The Philosophy of Carl G. Hempel

Carl G. Hempel (1905–1997)
Portrait of the Philosopher as a Young Man (ca. 1933)

The Philosophy of Carl G. Hempel

Studies in Science, Explanation, and Rationality

CARL G. HEMPEL

Edited by
JAMES H. FETZER

2001

OXFORD
UNIVERSITY PRESS

Oxford   New York
Athens   Auckland   Bangkok   Bogotá   Buenos Aires   Calcutta
Cape Town   Chennai   Dar es Salaam   Delhi   Florence   Hong Kong   Istanbul
Karachi   Kuala Lumpur   Madrid   Melbourne   Mexico City   Mumbai
Nairobi   Paris   São Paulo   Shanghai   Singapore   Taipei   Tokyo   Toronto   Warsaw

and associated companies in
Berlin   Ibadan

Copyright © 2001 by Oxford University Press

Published by Oxford University Press, Inc.
198 Madsion Avenue, New York, New York 10016

Oxford is a registered trademark of Oxford University Press

All rights reserved. No part of this publication may be reproduced,
stored in a retrieval system, or transmitted, in any form or by any means,
electronic, mechanical, photocopying, recording or otherwise,
without the prior permission of Oxford University Press.

Library of Congress Cataloging-in-Publication Data
Hempel, Carl Gustav, 1905–
The philosophy of Carl G. Hempel : studies in science, explanation, and rationality /
Carl G. Hempel ; edited by James H. Fetzer.
p.   cm.
Includes bibliographical references and indexes.
ISBN 0-19-512136-8; 0-19-514158-X (pbk.)
1. Knowledge, Theory of.   2. Science–Philosophy.   I. Title: Philosophy of Carl Gustav
Hempel. II. Fetzer, James H., 1940–   III. Title.

B945.H451   2000
191—dc21   99-087203

1 3 5 7 9 8 6 4 2

Printed in the United States of America
on acid-free paper

TO *DIANE*

Preface

Perhaps the three most important philosophers of science of the twentieth century have been Sir Karl Popper, Carl G. Hempel, and Thomas S. Kuhn. Popper exerted the most influence upon natural scientists and Kuhn upon social scientists and public alike. But Hempel's impact upon professional philosophers of science was unparalleled. His work, including the problems he addressed and the methods he employed, virtually defined the field, not just for a few years, but for decades. His most important book, *Aspects of Scientific Explanation* (1965a), has proven to be a rich and fertile source of philosophical reflections. That his influence should long endure remains of crucial interest to philosophy.

The papers included in this collection are intended to ensure that Hempel's work on these central problems will remain easily accessible for students and scholars alike. None of the pieces published in *Aspects* has been reprinted here; instead, an effort has been made to supplement that work by including his most important essays on these central problems before and after *Aspects*. A companion volume of new studies of Hempel's work, *Science, Explanation and Rationality: Aspects of the Philosophy of Carl G. Hempel*, shall soon appear from Oxford University Press. And Cambridge University Press is now publishing *Carl G. Hempel: Selected Philosophical Essays*, an independent anthology Richard Jeffrey has edited.

On a personal note, as a Princeton undergraduate (Class of 1962), it was my great good fortune to compose a thesis on the logical structure of explanations of human behavior for this remarkable man, who was not only the epitome of philosophical genius (with no doubt) but was also admired for his sterling character, including absolute honesty, almost infinite patience, and an abiding modesty that is truly rare among philosophers. Hempel dedicated *Aspects*, perhaps the most important landmark in the philosophical study of science, to his wife, Diane, with the words, "her sympathetic encouragement and unfaltering support would have deserved a better offering." This work is also for Diane. May it be found worthy.

Duluth, Minnesota J. H. F.
July 1999

Acknowledgments

1. "On the Nature of Mathematical Truth," *The American Mathematical Monthly*, Vol. 52 (December 1945), pp. 543–556. By permission of The Mathematical Association of America.
2. "Geometry and Empirical Science," *The American Mathematical Monthly*, Vol. 52 (January 1945), pp. 7–17. By permission of The Mathematical Association of America.
3. "Recent Problems of Induction," R. G. Colodny (ed.), *Mind and Cosmos* (Pittsburgh: University of Pittsburgh Press, 1966), pp. 112–134. By permission of The Pittsburgh Center for the Philosophy of Science.
4. "On the Structure of Scientific Theories," R. Suter (ed.), *The Isenberg Memorial Lecture Series 1965–66* (East Lansing: Michigan State University Press, 1969), pp. 11–38. By permission of Ronald Suter and of Diane Hempel.
5. "Explanation and Prediction by Covering Laws," B. Baumrin (ed.), *Philosophy of Science: The Delaware Seminar*, Vol. 1 (1961–62) (New York: Interscience Publishers, 1963), pp. 107–133. By permission of the University of Delaware Press.
6. "Deductive Nomological vs. Statistical Explanation," H. Feigl and G. Maxwell (eds.), *Minnesota Studies in the Philosophy of Science*, Vol. 3 (Minneapolis: University of Minnesota Press, 1962), pp. 98–169. By permission of the University of Minnesota Press.
7. "Maximal Specificity and Lawlikeness in Probabilistic Explanation," *Philosophy of Science* 35 (June 1968), pp. 116–133. By permission of The University of Chicago Press.
8. "Postscript 1976: More Recent Ideas on the Problem of Statistical Explanation," a translation of a new section on statistical explanation (by Hazel Maxian) of *Aspekete Wissenschaftlicher Erkldrung* (Berlin and New York: Walter de Gruyter, 1977). By permission of Walter de Gruyter and of Diane Hempel.
9. "Reduction: Ontological and Linguistic Facets," S. Morgenbesser, P. Suppes, and M. White (eds.), *Philosophy, Science, and Method: Essays in Honor of Ernest Nagel* (New York: St. Martin's Press, 1969), pp. 179–199. By permission of St. Martin's Press.

10. "The Meaning of Theoretical Terms: A Critique of the Standard Empiricist Construal," P. Suppes et al. (eds.), *Logical, Methodology and Philosophy of Science* 4 (North Holland Publishing Company, 1973), pp. 367–378. By permission of Patrick Suppes and of Diane Hempel.

11. "On the 'Standard Conception' of Scientific Theories," M. Radner and S. Winokur (eds.), *Minnesota Studies in the Philosophy of Science*, Vol. 4 (Minneapolis: University of Minnesota Press, 1970), pp. 142–163 . By permission of the University of Minnesota Press.

12. "Limits of a Deductive Construal of the Function of Scientific Theories," E. Ullmann-Margalit (ed.), *Science in Reflection, The Israel Colloquium*, Vol. 3 (Dordrecht: Kiuwer Academic Publishers, 1988), pp. 1–15. By permission of Kluwer Academic Publishers.

13. "Logical Positivism and the Social Sciences," in P. Achinstein and S. F. Barker (eds.), *The Legacy of Logical Positivism* (Baltimore: The Johns Hopkins Press, 1969), pp. 163–194. By permission of The Johns Hopkins Press.

14. "Explanation in Science and in History," Robert G. Colodny (ed.), *Frontiers of Science and Philosophy* (Pittsburgh: University of Pittsburgh Press, 1963), pp. 9–33. By permission of the University of Pittsburgh Press.

15. "Reasons and Covering Laws in Historical Explanation," S. Hook (ed.), *Philosophy and History: A Symposium* (New York: New York University Press, pp. 143–163. By permission of Ernest B. Hook.

16. "Rational Action," *Proceedings and Addresses of the American Philosophical Association*, Vol. 35 (October 1962), pp. 5–23. By permission of the University of Delaware Press.

17. "Science Unlimited?", *The Annals of the Japan Association for Philosophy of Science*, Vol. 4 (March 1973), pp. 187–202. By permission of the Japan Association for Philosophy of Science.

18. "Turns in the Evolution of the Problem of Induction," *Synthese* 46 (1981), pp. 389–404. By permission of Diane Hempel.

19. "Scientific Rationality: Normative vs. Descriptive Construals," in H. Berghel, A. Huebner, E. Koehler (eds.), *Wittgenstein, the Vienna Circle, and Critical Rationalism*. Proceedings of the Third International Wittgenstein Symposium, August 1978 (Vienna: Holder-Pichler-Tempsky, 1979), pp. 291–301. By permission of Holder-Pichler-Tempsky, Verlag.

20. "Valuation and Objectivity in Science," R. S. Cohen and L. Laudan (eds.), *Physics, Philosophy, and Psychoanalysis: Essays in Honor of Adolf Grunbaum* (Dordrecht: D. Reidel Publishing Company, 1983), pp. 73–100. By permission of Kluwer Academic Publishers.

Contents

Introduction   xiii

Part I: Cognitive Significance

1. On the Nature of Mathematical Truth   3

2. Geometry and Empirical Science   18

3. Recent Problems of Induction   29

4. On the Structure of Scientific Theories   49

Part II: Explanation and Prediction

5. Explanation and Prediction by Covering Laws   69

6. Deductive-Nomological versus Statistical Explanation   87

7. Maximal Specificity and Lawlikeness in Probabilistic Explanation   146

8. Postscript 1976: More Recent Ideas on the Problem of Statistical Explanation   165

Part III: Scientific Theories

9. Reduction: Ontological and Linguistic Facets   189

10. The Meaning of Theoretical Terms: A Critique of the Standard Empiricist Construal   208

11. On the "Standard Conception" of Scientific Theories   218

12. Limits of a Deductive Construal of the Function of Scientific Theories   237

Part IV: Explanations of Behavior

13. Logical Positivism and the Social Sciences   253

14. Explanation in Science and in History   276

15. Reasons and Covering Laws in Historical Explanation   297

16. Rational Action   311

Part V: Scientific Rationality

17. Science Unlimited?   329

18. Turns in the Evolution of the Problem of Induction   344

19. Scientific Rationality: Normative versus Descriptive Construals   357

20. Valuation and Objectivity in Science   372

A Bibliography of Carl G. Hempel   397

Index of Names   405

Index of Subjects   409

Introduction

Among the most influential philosophers of the twentieth century, Carl G. Hempel almost certainly exerted a more profound impact upon professional philosophy of science than did any of his contemporaries, including Rudolf Carnap, Sir Karl Popper, Hans Reichenbach, and Thomas S. Kuhn. In the first instance, his impact can be measured by the extent to which his work virtually defined the principal problems and alternative solutions that dominated philosophy of science during the century, including the problem of cognitive significance, the logic of confirmation, the structure of scientific theories, the logic of scientific explanation, and the problem of induction. Hempel focused on science, explanation, and rationality.

In the second instance, however, his legacy can be appraised by the success of his teaching and achievements of his students, including some of the most familiar names in the profession, who have made a great many highly varied contributions to philosophy of science. Among the most distinguished are Adolf Grünbaum, the author of impressive studies of space and time (and later of psychoanalysis), Nicholas Rescher, prolific editor and founder of *The American Philosophical Quarterly* and of *The History of Philosophy Quarterly* (and author/editor of nearly a hundred books), Larry Laudan, Robert Nozick, Lawrence Sklar, Jaegwon Kim, and Philip Kitcher, to cite only a handful. His influence has certainly been multiplied by his students.

Hempel sought solutions to difficult philosophical problems that were not only plausible and supported by suitable arguments but also were amenable to precise formulation by means of symbolic logic, especially utilizing the resources of sentential and predicate logic. Like his colleague, Gregory Vlastos, who utilized formal logic in the analysis of ancient philosophy, Hempel was committed to the highest standards of philosophical clarity and rigor. To the extent to which his solutions were susceptible to formal presentation, they were subject to the most demanding inspection and critical examination, which advanced philosophy. He cared more about finding the right solutions than whether his solutions were right.

This collection of articles and its companion volume of studies of Hempel's work are intended to ensure that students and scholars of the nature of knowledge and of the philosophy of science will continue to have ready access to some of his most important publications and have the opportunity to appreciate their meaning and import. While he can no longer present public lectures or publish new papers, his contributions to philosophy retain their vitality and remain far from finished. My purpose in this introduction is to provide a historical orientation and an analytical explanation of the reasons why Carl G. Hempel's contributions shall long endure.

1. Logical Atomism

Among the most important philosophical movements of the twentieth century were pragmatism, existentialism, phenomenology, and ordinary language philosophies, where the latter focused on the manifold use of ordinary language as a repository of solutions to problems of philosophy, many of which were widely supposed to be amenable to dissolution (to simply "disappear") when the use of language was properly understood. Other students of philosophy, however, considered ordinary language an imperfect instrument for the resolution of many conceptual difficulties, which required methods to revise, reform, or otherwise improve upon its resources. For them, ordinary language was a part of the problem, not the solution.

At least three movements of this kind would emerge between 1920 and 1960, each of which was strongly affected by the use of formal logic as a means for the exact solution to philosophical problems that ordinary language could not provide. The first was the offspring of Bertrand Russell and Ludwig Wittgenstein and was known as *logical atomism*. According to its tenets, the structure of language provides an access route to the structure of the world, where atomic sentences that have no other sentences as parts describe possible atomic states-of-affairs in the world, known as "atomic facts," while molecular sentences having other sentences as parts describe "molecular facts." The world itself thus has a logical structure.

In his *Tractatus Logico-Philosophicus* (1922), Wittgenstein confronted the conundrum of using language to talk about language, something that he believed was systematically impossible. In an introduction to that work, Russell, who advanced his own version of logical atomism in several papers, remarked that Wittgenstein apparently did not understand the crucial distinction between the use of language to talk about the world, for example, and the use of one language to talk about another language, known as the *object-language/metalanguage* distinction. Thus, in order to use some language to systematically discuss another, the metalanguage must be essentially richer than the object-language in several important respects.

The most important of these may be that, in order to formalize the conditions for the truth of a sentence in a language $L1$ (German, for example) in some other language $L2$, it is necessary to be able to have the ability to quote sentences from the object-language ("Schnee ist weis," say) and relate them to corresponding translations in the metalanguage ("Snow is white"), as Alfred Tarski (1935/36) explained. The sentence, "The sentence, 'Schnee ist weiss' is true-in-German if and only if snow is white," thus belongs to the metalanguage (enhanced English, let us say). Wittgenstein would not pursue formal syntax and semantics, however, but instead abandon logical atomism altogether for ordinary language philosophy, where his *Philosophical Investigations* (1953) is among its most important works.

2. Logical Positivism

Early Wittgenstein thought the meaning of a sentence could not be said but could only be shown. In his attempts to understand how language is related to the world, he offered *the picture theory of language*, which holds that sentences are true when they picture reality. The problem of understanding a language is thus a matter of understanding how the constituents of its sentences picture the constituents of the world. Sentences can be atomic or molecular and correspond to atomic or molecular constituents of the world as atomic and molecular facts. The truth conditions of molecular sentences depend only on the truth values (true, false) of their atomic components when formalized by truth-functional logic.

The picturing relation fundamental to logical atomism remained unexplained and was never really adequate to establish the isomorphic connection between parts of language and parts of the world this account required. Indeed, logical atomism appears to have been inspired by a misconception.[1] Logical properties are syntactical and semantical properties of languages rather than of the world, which may have a causal (or a lawful) structure but has no logical structure of its own. This realization is reinforced by the consideration that "facts" seem to be no more or less than *true sentences*. So sentences that correspond to facts are true because they are true sentences, suggesting that atomism is either trivial or false.

The leading movement of the 1930s and 1940s thus became *logical positivism*, an approach to understanding the nature of knowledge that was strongly affected by the philosophical legacy of David Hume, on the one hand, and by the formal technology of symbolic logic, on the other. While Hume maintained that all knowledge worthy of the name must reflect relations between ideas or derivations from experience and otherwise qualifies merely as metaphysical nonsense, logical positivism refashioned those views within a linguistic framework, where all knowledge worthy of the name must be either *analytic and a priori* or *synthetic and a posteriori* or else qualifies as cognitively insignificant. It is Hume with a linguistic turn.

In relation to a language L, sentences are said to be *analytic* (or *contradictory*) when their truth (or falsity) is determined by the properties of that language and logic alone; in particular, when their truth (or falsity) follows from the syntax (or grammar) and semantics (or vocabulary) of the language in which they are formulated. Since their truth (or falsity) does not depend on experience, knowledge of this kind is said to be *a priori*. Sentences are *synthetic* when their truth (or falsity) is not determined by the properties of that language and logic alone but also depends on the history of the world. Since information about the world's history cannot be acquired without experience, such knowledge is said to be *a posteriori*.

3. The Analytic/Synthetic Distinction

In relation to an ordinary language, such as English, sentences such as "If John is a bachelor, then John is unmarried," "Either Mary has gone to Paris or Mary has not gone to Paris," and so on, turn out to be true on the basis of their meaning or their grammar alone. So long as the same words mean the same thing when they occur in different contexts (the two instances of "John" in the first case or of "Mary has gone to Paris" in the second), these sentences cannot possibly be false. Whether the sentences, "John is a bachelor" and "Mary has gone to Paris" are true in turn, however, depends on the history of the world: their truth or falsity cannot be known on the basis of logic and language alone.

The position defined by logical positivism affirmed that all *a priori* knowledge is analytic and that all synthetic knowledge is *a posteriori*, thereby denying the existence of knowledge that is both synthetic and *a priori*. The logical-positivist position may therefore be represented as follows:

	A Priori	A Posteriori
Analytic	Yes	?
Synthetic	No	Yes

Figure 1. The Logical Positivist Position.

Thus, sentences whose truth (or falsity) could be ascertained on the basis of logic and language alone are *a priori* but provide no information or content about the world, while other sentences may provide information or content about the history of the world but their truth is only knowable *a posteriori*. In harmony with classic forms of empiricism exemplified by Hume, therefore, logical positivism denied the existence of synthetic knowledge that is *a priori*.

These distinctions, it should be emphasized, were only to be drawn relative to a presupposed language framework **L** that assigns meanings to at least some of the elements of its vocabulary and at least partially formalizes its grammar. As Carnap (1939) would explain, relative to an ordinary language *L*, semantical models of its vocabulary and syntactical models of its grammar are successively more and more abstract representations of the linguistic dispositions of its actual speakers that require simplification of the pragmatic phenomena. It follows that which class of sentences is "analytic-in-L" occurs as a deductive consequence of syntactical and semantical properties established by these modeling procedures.

Although knowledge that is both synthetic and *a priori* is rightly ruled out, to also discount the apparent possibility of metalinguistic knowledge about a language that is analytic but *a posteriori* would be wrong. The "?" in Figure 1 thus reflects the conception of empirical linguistics as an *a posteriori* discipline devoted to the discovery of analytic relations of synonymy-in-use and implicit definitional relations that are made explicit through the use of syntactical and semantical models. Of course, these relationships clearly presuppose the existence of metalanguages for the investigation of object-languages, but they could provide replies to questions that would eventually be raised about analyticity.

4. The Verifiabililty Criterion

Interestingly, logical positivists found inspiration in Wittgenstein's *Tractatus* for several reasons, including his concern with the relationship of language and the world, his contention that genuine propositions are truth-functions of elementary propositions, and his conclusion that logical truths are tautologies which say nothing about the world. Indeed, Wittgenstein made extensive use of truth-tables, which define connectives between sentences exclusively on the basis of the truth-values of their component sentences. In place of Wittgenstein's problematical notion of sentences picturing parts of the world, the logical positivists offered the notion of observation sentences that describe parts of the world as a more plausible approach relating language and knowledge through experience.

The precise manner in which empirical knowledge is related to experience is thus a crucial issue. Indeed, perhaps the most important principle defining logical positivism turns out to be the presumption that theoretical language—which makes reference to *nonobservables*, including dispositions as nonobservable properties of observable entities and *theoreticals* as nonobservable properties of nonobservable entities—is either reducible to *observational* or fails to be empirically meaningful. Thus, empirical knowledge is restricted to observations and their deductive consequences, a constraint that reduces the empirical content of scientific theories to logical constructions of observables.

Perhaps the most brilliant exposition of this position was presented in A.J. Ayer's *Language, Truth and Logic* (1936), which lucidly articulated the crucial principle that cognitively significant language consists only of assertions that are analytic or contradictory—in which case they may be said to have purely *logical meaning* or significance—or else are observation sentences or logically equivalent to finite sets of observation sentences—in which case they may be said to have *empirical meaning* or significance. Any sentences that could not satisfy this criterion, such as "God is omnipotent" or "The Absolute is perfect," were said to be cognitively insignificant as merely vacuous pseudostatements.

Distinctions were drawn between sentences of different grammatical forms, emphasizing the differences between declarative, imperative, interrogatory, and exclamatory sentences, where declarative make assertions, imperative issue commands, interrogatory ask questions, and exclamatory express attitudes.

xviii *Introduction*

Perhaps no aspect of logical positivism created more concern in the public mind than the attempt to account for maxims of morality as functions of attitude that might differ from person to person or from time to time. There appeared to be no more to right and wrong than differences in attitude toward different kinds of acts or, even worse, than expressions of attitude toward acts of various kinds.

5. Logical Empiricism

The classic core of logical positivism thus consisted in three basic premises, namely: (1) the analytic/synthetic distinction; (2) the observational/theoretical distinction; and (3) the verifiability criterion of meaningfulness. Although some outside the movement, such as Popper (1935), rejected the verifiability criterion and emphasized the importance of falsifiability in scientific practice without substituting an alternative falsifiability criterion in its place (because he believed that many forms of language were meaningful, even if they were not scientific), logical positivism became identified in the popular mind with scientific philosophy, which appeared to be rigorous, dry, and dehumanizing.

The transition from logical positivism to what would later be referred to as *logical empiricism* was promoted by several severe critiques of its basic tenets. In a set of influential articles (Hempel 1950, 1951, and 1965b), for example, Hempel persuasively demonstrated that the verifiability criterion of meaning implied that *existential generalizations* (which assert the existence of at least one thing of some kind and can be verified by a single case) are meaningful, while *universal generalizations* (which assert something about everything of a certain kind) are meaningless, even though they include general laws, such as Newton's laws of motion and of gravitation, as instances.

Indeed, on the assumption that a sentence is meaningful if and only if its negation is meaningful, Hempel easily demonstrated that implementing the verifiability criterion generated inconsistent consequences. The sentence, "At least one stork is red-legged," for example, is *meaningful* because it can be verified by observing a single red-legged stork, yet its negation, "Not even one stork is red-legged," cannot be shown to be true by any finite number of observations and is therefore *meaningless*. And descriptions of relative frequencies in finite classes are meaningful, but assertions about limits in infinite sequences are not. Many of the most important results of scientific inquiry thus were meaningless.

These realizations suggested that the logical relationship between scientific theories and empirical evidence cannot be exhausted by observation sentences and their *deductive consequences* alone, but must be expanded to include observation sentences and their *inductive consequences* as well. The concepts of confirmation and disconfirmation (as forms of partial verification and partial falsification) received renewed interest, where the crucial feature of scientific hypotheses appeared to be their empirical testability rather than verifiability. And Hempel proposed that cognitive significance was better envisioned as a matter of degree that could only be evaluated in relation to multiple criteria.

6. "Two Dogmas of Empiricism"

Exchanging the principle of verifiability for a criterion of testability was not entirely welcome news. There were those who supposed that inductive arguments—where, unlike deductive arguments, the truth of their premises provides support for, but does not guarantee, the truth of their conclusions—were inherently defective. A searing witticism of the time lampooned one widely used text on inductive and deductive logic by remarking that, in the first part of the book (on deduction), the fallacies were explained, and in the second part (on induction), they were committed. Reichenbach, Hempel, and Carnap thus made strenuous attempts to render inductive logic respectable.

While Hempel was proposing that cognitive significance should be understood as a property of whole theoretical systems—which are subject to comparative evaluation relative to the clarity and precision of their respective language frameworks, their respective degrees of systematic power for explaining and predicting the phenomena of experience, the formal simplicity, economy, or elegance with which they attain their degrees of systematic power, and the extent to which they have been confirmed by experiential finding—W. V. O. Quine was launching an assault on what was, if possible, an even more fundamental tenet of logical positivism, namely: the analytic/synthetic distinction.

In "Two Dogmas of Empiricism" (Quine 1953), one of the most influential philosophical articles of the twentieth century, Quine contended that implementing such a distinction depends on prior understanding of definitions, which consist of the word being defined (the *definiendum*) and its meaning (*definiens*). But that presupposes we already know which words possess which meanings. He suggested that appeals to synonymy (the sameness of meaning) or to meaning postulates or to truth-preserving interchangeability of expressions to elucidate the distinction are circular because they presuppose the existence of the very relationships they are being advanced to explain, concluding that the analytic/synthetic distinction is but "a metaphysical dogma" that should be abandoned.

Quine, a logician, however, allowed one class of narrowly analytic sentences whose truth depends only on their form, *logical truths* (including "$p \vee -p$" in sentential logic), and another class of broadly analytic sentences, *nominal definitions*, whose truth depends on explicit agreements in the form of conventions. And it was never clear why empirical reports on language-in-use by linguists, which are known as *meaning analyses*, were inadequate for their purpose, even though, as inductive generalizations based on experience, their status as *dictionary entries* was contingent. Quine's critique was subject to extensive—even voluminous—criticism, but it exerted an enormous impact.

7. Definitions of Dispositions

Quine may have been overly influenced by what is called "the paradox of analysis," according to which, in the process of relating words to definitions, either you know what they mean (in which case, analysis is unnecessary) or you do not (in which case, analysis is impossible). With respect to meaning, therefore, analytical philosophy is either unnecessary or impossible, an argument that overlooks the evident possibility that we might begin with words whose meanings are vague or ambiguous and only partially understood, but proceed through a process of conceptual clarification to advance recommendations and proposals about their meaning that enhance our understanding.

In his contribution to *The International Encyclopedia of Unified Science*, a series of volumes devoted to scientific philosophy, Hempel clearly advocated an approach of this kind as the method of *explication*, which had previously been introduced by Carnap (Hempel 1952). Both Carnap and Reichenbach had published detailed explications about the meaning of "probability," one devoted to its meaning within the context of *empirical hypotheses* about the relative frequency with which various events occur in the world, the other to its meaning within the context of *inductive reasoning* as a logical measure of evidential support for empirical hypotheses (Reichenbach 1949; Carnap 1950).

It was therefore striking when Hempel reported arguments due to Carnap that appeared to weaken the tenability of the analytic/synthetic distinction. In *Fundamentals of Concept Formation in Empirical Science* (Hempel 1952), he considered "magnetic" as an example of a *dispositional predicate* designating, not a directly observable property, but rather a disposition on the part of some things to display specific reactions (such as attracting small iron objects) under certain specific circumstances (such as the presence of small iron objects in the vicinity). On first consideration, it might seem appropriate to define this predicate by means of a conditional formulation employing the following form:

(D1) x is magnetic at t =df if, at t a small iron object is close to x, then it moves toward x.

The problem Hempel detected with such formulations is that the "if . . . then ——" conditional upon which it depends is normally interpreted in formal logic as a *material conditional* whose meaning is synonymous with "either not . . . or ——." Given this understanding of conditionality, the proposed definition would be logically equivalent with and have the same meaning as the alternative form:

(D2) x is magnetic at t =df at t, either it is not the case that a small iron object is close to x or else it moves toward x.

The problem here, however, is that this definiens would be satisfied not only by things for which, at t, small iron objects are close and move toward them, but also by things for which, at t, no such objects are nearby, which implies that anything (such as a yellow cow or a red rose) turns out to be magnetic at any time t during which no small iron objects are close, an obviously unacceptable result.

8. Extensional Formulations

Hempel admitted that the problem might be overcome by reformulating the definiens by means of *subjunctive conditionals* of the form, "If . . . were the case, then —— would be the case" (which make assertions about what would occur in their consequents on the hypothetical assumption that the conditions that they specify in their antecedents are satisfied) employing an alternative formulation:

(D3) x is magnetic at t =df if, at t, a small iron object were close to x, then that object would move toward x.

Hempel could not embrace this approach, however, because "despite considerable analytic efforts and significant partial results, no fully satisfactory explication seems available at present, and the formulation [here (D3)] represents a program rather than a solution" (Hempel 1952, p. 25). He therefore rejected it.

Hempel instead appealed to a technique Carnap had proposed to cope with this problem while remaining within the framework of *extensional (or truth-functional) logic*, where the truth values of conditionals and other connectives are completely determined by the truth values of their component sentences. He thus suggested that partial definitions could be formulated where the satisfaction of the predicate by x depended on its response to the test condition:

(D4)   If a small iron object is close to x at t, then x is magnetic at t if and only if that object moves toward x at t.

The problem encountered by (D1) and (D2) is thereby circumvented insofar as, when no small iron objects are in the vicinity of x, although the whole sentence is true of x, it does not support the troublesome inference that x is magnetic at t. Sentences of this kind were called "reduction sentences" rather than "definitions."

The possibility that more than one reduction sentence might be introduced as partial definitions for different manifestations of the underlying property, however, brought with it a rather unexpected consequence. Things that are magnetic could also satisfy certain additional conditions, such as the following:

(D5)   If x moves through a closed wire loop at t, then x is magnetic at t if and only if an electric current flows in the loop at t.

Sentences (D4) and (D5) thus yield conditions of applicability for the predicate "magnetic" that are nonexhaustive, since magnetism might have no end of displays. Hempel observed that anything that satisfies one of their test-condition/outcome-response pairs must also satisfy the second, on pain of contradiction.

Thus, by (D4), any object x that attracts some small iron body when near x is magnetic, and, by (D5), anything that is magnetic and moves through a closed wire loop will generate a current in that loop. They therefore jointly imply the consequence that any object x that attracts some small iron body when near x and moves through a closed wire loop will generate a current, a generalization that does not have the character of a definitional truth but rather that of an empirical law. Hempel drew the inference that concept formation by means of reduction sentences depends on the truth of certain empirical generalizations, which substantially undermined the analytic/synthetic distinction, since no one would want to suppose that analytic sentences can have synthetic consequences.

9. The Observational/Theoretical Distinction

The apparent demise of the verifiability criterion of meaningfulness and of the analytic/synthetic distinction were complemented by criticism of the observational/theoretical distinction, which was as fundamental to logical positivism as atomic states-of-affairs were to logical atomism. They shared a commitment to truth-functional language, where atomism employed it for the purpose of exposing the logical structure of the world as *an exercise in ontology*, while positivism relied upon it for the purpose of elaborating the logical structure of our knowledge of the world as *an exercise in epistemology*. Observation sentences were as essential to logical positivism as atomic facts were to logical atomism.

They were generally understood as sentences whose truth could be ascertained, under suitable conditions, on the basis of direct observation and relatively simple measurement. Observation sentences were synthetic and thus represented the results of causal interactions between observers and the observed. They were distinguished from other varieties of scientific language, where *observable predicates* designate observable properties of observable entities, *dispositional predicates* nonobservable properties of observable entities, and *theoretical predicates* nonobservable properties of nonobservable entities, reflecting their ease of accessibility by means of direct experience.

In order to facilitate scientific objectivity, *physicalistic* language was preferred to *phenomenalistic*, where the former envisions the contents of experience as properties of objects in the world, while the latter envisions them as properties of perceivers of those objects. Thus, the language of colors, shapes, and sizes was advocated over the language of color-sensations, shape-perceptions, and the like, which would allow for attributions of colors, shapes, and such on the basis of inferences from experience. The adoption of physicalistic language in lieu of phenomenalistic language was generally supposed (no doubt, correctly) to contribute to the attainment of the objectivity of science.

Thus, objectivity, as a property of scientific inquiries, does not require any privileged access to truth but instead depends on *intersubjective reliability*, which, properly understood, reflects the notion that, under similiar conditions of investigation involving similar hypotheses and similar evidence, different investigators ordinarily ought to accept, reject, and hold in suspense all and only the same conclusions. This objective is advanced by the use of special terms defined for scientific purposes as forms of technical language, instilled by formal instruction under controlled conditions to ensure a high degree of uniform usage, which is far more readily attainable by means of physicalistic language, which is public, than of phenomenalistic language, which is private.

10. Theory-Laden Observations

Surprisingly, some of the most powerful arguments against the observational/theoretical distinction originated with Popper (1935), which gained influence with its publication in English as Popper (1959). Popper argued that the distinction could not be sustained because both observable and theoretical *properties* are properly understood as *dispositions*. Leaving the logical properties of abstract entities to one side, if Popper were correct, then since dispositional properties are obviously dispositions, too, all the physical properties of the world—all of the properties that are capable of exercising causal influence within space/time—would be dispositional as tendencies to bring about specific outcomes under specific conditions, possibly with varying strengths.

Popper's position was especially persuasive because he argued in terms of very familiar examples. Thus, he maintained that the use of universals such as "glass" and "water" in sentences like, "Here is a glass of water'" necessarily transcend experience, because in describing something as *glass* or as *water*, we are attributing to it innumerable tendencies to behave in innumerable ways under various test conditions. Thus, consider the difference if the contents of a glass were mistaken for water but consisted of alcohol instead with respect to watering plants, quenching thirst, and extinguishing fires. What Popper uncovered was a deeper ontological distinction beneath the surface of epistemology.

In relation to *universals* understood as properties that can have more than one instance during the history of the world, moreover, Popper insisted that all universals are dispositions and, as a consequence, none of them can be adequately defined by means of extensional logic and observational predicates alone. On the contrary, *subjunctives and counterfactuals* (understood as subjunctives with false antecedents) would be required to properly unpack the meaning of universals of either the observational, the dispositional, or the theoretical kind. And in adopting this attitude, Popper received support from an unexpected direction; for Carnap would endorse the use of nonextensional logic for this role (Carnap 1963).

Carnap believed that, once the logical structure of lawlike sentences had been resolved and a suitable logic of causal modalities had been developed, it would be possible to explicate subjunctive and counterfactual conditionals and to provide explicit definitions for dispositional predicates. Recent efforts at developing the formal semantics of conditionals of each of these kinds on the basis of possible-world models appear to support Carnap's expectation.[2] If Carnap was correct, the demise of the observational/theoretical distinction may be attended by the restoration of the analytic/synthetic distinction, which may be the logical foundation, not just of philosophy of science, but of analytical philosophy in general.

11. Subjunctive Definitions

By employing subjunctive and counterfactual (and causal) conditionals, the explicit definition of dispositional predicates could satisfy the form (D3), where other manifestations under other test conditions might enhance the definiens:

(D6) x is magnetic at t =df if, at t, a small iron object were close to x,
then that object would move toward x; and,
if, at t, x were to move through a closed
wire loop, then an electric current would
flow through the loop; and, ...

for as many outcomes as that property may display under different conditions, where satisfying one of them would not logically guarantee satisfying the rest. Unlike material conditionals, subjunctives are never true merely because their antecedent conditions are not satisfied, which means that properties so defined might be attributed to objects whether or not they were being or ever had been subjected to those conditions, for example, on the basis of an inductive inference.

The difficulties encountered in understanding subjunctive and counterfactual conditionals were forcefully demonstrated by Nelson Goodman (1947 and 1955). Goodman argued that every predicate "is full of threats and promises" and that which should be relied upon to make predictions (called "projections") about the future should be evaluated on the basis of their success in past projections as a measure of their "degree of entrenchment." While acknowledging that the definition of dispositions, the nature of laws of nature, and the logic of inductive reasoning are inextricably intertwined, Goodman thereby offered a pragmatic solution to what always appeared to be a syntactical and semantical problem.

Hempel was profoundly impressed by Goodman's analysis, which he embraced as affording a partial solution to understanding the character of general laws. He adopted Goodman's conception of *lawlike sentences* as sentences that would be laws if they were true, where the capacity to support subjunctive and counterfactual conditionals was now assumed to be a condition that lawlike sentences must satisfy. This carried him beyond the resources of extensional logic, which only supported the conception of lawlike sentences as unrestrictedly general material conditionals. The pragmatic character of this condition was rather reminiscent of the attitude of Hume, who regarded our tendency to expect the future to resemble the past as merely a habit of mind.

12. Explications of Explanation

Hempel's monumental contribution to the philosophy of science, however, was his masterful analysis of the logical structure of *scientific explanations*, understood as arguments whose premises ("the explanans") include at least one general law L, which explain why the event described by the conclusion ("the explanandum") occurred by showing that it was to be expected relative to its initial (or antecedent) conditions C1-Cm (Hempel and Oppenheim 1948).

$$
\begin{array}{ll}
\text{General Law(s):} & L1, L2, \ldots, Ln \\
\text{Initial Conditions:} & C1, C2, \ldots, Cm
\end{array} \right\} \text{Explanans}
$$

$$
\text{Event Description:} \quad E \qquad \text{Explanandum}
$$

Figure 2. The Covering-Law Model.

A simple example might explain why a small coin expanded when heated by invoking the law that copper expands when heated and noting it was copper. Hempel considered a vast variety of modes of explanation, contending that those that—implicitly or explicitly—conform to this conception are scientific.

Hempel included the explanation of empirical generalizations and lawlike sentences through their subsumption by scientific theories within the scope of his approach, but devoted most of his attention to the elaboration of very precise and detailed accounts of the scientific explanation of singular events. And he advanced *deductive-nomological* and *inductive-probabilistic* versions to account for differences between subsumption by universal and by statistical covering laws. The differences between them, especially the peculiar difficulties generated by probabilistic explanations, would preoccupy much of his efforts over nearly two decades, as essays in this volume clearly reflect.

The crucial problem turned out to be explaining the nature of the logical link between explanans and explanandum when the covering laws were not universal but statistical. Suppose, for example, that a statistical law of the form, $P(B/A) = r$, assigned probability of value r to the occurrence of an outcome of kind B, given conditions of kind A. That might be understood as a relative frequency in the sense of Reichenbach. An explanation of the form,

$$
\begin{array}{ll}
\text{The Statistical Law:} & P(B/A) = r \\
\text{Initial Conditions:} & Axt \\
\hline
\text{The Explanandum:} & Bxt
\end{array} \quad [r]
$$

Figure 3. Inductive-Probabilistic Explanation.

invites the presumption that the bracketed variable [r] should be understood as a logical probability in Carnap's sense. Hempel initially adopted this approach, which reflects the epistemic conception of [r] as a degree of evidential support, but he would subsequently reject it (Hempel 1968).

While the covering-law approach dominated the philosophy of science in the '50s and the '60s, such difficulties, which were rooted in deep problems about the nature of probabilistic laws, stimulated other investigations, the most important being the *statistical relevance* model of Wesley Salmon (1971), which denied explanations were arguments and captivated the interest of the discipline in the '70s. Salmon would later abandon the interpretation of probabilities as relative frequencies for the Popperian alternative of causal propensities as probabilistic dispositions within the context of probabilistic explanation (Salmon 1984). But during the '80s and the '90s, no approach exercised its grip on the discipline as had Hempel's covering-law conception.

13. Cognitive Significance

Hempel's contributions to the philosophy of science are well displayed by the articles collected in this volume. "On the Nature of Mathematical Truth" provides a classic exposition of the thesis that *the truths of mathematics are analytic sentences,* which are derivable as theorems from specific premises as the axioms of a formal calculus. He explains why they cannot be suitably envisioned as self-evident truths or as empirical generalizations and employs Peano's formalization of arithmetic to illustrate the conception of mathematics as axiomatized deductive systems. But he emphasizes that these results only apply to pure mathematics, since truths of applied mathematics are synthetic.

"Geometry and Empirical Science" pursues this theme and elaborates the conception of scientific theories as *abstract calculi provided with empirical interpretations.* He illustrates this process with reference to Euclidean and non-Euclidean geometries, which make no assertions about the mathematical structure of physical space without adopting *coordinating definitions,* for example, which identify "straight lines" with paths of light rays and "points" with their intersections. And he persuasively explains the reasons why, as Einstein once succinctly observed, "As far as the propositions of mathematics refer to reality, they are not certain, and as far as they are certain, they do not refer to reality."

"Recent Problems of Induction" considers how the classic problem of jusifying inferences about the future on the basis of information about the past first raised by Hume has now received a *linguistic reformulation* by Goodman, who has raised "the new riddle of induction" concerning the character of projectible predicates, understood as kinds of predicates that should be relied on in the process of anticipating the future based on information about the past. The *paradoxes of confirmation,* which arise from the use of logic to formalize hypotheses, are reviewed and the *ambiguity of induction,* which will also become important for understanding the logic of statistical explanation, is discussed.

"On the Structure of Scientific Theories" advances a critique of the standard conception of scientific theories as formal calculi with empirical interpretations, which he previously championed. In order to overcome the untenability of the observational/theoretical distinction, he introduces an alternative conception of scientific theories as the union of *internal principles* distinctive to those theories with *bridge principles* that indicate how the processes invoked by the theory are related to the phenomena the theories are intended to explain. He disposes of unsound criticisms of the standard conception, but discovers suitable ground for proposing this clarification of the relations between theories and the world.

14. Explanation and Prediction

"Explanation and Prediction by Covering Laws" supplies an introduction to the account of explanation as subsumption by means of covering laws, elaborating *deductive-nomological* and *inductive-probabilistic* variants while defending this approach against criticisms from Michael Scriven. The logical link between the explanans and explanandum is assumed to satisfy Carnap's *requirement of total evidence* as an epistemic conception. Significantly, Hempel explicates the nomic expectability desideratum by insisting that explanations are only adequate provided knowledge of their explanans would have enabled us to *predict* the occurrence of their explanandum events, which induces a requirement that the value of the logical probability [r] must be equal to .5 or more. This condition makes the model vulnerable to the problem, which Salmon would subsequently exploit, that it is logically impossible to explain events that occur with low probability.

"Deductive-Nomological vs. Statistical Explanation," one of his most extensive studies, separates *potential* explanations from *true* explanations and *confirmed* explanations, where a potential explanation is one that would be adequate if it were true. Hempel distinguishes laws of coexistence from laws of succession and discusses differences between explanations, predictions, and retrodictions. Most important, he investigates *the ambiguity of statistical explanations* that arises from the possibility that alternative explanations that have true premises might confer different degrees of nomic expectability on the same explanandum, appealing to Carnap's conception of logical probability while proposing a *rough criterion of evidential adequacy* that is intended to resolve this problem.

"Maximal Specificity and Lawlikeness in Probabilistic Explanation" reports that Hempel no longer regards [r] as an inductive probability in Carnap's sense that represents the degree of evidential support for the thruth of an explanandum relative to our total evidence but rather as having a distinct rationale of its own; "for *the point of an explanation is not to provide evidence for the occurrence of the explanandum phenomenon, but to exhibit it as nomically expectable*." And he proposes a novel condition known as *the requirement of maximal specificity* that is intended to ensure that the lawlike premises of probabilistic explanations take into account all of the initial conditions whose presence made a difference to the occurrence of the explanandum event and thereby prohibit ambiguities; but the implementation of this requirement is related to an epistemic context K.

"More Recent Ideas on the Problem of Statistical Explanation" appeared as the "Postscript 1976" to a new translation of Hempel (1965a) in German. This essay, which has not been published in English before, explores criticism by Richard Jeffrey, Wesley Salmon, and others. He agrees with Jeffrey that the adequacy of an explanation does *not* depend on the high probability of its explanandum, but rather consists in a characterization of the stochastic process and initial conditions that produced the event, which can explain the occurrence of an improbable outcome "no less well" than the occurrence of a very probable event. He also accepts Salmon's criticism of the high probability requirement, which he now abandons. He apologizes for the epistemic relativity of the maximal specificity condition, but expresses skepticism about the potential for development of an ontic counterpart.

15. Scientific Theories

"Reduction: Ontological and Linguistic Facets" explains why the problem of reducing theories and laws at one level to those of another, lower level may be more adequately addressed by resorting to metalinguistic distinctions than by attempting to deal with these issues of ontology more directly. Using examples drawn from mechanistic biology, psychophysical associations, and the history of physics, he suggests that, even though reducing theories tend to be inconsistent with their predecessors, their reductions tend to preserve them as special cases. The construal of the relationship involved here as purely deductive, therefore, cannot be sustained. But characterizations in terms of incompatibility or incommensurability relations similarly oversimplify the nature of reduction in science.

"The Meaning of Theoretical Terms: A Critique of the Standard Empiricist Construal" extends Hempel's critique of the standard conception of scientific theories as abstract calculi with empirical interpretations by challenging the assumption that theoretical terms within them must have their meaning specified by sets of sentences interpreting them in an empirical vocabulary that is fully and clearly understood. Hempel considers various alternatives—including the methods of *implicit definition, meaning postulates, correspondence rules,* and *explicit definitions by observation predicates*—explaining why theoretical terms are better envisioned as connected to experience by an antecedently available vocabulary. Some readers might want to compare Hempel's arguments with those of Quine.

"On the 'Standard Conception' of Scientific Theories" continues his defense of the internal principle/bridge principle conception as an alternative to the standard conception. The internal principle/bridge principle account does not hinge upon the analytic/synthetic distinction, and bridge principles are on a par with internal principles as parts of a theory. He concedes that there is no precise criterion for distinguishing between them and that principles of both kinds may include theoretical as well as antecedently available language. Nor is their difference a matter of epistemic standing. The history of logical empiricism, he submits, reflects "a steady retrenchment of the initial belief in, or demand for, full definability of all scientific terms by means of some antecedent vocabulary consisting of observational predicates or the like." His account offers an alternative.

"Limits of a Deductive Construal of the Function of Scientific Theories" marks an important development in thinking about theories. Hempel now shows that the desire to formalize theories and their application for the purpose of explanation, prediction, and retrodiction *employing exclusively deductive principles* cannot be sustained for at least two important reasons: first, the necessity for "inductive ascent" from experiential data to explanatory hypotheses, which are not the only logically possible alternatives that might explain them; and, second, "the problem of provisos" in drawing inferences from theories and laws whose application in each instance depends upon various ontic assumptions about the presence or absence of boundary conditions that are not explicitly encompassed by those theories or laws themselves.[3] This realization undermines programs for the elimination of theoretical terms and casts grave doubt upon the tenability of the instrumentalist position.

16. Explanations of Behavior

"Logical Positivism and the Social Sciences" provides a historical and analytical exploration of contributions, especially by Otto Neurath, Rudolf Carnap, and Herbert Feigl, to understanding social phenomena and records the successive weakening of *empiricist criteria of cognitive significance* during the course of development from the earlier (logical postivist) stage to the later (logical empiricist) stage. Reducibility to observables thus gives way to testability; definability gives way to partial definitions; and the construction of psychological properties as dispositions gives way to their interpretation as theoretical properties instead. This essay affords a fascinating perspective on relations between this movement and physicalism, behaviorism, and mentalism, including illuminating remarks on the unity of language, unity of laws, and unity of methods of the various sciences.

"Explanation in Science and in History" manifests Hempel's long-standing interest in explanations of human behavior as well as of inanimate phenomena, a concern displayed already in his first important paper on explanation (Hempel 1942). He elaborates the notions of *elliptical* and *partial* explanations, remarking that the deductive-nomological and inductive-statistical models represent *ideal types* or *idealized standards* or *rational reconstructions* of the conditions of explanatory adequacy. They are not intended to describe the way in which scientists or historians present their findings and discoveries, which may satisfy those idealized standards only relative to implicit premises and background assumptions, whose content and truth may be extremely difficult to ascertain.

"Reasons and Covering Laws in Historical Explanation" addresses the common tendency to assume that explanations of behavior that appeal to reasons are not covered by the covering-law model because *reasons are not causes*. He explores the notion of rationality as a normative concept, which implies standards of conduct that a "rational agent" would be expected to satisfy, observing that whether or not any specific person qualifies as a rational agent in that sense is an empirical matter that cannot be known *a priori*. That an agent A was in a situation of kind C and that "the thing to do in a situation of kind C is X," for example, only implies that A did X on the additional assumption that A was a rational agent at the time, a crucial clarification that reaffirms the applicability of covering-law explanation.

"Rational Action" draws a distinction between rationality of action and rationality of belief, where agents are highly *rational with respect to actions* when they are strongly disposed to take actions that are effective, efficient, or reliable means for accomplishing their goals and highly *rational with respect to beliefs* when they are strongly disposed to accept, reject, or hold beliefs in suspense in proportion to the strength of the available relevant evidence in their support. These are logically independent traits: agents might be high in one and low in the other. They are "broadly dispositional" characteristics, whose manifestations in specific cases are brought about (possibly probabilistically) by causal interactions between an agents' motives and beliefs. Hypotheses about an agents' motives or beliefs are therefore *epistemically interdependent*, since inferences about specific motives based on observed behavior require assumptions about beliefs, and conversely.

17. Scientific Rationality

"Science Unlimited?" investigates the possible existence of boundaries or constraints upon what can be known by science that are logical, theoretical, or methodological in contrast to limitations imposed by money, time, and effort. These include the *inductive uncertainty* of scientific knowledge, the *potential scope* of scientific knowledge, limitations inherent to scientific explanations of individual events or of general regularities, whether science can explain *why* or only *how* events occur, and whether explaining an event entails explaining its premises, concluding with a fascinating analysis of whether science ought to be able to answer the question, "Why is there something rather than nothing?"

"Turns in the Evolution of the Problem of Induction" explores varied conceptions of the problem of justifying our expectations about the future on the basis of knowledge of the past. Deductive rules of inference are too strong, and probabilistic rules are too weak. The problem should be divided into rules that determine the strength of the available evidence and rules that determine when sufficient evidence is available. Inductive acceptance within the context of *pure science* (as a function of rationality of belief) must be distinguished from inductive acceptance in the context of *applied science* (as a function of rationality of action). While acting on hypotheses may require moral value judgments within applied science, accepting hypotheses within pure science only requires epistemic value judgments.

"Scientific Rationality: Normative vs. Descriptive Construals" pursues the nature of alternative approaches to the study of science. The *analytic-empiricist approach* attempts to identify normative standards relative to which the rationality of specific practices or procedures may be appraised as means that are suitable to attain their intended ends. The *sociological-historical approach*, by contrast, attempts to describe the actual practices and procedures that have been displayed during the history of science. The normative approach aims at an account that can provide a *justification* for specific practices as scientific, while the descriptive approach aims at an account that affords an *explanation* for the actual practices of scientists. Even descriptivists, however, take for granted that the practices they are describing are "rational." When they qualify as suitable means to attain specified ends, that may be correct; but to the extent to which they involve merely subjective judgments and personal preferences that cannot be so characterized, the conception of science they support is arational.

"Valuation and Objectivity in Science" provides a searching exploration of differences between *methodological rationalism*, which attempts to produce normative standards for critical evaluation of actual practice, and *methodological naturalism*, which seeks to describe the actual research behavior that scientists have displayed. The success of explication crucially depends on shared intuitions about aspects of the phenomena under consideration and thus presupposes the existence of widely shared agreement, for example, about clear cases of good practice and clear cases of bad practice. It would therefore be wrong to suppose that the production of normative standards of scientific practice is simply an *a priori* exercise, since the study of the history of science can make a valuable, but not determinative, contribution. The rationality of science precludes conceptions according to which "anything goes."

18. Concluding Reflections

The historical evolution of logical empiricism from its roots in logical positivism and logical atomism reflects a long-standing methodological preoccupation with logic and language. Even though the verifiability criterion has long since been abandoned and criteria of cognitive significance have undergone considerable weakening, especially in view of the untenability of the observational/theoretical distinction, the situation with respect to the analytic/synthetic distinction does not appear to be the same. Differences between sentences whose truth follows from logic and language alone as opposed to those whose truth does not are difficult to deny, even if dictionary entries and grammars result from empirical investigations of linguistic behavior in language-using communities and remain semantical and syntactical models.

Indeed, if physical properties are ontically dispositional and the kind of property dispositions are does not depend on the epistemic ease with which their presence or absence might be ascertained on the basis of experiential findings, it should come as no surprise that observations are theory-laden. And if more recent developments in formal semantics can accommodate nonextensional conditionals—including subjunctive, counterfactual, and probabilistic causal conditionals—and thereby fulfill Carnap's conception, then the demise of the observational/theoretical distinction may occur concurrently with the restitution of the analytic/synthetic distinction on equally good grounds.

The publication of *The Structure of Scientific Revolutions* (Kuhn 1962)—ironically, as the final volume of *The International Encyclopedia of Unified Science*—was among the most important developments that contributed to a dramatic diminution in enthusiasm for logical empiricism. Kuhn's work, which turned "paradigm" into a household word, was widely interpreted as likening revolutions in science to revolutions in politics, where one theory may succeed another only upon the death of its adherents. It was also supposed to have destroyed the myth that philosophers possessed any special wisdom with respect to the nature of science, as though every opinion were on a par with every other. A close reading of Kuhn's work shows that these are not *his* conclusions, but they were enormously influential, nevertheless.

If a new generation of sophisticated thinkers has now declared that philosophy of science has been misconceived, at least since the time of Aristotle, and requires "a naturalistic turn" as a kind of *science of science* that is more akin to history and sociology (Callebaut 1993), then perhaps they should consider how to distinguish frauds, charlatans, and quacks from "the real thing." Without knowing which members of "the scientific community" are practicing *science*, sociologists of science would not know whose behavior it should study. The "standards of science" can be identified with those that are upheld by the members of a scientific community only if they adhere to *scientific* standards. Among the most important lessons of Hempel's enduring legacy must be that scientific standards cannot be derived from descriptions of its practice alone.

NOTES

1. From the perspective of the theory of signs advanced by Charles S. Peirce, Wittgenstein was treating sentences in ordinary languages as if they were *icons*, which resemble what they stand for, rather than *symbols*, which are merely habitually associated with that for which they stand. From this point of view, therefore, he appears to have committed a category mistake. [See J. H. Fetzer, *Artificial Intelligence: Its Scope and Limits* (Dordrecht, 1990); or J. H. Fetzer, *Philosophy and Cognitive Science*, 2nd ed. (New York, 1996).]
2. A formal semantics for specifically scientific as opposed to ordinary language conditionals, including possible-world semantics for subjunctive conditionals and for causal conditionals of universal and probabilistic strength—which I developed (at Hempel's suggestion) in collaboration with Donald E. Nute—may be found in J. H. Fetzer, *Scientific Knowledge* (Dordrecht, 1981). [See J. H. Fetzer and D. E. Nute, "Syntax, Semantics, and Ontology: A Probabilistic Causal Calculus," *Synthese* (March 1979), pp. 453–495; J. H. Fetzer and D. E. Nute, "A Probabilistic Causal Calculus: Conflicting Conceptions," *Synthese* (June 1980), pp. 241–246; and "Errata," *Synthese* (September 1981), p. 493.
3. Lawlike sentences cannot be true unless their antecedents are complete in specifying the presence or absence of *every property* that makes a difference to the occurrence of their consequents. Since these properties may be theoretical, dispositional, or observational, this ontic condition locates the source of the epistemic problem of ascertaining whether or not every property that matters to their application has been taken into account, which will be the case only if there are no other relevant initial conditions. [See J. H. Fetzer, *Scientific Knowledge* (Dordrecht, 1981) and *Philosophy of Science* (New York, 1993).]

REFERENCES

Ayer, A. J. (1936), *Language, Truth, and Logic*. London. 2nd ed. 1946.
Callebaut, W., ed. (1993), *Taking the Naturalistic Turn: How Real Philosophy of Science is Done*. Chicago.
Carnap, R. (1939), *Foundations of Logic and Mathematics*. Chicago.
Carnap, R. (1950), *Logical Foundations of Probability*. Chicago. 2nd ed. 1962.
Carnap, R. (1963), "Replies and Systematic Expositions," in P. A. Schilpp, ed., *The Philosophy of Rudolf Carnap* (La Salle), pp. 859–1013.
Goodman, N. (1947), "The Problem of Counterfactual Conditionals," *Journal of Philosophy* 44, pp. 113–128.
Goodman, N. (1955), *Fact, Fiction, and Forecast*. Harvard.
Hempel, C. G. (1942), "The Function of General Laws in History," *Journal of Philosophy* 39, pp. 35–48.
Hempel, C. G. (1950), "Problems and Changes in the Empiricist Criterion of Meaning," *Revue internationale de Philosopie* No. 11, pp. 41–63.
Hempel, C. G. (1951), "The Concept of Cognitive Significance: A Reconsideration," *Proceedings of the American Academy of Arts and Sciences* 80/1 pp. 61–77.
Hempel, C. G. (1952), *Fundamentals of Concept Formation in Empirical Science*. Chicago.
Hempel, C. G. (1965a), *Aspects of Scientific Explanation*. New York.
Hempel, C. G. (1965b), "Empiricist Criteria of Cognitive Significance: Problems and Changes," in Hempel (1965a), pp. 101–119.
Hempel, C. G. (1968), "Maximal Specificity and Lawlikeness in Probabilistic Explanation," *Philosophy of Science* 35 (June), pp. 116–133.
Hempel, C. G. and P. Oppenheim (1948), "Studies in the Logic of Explanation," *Philosophy of Science* 15, pp. 135–175.
Kuhn, T. S. (1962), *The Structure of Scientific Revolutions*. Chicago.
Popper, K. R. (1935), *Logik der Forschung*. Wien.
Popper, K. R. (1959), *The Logic of Scientific Discovery*. London.
Quine, W.V. O. (1953), "Two Dogmas of Empiricism," in W. V. O. Quine, *From a Logical Point of View* (Harvard), pp. 20–46.
Reichenbach, H. (1949), *Theory of Probability*. Berkeley and Los Angeles.
Salmon, W. C. (1971), *Statistical Explanation and Statistical Relevance*. Pittsburgh.
Salmon, W. C. (1984), *Scientific Explanation and the Causal Structure of the World*. Princeton.
Tarski, A. (1935/36), "Der Wahrheitsbegriff in den formalisierten Sprachern," *Studia Philosophica*, Vol. 1 (1935/36), pp. 261–405.
Wittgenstein, L. (1922), *Tractatus Logico-Philosophicus*. London.
Wittgenstein, L. (1953), *Philosophical Investigations*. Oxford.

I

COGNITIVE SIGNIFICANCE

1

On the Nature of Mathematical Truth

1. The Problem

It is a basic principle of scientific inquiry that no proposition and no theory is to be accepted without adequate grounds. In empirical science, which includes both the natural and the social sciences, the grounds for the acceptance of a theory consist in the agreement of predictions based on the theory with empirical evidence obtained either by experiment or by systematic observation. But what are the grounds which sanction the acceptance of mathematics? That is the question I propose to discuss in the present essay. For reasons which will become clear subsequently, I shall use the term "mathematics" here to refer to arithmetic, algebra, and analysis—to the exclusion, in particular, of geometry.[1]

2. Are the Propositions of Mathematics Self-Evident Truths?

One of the several answers which have been given to our problem asserts that the truths of mathematics, in contradistinction to the hypotheses of empirical science, require neither factual evidence nor any other justification because they are "self-evident." This view, however, which ultimately relegates decisions as to mathematical truth to a feeling of self-evidence, encounters various difficulties. First of all, many mathematical theorems are so hard to establish that even to the specialist in the particular field they appear as anything but self-evident. Secondly, it is well known that some of the most interesting results of mathematics—especially in such fields as abstract set theory and topology—run counter to deeply ingrained intuitions and the customary kind of feeling of self-evidence. Thirdly, the existence of mathematical conjectures such as those of Goldbach and of Fermat, which are quite elementary in content and yet undecided up to this day, certainly shows that not all mathematical truths can be self-evident. And finally, even if self-evidence were attributed only to the basic postulates of mathematics, from which all other mathematical propositions can be deduced, it would be pertinent to remark that judgments as to what may be considered as self-evident are subjective; they may vary from person to person and certainly cannot constitute an adequate basis for decisions as to the objective validity of mathematical propositions.

3. Is Mathematics the Most General Empirical Science?

According to another view, advocated especially by John Stuart Mill, mathematics is itself an empirical science which differs from the other branches such as astronomy, physics, chemistry, etc., mainly in two respects: its subject matter is more general than that of any other field of scientific research, and its propositions have been tested and confirmed to a greater extent than those of even the most firmly established sections of astronomy or physics. Indeed, according to this view, the degree to which the laws of mathematics have been borne out by the past experiences of mankind is so overwhelming that—unjustifiably—we have come to think of mathematical theorems as qualitatively different from the well-confirmed hypotheses or theories of other branches of science: we consider them as certain, while other theories are thought of as at best "very probable" or very highly confirmed.

But this view, too, is open to serious objections. From a hypothesis which is empirical in character—such as, for example, Newton's law of gravitation—it is possible to derive predictions to the effect that under certain specified conditions certain specified observable phenomena will occur. The actual occurrence of these phenomena constitutes confirming evidence, their nonoccurrence disconfirming evidence for the hypothesis. It follows in particular that an empirical hypothesis is theoretically disconfirmable; i.e., it is possible to indicate what kind of evidence, if actually encountered, would disconfirm the hypothesis. In the light of this remark, consider now a simple "hypothesis" from arithmetic: $3+2 = 5$. If this is actually an empirical generalization of past experiences, then it must be possible to state what kind of evidence would oblige us to concede the hypothesis was not generally true after all. If any disconfirming evidence for the given proposition can be thought of, the following illustration might well be typical of it: We place some microbes on a slide, putting down first three of them and then another two. Afterward we count all the microbes to test whether in this instance 3 and 2 actually added up to 5. Suppose now that we counted 6 microbes altogether. Would we consider this as an empirical disconfirmation of the given proposition, or at least as a proof that it does not apply to microbes? Clearly not; rather, we would assume we had made a mistake in counting or that one of the microbes had split in two between the first and the second count. But under no circumstances could the phenomenon just described invalidate the arithmetical proposition in question; for the latter asserts nothing whatever about the behavior of microbes; it merely states that any set consisting of $3 + 2$ objects may also be said to consist of 5 objects.

And this is so because the symbols "3 + 2" and "5" denote the same number: they are synonymous by virtue of the fact that the symbols "2," "3," "5," and "+" are *defined* (or tacitly understood) in such a way that the above identity holds as a consequence of the meaning attached to the concepts involved in it.

4. The Analytic Character of Mathematical Propositions

The statement that 3 + 2 = 5, then, is true for similar reasons as, say, the assertion that no sexagenarian is 45 years of age. Both are true simply by virtue of definitions or of similar stipulations which determine the meaning of the key terms involved. Statements of this kind share certain important characteristics: Their validation naturally requires no empirical evidence; they can be shown to be true by a mere analysis of the meaning attached to the terms which occur in them. In the language of logic, sentences of this kind are called analytic or true a priori, which is to indicate that their truth is logically independent of, or logically prior to, any experiential evidence.[2] And while the statements of empirical science, which are synthetic and can be validated only a posteriori, are constantly subject to revision in the light of new evidence, the truth of an analytic statement can be established definitely, once and for all. However, this characteristic "theoretical certainty" of analytic propositions has to be paid for at a high price: An analytic statement conveys no factual information. Our statement about sexagenarians, for example, asserts nothing that could possibly conflict with any factual evidence: it has no factual implications, no empirical content; and it is precisely for this reason that the statement can be validated without recourse to empirical evidence.

Let us illustrate this view of the nature of mathematical propositions by reference to another, frequently cited, example of a mathematical—or rather logical—truth, namely the proposition that whenever $a = b$ and $b = c$ then $a = c$. On what grounds can this so-called "transitivity of identity" be asserted? Is it of an empirical nature and hence at least theoretically disconfirmable by empirical evidence? Suppose, for example, that $a, b, c,$ are certain shades of green, and that as far as we can see, $a = b$ and $b = c$, but clearly $a \neq c$. This phenomenon actually occurs under certain conditions; do we consider it as disconfirming evidence for the proposition under consideration? Undoubtedly not; we would argue that if $a \neq c$, it is impossible that $a = b$ and also $b = c$; between the terms of at least one of these latter pairs, there must obtain a difference, though perhaps only a subliminal one. And we would dismiss the possibility of empirical disconfirmation, and indeed the idea that an empirical test should be relevant here, on the grounds that identity is a transitive relation by virtue of its definition or by virtue of the basic postulates governing it.[3] Hence, the principle in question is true a priori.

5. Mathematics as an Axiomatized Deductive System

I have argued so far that the validity of mathematics rests neither on its alleged self-evident character nor on any empirical basis, but derives from the stipulations which determine the meaning of the mathematical concepts, and that the propositions of mathematics are therefore essentially "true by definition." This latter statement, however, is obviously oversimplified and needs restatement and a more careful justification.

For the rigorous development of a mathematical theory proceeds not simply from a set of definitions but rather from a set of nondefinitional propositions which are not proved within the theory; these are the postulates or axioms of the theory.[4] They are formulated in terms of certain basic or primitive concepts for which no definitions are provided within the theory. It is sometimes asserted that the postulates themselves represent "implicit definitions" of the primitive terms. Such a characterization of the postulates, however, is misleading. For while the postulates do limit, in a specific sense, the meanings that can possibly be ascribed to the primitives, any self-consistent postulate system admits, nevertheless, many different interpretations of the primitive terms (this will soon be illustrated), whereas a set of definitions in the strict sense of the word determines the meanings of the definienda in a unique fashion.

Once the primitive terms and the postulates have been laid down, the entire theory is completely determined; it is derivable from its postulational basis in the following sense: Every term of the theory is definable in terms of the primitives, and every proposition of the theory is logically deducible from the postulates. To be entirely precise, it is necessary also to specify the principles of logic which are to be used in the proof of the propositions, i.e., their deduction from the postulates. These principles can be stated quite explicitly. They fall into two groups : primitive sentences, or postulates, of logic (such as: If p and q is the case, then p is the case), and rules of deduction or inference (including, for example, the familiar *modus ponens* rule and the rules of substitution which make it possible to infer, from a general proposition, any one of its substitution instances). A more detailed discussion of the structure and content of logic would, however, lead too far afield in the context of this article.

6. Peano's Axiom System as a Basis for Mathematics

Let us now consider a postulate system from which the entire arithmetic of the natural numbers can be derived. This system was devised by the Italian mathematician and logician G. Peano (1858–1932). The primitives of this system are the terms "0," "number," and "successor." While, of course, no definition of these terms is given within the theory, the symbol "0" is intended to designate the number 0 in its usual meaning, while the term "number" is meant to refer to the natural numbers 0, 1, 2, 3 . . . exclusively. By the successor of a natural number n, which will sometimes briefly be called n', is meant the natural number immediately following n in the natural order. Peano's system contains the following 5 postulates:

P1. 0 is a number
P2. The successor of any number is a number
P3. No two numbers have the same successor
P4. 0 is not the successor of any number
P5. If P is a property such that (a) 0 has the property P, and (b) whenever a number n has the property P, then the successor of n also has the property P, then every number has the property P.

The last postulate embodies the principle of mathematical induction and illustrates in a very obvious manner the enforcement of a mathematical "truth" by stipulation. The construction of elementary arithmetic on this basis begins with the definition of the various natural numbers. 1 is defined as the successor of 0, or briefly as $0'$; 2 as $1'$, 3 as $2'$, and so on. By virtue of P2, this process can be continued indefinitely; because of P3 (in combination with P5), it never leads back to one of the numbers previously defined, and in view of P4, it does not lead back to 0 either.

As the next step, we can set up a definition of addition which expresses in a precise form the idea that the addition of any natural number to some given number may be considered as a repeated addition of 1; the latter operation is readily expressible by means of the successor relation. This definition of addition runs as follows:

D1. (a) $n + 0 = n$; (b) $n + k' = (n + k)'$.

The two stipulations of this recursive definition completely determine the sum of any two integers. Consider, for example, the sum $3 + 2$. According to the definitions of the numbers 2 and 1, we have $3 + 2 = 3 + 1' = 3 + (0')'$; by D1 (b), $3 + (0')' = (3 + 0')' = ((3 + 0)')'$; but by D1 (a), and by the definitions of the numbers 4 and 5, $((3 + 0)')' = (3')' = 4' = 5$. This proof also renders more explicit and precise the comments made earlier in this paper on the truth of the proposition that $3 + 2 = 5$: Within the Peano system of arithmetic, its truth flows not merely from the definition of the concepts involved, but also from the postulates that govern these various concepts. (In our specific example, the postulates P1 and P2 are presupposed to guarantee that 1, 2, 3, 4, 5 are numbers in Peano's system; the general proof that D1 determines the sum of any two numbers also makes use of P5.) If we call the postulates and definitions of an axiomatized theory the "stipulations" concerning the concepts of that theory, then we may say now that the propositions of the arithmetic of the natural numbers are true by virtue of the stipulations which have been laid down initially for the arithmetical concepts. (Note, incidentally, that our proof of the formula "$3 + 2 = 5$" repeatedly made use of the transitivity of identity; the latter is accepted here as one of the rules of logic which may be used in the proof of any arithmetical theorem; it is, therefore, included among Peano's postulates no more than any other principle of logic.)

Now, the multiplication of natural numbers may be defined by means of the following recursive definition, which expresses in a rigorous form the idea that a product nk of two integers may be considered as the sum of k terms each of which equals n.

D2. (a) $n \cdot 0 = 0$; (b) $n \cdot k' = n \cdot k + n$.

It now is possible to prove the familiar general laws governing addition and multiplication, such as the commutative, associative, and distributive laws ($n + k = k + n$, $n \cdot k = kn$; $n + (k + l) = (n + k) + l$, $n \cdot (k \cdot l) = (n \cdot k) \cdot l$; $n \cdot (k + l) = (n \cdot k) + (n \cdot l)$). In terms of addition and multiplication, the inverse operations of subtraction and division can then be defined. But it turns out that these "cannot always be performed"; i.e., in contradistinction to the sum and the product, the difference and the quotient are not defined for every couple of numbers; for example, 7 − 10 and 7 ÷ 10 are undefined. This situation suggests an enlargement of the number system by the introduction of negative and of rational numbers.

It is sometimes held that in order to effect this enlargement, we have to "assume" or else to "postulate" the existence of the desired additional kinds of numbers with properties that make them fit to fill the gaps of subtraction and division. This method of simply postulating what we want has its advantages; but, as Bertrand Russell[5] puts it, they are the same as the advantages of theft over honest toil; and it is a remarkable fact that the negative as well as the rational numbers can be obtained from Peano's primitives by the honest toil of constructing explicit definitions for them, without the introduction of any new postulates or assumptions whatsoever. Every positive and negative integer—in contradistinction to a natural number which has no sign—is definable as a certain set of ordered couples of natural numbers; thus, the integer +2 is definable as the set of all ordered couples (m, n) of natural numbers where $m = n + 2$; the integer −2 is the set of all ordered couples (m, n) of natural numbers with $n = m + 2$. Similarly, rational numbers are defined as classes of ordered couples of integers. The various arithmetical operations can then be defined with reference to these new types of numbers, and the validity of all the arithmetical laws governing these operations can be proved by virtue of nothing more than Peano's postulates and the definitions of the various arithmetical concepts involved.

The much broader system thus obtained is still incomplete in the sense that not every number in it has a square root, and more generally, not every algebraic equation whose coefficients are all numbers of the system has a solution in the system. This suggests further expansions of the number system by the introduction of real and finally of complex numbers. Again, this enormous extension can be affected by mere definition, without the introduction of a single new postulate.[6] On the basis thus obtained, the various arithmetical and algebraic operations can be defined for the numbers of the new system, the concepts of function, of limit, of derivative and integral can be introduced, and the familiar theorems pertaining to these concepts can be proved, so that finally the huge system of mathematics as here delimited rests on the narrow basis of Peano's system: Every concept of mathematics can be defined by means of Peano's three primitives, and every proposition of mathematics can be deduced from the five postulates enriched by the definitions of the nonprimitive terms.[7] These deductions can be carried out, in most cases, by means of nothing more than the principles of formal logic; the proof of some theorems concerning real numbers, however, requires one assumption which is not usually included among the latter. This is the so-called axiom of choice. It asserts that given a class of mutually exclusive classes, none of which is empty, there exists at least one class which has exactly one element in common with each of the given classes. By virtue of this principle and the rules of formal logic, the content of all of mathematics can thus be derived from Peano's modest system—a remarkable achievement in systematizing the content of mathematics and clarifying the foundations of its validity.

7. Interpretations of Peano's Primitives

As a consequence of this result, the whole system of mathematics might be said to be true by virtue of mere definitions (namely, of the nonprimitive mathematical terms) provided that the five Peano postulates are true. However, strictly speaking, we cannot, at this juncture, refer to the Peano postulates as propositions which are either true or false, for they contain three primitive terms which have not been assigned any specific meaning. All we can assert so far is that any specific interpretation of the primitives which satisfies the five postulates—i.e., turns them into true statements—will also satisfy all the theorems deduced from them. But for Peano's system, there are several—indeed, infinitely many—interpretations which will do this. For example, let us understand by 0 the origin of a half-line, by the successor of a point on that half-line the point 1 cm behind it, counting from the origin, and by a number any point which is either the origin or can be reached from it by a finite succession of steps each of which leads from one point to its successor. It can then readily be seen that all the Peano postulates as well as the ensuing theorems turn into true propositions, although the interpretation given to the primitives is certainly not the customary one, which was mentioned earlier. More generally, it can be shown that every progression of elements of any kind provides a true interpretation, or a "model," of the Peano system. This example illustrates our earlier observation that a postulate system cannot be regarded as a set of "implicit definitions" for the primitive terms: The Peano system permits many different interpretations, whereas in everyday as well as in scientific language, we attach one specific meaning to the concepts of arithmetic. Thus, e.g., in scientific and in everyday discourse, the concept 2 is understood in such a way that from the statement "Mr. Brown as well as Mr. Cope, but no one else is in the office, and

Mr. Brown is not the same person as Mr. Cope," the conclusion "Exactly two persons are in the office" may be validly inferred. But the stipulations laid down in Peano's system for the natural numbers, and for the number 2 in particular, do not enable us to draw this conclusion; they do not "implicitly determine" the customary meaning of the concept 2 or of the other arithmetical concepts. And the mathematician cannot acquiesce at this deficiency by arguing that he is not concerned with the customary meaning of the mathematical concepts; for in proving, say, that every positive real number has exactly two real square roots, he is himself using the concept 2 in its customary meaning, and his very theorem cannot be proved unless we presuppose more about the number 2 than is stipulated in the Peano system.

If therefore mathematics is to be a correct theory of the mathematical concepts in their intended meaning, it is not sufficient for its validation to have shown that the entire system is derivable from the Peano postulates plus suitable definitions; rather, we have to inquire further whether the Peano postulates are actually true when the primitives are understood in their customary meaning. This question, of course, can be answered only after the customary meaning of the terms "0," "natural number," and "successor" has been clearly defined. To this task we now turn.

8. Definition of the Customary Meaning of the Concepts of Arithmetic in Purely Logical Terms

At first blush, it might seem a hopeless undertaking to try to define these basic arithmetical concepts without presupposing other terms of arithmetic, which would involve us in a circular procedure. However, quite rigorous definitions of the desired kind can indeed be formulated, and it can be shown that for the concepts so defined, all Peano postulates turn into true statements. This important result is due to the research of the German logician G. Frege (1848–1925) and to the subsequent systematic and detailed work of the contemporary English logicians and philosophers B. Russell and A. N. Whitehead. Let us consider briefly the basic ideas underlying these definitions.[8]

A natural number—or, in Peano's term, a number—in its customary meaning can be considered as a characteristic of certain *classes* of objects. Thus, e.g., the class of the apostles has the number 12, the class of the Dionne quintuplets the number 5, any couple the number 2, and so on. Let us now express precisely the meaning of the assertion that a certain class C has the number 2, or briefly, that $n(C) = 2$. Brief reflection will show that the following definiens is adequate in the sense of the customary meaning of the concept 2: There is some object x and some object y such that (1) $x \varepsilon C$ (i.e., x is an element of C) and $y \varepsilon C$, (2) $x \neq y$, and (3) if z is any object such that $z \varepsilon C$, then either $z = x$ or $z = y$. (Note that on the basis of this definition it becomes indeed possible to infer the statement "The number of persons in the office is 2" from "Mr. Brown as well as Mr. Cope, but no one else is in the office, and Mr. Brown is not identical with

Mr. Cope"; C is here the class of persons in the office.) Analogously, the meaning of the statement that $n(C) = 1$ can be defined thus: There is some x such that $x \varepsilon C$, and any object y such that $y \varepsilon C$, is identical with x. Similarly, the customary meaning of the statement that $n(C) = 0$ is this: There is no object such that $x \varepsilon C$.

The general pattern of these definitions clearly lends itself to the definition of any natural number. Let us note especially that in the definitions thus obtained, the definiens never contains any arithmetical term, but merely expressions taken from the field of formal logic, including the signs of identity and difference. So far, we have defined only the meaning of such phrases as "$n(C) = 2$," but we have given no definition for the numbers 0, 1, 2, . . . apart from this context. This desideratum can be met on the basis of the consideration that 2 is that property which is common to all couples, i.e., to all classes C such that $n(C) = 2$. This common property may be conceptually represented by the class of all those classes which share this property. Thus, we arrive at the definition: 2 is the class of all couples, i.e., the class of all classes C for which $n(C) = 2$. This definition is by no means circular because the concept of couple—in other words, the meaning of "$n(C) = 2$"—has been previously defined without any reference to the number 2. Analogously, 1 is the class of all unit classes, i.e., the class of all classes C for which $n(C) = 1$. Finally, 0 is the class of all null classes, i.e., the class of all classes without elements. And as there is only one such class, 0 is simply the class whose only element is the null class. Clearly, the customary meaning of any given natural number can be defined in this fashion.[9] In order to characterize the intended interpretation of Peano's primitives, we actually need, of all the definitions here referred to, only that of the number 0. It remains to define the terms "successor" and "integer."

The definition of "successor," whose precise formulation involves too many niceties to be stated here, is a careful expression of a simple idea which is illustrated by the following example: Consider the number 5, i.e., the class of all quintuplets. Let us select an arbitrary one of these quintuplets and add to it an object which is not yet one of its members. 5', the successor of 5, may then be defined as the number applying to the set thus obtained (which, of course, is a sextuplet). Finally, it is possible to formulate a definition of the customary meaning of the concept of natural number; this definition, which again cannot be given here, expresses, in a rigorous form, the idea that the class of the natural numbers consists of the number 0, its successor, the successor of that successor, and so on.

If the definitions here characterized are carefully written out—this is one of the cases where the techniques of symbolic, or mathematical, logic prove indispensable—it is seen that the definiens of every one of them contains exclusively terms from the field of pure logic. In fact, it is possible to state the customary interpretation of Peano's primitives, and thus also the meaning of every concept definable by means of them—and that includes every concept of mathematics—in terms of the following 7 expressions, in addition to variables such as "x" and "C": *not; and; if—then; for every object x it is the case that . . . ; there is some object x such that . . . ; x is an element of class C; the* *class of all things x such that.* . . . And it is even possible to reduce the number of logical concepts needed to a mere four: The first three of the concepts just mentioned are all definable in terms of *"neither—nor,"* and the fifth is definable by means of the fourth and *"neither—nor."* Thus, all the concepts of mathematics prove definable in terms of four concepts of pure logic. (The definition of one of the more complex concepts of mathematics in terms of the four primitives just mentioned may well fill hundreds or even thousands of pages; but clearly this affects in no way the theoretical importance of the result just obtained; it does, however, show the great convenience and indeed practical indispensability for mathematics of having a large system of highly complex defined concepts available.)

9. The Truth of Peano's Postulates in Their Customary Interpretation

The definitions characterized in the preceding section may be said to render precise and explicit the customary meaning of the concepts of arithmetic. Moreover—and this is crucial for the question of the validity of mathematics—it can be shown that the Peano postulates all turn into true propositions if the primitives are construed in accordance with the definitions just considered.

Thus, P1 (0 is a number) is true because the class of all numbers—i.e., natural numbers—was defined as consisting of 0 and all its successors. The truth of P2 (The successor of any number is a number) follows from the same definition. This is true also of P5, the principle of mathematical induction. To prove this, however, we would have to resort to the precise definition of "integer" rather than the loose description given of that definition above. P4 (0 is not the successor of any number) is seen to be true as follows: By virtue of the definition of "successor," a number which is a successor of some number can apply only to classes which contain at least one element; but the number 0, by definition, applies to a class if and only if that class is empty. While the truth of P1, P2, P4, P5 can be inferred from the above definitions simply by means of the principles of logic, the proof of P3 (No two numbers have the same successor) presents a certain difficulty. As was mentioned in the preceding section, the definition of the successor of a number n is based on the process of adding, to a class of n elements, one element not yet contained in that class. Now if there should exist only a finite number of things altogether then this process could not be continued indefinitely, and P3, which (in conjunction with P1 and P2) implies that the integers form an infinite set, would be false. Russell's way of meeting this difficulty[10] was to introduce a special "axiom of infinity," which stipulates, in effect, the existence of infinitely many objects and thus makes P3 demonstrable. The axiom of infinity can be formulated in purely logical terms and may therefore be considered as a postulate of logic; however, it certainly does not belong to the generally recognized principles of logic; and it thus introduces a foreign element into the otherwise unexceptionable derivation of the

Peano postulates from pure logic. Recently, however, it has been shown[11] that a suitable system of logical principles can be set up which is even less comprehensive than the rules of logic which are commonly used,[12] and in which the existence of infinitely many objects can be proved without the need for a special axiom.

10. Mathematics as a Branch of Logic

As was pointed out earlier, all the theorems of arithmetic, algebra, and analysis can be deduced from the Peano postulates and the definitions of those mathematical terms which are not primitives in Peano's system. This deduction requires only the principles of logic plus, in certain cases, the axiom of choice. By combining this result with what has just been said about the Peano system, the following conclusion is obtained, which is also known as *the thesis of logicism concerning the nature of mathematics*:

Mathematics is a branch of logic. It can be derived from logic in the following sense:

a. All the concepts of mathematicst, i.e., of arithmetic, algebra, and analysis, can be defined in terms of four concepts o! pure logic.
b. All the theorems of mathematics can be deduced from those definitions by means of the principles of logic (including the axiom of choice).

In this sense it can be said that the propositions of the system of mathematics as here delimited are true by virtue of the definitions of the mathematical concepts involved, or that they make explicit certain characteristics with which we have endowed our mathematical concepts by definition. The propositions of mathematics have, therefore, the same unquestionable certainty which is typical of such propositions as "All bachelors are unmarried," but they also share the complete lack of empirical content which is associated with that certainty: The propositions of mathematics are devoid of all factual content; they convey no information whatever on any empirical subject matter.

11. On the Applicability of Mathematics to Empirical Subject Matter

This result seems to be irreconcilable with the fact that after all mathematics has proved to be eminently applicable to empirical subject matter, and that indeed the greater part of present-day scientific knowledge has been reached only through continual reliance on and application of the propositions of mathematics. Let us try to clarify this apparent paradox by reference to some examples.

Suppose that we are examining a certain amount of some gas, whose volume v, at a certain fixed temperature, is found to be 9 cubic feet when the pressure p is 4 atmospheres. And let us assume further that the volume of the gas for the same temperature and $p = 6$ at., is predicted by means of Boyle's law. Using elementary arithmetic we reason thus: For corresponding values of v and p, $vp = c$, and $v = 9$ when $p = 4$; hence $c = 36$: Therefore, when $p = 6$, then $v = 6$. Suppose that this prediction is borne out by subsequent test. Does that show that the arithmetic used has a predictive power of its own, that its propositions have factual implications? Certainly not. All the predictive power here deployed, all the empirical content exhibited stems from the initial data and from Boyle's law, which asserts that $vp = c$ for *any* two corresponding values of v and p, hence also for $v = 9$, $p = 4$, and for $p = 6$ and the corresponding value of v.[13] The function of the mathematics here applied is not predictive at all; rather, it is analytic or explicative: it renders explicit certain assumptions or assertions which are included in the content of the premises of the argument (in our case, these consist of Boyle's law plus the additional data); mathematical reasoning reveals that those premises contain—hidden in them, as it were,—an assertion about the case as yet unobserved. In accepting our premises—so arithmetic reveals—we have—knowingly or unknowingly—already accepted the implication that the p-value in question is 6. Mathematical as well as logical reasoning is a conceptual technique of making explicit what is implicitly contained in a set of premises. The conclusions to which this technique leads assert nothing that is *theoretically new* in the sense of not being contained in the content of the premises. But the results obtained may well be *psychologically new*: we may not have been aware, before using the techniques of logic and mathematics, what we committed ourselves to in accepting a certain set of assumptions or assertions.

A similar analysis is possible in all other cases of applied mathematics, including those involving, say, the calculus. Consider, for example, the hypothesis that a certain object, moving in a specified electric field, will undergo a constant acceleration of 5 feet/sec². For the purpose of testing this hypothesis, we might derive from it, by means of two successive integrations, the prediction that if the object is at rest at the beginning of the motion, then the distance covered by it at any time t is $\frac{5}{2}t^2$ feet. This conclusion may clearly be psychologically new to a person not acquainted with the subject, but it is not theoretically new; the content of the conclusion is already contained in that of the hypothesis about the constant acceleration. And indeed, here as well as in the case of the compression of a gas, a failure of the prediction to come true would be considered as indicative of the factual incorrectness of at least one of the premises involved (e.g., of Boyle's law in its application to the particular gas), but never as a sign that the logical and mathematical principles involved might be unsound.

Thus, in the establishment of empirical knowledge, mathematics (as well as logic) has, so to speak, the function of a theoretical juice extractor: the techniques of mathematical and logical theory can produce no more juice of factual information than is contained in the assumptions to which they are applied; but they may produce a great deal more juice of this kind than might have been anticipated upon a first intuitive inspection of those assumptions which form the raw material for the extractor.

At this point, it may be well to consider briefly the status of those mathematical disciplines which are not outgrowths of arithmetic and thus of logic; these include, in particular, topology, geometry, and the various branches of abstract algebra, such as the theory of groups, lattices, fields, etc. Each of these disciplines can be developed as a purely deductive system on the basis of a suitable set of postulates. If P be the conjunction of the postulates for a given theory, then the proof of a proposition T of that theory consists in deducing T from P by means of the principles of formal logic. What is established by the proof is therefore not the truth of T, but rather the fact that T is true provided that the postulates are. But since both P and T contain certain primitive terms of the theory, to which no specific meaning is assigned, it is not strictly possible to speak of the truth of either P or T; it is therefore more adequate to state the point as follows: If a proposition T is logically deduced from P, then every specific interpretation of the primitives which turns all the postulates of P into true statements, will also render T a true statement. Up to this point, the analysis is exactly analogous to that of arithmetic as based on Peano's set of postulates. In the case of arithmetic, however, it proved possible to go a step further, namely to define the customary meanings of the primitives in terms of purely logical concepts and to show that the postulates—and therefore also the theorems—of arithmetic are unconditionally true by virtue of these definitions. An analogous procedure is not applicable to those disciplines which are not outgrowths of arithmetic: The primitives of the various branches of abstract algebra have no specific "customary meaning"; and if geometry in its customary interpretation is thought of as a theory of the structure of physical space, then its primitives have to be construed as referring to certain types of physical entities, and the question of the truth of a geometrical theory in this interpretation turns into an *empirical* problem.[14] For the purpose of applying any one of these nonarithmetical disciplines to some specific field of mathematics or empirical science, it is therefore necessary first to assign to the primitives some specific meaning and then to ascertain whether in this interpretation the postulates turn into true statements. If this is the case, then we can be sure that all the theorems are true statements too, because they are logically derived from the postulates and thus simply explicate the content of the latter in the given interpretation. In their application to empirical subject matter, therefore, these mathematical theories no less than those which grow out of arithmetic and ultimately out of pure logic, have the function of an analytic tool, which brings to light the implications of a given set of assumptions but adds nothing to their content.

But while mathematics in no case contributes anything to the content of our knowledge of empirical matters, it is entirely indispensable as an instrument for the validation and even for the linguistic expression of such knowledge: The majority of the more far-reaching theories in empirical science—including those which lend themselves most eminently to prediction or to practical application—are stated with the help of mathematical concepts; the formulation of these theories makes use, in particular, of the number system, and of functional relationships among different metrical variables. Furthermore, the scientific test of these theories, the establishment of predictions by means of them, and finally their practical application, all require the deduction, from the general theory, of certain specific consequences; and such deduction would be entirely impossible without the techniques of mathematics which reveal what the given general theory implicitly asserts about a certain special case.

Thus, the analysis outlined on these pages exhibits the system of mathematics as a vast and ingenious conceptual structure without empirical content and yet an indispensable and powerful theoretical instrument for the scientific understanding and mastery of the world of our experience.

NOTES

1. A discussion of the status of geometry is given in my article, "Geometry and Empirical Science," *American Mathematical Monthly*, vol. 52, pp. 7–17, 1945.
2. The objection is sometimes raised that without certain types of experience, such as encountering several objects of the same kind, the integers and the arithmetical operations with them would never have been invented, and that therefore the propositions of arithmetic do have an empirical basis. This type of argument, however, involves a confusion of the logical and the psychological meaning of the term "basis." It may very well be the case that certain experiences occasion psychologically the formation of arithmetical ideas and in this sense form an empirical "basis" for them; but this point is entirely irrelevant for the logical questions as to the *grounds* on which the propositions of arithmetic may be accepted as true. The point made above is that no empirical "basis" or evidence whatever is needed to establish the truth of the propositions of arithmetic.
3. A precise account of the definition and the essential characteristics of the identity relation may be found in A. Tarski, *Introduction to Logic*, New York, 1941, Ch. III.
4. For a lucid and concise account of the axiomatic method, see A. Tarski, ibid., Ch. VI.
5. Bertrand Russell, *Introduction to Mathematical Philosophy*, New York and London, 1919, p. 71.
6. For a more detailed account of the construction of the number system on Peano's basis, cf. Bertrand Russell, ibid., esp. Chs. I and VII. A rigorous and concise presentation of that construction, beginning, however, with the set of all integers rather than that of the natural numbers, may be found in G. Birkhoff and S. MacLane, *A Survey of Modern Algebra*, New York 1941, Chs. I, II, III, V. For a general survey of the construction of the number system, cf. also J. W. Young, *Lectures on the Fundamental Concepts of Algebra and Geometry*, New York, 1911, esp. lectures X, XI, XII.
7. As a result of very deep-reaching investigations carried out by K. Godel it is known that arithmetic, and a fortiori mathematics, is an incomplete theory in the following sense: While all those propositions which belong to the classical systems of arithmetic, algebra, and analysis can indeed be derived, in the sense characterized above, from the Peano postulates, there exist nevertheless other propositions which can be expressed in purely arithmetical terms, and which are true, but which cannot be derived from the Peano system. And more generally: For any postulate system of arithmetic (or of mathematics for that matter) which is not self-contradictory, there exist propositions which are true, and which can be stated in purely arithmetical terms, but which cannot be derived from that postulate system. In other words, it is impossible to con-

struct a postulate system which is not self-contradictory, and which contains among its consequences all true propositions which can be formulated within the language of arithmetic.

This fact does not, however, affect the result outlined above, namely, that it is possible to deduce, from the Peano postulates and the additional definitions of nonprimitive terms, all those propositions which constitute the classical theory of arithmetic, algebra, and analysis; and it is to these propositions that I refer above and subsequently as the propositions of mathematics.

8. For a more detailed discussion, cf. Russell, *Introduction to Mathematical Philosophy*, Chs. II, III, IV. A complete technical development of the idea can be found in the great standard work in mathematical logic, A. N. Whitehead and B. Russell, *Principia Mathematica*, Cambridge, England, 1910–1913. For a very precise development of the theory, see W. V. O. Quine, *Mathematical Logic*, New York 1940. A specific discussion of the Peano system and its interpretations from the viewpoint of semantics is included in R. Carnap, *Foundations of Logic and Mathematics*, International Encyclopedia of Unified Science, vol. I, no. 3, Chicago, 1939; especially sections 14, 17, 18.

9. The assertion that the definitions given above state the "customary" meaning of the arithmetical terms involved is to be understood in the logical, not the psychological sense of the term "meaning." It would obviously be absurd to claim that the above definitions express "what everybody has in mind" when talking about numbers and the various operations that can be performed with them. What is achieved by those definitions is rather a "logical reconstruction" of the concepts of arithmetic in the sense that if the definitions are accepted, then those statements in science and everyday discourse which involve arithmetical terms can be interpreted coherently and systematically in such a manner that they are capable of objective validation. The statement about the two persons in the office provides a very elementary illustration of what is meant here.

10. Cf. Bertrand Russell, *Introduction to Mathematical Philosophy*, p. 24 and Ch. XIII.

11. This result has been obtained by W. V. O. Quine; cf. his *Mathematical Logic*.

12. The principles of logic developed in Quine's work and in similar modern systems of formal logic embody certain restrictions as compared with those logical rules which had been rather generally accepted as sound until about the turn of the twentieth century. At that time, the discovery of the famous paradoxes of logic, especially of Russell's paradox (cf. Russell, *Introduction to Mathematical Philosophy*, Ch. XIII) revealed the fact that the logical principles implicit in customary mathematical reasoning involved contradictions and therefore had to be curtailed in one manner or another.

13. Note that we may say "hence" by virtue of the rule of substitution, which is one of the rules of logical inference.

14. For a more detailed discussion of this point, cf. the article mentioned in note 1.

2

Geometry and Empirical Science

1. Introduction

The most distinctive characteristic which differentiates mathematics from the various branches of empirical science, and which accounts for its fame as the queen of the sciences, is no doubt the peculiar certainty and necessity of its results. No proposition in even the most advanced parts of empirical science can ever attain this status; a hypothesis concerning "matters of empirical fact" can at best acquire what is loosely called a high probability or a high degree of confirmation on the basis of the relevant evidence available; but however well it may have been confirmed by careful tests, the possibility can never be precluded that it will have to be discarded later in the light of new and disconfirming evidence. Thus, all the theories and hypotheses of empirical science share this provisional character of being established and accepted "until further notice," whereas a mathematical theorem, once proved, is established once and for all; it holds with that particular certainty which no subsequent empirical discoveries, however unexpected and extraordinary, can ever affect to the slightest extent. It is the purpose of this essay to examine the nature of that proverbial "mathematical certainty" with special reference to geometry, in an attempt to shed some light on the question as to the validity of geometrical theories, and their significance for our knowledge of the structure of physical space.

The nature of mathematical truth can be understood through an analysis of the method by means of which it is established. On this point I can be very brief: it is the method of mathematical demonstration, which consists in the logical deduction of the proposition to be proved from other propositions, previously established. Clearly, this procedure would involve an infinite regress unless some propositions were accepted without proof; such propositions are indeed found in every mathematical discipline which is rigorously developed; they are the *axioms* or *postulates* (we shall use these terms interchangeably) of the theory. Geometry provides the historically first example of the axiomatic presentation of a mathematical discipline. The classical set of postulates, however, on which Euclid based his system, has proved insufficient for the deduction of the well-known theorems of so-called Euclidean geometry; it has therefore been revised and supplemented in modern times, and at present various adequate systems of postulates for euclidean geometry are available; the one most closely related to Euclid's system is probably that of Hilbert.

2. The Inadequacy of Euclid's Postulates

The inadequacy of Euclid's own set of postulates illustrates a point which is crucial for the axiomatic method in modern mathematics: Once the postulates for a theory have been laid down, every further proposition of the theory must be proved exclusively by logical deduction from the postulates; any appeal, explicit or implicit, to a feeling of self-evidence, or to the characteristics of geometrical figures, or to our experiences concerning the behavior of rigid bodies in physical space, or the like, is strictly prohibited; such devices may have a heuristic value in guiding our efforts to find a strict proof for a theorem, but the proof itself must contain absolutely no reference to such aids. This is particularly important in geometry, where our so-called intuition of geometrical relationships, supported by reference to figures or to previous physical experiences, may induce us tacitly to make use of assumptions which are neither formulated in our postulates nor provable by means of them. Consider, for example, the theorem that in a triangle the three medians bisecting the sides intersect in one point which divides each of them in the ratio of 1:2. To prove this theorem, one shows first that in any triangle ABC (see figure) the line segment MN which connects the centers of AB and AC is parallel to BC and therefore half as long as the latter side. Then the lines BN and CM are drawn, and an examination of the triangles MON and BOC leads to the proof of the theorem. In this procedure, it is usually taken for granted that BN and CM intersect in a point O which lies between B and N as well as between C and M. This assumption is based on geometrical intuition, and indeed, it cannot be deduced from Euclid's postulates; to make it strictly demonstrable and independent of any reference to intuition, a special group of postulates has been added to those of Euclid; they are the postulates of order. One of these—to give an example—asserts that if A, B, C are points on a straight line l, and if B lies between A and C, then B also lies between C and A. Not even as "trivial" an assumption as this may be taken for granted; the system of postulates has to

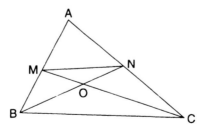

be made so complete that all the required propositions can be deduced from it by purely logical means.

Another illustration of the point under consideration is provided by the proposition that triangles which agree in two sides and the enclosed angle, are congruent. In Euclid's *Elements*, this proposition is presented as a theorem; the alleged proof, however, makes use of the ideas of motion and superimposition of figures and thus involves tacit assumptions which are based on our geometric intuition and on experiences with rigid bodies, but which are definitely not warranted by—i.e., deducible from—Euclid's postulates. In Hilbert's system, therefore, this proposition (more precisely: part of it) is explicitly included among the postulates.

3. Mathematical Certainty

It is this purely deductive character of mathematical proof which forms the basis of mathematical certainty: What the rigorous proof of a theorem—say the proposition about the sum of the angles in a triangle—establishes is not the truth of the proposition in question but rather a conditional insight to the effect that that proposition is certainly true *provided that* the postulates are true; in other words, the proof of a mathematical proposition establishes the fact that the latter is logically implied by the postulates of the theory in question. Thus, each mathematical theorem can be cast into the form

$$(P_1 \cdot P_2 \cdot P_3 \cdot \cdots \cdot P_N) \rightarrow T$$

where the expression on the left is the conjunction (joint assertion) of all the postulates, the symbol on the right represents the theorem in its customary formulation, and the arrow expresses the relation of logical implication or entailment. Precisely this character of mathematical theorems is the reason for their peculiar certainty and necessity, as I shall now attempt to show.

It is typical of any purely logical deduction that the conclusion to which it leads simply reasserts (a proper or improper) part of what has already been stated in the premises. Thus, to illustrate this point by a very elementary example, from the premise, "This figure is a right triangle," we can deduce the conclusion, "This figure is a triangle"; but this conclusion clearly reiterates part of the information already contained in the premise. Again, from the premises, "All primes different from 2 are odd" and "n is a prime different from 2," we can infer logically that n is odd; but this consequence merely repeats part (indeed a relatively small part) of the information contained in the premises. The same situation prevails in all other cases of logical deduction; and we may, therefore, say that logical deduction—which is the one and only method of mathematical proof—is a technique of conceptual analysis: it discloses what assertions are concealed in a given set of premises, and it makes us realize to what we committed ourselves in accepting those premises; but none of the results obtained by this technique ever goes by one iota beyond the information already contained in the initial assumptions.

Since all mathematical proofs rest exclusively on logical deductions from certain postulates, it follows that a mathematical theorem, such as the Pythagorean theorem in geometry, asserts nothing that is *objectively* or *theoretically new* as compared with the postulates from which it is derived, although its content may well be *psychologically new* in the sense that we were not aware of its being implicity contained in the postulates.

The nature of the peculiar certainty of mathematics is now clear: A mathematical theorem is certain *relatively* to the set of postulates from which it is derived; i.e., it is necessarily true *if* those postulates are true; and this is so because the theorem, if rigorously proved, simply reasserts part of what has been stipulated in the postulates. A truth of this conditional type obviously implies no assertions about matters of empirical fact and can, therefore, never get into conflict with any empirical findings, even of the most unexpected kind; consequently, unlike the hypotheses and theories of empirical science, it can never suffer the fate of being disconfirmed by new evidence: A mathematical truth is irrefutably certain just because it is devoid of factual, or empirical content. Any theorem of geometry, therefore, when cast into the conditional form described earlier, is analytic in the technical sense of logic, and thus true a priori; i.e., its truth can be established by means of the formal machinery of logic alone, without any reference to empirical data.

4. Postulates and Truth

Now it might be felt that our analysis of geometrical truth so far tells only half of the relevant story. For while a geometrical proof no doubt enables us to assert a proposition conditionally—namely, on condition that the postulates are accepted—, is it not correct to add that geometry also unconditionally asserts the truth of its postulates and thus, by virtue of the deductive relationship between postulates and theorems, enables us unconditionally to assert the truth of its theorems? Is it not an unconditional assertion of geometry that two points determine one and only one straight line that connects them, or that in any triangle, the sum of the angles equals two right angles? That this is definitely not the case, is evidenced by two important aspects of the axiomatic treatment of geometry which will now be briefly considered.

The first of these features is the well-known fact that in the more recent development of mathematics, several systems of geometry have been constructed which are incompatible with Euclidean geometry, and in which, for example, the two propositions just mentioned do not necessarily hold. Let us briefly recollect some of the basic facts concerning these *non-Euclidean geometries*. The postulates on which euclidean geometry rests include the famous postulate of the parallels, which, in the case of plane geometry, asserts in effect that through every point P not on a given line l there exists exactly one parallel to l, i.e., one straight line which does not meet l. As this postulate is considerably less simple than the others, and as it was also felt to be intuitively less plausible than the latter, many efforts were made in the history of geometry to prove that this proposition need not be accepted as an axiom, but that it can be deduced as a theorem from the remaining body of postulates. All attempts in this direction failed, however; and finally it was conclusively demonstrated that a proof of the parallel principle on the basis of the other postulates of Euclidean geometry (even in its modern, completed form) is impossible. This was shown by proving that a perfectly self-consistent geometrical theory is obtained if the postulate of the parallels is replaced by the assumption that through any point P not on a given straight line l there exist at least two parallels to l. This postulate obviously contradicts the euclidean postulate of the parallels, and if the latter were actually a consequence of the other postulates of euclidean geometry, then the new set of postulates would clearly involve a contradiction, which can be shown not to be the case. This first non-Euclidean type of geometry, which is called hyperbolic geometry, was discovered in the early '20s of the last century almost simultaneously, but independently by the Russian N. I. Lobatschefskij, and by the Hungarian J. Bolyai. Later, Riemann developed an alternative geometry, known as elliptical geometry, in which the axiom of the parallels is replaced by the postulate that no line has any parallels. (The acceptance of this postulate, however, in contradistinction to that of hyperbolic geometry, requires the modification of some further axioms of Euclidean geometry, if a consistent new theory is to result.) As is to be expected, many of the theorems of these non-Euclidean geometries are at variance with those of Euclidean theory; thus, e.g., in the hyperbolic geometry of two dimensions, there exist, for each straight line l, through any point P not on l, infinitely many straight lines which do not meet l; also, the sum of the angles in any triangle is less than two right angles. In elliptic geometry, this angle sum is always greater than two right angles; no two straight lines are parallel; and while two different points usually determine exactly one straight line connecting them (as they always do in Euclidean geometry), there are certain pairs of points which are connected by infinitely many different straight lines. An illustration of this latter type of geometry is provided by the geometrical structure of that curved two-dimensional space which is represented by the surface of a sphere, when the concept of a straight line is interpreted by that of a great circle on the sphere. In this space, there are no parallel lines since any two great circles intersect; the endpoints of any diameter of the sphere are points connected by infinitely many different "straight lines," and the sum of the angles in a triangle is always in excess of two right angles. Also, in this space, the ratio between the circumference and the diameter of a circle (not necessarily a great circle) is always less than π.

Elliptic and hyperbolic geometry are not the only types of non-Euclidean geometry; various other types have been developed; we shall later have occasion to refer to a much more general form of non-Euclidean geometry which was likewise devised by Riemann.

The fact that these different types of geometry have been developed in modern mathematics shows clearly that mathematics cannot be said to assert the truth of any particular set of geometrical postulates; all that pure mathematics is interested in, and all that it can establish, is the deductive consequences of given sets of postulates and thus the necessary truth of the ensuing theorems relatively to the postulates under consideration.

A second observation which likewise shows that mathematics does not assert the truth of any particular set of postulates refers to *the status of the concepts in geometry*. There exists, in every axiomatized theory, a close parallelism between the treatment of the propositions and that of the concepts of the system. As we have seen, the propositions fall into two classes: the postulates, for which no proof is given, and the theorems, each of which has to be derived from the postulates. Analogously, the concepts fall into two classes: the primitive or basic concepts, for which no definition is given, and the others, each of which has to be precisely defined in terms of the primitives. (The admission of some undefined concepts is clearly necessary if an infinite regress in definition is to be avoided.) The analogy goes farther: Just as there exists an infinity of theoretically suitable axiom systems for one and the same theory—say, Euclidean geometry—, so there also exists an infinity of theoretically possible choices for the primitive terms of that theory; very often—but not always—different axiomatizations of the same theory involve not only different postulates, but also different sets of primitives. Hilbert's axiomatization of plane geometry contains six primitives: point, straight line, incidence (of a point on a line), betweenness (as a relation of three points on a straight line), congruence for line segments, and congruence for angles. (Solid geometry, in Hilbert's axiomatization, requires two further primitives, that of plane and that of incidence of a point on a plane.) All other concepts of geometry, such as those of angle, triangle, circle, etc., are defined in terms of these basic concepts.

But if the primitives are not defined within geometrical theory, what meaning are we to assign to them? The answer is that it is entirely unnecessary to connect any particular meaning with them. True, the words "point," "straight line," etc., carry definite connotations with them which relate to the familiar geometrical figures, but the validity of the propositions is completely independent of these connotations. Indeed, suppose that in axiomatized Euclidean geometry, we replace the oversuggestive terms "point," "straight line," "incidence," "betweenness," etc., by the neutral terms "object of kind 1," "object of kind 2," "relation No. 1, "relation No. 2," etc., and suppose that we present this modified wording of geometry to a competent mathematician or logician who, however, knows nothing of the customary connotations of the primitive terms. For this logician, all proofs would clearly remain valid, for as we saw before, a rigorous proof in geometry rests on deduction from the axioms alone without any reference to the customary interpretation of the various geometrical concepts used. We see therefore that indeed no specific meaning has to be attached to the primitive terms of an axiomatized theory; and in a precise logical presentation of axiomatized geometry the primitive concepts are accordingly treated as so-called logical variables.

As a consequence, geometry cannot be said to assert the truth of its postulates, since the latter are formulated in terms of concepts without any specific meaning; indeed, for this very reason, the postulates themselves do not make any specific assertion which could possibly be called true or false! In the terminology of modern logic, the postulates are not sentences, but sentential functions with the primitive concepts as variable arguments. This point also shows that the postulates of geometry cannot be considered as "self-evident truths," because where no assertion is made, no self-evidence can be claimed.

5. Pure and Physical Geometry

Geometry thus construed is a purely formal discipline; we shall refer to it also as *pure geometry*. A pure geometry, then,—no matter whether it is of the euclidean or of a noneuclidean variety—deals with no specific subject-matter; in particular, it asserts nothing about physical space. All its theorems are analytic and thus true with certainty precisely because they are devoid of factual content. Thus, to characterize the import of pure geometry, we might use the standard form of a movie-disclaimer: No portrayal of the characteristics of geometrical figures or of the spatial properties or relationships of actual physical bodies is intended, and any similarities between the primitive concepts and their customary geometrical connotations are purely coincidental.

But just as in the case of some motion pictures, so in the case at least of Euclidean geometry, the disclaimer does not sound quite convincing: Historically speaking, at least, Euclidean geometry has its origin in the generalization and systematization of certain empirical discoveries which were made in connection with the measurement of areas and volumes, the practice of surveying, and the development of astronomy. Thus understood, geometry has factual import; it is an empirical science which might be called, in very general terms, the theory of the structure of physical space, or briefly, *physical geometry*. What is the relation between pure and physical geometry?

When the physicist uses the concepts of point, straight line, incidence, etc., in statements about physical objects, he obviously connects with each of them a more or less definite physical meaning. Thus, the term "point" serves to designate physical points, i.e., objects of the kind illustrated by pinpoints, cross hairs, etc. Similarly, the term "straight line" refers to straight lines in the sense of physics, such as illustrated by taut strings or by the path of light rays in a homogeneous medium. Analogously, each of the other geometrical concepts has a concrete physical meaning in the statements of physical geometry. In view of this situation, we can say that physical geometry is obtained by what is called, in contemporary logic, a semantical interpretation of pure geometry. Generally speaking, a semantical interpretation of a pure mathematical theory, whose primitives are not assigned any specific meaning, consists in giving each primitive (and thus, indirectly, each defined term) a specific meaning or designatum. In the case of physical geometry, this meaning is physical in the sense just illustrated; it is possible, however, to assign a purely arithmetical meaning to each concept of geometry; the possibility of such an arithmetical interpretation of geometry is of great importance in the study of the consistency and other logical characteristics of geometry, but it falls outside the scope of the present discussion.

By virtue of the physical interpretation of the originally uninterpreted primitives of a geometrical theory, physical meaning is indirectly assigned also to every defined concept of the theory; and if every geometrical term is now taken in its physical interpretation, then every postulate and every theorem of the theory under consideration turns into a statement of physics, with respect to which the question as to truth or falsity may meaningfully be raised—a circumstance which clearly contradistinguishes the propositions of physical geometry from those of the corresponding uninterpreted pure theory. Consider, for example, the following postulate of pure Euclidean geometry: For any two objects x, y of kind 1, there exists exactly one object l of kind 2 such that both x and y stand in relation No. 1 to l. As long as the three primitives occurring in this postulate are uninterpreted, it is obviously meaningless to ask whether the postulate is true. But by virtue of the above physical interpretation, the postulate turns into the following statement: For any two physical points x, y there exists exactly one physical straight line l such that both x and y lie on l. But this is a physical hypothesis, and we may now meaningfully ask whether it is true or false. Similarly, the theorem about the sum of the angles in a triangle turns into the assertion that the sum of the angles (in the physical sense) of a figure bounded by the paths of three light rays equals two right angles.

Thus, the physical interpretation transforms a given pure geometrical theory—euclidean or non-Euclidean—into a system of physical hypotheses which, if true, might be said to constitute a theory of the structure of physical space. But the question whether a given geometrical theory in physical interpretation is factually correct represents a problem not of pure mathematics but of empirical science; it has to be settled on the basis of suitable experiments or systematic observations. The only assertion the mathematician can make in this context is this: If all the postulates of a given geometry, in their physical interpretation, are true, then all the theorems of that geometry, in their physical interpretation, are necessarily true, too, since they are logically deducible from the postulates. It might seem, therefore, that in order to decide whether physical space is Euclidean or non-Euclidean in structure, all that we have to do is to test the respective postulates in their physical interpretation. However, this is not directly feasible; here, as in the case of any other physical theory, the basic hypotheses are largely incapable of a direct experimental test; in geometry, this is particularly obvious for such postulates as the parallel axiom or Cantor's axiom of continuity in Hilbert's system of Euclidean geometry, which makes an assertion about certain infinite sets of points on a straight line. Thus, the empirical test of a physical geometry no less than that of any other scientific theory has to proceed indirectly; namely, by deducing from the basic hypotheses of the theory certain consequences, or predictions, which are amenable to an experimental test. If a test bears out a prediction, then it constitutes confirming evidence (though, of course, no conclusive proof) for the theory; otherwise, it disconfirms the theory. If an adequate amount of confirming evidence for a theory has been established, and if no disconfirming evidence has been found, then the theory may be accepted by the scientist "until further notice."

It is in the context of this indirect procedure that pure mathematics and logic acquire their inestimable importance for empirical science: While formal logic and pure mathematics do not in themselves establish any assertions about matters of empirical fact, they provide an efficient and entirely indispensable machinery for deducing, from abstract theoretical assumptions, such as the laws of Newtonian mechanics or the postulates of Euclidean geometry in physical interpretation, consequences concrete and specific enough to be accessible to direct experimental test. Thus, e.g., pure Euclidean geometry shows that from its postulates there may be deduced the theorem about the sum of the angles in a triangle, and that this deduction is possible no matter how the basic concepts of geometry are interpreted; hence also in the case of the physical interpretation of Euclidean geometry. This theorem, in its physical interpretation, is accessible to experimental test; and since the postulates of elliptic and of hyperbolic geometry imply values different from two right angles for the angle sum of a triangle, this particular proposition seems to afford a good opportunity for a crucial experiment. And no less a mathematician than Gauss did indeed perform this test; by means of optical methods—and thus using the interpretation of physical straight lines as paths of light rays—he ascertained the angle sum of a large triangle determined by three mountain tops. Within the limits of experimental error, he found it equal to two right angles.

6. On Poincaré's Conventionalism Concerning Geometry

But suppose that Gauss had found a noticeable deviation from this value; would that have meant a refutation of euclidean geometry in its physical interpretation, or, in other words, of the hypothesis that physical space is euclidean in structure? Not necessarily; for the deviation might have been accounted for by a hypothesis to the effects that the paths of the light rays involved in the sighting process were bent by some disturbing force and thus were not actually straight lines. The same kind of reference to deforming forces could also be used if, say, the euclidean theorems of congruence for plane figures were tested in their physical interpretation by means of experiments involving rigid bodies, and if any violations of the theorems were found. This point is by no means trivial; Henri Poincaré, the great French mathematician and theoretical physicist, based on considerations of this type his famous *conventionalism concerning geometry*. It was his opinion that no empirical test, whatever its outcome, can conclusively invalidate the Euclidean conception of physical space; in other words, the validity of Euclidean geometry in physical science can always be preserved—if necessary, by suitable changes in the theories of physics, such as the introduction of new hypotheses concerning deforming or deflecting forces. Thus, the question as to whether physical space has a Euclidean or a non-Euclidean structure would become a matter of convention, and the decision to preserve Euclidean geometry at all costs would recommend itself, according to Poincaré, by the greater simplicity of Euclidean as compared with non-Euclidean geometrical theory.

It appears, however, that Poincaré's account is an oversimplification. It rightly calls attention to the fact that the test of a physical geometry G always presupposes a certain body P of nongeometrical physical hypotheses (including the physical theory of the instruments of measurement and observation used in the test), and that the so-called test of G actually bears on the combined theoretical system G·P rather than on G alone. Now, if predictions derived from G·P are contradicted by experimental findings, then a change in the theoretical structure becomes necessary. In classical physics, G always was Euclidean geometry in its physical interpretation, GE; and when experimental evidence required a modification of the theory, it was P rather than GE which was changed. But Poincaré's assertion that this procedure would always be distinguished by its greater simplicity is not entirely correct; for what has to be taken into consideration is the simplicity of the total system G·P, and not just that of its geometrical part. And here it is clearly conceivable that a simpler total theory in accordance with all the relevant empirical evidence is obtainable by going over to a non-Euclidean form of geometry rather than by preserving the Euclidean structure of physical space and making adjustments only in part P.

And indeed, just this situation has arisen in physics in connection with the development of the general theory of relativity: If the primitive terms of geometry are given physical interpretations along the lines indicated before, then certain findings in astronomy represent good evidence in favor of a total physical theory with a non-Euclidean geometry as part G. According to this theory, the physical universe at large is a three-dimensional curved space of a very complex geometrical structure; it is finite in volume and yet unbounded in all directions. However, in comparatively small areas, such as those involved in Gauss's experiment, euclidean geometry can serve as a good approximative account of the geometrical structure of space. The kind of structure ascribed to physical space in this theory may be illustrated by an analogue in two dimensions; namely, the surface of a sphere. The geometrical structure of the latter, as was pointed out before, can be described by means of elliptic geometry, if the primitive term "straight line" is interpreted as meaning "great circle," and if the other primitives are given analogous interpretations. In this sense, the surface of a sphere is a two-dimensional curved space of non-Euclidean structure, whereas the plane is a two-dimensional space of Euclidean structure. While the plane is unbounded in all directions, and infinite in size, the spherical surface is finite in size and yet unbounded in all directions: a two-dimensional physicist, traveling along "straight lines" of that space would never encounter any boundaries of his space; instead, he would finally return to his point of departure, provided that his life span and his technical facilities were sufficient for such a trip in consideration of the size of his "universe." It is interesting to note that the physicists of that world, even if they lacked any intuition of a three-dimensional space, could empirically ascertain the fact that their two-dimensional space was curved. This might be done by means of the method of traveling along straight lines; another, simpler test would consist in determining the angle sum in a triangle; again another in determining, by means of measuring tapes, the ratio of the circumference of a circle (not necessarily a great circle) to its diameter; this ratio would turn out to be less than π.

The geometrical structure which relativity physics ascribes to physical space is a three-dimensional analogue to that of the surface of a sphere, or, to be more exact, to that of the closed and finite surface of a potato, whose curvature varies from point to point. In our physical universe, the curvature of space at a given point is determined by the distribution of masses in its neighborhood; near large masses such as the sun, space is strongly curved, while in regions of low mass-density, the structure of the universe is approximately Euclidean. The hypothesis stating the connection between the mass distribution and the curvature of space at a point has been approximately confirmed by astronomical observations concerning the paths of light rays in the gravitational field of the sun.

The geometrical theory which is used to describe the structure of the physical universe is of a type that may be characterized as a generalization of elliptic geometry. It was originally constructed by Riemann as a purely mathematical theory, without any concrete possibility of practical application at hand. When Einstein, in developing his general theory of relativity, looked for an appropriate mathematical theory to deal with the structure of physical space, he found in Riemann's abstract system the conceptual tool he needed. This fact throws an interesting sidelight on the importance for scientific progress of that type of investigation which the "practical-minded" man in the street tends to dismiss as useless, abstract mathematical speculation.

Of course, a geometrical theory in physical interpretation can never be validated with mathematical certainty, no matter how extensive the experimental tests to which it is subjected; like any other theory of empirical science, it can acquire only a more or less high degree of confirmation. Indeed, the considerations presented in this article show that the demand for mathematical certainty in empirical matters is misguided and unreasonable; for, as we saw, mathematical certainty of knowledge can be attained only at the price of analyticity and thus of complete lack of factual content. Let me summarize this insight in Einstein's words: "As far as the laws of mathematics refer to reality, they are not certain; and as far as they are certain, they do not refer to reality."

3

Recent Problems of Induction

> In the application of inductive logic to a given knowledge situation, the total evidence available must be used as a basis for determining the degree of confirmation.
> —Rudolf Carnap, *Logical Foundations of Probability* (1962)

1. The Classical Problem of Induction

In the philosophical discussion of induction, one problem has long occupied the center stage—so much so, indeed, that it is usually referred to as *the* problem of induction. That is the problem of justifying the way in which, in scientific inquiry and in our everyday pursuits, we base beliefs and assertions about empirical matters on logically inconclusive evidence.

This classical problem of justification, raised by Hume and made famous by his skeptical solution, is indeed of great philosophical importance. But more recent studies, most of which were carried out during the past two or three decades, have given rise to new problems of induction, no less perplexing and important than the classical one, which are logically prior to it in the sense that the classical problem cannot even be clearly stated—let alone solved—without some prior clarification of the new puzzles.

In this essay, I propose to discuss some of these recent problems of induction.

Induction may be regarded as affecting a transition from some body of empirical information to a hypothesis which is not logically implied by it, and for this reason it is often referred to as nondemonstrative *inference*. This characterization has to be taken with a grain of salt; but it is suggestive and convenient, and in accordance with it, I will therefore sometimes refer to the sentences specifying the evidence as the *premises* and to the hypothesis based on it as the *conclusion* of an "inductive inference."

Among the simplest types of inductive reasoning are those in which the evidence consists of a set of examined instances of a generalization, and the hypothesis is either the generalization itself or a statement about some unexamined instances of it. A standard example is the inference from the evidence statement that all ravens so far observed have been black to the generalization that all ravens are black or to the prediction that the birds now hatching in a given clutch of raven eggs will be black or to the retrodiction that a raven whose skeleton was found at an archeological site was black. As these examples show, induction does not always proceed from the particular to the general or from statements about the past or present to statements about the future.

The inductive procedures of science comprise many other, more complex and circumstantial, kinds of nondemonstrative reasoning, such as those used in making a medical diagnosis on the basis of observed symptoms, in basing statements about remote historical events on presently available evidence, or in establishing a theory on the basis of appropriate experimental data.

However, most of the problems to be considered here can be illustrated by inductions of the simple kind that proceed from instances of a generalization, and in general I will use these as examples.

2. The Narrow Inductivist View of Scientific Inquiry

It should be stressed at the outset that what we have called inductive inference must not be thought of as an effective method of discovery, which by a mechanical procedure leads from observational data to appropriate hypotheses or theories. This misconception underlies what might be called the narrow inductivist view of scientific inquiry, a view that is well illustrated by the following pronouncement:

> If we try to imagine how a mind of superhuman power and reach, but normal so far as the logical processes of its thought are concerned . . . would use the scientific method, the process would be as follows: First, all facts would be observed and recorded, *without selection* or *a priori* guess as to their relative importance. Second, the observed and recorded facts would be analyzed, compared, and classified, *without hypothesis or postulates* other than those necessarily involved in the logic of thought. Third, from this analysis of the facts, generalization would be inductively drawn as to the relations, classificatory or causal, between them. Fourth, further research would be deductive as well as inductive, employing inferences from previously established generalizations.[1]

It need hardly be argued in detail that this conception of scientific procedure, and of the role induction plays in it, is untenable; the reasons have been set forth by many writers. Let us just note that an inquiry conforming to this idea would never go beyond the first stage, for—presumably to safeguard scientific objectivity—no initial hypotheses about the mutual relevance and interconnections of facts are to be entertained in this stage, and as a result, there would be no criteria for the selection of the facts to be recorded. The initial stage would therefore degenerate into an indiscriminate and interminable gathering of data from an unlimited range of observable facts, and the inquiry would be totally without aim or direction.

Similar difficulties would beset the second stage—if it could ever be reached—for the classification or comparison of data again requires criteria. These are normally suggested by hypotheses about the empirical connections between various features of the "facts" under study. But the conception just cited would prohibit the use of such hypotheses, and the second stage of inquiry as here envisaged would again lack aim and direction.

It might seem that the quoted account of inductive scientific procedure could be rectified by simply adding the observation that any particular scientific investigation is aimed at solving a specified problem, and that the initial selection of data should therefore be limited to facts that are relevant to that problem. But this will not do, for the statement of a problem does not generally determine what kinds of data are relevant to its solution. The question as to the causes of lung cancer does not by itself determine what sorts of data would be relevant—whether, for example, differences in age, occupation, sex, or dietary habits should be recorded and studied. The notion of "relevant" facts acquires a clear meaning only when some specific answer to the problem has been suggested, however tentatively, in the form of a hypothesis: an observed fact will then be favorably or unfavorably relevant to the hypothesis according as its occurrence is by implication affirmed or denied by the hypothesis. Thus, the conjecture that smoking is a potent causative factor in lung cancer affirms by implication a higher incidence of the disease among smokers than among nonsmokers. Data showing for a suitable group of subjects that this is the case or that it is not would therefore constitute favorably relevant (confirming) or unfavorably relevant (disconfirming) evidence for the hypothesis. Generally, then, those data are relevant and need to be gathered which can support or disconfirm the contemplated hypothesis and which thus provide a basis for testing it.

Contrary to the conception quoted above, therefore, hypotheses are put forward in science as tentative answers to the problem under investigation. And contrary to what is suggested by the description of the third stage of inquiry above, such answers in the form of hypotheses or theories cannot be inferred from empirical evidence by means of some set of mechanically applicable rules of induction. There is no generally applicable mechanical routine of "inductive inference" which leads from a given set of data to a corresponding hypothesis or theory somewhat in the way in which the familiar routine of multiplication leads from any two given integers, by a finite number of mechanically performable steps, to the corresponding product.

To be sure, mechanical induction routines can be specified for certain special kinds of cases, such as the construction of a curve, and of an analytic expression for the corresponding function, which will fit a finite set of points. Given a finite set of measurements of associated values of temperature and volume for a given body of gas under constant pressure, this kind of procedure could serve mechanically to produce a tentative general law connecting temperature and volume of the gas. But for generating scientific theories, no such procedure can be devised.

Consider, for example, a theory, such as the theory of gravitation or the atomic theory of matter, which is introduced to account for certain previously established empirical facts, such as regularities of planetary motion and free fall, or certain chemical findings such as those expressed by the laws of constant and of multiple proportions. Such a theory is formulated in terms of certain concepts (those of gravitational force, of atom, of molecule, etc.) which are novel in the sense that they had played no role in the description of the empirical facts which the theory is designed to explain. And surely, no set of induction rules could be devised which would be generally applicable to just any set of empirical data (physical, chemical, biological, etc.) and which, in a sequence of mechanically performable steps, would generate appropriate novel concepts, functioning in an explanatory theory, on the basis of a description of the data.[2]

Scientific hypotheses and theories, then, are not mechanically inferred from observed "facts": *They are invented by an exercise of creative imagination.* Einstein, among others, often emphasized this point, and more than a century ago William Whewell presented the same basic view of induction. Whewell speaks of scientific discovery as a "process of invention, trial, and acceptance or rejection" of hypotheses and refers to great scientific advances as achieved by "Happy Guesses," by "felicitous and inexplicable strokes of inventive talent," and he adds: "No rules can ensure to us similar success in new cases; or can enable men who do not possess similar endowments, to make like advances in knowledge."[3] Similarly, Karl Popper has characterized scientific hypotheses and theories as conjectures, which must then be subjected to test and possible falsification.[4] Such conjectures are often arrived at by anything but explicit and systematic reasoning. The chemist Kékulé, for example, reports that his ring formula for the benzene molecule occurred to him in a reverie into which he had fallen before his fireplace. Gazing into the flames, he seemed to see snakes dancing about; and suddenly one of them moved into the foreground and formed a ring by seizing hold of its own tail. Kékulé does not tell us whether the snake was forming a *hexagonal* ring, but that was the structure he promptly ascribed to the benzene molecule.

Although no restrictions are imposed upon the *invention* of theories, scientific objectivity is safeguarded by making their *acceptance* dependent upon the outcome of careful tests. These consist in deriving, from the theory, consequences that admit of observational or experimental investigation, and then checking them by suitable observations or experiments. If careful testing bears out the consequences, the hypothesis is accordingly supported. But normally a scientific hypothesis asserts more than (i.e., cannot be inferred from) some finite set of consequences that may have been put to test, so that even strong evidential support affords no conclusive proof. It is precisely this fact, of course, that makes inductive "inference" nondemonstrative and gives rise to the classical problem of induction.

Karl Popper, in his analysis of this problem, stresses that the inferences involved in testing a scientific theory always run deductively from the theory to implications about empirical facts, never in the opposite direction; and he argues that therefore "Induction, i.e., inference based on many observations, is a myth. It is neither a psychological fact, nor a fact of ordinary life, nor one of scientific procedure";[5] and it is essentially this observation which, he holds, "solves . . . Hume's problem of induction."[6] But this is surely too strong a claim, for although the procedure of empirical science is not inductive in the narrow sense we have discussed and rejected, it still may be said to be *inductive in a wider sense*, referred to at the beginning of this essay: While scientific hypotheses and theories are not *inferred* from empirical data by means of some effective inductive procedure, they are *accepted* on the basis of observational or experimental findings which afford no deductively conclusive evidence for their truth. Thus, the classical problem of induction retains its import: What justification is there for accepting hypotheses on the basis of incomplete evidence?

The search for an answer to this question will require a clearer specification of the procedure that is to be justified; for while the hypotheses and theories of empirical science are not deductively implied by the evidence, it evidently will not count as inductively sound reasoning to accept a hypothesis on the basis of just any inconclusive evidence. Thus, there arises the logically prior problem of giving a more explicit characterization and precise criteria of what counts as sound inductive reasoning in science.

It may be instructive briefly to consider the analogue to this problem for deductive reasoning.

3. Deduction and Induction; Discovery and Validation

Deductive soundness, of course, is tantamount to deductive validity. This notion can be suggestively although imprecisely characterized by saying that an argument is deductively valid if its premises and its conclusion are so related that if all the premises are true, then the conclusion cannot fail to be true as well.[7]

As for *criteria* of deductive validity, the theory of deductive logic specifies a variety of forms of inference which are deductively valid, such as, for example, *modus ponens*:

$$\frac{p \supset q}{q}$$

or the inference rules of quantificational logic. Each of these represents a sufficient but not necessary condition of deductive validity. These criteria have the important characteristic of being expressible by reference to the syntactical structure of the argument, and thus without any reference to the meanings of the extralogical terms occurring in premises and conclusion. As we will see later, criteria of inductive soundness cannot be stated in purely syntactical terms.

We have already noted that whatever the rules of induction may be, they cannot be expected to specify mechanical routines leading from empirical evidence to appropriate hypotheses. Are the rules of deductive inference superior in this respect? Consider their role in logic and mathematics.

A moment's reflection shows that no interesting theorem in these fields is discovered by a mechanical application of the rules of deductive inference. Unless a putative theorem has first been put forward, such application would lack direction. Discovery in logic and mathematics, no less than in empirical science, *calls for imagination and invention*; it does not follow any mechanical rules.

Next, even when a putative theorem has been proposed, the rules of deduction do not, in general, provide a mechanical routine for proving or disproving it. This is illustrated by the famous arithmetical conjectures of Goldbach and of Fermat, which were proposed centuries ago but have remained undecided to this day. Mechanical routines for proving or disproving any given conjecture can be specified only for systems that admit of a decision procedure; and even for first-order quantificational logic and for elementary arithmetic, it is known that there can be no such procedure. In general, then, the construction of a proof or a disproof for a given logical or mathematical conjecture requires ingenuity.

But when a putative theorem has been proposed and a step-by-step argument has been offered as a presumptive proof for it, then the rules of deductive logic afford a means of establishing the validity of the argument: If each step conforms to one of those rules—a matter which can be decided by mechanical check—then the argument is a valid proof of the proposed theorem.

In sum, the formal rules of deductive inference are not rules of discovery leading mechanically to correct theorems or even to proofs for conjectured theorems which are in fact provable; rather, they provide criteria of soundness or of validity for proposed deductive proofs.

Analogously, rules of inductive inference will have to be conceived, not as canons of discovery, but as criteria of validation for proposed inductive arguments; far from generating a hypothesis from given evidence, they will *presuppose* that, in addition to a body of evidence, a hypothesis has been put forward, and they will then serve to appraise the soundness of the hypothesis on the basis of the evidence.

Broadly speaking, inductive arguments might be thought of as taking one of these forms:

$$\frac{e}{h}$$ (i.e., evidence e supports hypothesis h)

$$\frac{e}{h}[r]$$ (i.e., evidence e supports hypothesis h to degree r)

Here, the double line is to indicate that the relation of e to h is not that of full deductive implication but that of partial inductive support.

The second of these schemata incorporates the construal of inductive support as a quantitative concept. Rules of induction pertaining to it would provide criteria determining the degree of support conferred on certain kinds of hypotheses by certain kinds of evidence sentences; these criteria might even amount to a general definition assigning a definite value of r to any given e and h; this is one objective of Carnap's inductive logic.[8]

The first schema treats inductive support or confirmation as a qualitative concept; the corresponding inference rules would specify conditions under which a given evidence sentence supports, or confirms, a given hypothesis.[9]

The formulation of rules of these or similar kinds will be required to explicate the concept of inductive inference in terms of which the classical problem of justification is formulated. And it is in this context of explication that the newer problems of induction arise. We now turn to one of those problems; it concerns the qualitative concept of confirmation.

4. The Paradoxes of Qualitative Confirmation

The most familiar rules of induction concern generalizations of the simple form "All F are G." According to one widely asserted rule, a hypothesis of this kind receives support from its positive instances—i.e., from cases of F that have been found also to be G. For example, the hypothesis "All ravens are black," or

(h) $\qquad (x)(Rx \supset Bx)$

is supported, or confirmed, by any object i such that

(I) $\qquad Ri \cdot Bi$

or, as we will say, by any evidence sentence of the form "$Ri \cdot Bi$." Let us refer to such instances as *positive instances of type I for h*. Similarly, h is disconfirmed (invalidated) by any evidence sentence of the form $Ri \cdot -Bi$. This criterion was explicitly discussed and advocated by Jean Nicod;[10] I will therefore call it Nicod's criterion.

Now, the hypothesis h is logically equivalent to, and thus makes exactly the same assertion as, the statement that all nonblack things are nonravens, or

(h') $\qquad (x)(-Bx \supset -Rx)$

According to Nicod's criterion, this generalization is confirmed by *its* instances— i.e., by any individual j such that

(II) $\qquad -Bj \cdot -Rj$

But since h' expresses exactly the same assertion as h, any such individual will also confirm h. Consequently, such things as a yellow rose, a green caterpillar, or a red herring confirm the generalization "All ravens are black," by virtue of being nonblack nonravens. I will call such objects *positive instances of type II for h*.

Next, the hypothesis h is logically equivalent also to the following statement:

(h") $\qquad (x)[(Rx \vee -Rx) \supset (-Rx \vee Bx)]$

in words: Anything that is a raven or not a raven—i.e., anything at all—either is not a raven or is black. Confirmatory instances for this version, which I will call *positive instances of type III for h*, consist of individuals k such that

(III) $\qquad\qquad\qquad -Rk \vee Bk$

This condition is met by any object k that is not a raven (no matter whether it is black) and by any object k that is black (no matter whether it is a raven). Any such object, then, affords a confirmatory instance in support of the hypothesis that all ravens are black.

On the other hand, the hypothesis h can be equivalently expressed by the sentence

(h''') $\qquad\qquad\qquad (x)[(Rx \cdot -Bx) \supset (Rx \cdot -Rx)]$

for which nothing can possibly be a confirmatory instance in the sense of Nicod's criterion, since nothing can be both a raven and not a raven.

These peculiarities, and some related ones, of the notion of confirmatory instance of a generalization have come to be referred to as the *paradoxes of confirmation*.[11] And indeed, at first glance they appear to be implausible and perhaps even logically unsound. But on further reflection one has to conclude, I think, that they are perfectly sound, that it is our intuition in the matter which leads us astray, so that the startling results are paradoxical only in a psychological, but not in a logical sense.

To see this, let us note first that the results in question follow deductively from two simple basic principles, namely: (A) A generalization of the form "All F are G" is confirmed by its positive instances—i.e., by cases of F that have been found also to be cases of G. (B) Whatever confirms a hypothesis also confirms any logically equivalent one.

Principle (A) is, in effect, part of Nicod's criterion, of which Nicod himself remarks that it "cannot claim the force of an axiom. But it offers itself so naturally and introduces such great simplicity, that reason welcomes it without feeling any imposition."[12] We will encounter some surprising exceptions to it in Sections 5 and 6, but it does indeed seem very reasonable in cases of the kind we have considered so far—i.e., in reference to generalizations of universal conditional form containing exclusively property terms (one-place predicates).

Principle (B) may be called the equivalence condition. It simply reflects the idea that whether given evidence confirms a hypothesis must depend only on the content of the hypothesis and not on the way in which it happens to be formulated.

And once we accept these principles, we must also accept their surprising logical consequences.

Let us look at these consequences now from a different point of view, which will support the claim that they are sound. Suppose we are told that in the next room there is an object i which is a raven. Our hypothesis h then tells us about i that it is black, and if we find that this is indeed the case, so that we have $Ri \cdot Bi$, then this must surely count as bearing out, or confirming, the hypothesis.

Next, suppose we are told that in the adjoining room there is an object j that is not black. Again, our hypothesis tells us something more about it, namely, that it is not a raven. And if we find that this is indeed so—i.e., that $-Bj \cdot -Rj$, then this bears out, and thus supports, the hypothesis.

Finally, even if we are told only that in the next room there is an object k, the hypothesis still tells us something about it, namely, that either it is no raven or it is black—i.e., that $-Rk \lor Bk$; and if this is found to be the case, it again bears out the hypothesis.

Thus, our three types of positive instance must indeed be counted as confirmatory or supporting evidence for the generalization that all ravens are black.

Finally, the fact that the formulation h''' of our generalization admits of no confirming instances in the sense of Nicod's criterion presents no serious problem if, as here has been done, that criterion is stated as a sufficient but not necessary condition of confirmation.

But why does it seem implausible or paradoxical in the first place that positive instances of types *II* and *III* should be confirmatory for the generalization h? One important reason seems to lie in the assumption that the hypothesis "All ravens are black" is a statement about ravens and not about nonravens, let alone about all things in general. But surely, such a construal is untenable; anyone who accepts h would be bound to accept also the sentences h' and h'', which by the same token would have to be viewed as statements about nonravens and about all things, respectively. The use made of some statements of the form "All F are G" illustrates the same point. The Wassermann test, for example, is based, roughly speaking, on the generalization that any person infected with syphilis has a positive Wassermann reaction; but in view of its diagnostic implications for cases yielding a negative test result, this generalization surely cannot be said to be about syphilitically infected persons only.

To say that positive instances of types *I II*, and *III* all confirm the hypothesis h is not to say, however, that they confirm the generalization to the same extent. Indeed, several writers have argued that the different types differ greatly in this respect and that, in particular, a positive instance of type *I*, i.e., a black raven, lends much stronger support to our generalization than a positive instance of type *II*, i.e., a nonblack object that is not a raven; and they have suggested that this is the objective basis for the first impression that instances of type *I* alone can count as confirmatory for our hypothesis.

This view can be made plausible by the following suggestive but imprecise consideration: Let k be the hypothesis "All marbles in this bag are red," and suppose that there are twenty marbles in the bag. Then the generalization k has twenty instances of type *I*, each being provided by one of the marbles. If we have checked each of the twenty objects that are marbles in the bag, we have exhaustively tested the hypothesis. And roughly speaking we might say that if we have examined one of the marbles and found it red, we have shown one twentieth of the total content of the hypothesis to be true.

Now consider the contrapositive of our generalization—i.e., the statement, "Any object that is not red is not a marble in this bag." Its instances are provided by all nonred objects. There are a large number of these in the world—perhaps infinitely many of them. Examining one of them and averring that it is not a marble in the bag is therefore to check, and corroborate, only a tiny portion of all that the hypothesis affirms. Hence, a positive finding of type II would indeed support our generalization, but only to a very small extent.

Analogously in the case of the ravens. If we may assume that there are vastly more nonblack things than there are ravens, then the observation of one nonblack thing that is not a raven would seem to lend vastly less support to the generalization that all ravens are black than would the observation of one raven that *is* black.

This argument might serve to mitigate the paradoxes of confirmation.[13] But I have stated it here only in an intuitive fashion. A precise formulation would require an explicit quantitative theory of degrees of confirmation or of inductive probability, such as Carnap's. Even within the framework of such a theory, the argument presupposes further assumptions, and the extent to which it can be sustained is not fully clear as yet.

Let us now turn to another perplexing aspect of induction. I will call it Goodman's riddle, because it was Nelson Goodman who first called attention to this problem and proposed a solution for it.[14]

5. Goodman's Riddle: A Failure of Confirmation by "Positive Instances"

One of the two basic principles from which we deduced the paradoxes of confirmation stated that a generalization of the form "All F are G" is confirmed, or supported, by its positive instances of type I—i.e., by objects which are F and also G. Although this principle seems entirely obvious, Goodman has shown that there are generalizations that derive no support at all from their observed instances. Take for example the hypothesis

(h)   All ravens are blite

where an object is said to be blite if it is either examined before midnight tonight and is black or is not examined before midnight and is white.

Suppose now that all the ravens examined so far have been found to be black; then, by definition, all ravens so far examined are also blite. Yet this latter information does not support the generalization h, for that generalization implies that all ravens examined after midnight will be white—and surely our evidence must be held to militate against this forecast rather than to support it.

Thus, some generalizations do derive support from their positive instances of type I; for example, "All ravens are black," "All gases expand when heated," "In all cases of free fall from rest, the distance covered is proportional to the square of the elapsed time," and so forth; but other generalizations, of which "All ravens are blite" is an example, are not supported by their instances. Goodman expresses this idea by saying that the former generalizations can, whereas the latter cannot, be *projected* from examined instances to as yet unexamined ones.

The question then arises how to distinguish between projectible and nonprojectible generalizations. Goodman notes that the two differ in the character of the terms employed in their formulation. The term "black," for example, lends itself to projection; the term "blite" does not. He traces the difference between these two kinds of terms to what he calls their *entrenchment*—i.e., the extent to which they have been used in previously projected hypotheses. The word "blite," for example, has never before been used in a projection, and is thus much less entrenched than such words as "black," "raven," "gas," "temperature," "velocity," and so on, all of which have served in many previous inductive projections—successful as well as unsuccessful ones. What Goodman thus suggests is that our generalizations are chosen not only in consideration of how well they accord with the available evidence, but also in consideration of how well entrenched are their constituent extralogical terms.

By reference to the relative entrenchment of those terms, Goodman then formulates criteria for the comparison of generalizations in regard to their projectibility, and he thus constructs the beginnings of a theory of inductive projection.

I cannot enter into the details of Goodman's theory here, but I do wish to point out one of its implications which is, I think, of great importance for the conception of inductive inference.

As we noted earlier, the standard rules of deductive inference make reference only to the syntactical form of the sentences involved; the inference rules of quantification theory, for example, apply to all premises and conclusions of the requisite form, no matter whether the extralogical predicates they contain are familiar or strange, well entrenched or poorly entrenched. Thus,

	All ravens are blite
and	
	r is a raven
deductively implies	
	r is blite
no less than	
	All ravens are black
and	
	r is a raven
deductively implies	
	r is black

But on Goodman's conception of projectibility, even elementary rules of induction cannot be similarly stated in purely syntactical terms. For example, the rule that a positive instance confirms a generalization holds only for generalizations with adequately entrenched predicates; and entrenchment is neither a syntactical nor even a semantic property of terms, but a pragmatic one; it pertains to the actual use that has been made of a term in generalizations projected in the past.

6. A Further Failure of Confirmation by "Positive Instances"

Goodman's riddle shows that Nicod's criterion does not offer a generally adequate sufficient condition of confirmation: Positive instances do not confirm nonprojectible hypotheses.

But the criterion fails also in cases of a quite different kind, which do not hinge on the use of predicates such as "blite." Consider the hypothesis, "If for any two persons x,y it is not the case that each likes the other, then the first likes the second, but not vice versa"; in symbolic notation:

$$(h) \qquad (x)(y)[-(Lxy \cdot Lyx) \supset (Lxy \cdot -Lyx)]$$

Let e be the information that a,b are two persons such that a likes b but not vice versa, i.e. that

$$(e) \qquad Lab \cdot -Lba$$

This information can equivalently be stated as follows:

$$(e') \qquad -(Lab \cdot Lba) \text{ and } (Lab \cdot -Lba)$$

for the first of these two sentences is a logical consequence of the second one. The sentence e' then represents a positive instance of type I for h; hence, on Nicod's criterion, e' should confirm h.[15]

But e' is equivalent to

$$(e'') \qquad -(Lba \cdot Lab) \text{ and } (-Lba \cdot Lab)$$

and this, on Nicod's criterion, disconfirms h. In intuitive terms, the preceding argument is to this effect: If a is counted as the first person and b as the second, then the information provided by e shows that, as e' makes explicit, a and b satisfy both the antecedent and the consequent of h and thus confirm the hypothesis; but if b is counted as the first person and a as the second one, then by virtue of the same information, b and a satisfy the antecedent but not the consequent of h, as is made explicit in e''. Thus, on Nicod's criterion, e constitutes both confirming and invalidating evidence for h.

Incidentally, h can be thrown into the form

$$(h') \qquad (x)(y)(Lxy \cdot Lyx),$$

which makes it obvious that the evidence e logically contradicts the given hypothesis; hence, the same is true of e', although Nicod's criterion qualifies e' as confirming h.[16]

Hypotheses of the form illustrated by h can be formulated in terms of well-entrenched predicate expressions, such as "x likes y" and "x is soluble in y";

the difficulty here illustrated does not, therefore, spring from the use of ill-behaved predicates of the Goodmanian variety.

The difficulty rather shows that the intuition which informs the Nicod criterion simply fails when the hypotheses under consideration include relational terms rather than only property terms. If one considers, in addition, that the Nicod criterion is limited to hypotheses of universal conditional form, then it becomes clear that it would be of great interest to develop a general characterization of qualitative confirmation which (1) affords a full definition rather than only partial criteria for the confirmation of a hypothesis h by an evidence sentence e, (2) is applicable to any hypothesis, of whatever logical form, that can be expressed within a specified language, and (3) avoids the difficulties of the Nicod criterion which have just been pointed out.

An explicit definition of this kind for the concept "h qualitatively confirms e" has in fact been constructed for the case where h and e are formulated in a formalized language that has the structure of a first-order functional calculus without identity; h may be any sentence whatsoever in such a language, and e may be any consistent sentence containing no quantifiers. The concept thus defined demonstrably avoids the difficulties encountered by the Nicod criterion in the case of hypotheses with relational predicates; and it implies the Nicod criterion in reference to those hypotheses of universal conditional form which contain only property terms. It has been argued, however, that the concept thus arrived at is not fully satisfactory as an explication of the vague idea of qualitative confirmation because it fails to capture certain characteristics which might plausibly be attributed to the relation of qualitative confirmation.[17]

7. The Ambiguity of Induction

I now turn to a further basic problem, which I will call the problem of inductive ambiguity. This facet of induction, unlike those we have considered so far, is not a recent discovery; both the problem and a possible solution of it have been recognized, if not always very explicitly, by several writers on probability, past as well as contemporary. But certain aspects of the problem are of special interest in the context of our discussion, and I will therefore consider them briefly.

Suppose that we have the following information:

(e_1) Jones, a patient with a sound heart, has just had an appendectomy, and of all persons with sound hearts who underwent appendectomy in the past decade, 93% had an uneventful recovery.

This information, taken by itself, would clearly lend strong support to the hypothesis

(h_1) Jones will have an uneventful recovery.

But suppose that we also have the information:

(e_2) Jones is a nonagenarian with serious kidney failure; he just had an appendectomy after his appendix had ruptured; and in the past decade, of all cases of appendectomy after rupture of the appendix among nonagenarians with serious kidney failure only 8% had an uneventful recovery.

This information by itself lends strong support to the contradictory of h_1:

($-h_1$) Jones will not have an uneventful recovery.

But e_1 and e_2 are logically compatible and may well both be part of the information available to us and accepted by us at the time when Jones's prognosis is being considered. In this case, our available evidence provides us with a basis for two rival arguments, both of them inductively sound, whose "conclusions" contradict each other. This is what I referred to above as the ambiguity of inductive reasoning: Inductively sound reasoning based on a consistent, and thus possibly true, set of "premises" may lead to contradictory "conclusions."

This possibility is without parallel in deductive reasoning: The consequences deducible from any premises selected from a consistent set of sentences form again a consistent set.

When two sound inductive arguments thus conflict, which conclusion, if any, is it reasonable to accept, and perhaps to act on? The answer, which has long been acknowledged, at least implicitly, is this: If the available evidence includes the premises of both arguments, it is irrational to base our expectations concerning the conclusions exclusively on the premises of one or the other of the arguments; the credence given to any contemplated hypothesis should always be determined by the support it receives from the *total* evidence available at the time. (Parts may be omitted if they are irrelevant in the sense that their omission leaves the inductive support of the contemplated hypothesis unchanged.) This is what Carnap has called the *requirement of total evidence*. According to it, an estimate of Jones's prospects of recovery should be based on all the relevant evidence at our disposal; and clearly, a physician trying to make a reasonable prognosis will try to meet this requirement as best he can.

What the requirement of total evidence demands, then, is that the credence given to a hypothesis h in a given knowledge situation should be determined by the inductive support, or confirmation, which h receives from the total evidence e available in that situation. Let us call this confirmation $c(h,e)$. Now for some brief comments on this maxim.

1. In the form just stated, the requirement presupposes a quantitative concept of the degree, $c(h,e)$, to which the evidence e confirms or supports the hypothesis h. This raises the question how such a concept might be defined and whether it can be characterized so generally that $c(h,e)$ is determined for *any* hypothesis h that might be proposed, relative to *any* body of evidence e that might be available. This issue has been much discussed in recent decades. Carnap, in his theory of inductive logic, has developed an explicit and completely gen-

eral definition of the concept for the case where e and h are any two sentences expressible in one or another of certain formalized languages of relatively simple logical structure.[18] Others have argued that the concept in question can be satisfactorily defined at best for certain special types of hypotheses and of evidential information. For example, if the total relevant evidence consists just of the sentences e_1 and e_2 listed above, certain analysts would hold that no probability or degree of confirmation can be significantly assigned to the hypothesis, "Jones will have an uneventful recovery," since the evidence provides no information about the percentage of uneventful recoveries among nonagenarians with sound hearts but seriously defective kidneys who undergo appendectomy after rupture of the appendix.

2. Next, let us note that while the requirement of total evidence is a principle concerning induction, it is not a rule of inductive inference or, more precisely, of inductive support, for it does not concern the question whether, or how strongly, a given hypothesis is supported by given evidence. The requirement is concerned rather with the rational use, or application, of inductive reasoning in the formation of empirical beliefs. This observation suggests a distinction between two kinds of rules pertaining to inductive reasoning:

  a. *Rules of inductive support, or of valid inductive inference.* These would encompass, for example, all criteria concerning the qualitative confirmation or disconfirmation of generalizations by positive or negative instances; criteria determining degrees of confirmation; and also all general principles connecting degrees of confirmation with each other, such as the law, that the degrees of confirmation of a hypothesis and of its contradictory on the same evidence add up to unity.
  b. *Rules of application.* These concern the use of rules of the former kind in the rational formation of empirical beliefs. The requirement of total evidence is one such rule of application, but not the only one, as will soon be seen.

The distinction between rules of inference and rules of application can be made also in reference to deductive reasoning. The rules of inference, as we noted earlier, provide criteria of deductive validity; but they qualify as deductively valid many particular arguments whose conclusions are false, and they do not concern the conditions under which it is reasonable to believe, or to accept, the conclusion of a deductively valid argument. To do so would be the task of rules for the rational application of deductive inference.

One such rule would stipulate, for example, that if we have accepted a set of statements as presumably true, then any logical consequence of that set (or, perhaps rather, any statement that is known to be such a consequence) should equally be accepted as presumably true.

The two kinds of rules for deduction call for quite different kinds of justification. An inference rule such as *modus ponens* might be justified by showing that when applied to true premises it will invariably yield a true conclusion—which is what is meant by the claim that an argument conforming to the rule is deductively valid.

But in order to justify a rule of application, we will have to consider what ends the acceptance or rejection of deductive conclusions is to serve. For example, if we are interested in a set of accepting statements, or of corresponding beliefs, which will afford us an emotionally reassuring or esthetically satisfying account of the world, then it will not always be reasonable to accept, or to believe, the logical consequences of what we have previously accepted. If, on the other hand, truth is what we value in our accepted statements, and if we are accordingly concerned to give credence to all statements that are true as far as our information enables us to tell, then indeed we have to accept all the consequences of previously accepted statements; thus, justification of our rule of application requires reference to the objectives, or the values, that our acceptance procedure is meant to achieve.

8. Induction and Valuation

Similarly, if we wish to devise rules for the rational application of valid inductive reasoning, or if we wish to appraise or justify such rules, we will have to take into account the objectives to be achieved by the inductive acceptance procedure, or the values or disvalues of the consequences that might result from correct or from incorrect acceptance decisions. In this sense, the construction and the justification of inductive acceptance rules for empirical statements presupposes judgments of value.

This is especially obvious when we wish to decide whether a given hypothesis is to be accepted in the strong sense of being relied on as a basis for practical action. Suppose, for example, that a new vaccine has been developed for immunization against a serious infectious disease that can afflict humans as well as chimpanzees. Let h be the hypothesis that the vaccine is both safe and effective in a sense specified by suitable operational criteria, and suppose that the hypothesis has been tested by examining a number of samples of the vaccine for safety and effectiveness. Let e be the evidence thus obtained.

Our rules of inductive support may then tell us how strongly the hypothesis is confirmed by the evidence; but in deciding whether to act on it we will have to consider, besides the strength of confirmation, also the kind of action that is contemplated, and what benefits might result from a correct decision, what harm from a mistaken one. For example, our standards of acceptance are likely to differ according as humans or chimpanzees are to be treated with the vaccine; and it may well happen that *on the same evidence* the given hypothesis is accepted as a basis of action in one case but rejected in the other.

Inductive decisions of this kind have been extensively studied in the mathematical theory of testing and decision-making. This theory deals in particular with the case where the values or disvalues attached to the possible consequences of the available decisions are expressible in numerical terms as so-called utilities. For such situations, the theory has developed a number of specific decision rules, which are rules of application in our sense. These rules—maximin, maximax, maximizing the expectable utility of the outcome, and others—make the acceptance or the rejection of the hypothesis contingent on the utilities assigned to the different possible consequences of acceptance or rejection; and when a measure for the evidential support of the hypothesis is available, that support is likewise taken into consideration.[19] In this fashion, the inductive decision rules combine empirical considerations with explicitly valuational ones.

That rules for the acceptance or rejection of empirical hypotheses thus presuppose valuational considerations has been emphasized by several writers. Some of these have made the stronger claim that the values in question are ethical values. Thus, Churchman asserts that "the simplest question of fact in science requires for even an approximation, a judgment of value," and that "the science of ethics . . . is *basic* to the meaning of any question the experimental scientist raises."[20] And in the context of a detailed study of the logic of testing statistical hypotheses, Braithwaite asserts, in a similar vein: "To say that it is 'practically certain' that the next 1000 births in Cambridge will include the birth of at least one boy includes a hedonic or ethical assessment."[21]

But while it is true that the justification of rules of acceptance for statements of fact requires reference to judgments of preference or of valuation, the claim that the values concerned are ethical values is, I think, open to question. Our argument about valuational presuppositions has so far been concerned only with the acceptance of hypotheses as a basis of specific *actions*, and in this case the underlying valuations may indeed be ethical in character. But what standards will govern the acceptance and rejection of hypotheses for which no practical application is contemplated? Braithwaite's statement about male births in Cambridge might well belong in that category, and surely so do the hypotheses examined in pure, or basic, scientific research; these might concern, for example, the rate of recession of distant galaxies or the spontaneous creation of hydrogen atoms in empty space. In such cases, it seems, we simply wish to decide, in consideration of the available evidence, whether to believe a proposed hypothesis; whether to record it, so to speak, in our book of tentative scientific knowledge, without envisaging any technological application. Here, we cannot relevantly base our decisions on any utilities or disutilities attached to practical consequences of acceptance or rejection and, in particular, ethical considerations play no part.

What will have to be taken into account in constructing or justifying inductive acceptance rules for pure scientific research are the objectives of such research or the importance attached in pure science to achieving certain kinds of results. What objectives does pure scientific research seek to achieve? Truth of the accepted statements might be held to be one of them. But surely not truth at all costs. For then, the only rational decision policy would be never to accept any hypothesis on inductive grounds since, however well supported, it might be false.

Scientific research is not even aimed at achieving very high probability of truth, or very strong inductive support, at all costs. Science is willing to take considerable chances on this score. It is willing to accept a theory that vastly outreaches its evidential basis if that theory promises to exhibit an underlying order, a system of deep and simple systematic connections among what had previously been a mass of disparate and multifarious facts.

It is an intriguing but as yet open question whether the objectives, or the values, that inform pure scientific inquiry can all be adequately characterized in terms of such theoretical desiderata as confirmation, explanatory power, and simplicity and, if so, whether these features admit of a satisfactory combination into a concept of purely theoretical or scientific utility that could be involved in the construction of acceptance rules for hypotheses and theories in pure science. Indeed, it is by no means clear whether the conception of basic scientific research as leading to the provisional acceptance or rejection of hypotheses is tenable at all. One of the problems here at issue is whether the notion of accepting a hypothesis independently of any contemplated action can be satisfactorily explicated within the framework of a purely logical and methodological analysis of scientific inquiry[22] or whether, if any illuminating construal of the idea is possible at all, it will have to be given in the context of a psychological, sociological, and historical study of scientific research.[23]

To conclude with a summary that centers about the classical problem of induction: For a clear statement of the classical problem of justification, two things are required. First, the procedure to be justified must be clearly characterized—this calls for an explication of the rules governing the inductive appraisal of hypotheses and theories; second, the intended objectives of the procedure must be indicated, for a justification of any procedure will have to be relative to the ends it is intended to serve. Concerning the first of these tasks, we noted that while there are no systematic mechanical rules of inductive discovery, two other kinds of rule have to be envisaged and distinguished, namely, rules of support and rules of application. And in our discussion of the objectives of inductive procedures we noted certain connections between rational belief on the one hand and valuation on the other.

Whatever insights further inquiry may yield, the recognition and partial exploration of these basic problems has placed the classical problem of induction into a new and clearer perspective and has thereby advanced its philosophical clarification.

NOTES

1. A. B. Wolfe, "Functional Economics," *The Trend of Economics,* ed. R. G. Tugwell (New York: Knopf, 1924), p. 450 (author's italics).
2. This argument does not presuppose a fixed division of the vocabulary of empirical science into observational and theoretical terms; it is quite compatible with acknowledging that as a theory becomes increasingly well established and accepted, certain statements couched in terms of its characteristic concepts may come to be qualified as descriptions of "observed facts."

3. William Whewell, *The Philosophy of the Inductive Sciences*, 2d ed. (London: John W. Parker, 1847), II, 41 (author's italics).
4. See, for example, Popper's essay, "Science: Conjectures and Refutations," in his book, *Conjectures and Refutations* (New York and London: Basic Books, 1962).
5. Karl Popper, "Philosophy of Science: A Personal Report," *British Philosophy in the Mid-Century*, ed. C. A. Mace (London: Allen and Unwin, 1957), pp. 155–91, quotation from p. 181.
6. Popper, "Philosophy of Science," p. 183.
7. Precise general characterizations of deductive validity, for arguments in languages of certain specified forms, will be found, e.g., in W. V. O. Quine, *Methods of Logic*, rev. ed. (New York: Holt, Rinehart & Winston, 1959).
8. See especially the following publications by Rudolf Carnap: *Logical Foundations of Probability*, 2d ed. (Chicago: U. of Chicago Press, 1962); "The Aim of Inductive Logic," *Logic, Methodology and Philosophy of Science: Proceedings of the 1960 International Congress*, eds. E. Nagel, P. Suppes, and A. Tarski (Stanford: Stanford U. Press, 1962), pp. 303–18.
9. It seems to me, therefore, that Popper begs the question when he declares: "But it is obvious that this rule or craft of 'valid induction' . . . simply does not exist. No rule can ever guarantee that a generalization inferred from true observations, however often repeated, is true" ("Philosophy of Science," p. 181). That inductive reasoning is not *deductively* valid is granted at the outset; the problem is that of constructing a concept of *inductive* validity.
10. Jean Nicod, *Foundations of Geometry and Induction* (New York: Harcourt, Brace & World, 1930), p. 219. Nicod here speaks of "truths or facts," namely, "the presence or absence of B in a case of A," as confirming or invalidating "the law A entails B" (author's italics). Such confirmatory and disconfirmatory facts can be thought of as described by corresponding evidence sentences. Nicod remarks about his criterion: "We have not seen it stated in any explicit manner. However, we do not think that anything ever written on induction is incompatible with it" (p. 220). Whether Nicod regards the specified conditions as necessary and sufficient or merely as sufficient for confirmation or invalidation is not entirely clear, although he does say: "It is conceivable that we have here the only two direct modes in which a fact can influence the probability of a law" (p. 219). We will construe his criteria simply as *sufficient* conditions of confirmation and invalidation.
11. These paradoxes were first noted in my essay "Le problème de la vérité," *Theoria* (Göteborg), 3 (1937), 206–46 (see especially p. 222) and were discussed in greater detail in my articles "Studies in the Logic of Confirmation," *Mind*, 54 (1945), 1–26, 97–121, and "A Purely Syntactical Definition of Confirmation," *The J. of Symbolic Logic*, 8 (1943), 122–43.
12. Nicod, *Geometry and Induction*, pp. 219–20.
13. It was first offered by Janina Hosiasson-Lindenbaum in her article "On Confirmation," *The J. of Symbolic Logic*, 5 (1940), 133–48. Similar ideas were proposed by, among others, D. Pears, "Hypotheticals," *Analysis*, 10 (1950), 49–63; I. J. Good, "The Paradoxes of Confirmation," Pts. I and II, *The British J. for the Philosophy of Science*, 11 (1960), 145–48; 12 (1961) 63–64. A detailed and illuminating study of qualitative confirmation and its paradoxes is offered in sec. 3, Pt. I of Israel Scheffler, *The Anatomy of inquiry* (New York: Knopf, 1963).
14. Nelson Goodman, *Fact, Fiction, and Forecast* (Cambridge: Harvard U. Press, 1955); 2d, rev. ed. (Indianapolis: Bobbs-Merrill, 1965).

15. Nicod does not explicitly deal with hypotheses which, like h, contain relational terms rather than only property terms such as "raven" and "black"; but the application here suggested certainly seems to be in full accord with his basic conception.
16. This further paradox of qualitative confirmation was briefly noted in my article, "Studies in the Logic of Confirmation," p. 13.
17. The general definition is developed in "A Purely Syntactical Definition of Confirmation"; the gist of it is presented in sec. 9 of my article essay, "Studies in the Logic of Confirmation." The objections in question were raised especially by R. Carnap in *Logical Foundations of Probability*, secs. 86–88. Briefly, Carnap's principal objection is to the effect that under an adequate definition of qualitative confirmation, e should confirm h only if, in the sense of inductive probability theory, e raises the prior probability of h; and my definition of confirmation is not compatible with such a construal.
18. See especially the following publications: "On Inductive Logic," *Philosophy of Science*, 12 (1945), 72–97; *Logical Foundations of Probability*; *The Continuum of inductive Methods* (Chicago: U. of Chicago Press, 1952).
19. A lucid account of these rules and of their theoretical use will be found in R. D. Luce and H. Raiffa, *Games and Decisions* (New York: Wiley, 1957).
20. C. W. Churchman, *Theory of Experimental Inference* (New York: Macmillan, 1948), pp. vii, viii (author's italics).
21. R. B. Braithwaite, *Scientific Explanation* (Cambridge: Cambridge U. Press, 1953), p. 251.
22. For a fuller discussion and bibliographic references concerning these issues, see, e.g., sec. 12 of C. G. Hempel, "Deductive-Nomological *vs*. Statistical Explanation" in *Scientific Explanation, Space, and Time*, eds. H. Feigl and G. Maxwell, Minnesota Studies in the Philosophy of Science, III (Minneapolis: U. of Minnesota Press, 1962), 98–169. Some of the basic issues are examined in R. B. Braithwaite's paper, "The Role of Values in Scientific Inference," and especially the discussion of that paper in *Induction: Some Current Issues*, eds. H. E. Kyburg, Jr., and E. Nagel (Middletown, Conn.: Wesleyan U. Press, 1963), pp. 180–204.
23. Such an alternative conception is represented, e.g., by T. S. Kuhn's work, *The Structure of Scientific Revolutions* (Chicago: U. of Chicago Press, 1962).

4

On the Structure of Scientific Theories

1. Two Schematic Construals of a Theory

The questions to be examined in this essay concern, broadly speaking, the logical structure and the epistemic status of scientific theories. To a large extent, my discussion will deal with the well-known conception of a theory as consisting of, or being decomposable into, two principal components:

(1) An uninterpreted deductive system, usually thought of as an axiomatized calculus C, whose postulates correspond to the basic principles of the theory and provide implicit definitions for its constitutive terms;

(2) A set R of statements that assign empirical meaning to the terms and the sentences of C by linking them to potential observational or experimental findings and thus interpreting them, ultimately, in terms of some observational vocabulary. Such statements, conceived as subject to more or less stringent syntactic and semantic requirements, have been variously referred to as the entries of a "dictionary" for translating theoretical into experimental language, as "rules of correspondence" or "rules of interpretation," as "coordinating definitions" and, of course, as "operational definitions."

Some writers have put considerable emphasis on the need for a third constituent of a good scientific theory, namely, a model (in an appropriate sense of the word) which interprets the abstract calculus C in terms of concepts or principles with which we are acquainted by previous experience. This point is not of immediate relevance to the issues with which I propose to deal first; it will, however, be considered later on.

The conception of a theory as consisting of an abstract calculus C and a set R of rules of correspondence will be briefly represented by the following schema, which construes a theory as the sum of two classes of sentences:

(Schema I) $\qquad T = C \cup R$

I have myself repeatedly used a construal of this kind in discussing the status of theories and of theoretical entities; but it seems to me now that, while not strictly untenable, it is misleading in several respects. I will point out what seem to me the inadequacies of Schema I by contrasting it with an alternative construal.

As a rule, theories are introduced in a field of scientific inquiry after prior research has yielded a body of initial knowledge—usually in the form of general laws or empirical generalizations—concerning the phenomena under investigation. Theories are then developed in an effort to achieve a deeper and more comprehensive understanding of those phenomena by presenting them as manifestations or resultants of certain underlying processes, and by exhibiting the previously established laws or generalizations as consequences (more accurately: as approximations of consequences) of certain basic general principles assumed to govern the underlying occurrences. This broad characterization applies equally, I think, to the two types of theory which Nagel, following Rankine, distinguishes in his penetrating study of the subject;[1] namely, "abstractive" theories, such as the Newtonian theory of gravitation and motion, and "hypothetical" theories, such as the kinetic theory of heat or the undulatory and corpuscular theories of light.

Prima facie, therefore, it seems reasonable to think of a theory as consisting of statements, or principles, of two kinds; let us call them *internal principles* and *bridge principles* for short. The internal principles will specify the "theoretical scenario"; they will characterize the basic processes posited by the theory, the kinds of entities they involve, and the laws to which they are held to conform. The bridge principles will indicate the ways in which the processes envisaged by the theory are related to the previously investigated empirical phenomena which the theory is intended to explain.

This conception will be indicated by the following schema:

(Schema II) $\qquad T = I \cup B$

Here, I is the class of internal principles, B the class of bridge principles.

There are obvious similarities to the conception indicated by Schema I, but there are also considerable differences; this will become clear in the following amplification and comparison of the two construals.

2. Internal and Bridge Principles; Theoretical and Pretheoretical Terms

Consider first the vocabularies (apart from logical and mathematical terms) in which the two sets of principles, I and B, are expressed. The formulation of the internal principles will normally make use of a "theoretical vocabulary"; V_T, i.e., a set of terms introduced specifically to characterize the various constituents of the theoretical scenario and the laws assumed to govern them. The formulation of the bridge principles will require, in addition to the theoretical vocabulary, a set of terms suited to describe the empirical phenomena and uniformities that the theory is intended to explain. The terms of this second kind are therefore available and understood prior to the formulation of the theory; they may therefore be said to constitute, relative to the theory in question, an antecedently understood, or pretheoretical, vocabulary V_A. For the terms of this vocabulary, there are objective rules of use which, at least initially, are independent of the theory. A vocabulary of this kind is equally presupposed, of course, by the conception of correspondence rules invoked in Schema I, for those rules are assumed to impart empirical meanings upon certain expressions of the calculus C by linking them to an appropriate experimental or observational subject matter.

Thus, the pretheoretical vocabulary has often been construed as consisting of observational terms. Broadly speaking, an observational term is one that stands for some characteristic of things or events that is directly observable, i.e., whose presence or absence in a particular case can, under suitable circumstances (such as normal lighting or the like), be ascertained by direct observation, without the use of special instruments. Thus, such words as 'blue', 'liquid', 'hot', 'acrid-smelling', 'longer than', 'contiguous with', might count as observational terms.

Closer studies of this idea strongly suggest, however, that the usual characterizations of observability fall short of determining a sufficiently clear and analytically useful dividing line between observational and nonobservational terms. It could be argued, for example, that under the rough criterion just mentioned, such terms as 'electrically charged', 'acid', and 'has a higher refractive index than' qualify as observational, since for each of them, circumstances can be specified in which the presence or absence of the characteristic in question can be ascertained very reliably by means of a few direct observations.[2]

The notion of observability presents another puzzling aspect, which does not seem to have been widely noticed, and which therefore might deserve brief mention here. The distinction between "directly observable" and "not directly observable" has been applied by many of its proponents not only to attributes (i.e., properties and relations) of physical objects and processes, but also to those objects and processes themselves. Thus, such things as mountains, chairs, and apples are said to be directly observable; electrons, photons, and mesons are not. Similarly, events or processes such as a thunderstorm or the motion of a car might be said to be observable, whereas quantum jumps or the mutual annihilation of a particle and a corresponding antiparticle might be qualified as not directly observable. But is the notion of observability as invoked in these contexts identical with that applied to attributes, or can it at least be explicated in terms of the latter? For brevity, I will limit my comments on this question to the case of "thing-like" entities; they can readily be extended to processes and events.

As is illustrated by the example of mountains, chairs, electrons, mesons, and the like, thing-like entities are often referred to by means of descriptive predicate terms. Now, under the rough criterion stated above, predicates like 'chair' (or 'is a chair'), 'mountain', and 'apple' would count as observational, whereas predicates like 'electron', 'meson', and 'photon' would not. Hence, it might seem that observability for thing-like entities can be explicated as follows in terms of observability for attributes of things: When 'P' is a predicate term, then the entities describable as P's are observable if and only if 'P' is an observational predicate. Thus, apples are observable, and indeed the word 'apple' is an observational predicate; it stands for a directly observable characteristic of physical bodies. And mesons are nonobservable; and indeed, the word 'meson' is not an observational predicate.

But the proposed explication is quite incorrect. Consider such entities as diamonds, Stradivarius violins, precision chronometers, purebred Pekinese dogs, or philosophers. All of them are objects that can be directly seen, felt, and in most cases also heard—in contrast to electrons and the like, which cannot. Yet the expressions 'diamond', 'Stradivarius violin', 'precision chronometer', 'purebred Pekinese' and 'philosopher' are none of them observational predicates: the presence or absence of the characteristics they stand for cannot be ascertained by direct observation. Hence, it is not the case that objects of kind P, or P's for short, are observable if and only if 'P' is an observational predicate: it may be that P's are observable, but not *qua* P's, not under that description.

The following alternative explication may appear plausible: An object is directly observable just in case it has at least one directly observable attribute. Thus, Pekinese dogs are furry; Stradivarius violins, like other violins, are hard objects with smooth surfaces of reddish-brown color; and so forth. But the contemplated criterion is too inclusive: an object—a flea's egg, perhaps—may be too small to be directly observable, and may yet have a certain color, say brown, which is observable in the sense that under suitable circumstances—namely, when it occurs on a large enough surface and in natural light—its presence can be quite reliably ascertained by direct observation.

At present, I am unable to offer a satisfactory explication of the concept of observability referring to thing-like entities in terms of the concept of observability as applied to attributes; and the point of the preceding remarks is mainly to note that if in the characterization of theories one wished to make use of the notion of observability, one would presumably need two concepts, both of which are rather vague.

Fortunately, however, the notion of observability is not required for our purposes; for the elements of the pretheoretical vocabulary V_A need not, and indeed should not, generally be conceived as observational terms in the sense adumbrated above: in many cases, V_A will contain terms originally introduced by an earlier theory that is empirically well supported, and whose internal and bridge principles provide rules for their use. Consider some examples.

In the classical kinetic theory of gases, the internal principles are assumptions about the gas molecules; they concern their size, their mass, their number; and they include also various laws, partly taken over from classical mechanics, partly statistical in nature, pertaining to the motions and collisions of the molecules, and to the resulting changes in their momenta and energies. The bridge principles include statements such as that the temperature of a gas is proportional to the mean kinetic energy of its molecules, and that the rates at which different gases diffuse through the walls of a container are proportional to the numbers of molecules of the gases in question, and to their average speeds. By means of such bridge principles, certain microcharacteristics of a gas, which belong to the scenario of the kinetic theory, are linked to macroscopic features such as temperature, pressure, and diffusion rate; these can be described, and generalizations concerning them can be formulated, in terms of an antecedently available vocabulary, namely, that of classical thermodynamics. And some of the macroscopic features in question might perhaps be regarded as fairly directly observable or measurable (although, of course, the relevant measurements will require the use of instruments).

But the features to which certain aspects of the theoretical scenario are linked by the bridge principles are not always as close to observability as the volume, temperature, and pressure of a body of gas. Consider, for example, the explanation provided by Bohr's model of the hydrogen atom for the fact that the light emitted by glowing hydrogen vapor is restricted to certain discrete wave lengths, which appear in the hydrogen spectrum as a corresponding set of lines. According to Bohr's model, the hydrogen atom contains one electron, which circles the nucleus in one or another of a series of discrete orbits that are available to it. When a mass of hydrogen gas is electrically or thermally excited, the electrons in some of the atoms are shifted to outer orbits, which represent higher energy states. Eventually, such electrons will jump back from some outer to some inner orbit; in the process, they emit monochromatic radiation whose wave length is uniquely determined by the released energy, which in turn is fully determined by the two orbits between which the jump takes place. Consequently, the radiation emitted by excited hydrogen vapor can assume only certain specific and discrete wave lengths. Moreover, the quantitative details of Bohr's model account for the specific wave lengths associated with the lines in the hydrogen spectrum. In particular, they imply Balmer's formula, an ingenious empirical generalization, which had been previously established, and which specifies the wave lengths of one series of discrete lines, called the Balmer series, in the hydrogen spectrum:

$$\lambda = b \frac{n^2}{n^2 - 4}$$

Here, b is a numerically specified constant; and if n is given the values 3, 4, 5, ..., the resulting values of λ give the wave lengths of the lines in the Balmer series.

In this case, the internal principles comprise the various assumptions that characterize Bohr's model of the hydrogen atom. The bridge principles, on the other hand, include such statements as these:

(a) the light emitted by excited hydrogen vapor results from the energy released when electrons jump from outer to inner orbits;
(b) An electronic jump that releases the energy E results in the emission of light with the wave length

$$\lambda = (h \cdot c) / E$$

where h is Planck's constant and c the velocity of light.

These and other bridge principles connect the entities and processes posited by the theory with certain fetitures of the subject matter to be explained, namely, the wave lengths associated with the lines in the hydrogen spectrum. But while the lines, and the discrete patterns they form, might be counted as observables, the wave lengths and the corresponding frequencies surely cannot; and the terms that serve to characterize them are not observational terms. But they are antecedently understood nonetheless: when Bohr proposed his model, rules for their use, including principles for the measurement of wave lengths, had already been established: these were based on antecedent theories, especially wave optics.

In characterizing scientific theories, it is not necessary, therefore, to presuppose one partition of all scientific terms into observational and theoretical ones. It suffices to note that whatever characteristic terms a new theory introduces will have to be linked by bridge principles to an antecedently understood vocabulary that serves to describe, and to express generalizations about, the subject matter that the theory is to explain.

This observation marks no great departure from the ideas embodied in Schema I; for while the rules of correspondence, R, are often said to connect theoretical terms with observational ones, several writers who have made expository use of that schema have clearly envisaged a less stringent conception that is quite compatible with the one just outlined.

3. On the Notion of an Axiomatized Theoretical Calculus

But Schema I presents more serious problems. One of them concerns the rationale of dividing a theory into an axiomatized uninterpreted calculus and a set of correspondence rules. Before indicating the difficulties I find with this conception, I want to mention one objection to it that I do not think pertinent. This is the observation that the proposed division into calculus and interpretation has no counterpart in the ways in which theories are actually formulated and used by scientists, and that, moreover, a rigid system C can represent at best one momentary stage of what is in fact a continually developing and changing system of ideas. This objection seems to me to miss its mark. For Schema I is meant to reflect certain logical and epistemological features of scientific theories rather than the heuristic and pragmatic aspects of their formulation and use, however interesting and important these may be in their own right.

But this remark immediately raises the question: What are the logical or epistemological characteristics of theories that are pointed up or illuminated by a construal in the manner of Schema I? Consider first the system C, which seems clearly intended to include those sentences which, in Schema II, are referred to as the internal principles. This system is conceived, in Schema I, as an axiomatized and uninterpreted calculus. Let us examine the significance of these two characteristics in turn.

As for axiomatization, its great importance in logic and mathematics and their metadisciplines need not be emphasized here. Under certain conditions, it can undoubtedly provide also philosophical illumination for questions concerning scientific theories; but several among the extant axiomatizations of such theories seem to me to have very little significance either for the concerns of empirical science or for those of philosophy—however ingenious they may be as logical accomplishments.

One outstanding example of a philosophically illuminating use of axiomatization is Reichenbach's work on the philosophical significance of relativistic physics; and his division of physical geometry into an axiomatized mathematical system and a set of "coordinative definitions" is an early instance of a construal of a scientific theory in the manner of Schema I.[3] While technically inferior to more recent examples, Reichenbach's axiomatization of the relativistic theory of spacetime is distinguished by aiming at definite philosophical objectives: among other things, it was meant to separate what Reichenbach regarded as conventional or stipulative truths from empirical hypotheses, and more generally, it was to provide a basis for a critique of Kant's conception of synthetic *a priori* knowledge.

But while axiomatization, if appropriately used, may be helpful, it is not necessary for this purpose—nor is it sufficient. For axiomatization is basically an expository device; it exhibits logical relationships between statements, but not their epistemic grounds or connections. A given scientific theory admits of many different axiomatizations, and the postulates chosen in a particular axiomatization need not, therefore, correspond to what in some more substantial sense might count as the basic assumptions of the theory; nor need the terms chosen as undefined, or primitive, in a given axiomatization represent what on epistemological or other grounds might qualify as the basic concepts of the theory; nor need the formal definitions of other theoretical terms by means of the chosen primitives correspond to statements which in science would be regarded as definitionally true and thus analytic. In an axiomatization of Newtonian mechanics, the second law of motion can be given the status of a definition, a postulate, or a theorem, as one pleases; but the role it is thus assigned within the axiomatized system does not indicate whether in its scientific use it functions as a definitional truth, as a basic theoretical law, or as a derivative one (if indeed it may be said to have just one of these functions).

Hence, whatever philosophical insights may be obtainable by considering a theory in an axiomatized form will require certain appropriate specific axiomatizations and not just any axiomatization; let alone simply the assumption that the theory has been put into some unspecified axiomatic form. But this last assumption alone enters into Schema I, and I do not see what philosophical illumination it could provide.

4. The Role of Pretheoretical Concepts in Internal Principles

The system C envisaged in Schema I is, moreover, conceived as uninterpreted in the sense that, within C, the extralogical terms have been assigned no meanings other than those accruing to them by reason of their role in the postulates, which thus constitute incomplete "implicit definitions" of the terms. One root of this idea is doubtless the realization that the description of the theoretical scenario of a theory normally makes use of characteristic new terms, which acquire empirical significance and applicability only when they are suitably linked to pretheoretical terms, antecedently used in the field of inquiry. But the conception of C as an uninterpreted system suggests that the internal principles of a theory are formulated exclusively in terms of the "new" theoretical vocabulary V_T (plus logical and mathematical symbols), which logically play the role of variables or dummies. All the sentences of C would then have the character of schemata such as these: 'All X's are Y's or Z's'; 'Every Y has three quantitative characteristics, q, r, s, such that $q = c \cdot r \cdot s$, where c is a constant', and so forth.[4] Actually, however, the internal principles of most, if not all, scientific theories contain not only the terms of the theoretical vocabulary; but also pretheoretical terms, taken over from the antecedently available vocabulary. To state the point more intuitively: The theoretical scenario is normally described to a large extent by means of terms which already have definite empirical content, and which are already understood, prior to, and independently of, the theory. For example, the theoretical assumptions of the classical kinetic theory of gases attribute masses, volumes, velocities, momenta, kinetic energies, and other characteristics familiar from the study of macroscopic objects, to molecules and to atoms; and the classical wave and particle theories of light characterize the processes in their theoretical scenarios by means of the concepts of wave length, frequency, velocity of a wave; path of a particle; deflection of a particle as a result of attraction—all of which are antecedently understood.

Hence, the internal principles of a theory, and thus, the system C, must be viewed as containing two kinds of extralogical terms, as indicated by the following schematization:

$$I = \Phi(t_1, t_2, \ldots t_k; p_1, p_2, \ldots, p_m)$$

Here, the t's are terms belonging to the theoretical vocabulary of T, whereas the p's are pretheoretical terms.

Thus, the basic principles of a theory cannot be conceived as an uninterpreted calculus, whose formulas contain, apart from logical or mathematical expressions, only theoretical terms which, formally, play the role of variables, and which obtain empirical relevance by way of correspondence principles.

It might be objected, in a spirit akin to that of operationism, that in this new context, the "old" terms $p_1, p_2, \ldots p_m$ represent new concepts, quite different from those they signify in their pretheoretical employment. For the use of such terms as 'mass', 'velocity', 'energy', etc., in reference to atoms or subatomic particles requires entirely new operational criteria of application, since at the atomic and subatomic levels, the quantities in question cannot be measured by means of scales, electrometers, and the like, which afford operational criteria for their measurement at the pretheoretical level. On the strict operationist maxim that different criteria of application determine different concepts, we would therefore have to conclude that, when used in internal principles, the terms $p_1, p_2, \ldots, p_m$ stand for new concepts, quite different from those they signify in their pretheoretical use. And we would have to add that it is therefore improper to use the old pretheoretical terms in theoretical contexts: they should be replaced here by appropriate new terms, which, along with $t_1, t_2, \ldots, t_k$, would then belong to the theoretical vocabulary.

This does not seem to me a compelling argument, however. For, first of all, the operationist conception on which it is based is itself untenable. Suppose we tried to adhere strictly to the maxim that different operational criteria of application determine different concepts which, in principle, should be distinguished by the use of different terms such as 'tactual length' and 'optical length' in one of Bridgman's examples;[5] we would then be led to an infinite proliferation of concepts and corresponding terms, which would totally defeat the scientific effort to find conceptually simple and economical theories that account for a great variety of empirical phenomena. For we would be obliged to say that the measurement of length, not only by means of rigid rods and by optical means, but even by rigid rods of different chemical composition or mass or date of manufacture, determines so many different concepts of length; and that weighing by means of two balances of however "identical" construction similarly determines different concepts of weight; for the two balances will always differ in some aspects, and will thus give rise to two different weighing operations. Hence, the operationist maxim we have considered is self-defeating. And in science any term is regarded, of course, as allowing for many alternative criteria of application. Each of these criteria is based on laws that connect the characteristic the term stands for (e.g., temperature) with observable phenomena of a certain kind (e.g., the readings of a mercury thermometer, a gas thermometer, a thermoelectric measuring device, etc.), which thus become indicators of the characteristic in question.

Moreover, the criteria of application are only part of the rules which govern the use of scientific terms, and which thus have a bearing on their "meanings": to a large extent, the use of a set of terms is determined by the general principles in which they function. Concerning the pretheoretical and the theoretical uses of the terms $p_1, p_2, \ldots, p_m$, it seems therefore interesting and important to note that some of the basic theoretical laws that are applied to the concepts of mass, volume, velocity, energy, etc., in their internal, theoretical use are taken over from their pretheoretical use. For example, in the classical kinetic theory, mass is taken to be additive in the sense that the mass of several molecules taken together is assumed to equal the sum of the masses of the individual molecules, just as the mass of a system consisting of several macroscopic physical bodies is taken to equal the sum of the masses of those bodies.[6] In fact, as a rule, this is not even explicitly stated: unless mention to the contrary is made, the attribution of mass to atoms and molecules is tacitly understood to imply the applicability of such fundamental laws in the new domain. Similarly, the conservation principles for mass, energy, and momentum are—at least initially—assumed to hold also for the atomic and molecular constituents of a gas, and so are the laws of motion.

In fact, the additivity principle for masses is used in three roles: initially as a pretheoretical law governing the concept of mass; then, in the context of the kinetic theory, as an internal principle, as has just been noted; and finally, as a bridge principle. In this last role it implies, for example, that the mass of a body of gas equals the sum of the masses of its constituent molecules; it thus connects certain features of the theoretical scenario with corresponding features of macroscopic systems that can be described in pretheoretical terms. This micro-macro applicability of the additivity principle is clearly presupposed in the explanation of the laws of multiple proportions and similarly in certain methods of determining Avogadro's number. These considerations strongly suggest that the term 'mass' (and others) can hardly be taken to stand for quite different concepts, depending on the kind of entity to which it is applied.

The transfer of basic pretheoretical laws to the domain of the theoretical scenario is reminiscent of a principle normally observed when mathematical concepts are extended to a wider range of application; namely, the maxim that for the wider range, the concepts are to be construed in such a way as to ensure, as far as possible, the continued validity of the basic laws that govern them in their original area of application.

Thus, the notion of exponentiation for real numbers as arguments might first be introduced for the case where the exponents are natural numbers, a^m being defined, in effect, as a product of m factors, all of which equal a. For the operation thus specified, certain basic laws can then be proved, such as the following:

(1) $\quad$ If $m > n$, then $\dfrac{a^m}{a^n} = a^{m-n}$ (for $a \neq 0$)

(2) $\quad$ If m is an integral multiple of n, then

$$\sqrt[n]{a^m} = a^{\frac{m}{n}}$$

When the question arises of extending the notion of exponentiation to a wider class of exponents—including 0, the negative integers, and all rational numbers, for example—the requisite definitions are chosen in such a way that the basic laws of exponentiation continue to hold in the new domain. Application of this requirement to (1) yields the following two formulas, which immediately provide the explicit definitions for a^0 and a^{-m}:

(3) $\quad a^0 = a^{m-m} = \dfrac{a^m}{a^m} = 1$ (for $a \neq 0$)

(4) $\quad a^{-m} = a^{0-m} = \dfrac{a^0}{a^m} = \dfrac{1}{a^m}$ (for $a \neq 0$)

Similarly, in consideration of (2) and (3), the principle requires that, for any two positive integers m and n, we set

(5) $\quad a^{\frac{m}{n}} = \sqrt[n]{a^m}$

(6) $$a^{-\frac{m}{n}} = \frac{1}{\sqrt[n]{a^m}} \quad \text{(for } a \neq 0\text{)}$$

There is a significant analogy between this way of extending mathematical concepts from a narrower to a wider domain of application and the manner in which the use of terms such 'mass', 'momentum', and 'kinetic energy' is extended to a domain that includes atoms and molecules: in both cases, certain basic laws which hold in the original domain are carried over to the extended domain. And this preservation of laws surely affords grounds for saying that when the terms in question are applied in the new domain, they do not stand for entirely new and different concepts.

In our mathematical illustration, the narrower and the wider concepts of exponentiation are closely related; the former may, in fact, be considered a subconcept of the latter. But the two are not identical: the "new" concept is applicable to a wider class of arguments, and some of the general principles that hold for the original concept (e.g., that a^m is a single-valued function of its two arguments) fail for the "new" one (in the class of real numbers, $a^{1/2}$ has two, one, or no values according as a is positive, zero, or negative).

In regard to the concepts of mass, velocity, etc., as used in the classical kinetic theory, however, a strong case could be made for holding that they are simply the same as those applied, in classical mechanics, to medium-sized objects. Classical mechanics imposes no lower bounds on the size or the mass of the bodies to which the concepts of mass, velocity, and kinetic energy can be significantly applied (in fact, Newtonian mechanics makes use of the concept of point masses); and the laws governing these concepts are subject to no such restrictions, either. This latter fact, incidentally, yields also a more direct retort to the operationist claim that the concept of mass as applied to atoms must be different from that applied to macroscopic bodies since the two cases require different operations of measurement: the application of classical mechanical considerations to the case of objects of the size of atoms shows that macroscopic scales are not sufficiently sensitive for weighing them, but that certain indirect procedures will provide an operational basis for determining their masses. It seems reasonable, therefore, to say that the need for different methods of determining masses does not indicate a difference in the meaning of 'mass' in the two cases, but only a large difference in the masses of the objects concerned. Of the classical kinetic theory of gases, then, it may be said that its internal principles characterize its theoretical scenario to a large extent with the help of the same concepts of mass, velocity, momentum, energy, and so forth that were already available at the pretheoretical level. A fortiori, the internal principles do not just contain new theoretical terms, and it seems misleading, therefore, to view them, as is done in Schema I, as a system C of sentences containing only uninterpreted terms.

It is not so clear, however, whether an analogous argument applies in all other cases. More recent theories of the microstructure of matter, for example, acknowledge that the mass of any atomic nucleus is somewhat less than the sum of its constituent protons and neutrons taken separately; thus, these theories abandon the principle of additivity for the concept of mass, and they replace the principle of conservation of mass by the principle of conservation of mass-energy. Are we to say. then, that the term 'mass' as used in these theories has a different meaning—stands for a different concept—than the same word in its earlier scientific use (which is pretheoretical in relation to these theories)? Or are we to say that the basic meaning of 'mass' has remained the same, and what has been changed is simply a set of—admittedly very far-reaching and fundamental—empirical assumptions about the one concept of mass to which the different theories refer?

The issue illustrated by this question has been widely discussed in recent years, and both of the positions just suggested have had their proponents. Hanson and Feyerabend, among others, have eloquently urged that the meanings of scientific terms depend on, and change with, the theories in which they function, and that even the meanings of so-called observational terms, which serve to describe the empirical evidence for or against a theory, are informed by the theory and change with it.[7] Feyerabend holds, for example, that temperature as conceived by Galileo and as supposedly measured by his thermoscope, is quite different from the current concept of temperature, since the theory of the relevant measurements has undergone considerable change; moreover, Feyerabend reasons that even the concept of temperature as used in classical thermodynamics is not the same as the one based on the kinetic theory of heat since, for example, the second law of thermodynamics holds as a strict universal law for the former, but not for the latter.[8]

Putnam, on the other hand, holds that the Italian word for 'temperature' as used by Galileo and the word 'temperature' as used today have the same "meaning," since they refer to the same physical magnitude, namely, "the magnitude whose greater and lower intensities are measured by the human sensorium as *warmer* and *colder* respectively."[9]

Now, it does seem to me suggestive and plausible to say that, in a somewhat elusive sense, Galileo's thermoscopic studies and more recent physical research on temperature are concerned with the same quantitative characteristic of physical bodies; but I doubt that this statement admits of a very precise explication and defense. The argument offered by Putnam, for example, raises the question whether we are entitled to speak of "the" magnitude whose intensities are recorded by the human sensorium as warmer and colder, for this phrasing presupposes that there is one and only one such physical magnitude. But the intensity of the heat or cold we feel on touching a given object depends not only on its temperature, but also on its specific heat; it makes a great deal of difference whether our skin is in contact with air of 170° F or with water of the same temperature.

A clear analytic resolution of the issue would require an adequate explication of the notion of sameness of concepts or of the corresponding notion of the synonymy of terms; and it seems to me that no satisfactory general explications of these notions are currently available.

But no matter what position one may take concerning the extent to which pretheoretical terms change their meanings when they are take over by a theory, the considerations offered in this section show, I think, that it is misleading to view the internal principles of a theory as an uninterpreted calculus and the theoretical terms accordingly as variables, as markers of empty shells into which the juice of empirical content is pumped through the pipelines called correspondence rules.

5. Nagel's Notion of a Model as a Component of a Theory

Ernest Nagel, in his penetrating and carefully documented study of the structure of scientific theories, distinguishes three major components of a theory, namely (1) an abstract calculus that "implicitly defines" the basic notions of the system; (2) a set of "rules of correspondence," which assign an empirical content to the calculus by relating it to specific observational or experimental materials, and (3) a "model," which interprets the abstract calculus in terms of more or less familiar conceptual or visualizable materials."[10] In the case of the Bohr theory of the hydrogen atom, for example, the calculus would consist in the mathematical formulas constituting the basic equations of the theory; these will contain certain uninterpreted variables like 'r_i', 'E_j', 'λ'. The model specifies the conception, referred to earlier, of a hydrogen atom as consisting of a nucleus circled by an electron to which a series of discrete circular orbits are available, etc.; in this model, the variable 'r_i' is interpreted as the length of the radius of the ith orbit, 'E_j' as the total energy of the atom when the electron is in the jth orbit, 'λ' as the wave length of the radiation generated by an orbital jump, and so forth. The correspondence rules, finally, link the notion of specific orbital jumps of an electron to the experimental notion of corresponding particular spectral lines, and they establish other linkages of this kind.

The term 'model of a theory' has been used in a number of different senses. In one of its uses, it refers to such constructions as the mechanical models of electric currents or of the luminiferous ether that played a considerable role in the physics of the late nineteenth and early twentieth centuries. Models of this kind carry an implicit 'as if' clause with them; thus, electric currents in wires behave in certain respects as if they consisted in the flow of a liquid through pipes of various widths and under various pressures, etc. But these models clearly were not intended to represent the actual microstructure of the modeled phenomena: wires were not thought actually to be pipes, and the ether was not believed actually to consist of the components envisaged by one model, namely, sets of nested hollow spheres, separated by elastic springs. The kinetic theory of gases, on the other hand,—and, I think, similarly Bohr's model of atomic structure—does put forward a conjecture as to the actual microstructure of the objects under study: gases are claimed actually to consist of molecules moving about, and colliding, at various high speeds; atoms are claimed actually to have certain specific subatomic constituents and to be capable of certain discrete energy states.[11] To be sure, these claims, like those of any other empirical hypothesis, may subsequently be amplified and refined or modified or partly discarded (as happened to Bohr's model, but they form an integral part of the theory. Models of the former kind, on the other hand, consist in physical systems which exhibit certain material analogies to the phenomena under study or whose behavior is governed by laws having the same mathematical form as basic laws of the theory. Such models may be of considerable psychological interest and heuristic value; they may make it easier to grasp a new theory and to become familiar with it, and they may suggest various consequences and even extensions of the basic assumptions of the theory; but they are not part of the content of the theory and are, thus, logically dispensable.[12]

For this reason I will restrict the following remarks to those cases where the "model" expresses part of the content of the theory, as in the kinetic theory of gases, the classical wave and particle theories of light, the molecular-lattice theory of crystal structure, recent theories of the molecular structure of genes and the basis of the genetic code, and I take it, in Bohr's theory of atomic structure.

In regard to the latter, Nagel notes that, usually, the theory is not presented as an abstract set of postulates supplemented by rules of correspondence, but is embedded in the model mentioned above because, among other reasons, it can thus be understood with greater ease than can the inevitably more complex formal exposition.[13] But for the reasons outlined earlier, I think their significance goes further. By characterizing certain theoretical variables as masses, energies, and the like, the theory commits itself to extending to these variables certain laws characteristic of those concepts or, if some of these laws are suspended, to making the appropriate modifications explicit (as happens in Bohr's model, where, in contrast to the principles of classical electromagnetic theory, an orbiting electron is assumed not to radiate energy). Thus, the specification of the model determines in part what consequences can be inferred from the theory and hence, what the theory can explain or predict.

I will try to state the point at issue in yet another way. The postulates of the formal calculus of a theory are often said to constitute implicit definitions of its basic terms: they rule out all those conceivable interpretations of the primitive terms which would turn some of the postulates into false statements.

Thus, the single postulate that an—otherwise unspecified—relation R is asymmetrical limits the possible interpretations of 'R' and thereby "implicitly defines" it or partially specifies its meaning; for example, the postulate precludes the interpretation, by model or by correspondence rules, of 'Rxy' as 'x resembles y', 'x respects y', 'x is adjacent to y', and so forth. But the model of a theory in the sense here at issue imposes similar constraints on the empirical content that may be assigned to theoretical terms by means of correspondence rules. Thus, the "interpretation," by the "model," of certain theoretical magnitudes as masses and charges of particles precludes the adoption of correspondence rules that would assign to those magnitudes the physical dimensions of mass densities and temperatures. Moreover, the masses and charges of particles would have to be linked by suitable composition laws to inacroscopically ascertainable masses and charges; and so forth. Thus, the statements of such a model are not only of heuristic value: they have a systematic-logical function much like the formulas of the calculus. And this seems to me to remove one further reason against conceiving what I called the internal principles of a theory as divided into a calculus and a model.

6. The Status of Bridge Principles

In Schema I, the sentences in R are regarded as statements that assign empirical content to the terms of the calculus; and the use of such designations for them as 'operational *definition*', 'coordinating *definition*', '*rule* of correspondence' tends to convey the suggestion that they have the character of stipulations or, more accurately, of statements made true by definition or by terminological conventions of a more general kind; a similar impression is created by Campbell's construal of them as a "dictionary" that serves to relate the "hypothesis" of a theory to (pretheoretical) empirical propositions. This construal is definitely not intended or endorsed by all proponents of the idea of correspondence rules; and it is surely untenable for several reasons, among them the following:

As was said earlier, the basic principles of a theory often assert the existence of certain kinds of entities and processes, which are characterized, to a greater or lesser extent, by means of antecedently understood empirical concepts; and the so-called coordinating principles then have the character of hypotheses expressing specific connections between the entities and processes posited by the theory and certain phenomena that have been examined "pretheoretically." But not all these connecting hypotheses can have the nonempirical status of truths by terminological convention. For quite often, a theory provides us with several principles that link a theoretical notion to different potential observational or experimental findings. For example, physical theory offers several different ways of determining Avogadro's number or the charge of an electron or the velocity of light; this implies the claim that if one of the experimental methods yields a certain numerical value for the magnitude in question, then the alternative methods will yield the same value: but whether this is in fact the case surely is an empirical matter and cannot be settled by definitions, rules, or stipulations. It was precisely in order to avoid the intrusion of this empirical element, and the associated inductive "risk," that Bridgman regarded different operational procedures as specifying different concepts. On the other hand, when Carnap, in "Testability and Meaning," offered the basis for a logically much more subtle and supple restatement of the idea of an operational specification of meaning, he noted explicitly that when a term is introduced by means of several reduction sentences (the counterparts, in his theory, of operational definitions), then the latter usually have empirical consequences, expressible by a well-defined sentence which he calls the representative sentence of that set of reduction sentences.[14]

Still another consideration is relevant here. Even if a "rule of correspondence" is initially established by stipulative *fiat*, it may lose its status as true by convention, and become liable to modification in response to empirical evidence or theoretical developments. For example, to mark off, or "define" in experimental terms, equal intervals of time, some periodic process must be chosen to serve as a standard clock, such as the daily apparent motion of a star, or the swinging of a pendulum. The time intervals determined by the recurring phases of the chosen process are then equal by convention or stipulation. But what particular standard clock (and rule of correspondence specifying it) is chosen may make a great deal of difference for the possibility of formulating a system of simple and comprehensive laws and theoretical principles concerning the time dependence of empirical phenomena. For example, if the pulse beat of a particular person were chosen as the standard, then the "speed," the temporal rate of change, of all empirical phenomena would become causally linked to the state of health of the person chosen as the standard clock, and it would be impossible to formulate any simple laws of free fall, of planetary motion, of harmonic oscillation, of radioactive decay, etc. But even a perfectly plausible choice of standard clocks, such as the rotating earth or a pendulum device, leads to basically similar consequences: The rate of the earth's rotation is known to be slowly decreasing, and the period of a pendulum is dependent on several extraneous factors, such as location (since the gravitational force acting on the pendulum bob varies with the location). The theoretical principles and the empirical considerations that indicate these deviations and determine them quantitatively may well have grown out of findings obtained with the help of clocks of the incriminated kind; but theoretical consistency demands that those clocks, and hence the associated correspondence rules, now be qualified as only approximately correct, and that new, more accurate, methods of measurement be specified.

Thus, even though a sentence may originally be introduced by stipulation, and may thus reflect a rule of correspondence in the narrower sense of 'rule', it soon joins the club of all other member-statements of the theory and becomes subject to revision in response to further empirical findings and theoretical developments; and the bridge principles of Schema II are conceived as having this character.

It might seem that the preceding argument about the change of status of statements first established by stipulation, is flawed by confounding logical and epistemological questions with genetic-pragmatic ones. But the point can be argued without offering grounds for this suspicion: Even if we imagine a scientific theory "frozen," if we consider its form and status at one particular stage, there are no clear ways of distinguishing those statements of the theory which are made true by rule or convention from those which are not. The initially plausible idea that statements of the first kind should not be liable to modification in response to empirical evidence, will not serve the purpose, as we have noted: no statements other than the truths of logic and mathematics come with such a guarantee.

The considerations presented in support of Schema II have still another consequence; they show that within the class T of statements that constitute a theory, the dividing line between the two subclasses I and B is not very sharply determined. In particular, it cannot be characterized syntactically, by reference to the constituent terms; for, as we noted, sentences in either class contain theoretical terms characteristic of the theory as well as pretheoretical terms. Nor is the difference one of epistemic status, such as stipulational vs. empirical truth.

The distinction between the internal principles and the bridge principles of a theory must be understood in the somewhat intuitive manner in which it was introduced earlier in this essay. It is thus a vague distinction, but to the extent that the preceding considerations are sound, it can none the less provide philosophic illumination.

NOTES

1. E. Nagel, *The Structure of Science*; New York: Harcourt, Brace and World, Inc., 1961; pp. 125–129.
2. For a discussion of some of the difficulties besetting the distinction between observational and theoretical terms, and for further bibliographic references, see, for example, H. Putnam, "What Theories Are Not" in E. Nagel, P. Suppes, A. Tarski (eds.), *Logic, Methodology and Philosophy of Science*; Stanford: Stanford University Press, 1962; pp. 240–251; and P. Achinstein, "The Problem of Theoretical Terms," *American Philosophical Quarterly* 2, 193–203 (1965).
3. The axiomatization is given in H. Reichenbach, *Axiomatik der relativistischen Raum-Zeit-Lehre*. Braunschweig: Vieweg, 1924. This book forms the basis of many of the ideas developed in Reichenbach's *Philosophy of Space and Time*; New York: Dover Publications, 1958.
4. N. R. Campbell, whose construal of a scientific theory as consisting of a "hypothesis" and a "dictionary" is an early instance of a conception akin to that of Schema I, gives examples of just this type for the "hypotheses," i.e. calculi, of physical theories; see his *Foundations of Science* (formerly: *Physics, The Elements*); New York: Dover Publications, 1957, pp. 122–129.
5. P. W. Bridgman, *The Logic of Modern Physics*; New York: The Macmillan Co., 1948; p. 16.
6. This additivity is not the same thing as the conservation of mass: even if the total mass of an isolated system consisting of several physical bodies always equals the sum of the masses of the components, the total mass might conceivably decrease in the course of time, and thus not be conserved, as a result, say, of the spontaneous disappearance of some of the components of the system.
7. See, for example, P. K. Feyerabend, "Explanation, Reduction, and Empiricism," in H. Feigl and G. Maxwell (eds.), *Minnesota Studies in the Philosophy of Science*, vol. III (Minneapolis: University of Minnesota Press, 1962), pp. 28–97; P. K. Feyerabend, "Reply to Criticism," in R. S. Cohen and M. W. Wartofsky (eds.), *Boston Studies in the Philosophy of Science*, vol. II (New York: Humanities Press, 1965), pp. 223–261; N. R. Hanson, *Patterns of Discovery* (Cambridge, England: Cambridge University Press, 1958). For illuminating comments on the debate, and for further bibliographic references, see P. Achinstein, "The Problem of Theoretical Terms"; P. Achinstein, "On the Meaning of Scientific Terms," *The Journal of Philosophy* 61, 497–510 (1964), and D. Shapere, "Meaning and Scientific Change," in R. G. Colodny (ed.), *Mind and Cosmos*; (Pittsburgh: The University of Pittsburgh Press, 1966), pp. 4–85.

8. Cf. Feyerabend, op. cit., (1962), p. 37 and pp. 78–84; also Section VI of Feyerabend, "Problems o Empiricism," in R. G. Colodny (ed.), *Beyond the Edge of Certainty;* Englewood Cliffs, N.J.: Prentice-Hall, 1965, pp. 145–260.
9. H. Putnam, "How Not to Talk about Meaning," in R. S. Cohen and M. W. Wartofsky (eds.), *Boston Studies in the Philosophy of Science,* vol. II (New York: Humanities Press, 1965), pp. 205–222. Quoted passage, including italics, from p. 218.
10. See E. Nagel, *The Structure of Science,* especially chapter 5.
11. This idea is lucidly developed, in the context of a distinction of four senses of 'model of a theory', in M. Spector, "Models and Theories," *The British Journal for the Philosophy of Science* 16, 121–142 (1965).
12. For fuller discussions of these sues, see, for example, E. Nagel, *The Structure of Science,* pp. 107–117; M. B. Hesse, *Models and Analogies in Science;* London and New York: Sheed and Ward, 1963; C. G. Hempel, *Aspects of Scientific Explanation;* New York: The Free Press, 1965, pp. 433–447.
13. E. Nagel, *The Structure of Science,* p. 95.
14. See R. Carnap, "Testability and Meaning," *Philosophy of Science,* 3, 419–471 (1936) and 4, 1–40 (1937); especially p. 451. Carnap has since then developed his analysis of this topic further; for a recent statement, see section 24 (pp. 958–966) of his "Replies and Systematic Expositions," in P. A. Schilpp (ed.) *The Philosophy of Rudolf Carnap* (La Salle, Illinois; Open Court, 1963).

II

EXPLANATION AND PREDICTION

5

Explanation and Prediction by Covering Laws

1. The Two Covering-Law Models of Scientific Explanation

In this essay, I propose to present two models of scientific explanation; to comment on the range of their applicability and, briefly, on their qualifications as models of scientific prediction; and to examine some of the criticisms directed against these models and some alternatives to them that have been put forward in the recent literature. I will refer to the models in question as the deductive-nomological, or briefly the deductive, model, and the inductive-probabilistic, or briefly the probabilistic, model. Both have been described in considerable detail in earlier publications,[1] and I will therefore be brief in outlining their main characteristics here; but I will add some new amplificatory observations.

The two models are intended to exhibit the logical structure of two basic types of scientific explanation, of two ways in which empirical science answers the question as to the Why? of empirical phenomena. The two modes of explanation schematized by our models have one important feature in common: Both account for a given phenomenon—for example, a particular event—by showing that it came about in accordance with certain general laws or theoretical principles—say, $L_1, L_2, \ldots, L_m$—in the sense that its occurrence can be inferred from those laws taken in conjunction with a set of statements—say, $C_1, C_2, \ldots, C_n$,—which describe certain particular empirical circumstances. Thus, the two modes of explanation agree in accounting for a given phenomenon by reference to what, following Dray's example, I will call *covering laws*; I will accordingly refer to both of the corresponding models as *covering-law models* of scientific explanation.[2]

The difference between the two types of explanation represented by those models lies in the character of the laws invoked and, as a consequence, in the logical character of the inference that links the statement of the phenomenon in question to the explanatory information. A deductive-nomological explanation is based on laws which express unexceptional uniformities; such laws are of strictly universal form, of which the following is a simple example: 'In every case x, without exception, when the (more or less complex) conditions A are satisfied, an event or state of affairs of kind B comes about,' or, symbolically, '$(x)(Ax \supset Bx)$.' Schematically speaking, a law of this kind might then serve to explain the occurrence of B in a particular case i by reference to the information that i satisfies conditions A. Generally, the deductive-nomological model construes an explanation by means of strictly universal laws as a deductive argument of the form

$$\frac{\begin{array}{c} L_1, L_2, \ldots, L_m \\ \\ C_1, C_2, \ldots, C_n \end{array}}{E} \qquad (1.1)$$

The premises are said to form the explanans; the conclusion, i.e., the statement E describing the phenomenon to be explained, is called the explanandum-statement or briefly the explanandum. For convenience the word 'explanandum' will occasionally be used also to refer to the phenomenon described by E, i.e., the explanandum-phenomenon; this will be done only when the context makes it quite clear which is meant.

Usually—and especially in the case of causal explanation, which is one variety of deductive-nomological explanation,—the particular circumstances specified in the sentences $C_1, C_2, \ldots, C_n$ will be such that their occurrence is prior to, or at most simultaneous with, that of the event to be explained. But some scientific explanations also invoke, in the explanans, certain occurrences that are later than the explanandum-event. Suppose, for example, that the path taken by a light ray between two points, P_1 and P_2, is explained by deductive subsumption under Fermat's law that a light ray's path is always such that the time required to cover it is an extremum: then the statements $C_1, C_2, \ldots, C_n$ will include, in addition to a specification of the refractive characteristics of the optical media involved, statements to the effect that the path in question did contain both P_1 and P_2; hence, in explaining its passage through a certain intermediate point P_3, which then constitutes the explanandum event, reference is made in the explanans to some later event, namely, the ray's going through P_2.

The kind of explanation schematized in (1.1), then, deductively subsumes the explanandum under general laws and thus shows, to put it loosely, that according to those laws the explanandum-phenomenon "had to occur" in virtue of the particular circumstances described by $C_1, C_2, \ldots, C_n$. This procedure is illustrated, for example, by the explanation of mirror images, the broken appearance of an oar partly submerged under water; the size of images produced by lens systems, etc., with the help of the optical laws of reflection and refraction; and by the explanation of certain aspects of the motions of freely falling bodies, the moon, artificial satellites, the planets, etc., with the help of the laws of Newtonian mechanics. Professor Hanson's discussion of the HD (hypothetico-deductive) method in science suggests further illustrations; for example, his account of the HD analysis of the problem of perturbations can be used to construct another example of deductive-nomological explanation.

In an inductive-probabilistic explanation, on the other hand, at least some of the relevant laws are not of strictly universal, but of statistical character. The simplest statements of this kind have the form: 'the statistical probability (i.e., roughly, the long-run relative frequency) for the occurrence of an event of kind B under conditions of kind A is r' or, in symbols, '$p_s(B,A) = r$.' If the probability r is close to 1 then a law of this type may serve to explain the occurrence of B in a particular case i by reference to the information that i satisfies condition A or, briefly, that Ai. But this information, together with the statistical law invoked, does not, of course, deductively imply the explanandum-statement 'Bi,' which asserts the occurrence of B in the individual case i; rather, it lends to this statement strong inductive support; or, to use Carnap's terminology, it confers upon the explanandum-statement a high logical, or inductive, probability. The simplest kind of inductive-probabilistic explanation, then, may be schematized as follows:

$$\left.\begin{array}{l} Ai \\ p_s(B,A) = 1-\epsilon \quad \text{(where } \epsilon \text{ is small)} \\ Bi \end{array}\right\} \quad \begin{array}{l} \text{confers high} \\ \text{inductive} \\ \text{probability on} \end{array} \qquad (1.2)$$

Let me give a simple illustration of this type of argument. Suppose that a blind drawing made from an urn filled with marbles produces a white marble. The result might be probabilistically explained by pointing out that there were 1000 marbles in the urn, all but one of them white, and that—as a result of thorough stirring and shaking of the contents of the urn before each drawing—the statistical probability for a blind drawing to yield a white marble is $p_s(W,D) = 0.999$. This law, in combination with the information, Di, that the instance i in question was a drawing from the urn, does not deductively imply, but lends strong inductive support, or high inductive (logical) probability, to the presumption that in the case i, the result is a white ball, or briefly that Wi.

It might be objected that after all, the explanatory information offered in this case does not exclude the drawing of a nonwhite ball in the particular case i, and that therefore the probabilistic argument does not explain the given outcome. And indeed, this kind of objection is sometimes raised. But it presupposes what seems to me a too restrictive construal of explanation: while in deductive-nomological explanation, the explanans does indeed preclude the nonoccurrence of the explanandum event, probabilistic arguments, which make their conclusions no more than highly probable, are surely often offered as scientific explanations. In particular, such probabilistic accounts play a fundamentally important role in such fields as statistical mechanics, quantum theory, and genetic theory. For example, the explanation of certain uniformities that hold at the macro level in terms of assumptions about statistical uniformities at the level of the underlying micro phenomena has basically the logical character just considered, although, needless to say, the details are vastly more complex.

It is also often held that statistical laws can account only for what happens in large samples, but not for what happens in a single case. But this view is not literally correct. Even statements about what happens in large samples can at best be made (inductively) very probable by the relevant statistical laws, but they are never implied with deductive certainty by the latter. In our illustration, for example, the statistical law that $p_s(W,D) = 0.999$ does not conclusively imply any nonanalytic statement about the number of white marbles obtained in a given sequence of even 10,000 drawings (after each of which the marble is put back into the urn); rather, it assigns to every number of white marbles that could possibly be obtained in 10,000 drawings, and also to every possible interval—e.g., between 9000 and 10,000 white marbles—a certain, more or less high, probability. And some statistical laws, such as the one just mentioned, assign in the same manner a very high probability to a particular kind of outcome—in our illustration the drawing of a white marble—even for one single trial or test. The difference between a "sample of one" and a "large sample" is, after all, a matter of degree, and it is hardly to be expected that in the context of probabilistic explanation it should give rise to an essential logical difference.

A probabilistic explanation, then, shows that in view of the information provided in the explanans, which includes statistical probability laws, the explanandum phenomenon is to be expected with high inductive probability.

But suppose now that by way of an explanation for the occurrence of B in a particular case i we are offered the explanans considered earlier, consisting of the statements '$p_s(B,A) = 1 - \epsilon$' and 'Ai.' Then even if this information is true, it is still possible that i should also satisfy some further conditions, say A^*, under which, according to a further true statistical law, the occurrence of B has an extremely small probability. For example, in a particular instance of throat infection, it may well be the case that (i) the illness is a streptococcus infection, and the statistical probability of a favorable response of such infections to penicillin treatment is high, and also (ii) the streptococci in the case at hand are of a strain that is highly resistant to penicillin, and the statistical probability for an infection caused by this strain to respond favorably to penicillin treatment is extremely small. If both (i) and (ii) are known, then a quick recovery in the case at hand surely cannot be reasonably explained by simply adducing the information mentioned under (i), even though this does confer a high probability upon the explanandum. Generally speaking: If a probabilistic explanation is to be rationally acceptable, no further information must be available whose inclusion in the explanans would change the probability of the explanandum. In other words, a rationally acceptable probabilistic explanation must satisfy what Carnap has called the requirement of total evidence; i.e., its explanans must include all the available information that is inductively relevant to the explanandum. In a deductive explanation, whose explanans logically implies the explanandum and thus confers upon it the logical probability 1, and whose deductive conclusiveness remains unchanged if to the explanans any further available information is added, the requirement of total evidence is always trivially satisfied.[3]

These, then, are the two covering-law models of explanation. What is the scope of their applicability? Do they account for all kinds of scientific explanation? It seems to me that between them, they accommodate all the explanations typically provided by the physical sciences; at any rate, I am not aware of any explanation in this area that cannot be quite satisfactorily construed as an instance—which may, however, be elliptically formulated—of deductive or of probabilistic explanation as they are schematically construed in those models.

The issue becomes more controversial when we turn to other branches of empirical science. Indeed, it is widely held that in order to do justice to their characteristic subject matters and their peculiar objectives, some of those other branches must, and do, resort to explanatory methods *sui generis*. For example, biology has been said to require teleological concepts and hypotheses in order to be able to account for regeneration, reproduction, homeostasis, and various other phenomena typically found in living organisms; and the resulting explanations have been held to be fundamentally different from the kinds of explanation offered by physics and chemistry.

Now indeed, some kinds of teleological explanation whieli have been suggested for biological phenomena fit neither of the covering-law models. This is true, for example, of vitalistic and neovitalistic accounts couched in terms of vital forces, entelechies, or similar agents, which are assumed to safeguard or restore the normal functioning of a biological system as far as this is possible without violation of physical or chemical laws. The trouble with explanations of this type—in sharp contrast, for example, to explanations invoking gravitational or electromagnetic forces—is that they include no general statements indicating under what conditions, and in what specific manner, an entelechy will go into action, and within what range of possible interferences with a biological system it will succeed in safeguarding or restoring the system's normal way of functioning. Consequently, these explanations do not tell us—not even in terms of probabilistic laws—what to expect in any given case, and thus they give us no insight into biological phenomena, no understanding of them (even though they may have a certain intuitive appeal); and precisely for this reason, they are worthless for scientific purposes and have, in fact, been abandoned by biologists. The reason for their failure does not lie, of course, in the assumption that entelechies are invisible and indeed noncorporeal entities; for neither are the gravitational or electromagnetic fields of classical theory visible or corporeal, and yet they provide the basis for important scientific explanations. For those field concepts function in the context of a set of general laws and theoretical principles which specify, among other things, under what conditions fields of what characteristics will arise, and what their effects will be, say, on the motion of a given macrophysical object or on an elementary particle. It is precisely the lack of corresponding laws or theoretical principles for entelechies which deprives the latter concept of all explanatory force.

But while vitalistic accounts have had to be discarded, very promising advances toward a scientific understanding of some of the biological phenomena in question have been made with the help of physical and chemical theory, cybernetic and information-theoretical principles, and other means; but these approaches all lend themselves to analysis in terms of the covering-law models.

Several other branches of empirical science use interesting explanatory methods that appear to be distinctly their own. There is, for example, the method of functional analysis as used in cultural anthropology and sociology; there is the method, used by historians among others, of explaining deliberate human actions by reference to motivating reasons; there are the psychoanalytic explanations of dreams, phantasies, slips of pen, tongue, and memory, neurotic symptoms, etc., in terms of subconscious psychic processes; there is the method of genetic explanation. Do any of these modes of explanation differ logically from those we have considered so far? It is impossible, of course, to examine this vast problem in the present context; but I will at least state the answer which I am inclined to give, and for which I have tried to offer support in other essays [14, 16, 17].

Among the various explanations here referred to, there are some which lack the status of even a potential scientific explanation because they are entirely incapable of any objective test. This is true of neovitalistic and of some psychoanalytic explanations, among others; they do not tell us, even with probability, what to expect under any kind of circumstance. For this reason, they admit of no test and certainly do not offer an explanation of anything. Some other explanatory accounts—including certain instances of functional analysis—are couched in terms of testable empirical assertions which, however, offer no adequate grounds for expecting the occurrence of the explanandum-phenomenon; thus they violate what is surely a necessary, though not sufficient, condition for any scientific explanation. Still other "explanations" have the character of straightforward descriptions: they tell us that, rather than why, certain things are the case. But if we set aside accounts of these kinds, then the remaining explanations—notwithstanding great differenecs in the explicitness and precision with which they are formulated—can, I think, be accounted for as more or less complex concatenations of arguments each of which is of one or the other of the two basic types considered earlier; and all of them derive whatever explanatory power they possess from universal or statistical laws which they explicitly invoke or implicitly presuppose.

The assessment just outlined of the adequacy of the covering-law models for the various sorts of explanation offered in different areas of empirical science implies, of course, no claim whatever concerning the extent to which scientific explanations can actually be achieved for the phenomena studied in different branches of scientific inquiry; and even less does it imply or presuppose universal determinism.

2. Predictive Aspects of the Covering-Law Models

In the preceding section, I appealed, at certain points, to a requirement which must be satisfied, it seems to me, by any account that is to qualify as a scientific explanation. I will refer to it as a *condition of adequacy for scientific explanations* and will now state it somewhat more fully, and then comment on it. The requirement is that any adequate scientific answer to a question of the type 'Why is X the case?' must provide information which constitutes good grounds for believing or expecting that X is the case. Of course, by no means all information that offers such grounds does thereby provide an explanation. For example, a report on the favorable outcome of a thorough experimental test of the hypothesis that X is the case may constitute good inductive grounds for believing the hypothesis to be true, but it surely provides no explanation of why X is the case. Thus, the condition of adequacy just stated is a necessary, but not a sufficient one. Indeed, we might add that the request for an explanation normally arises only when the explanandum-event is known or presumed to have occurred; hence, offering grounds for believing that it has occurred is not the principal task of an explanation at all; but it is nevertheless a necessary condition for its adequacy.

Explanations by covering laws, whether deductive or probabilistic, clearly satisfy this condition; and they have a further feature which seems to me characteristic of all scientific explanation, namely, they rest on general laws. In the explanation of an individual event, these laws connect the explanandum-event with the particular circumstances cited in the explanans, and they thus confer explanatory significance upon the latter. To be sure, in the usual wording of an explanation the laws often are not all explicitly mentioned. An explanation for a particular event, for example, may sometimes be stated by saying that the event in question was *caused* by such and such other events and particular conditions. But a causal claim of this kind presupposes causal laws, and the assumption of such laws is thus implicit in the explanation.

One consequence of the condition of adequacy I have been discussing may be stated as follows: In the explanation of a given event, the explanans must be such that if it had been known before the occurrence of the explanandum-event, it would have enabled us to predict the latter deductively or inductively.[4] This principle seems to me obvious, and indeed almost trivial; yet it has met with objections. One of these, advanced by Scriven, holds that there are instances of perfectly good explanations which offer no basis for potential prediction. By way of illustration, Scriven [24], p. 480) refers to the case where "we have a proposition of the form 'The only cause of X is A' (I)—for example, 'The only cause of paresis is syphilis.' Notice that this is perfectly compatible with the statement that A is often not followed by X—in fact, very few syphilitics develop paresis. . . . Hence, when A is observed, we can predict that X is *more* likely to occur than without A, hut still extremely unlikely. So we must, on the evidence, still predict that it will *not* occur. But if it does, we can appeal to (I) to provide and guarantee our explanation." But surely, the information that patient B has previously suffered from syphilis does not suffice to explain his having developed paresis, precisely because paresis is such a rare sequel of syphilis. The specification of what might be called a nomically necessary antecedent of an event does not, in general, provide an explanation for the latter. Otherwise, we might argue, for example, that only individuals who are alive contract paresis, and we might adduce B's having been alive as explaining his having contracted paresis. Note that the probability for living individuals to develop paresis is positive, whereas for those who are not alive it is zero, just as in the case of the explanation by prior syphilitic infection.

It might be suspected that this counterexample simply is not an instance of specifying "the only cause" in the sense intended by Scriven. however, Scriven presents us with no precise criteria for what he wishes to understand by "the only cause" of a phenomenon. He does indeed indicate what he means by a cause: "Speaking loosely, we could say that a cause is a nonredundant member of a set of conditions jointly sufficient for the effect . . . , the choice between the several candidates that usually meet this requirement being based on considerations of context. However, many qualifications must be made to this analysis . . . , and in the end probably the best view of *a cause* is that it *is any physical explanation which involves reference to only one state or event (or a few)* other than the effect" ([25], pp. 215–216; italics mine). But—disregarding the point that, even according to Scriven, an explanation of an event is a certain kind of communication about it and must therefore surely be distinguished from its causes—the account thus achieved is simply circular, because the kind of *explanation* referred to in Scriven's paresis example is in turn characterized as specifying the only *cause* of an event. The same characteristic seems to me to vitiate Scriven's general analysis of causal explanation, according to which a causal explanation of a state or event X must specify "the cause of X," and must do so in a manner which meets certain specified requirements of intelligibility ([25], p. 204). For since, as has just been mentioned, Scriven's definition of 'cause' is in terms of 'explanation,' his analysis of the meanings of the two terms plainly moves in a circle.

Among the objections that have been raised against the attribution of potential predictive import to explanations, there are some, however, which are aimed at the stronger thesis, put forth in an article by Oppenheim and myself [18], that scientific explanation and scientific prediction have the same logical structure and differ only in pragmatic characteristics.[5] I agree with the observation made both by Scheffler ([22], Sec. 1) and by Scriven (e.g., [25], p. 177) that what is usually called a prediction is not an argument but a statement—or, more precisely, as pointed out by Scheffler, a statement-token, i.e., a concrete utterance or inscription of a statement,[6] purporting to describe some future event. Our thesis about the structural identity of scientific explanation and prediction Should be understood, of course, to refer to the logical structure, not of predictive statements, but of *predictive arguments* that serve to establish predictive statements in science. But I would now want to weaken the thesis so as to assert only that the two covering-law models represent the logical structure of two important types of predictive inference in empirical science, but not that these are the only types. That there are predictive arguments that could not qualify as explanations has been asserted, for example, by Scheffler ([22], p. 296) and by Scriven (e.g., [24], p. 480); the one type of argument in support of this claim that I find clear-cut and convincing is Scheffler's observation to the effect that scientific predictions may be inductively grounded on information that includes no statements in the form of general laws, but only a finite set of particular data, e.g., the results obtained in trying out a given die or a given roulette wheel a finite number of times for the purpose of predicting later results.

Predictive arguments of this type, from finite sample to finite sample, clearly are not of the covering-law form.

The preceding discussion cannot, of course, do justice to all the comments and criticisms that have been directed at the theses here considered concerning the predictive aspects of scientific explanation. . . . ; further responses to some of the criticisms have been put forward elsewhere [10, 15, 20].

3. Some Misunderstandings Concerning the Covering-Law Models

In this section, I will survey briefly certain objections against the covering-law models which completely miss their target because they involve misunderstandings or even outright misrepresentations.

(i) In a recent essay [23], Scriven speaks of "the deductive model, with its syllogistic form, where no student of elementary logic could fail to complete the inference, given time premise" (p. 462). He gives no reference to the literature to support this simple-minded and arbitrary construal, and indeed, this would be impossible; for the deductive model has never been conceived in this fashion by its advocates, and the general form (1.1) of the model clearly allows for the use of highly complex general laws—e.g., laws representing certain quantitative parameters as mathematical functions of others; and the deduction of the explanandum cannot, in such cases, be achieved by syllogistic methods, as any student of elementary logic knows.[7]

(ii) Referring to writings by Oppenheim and by myself on the subject of explanation, Scriven maintains, again without any documentation whatsoever, that the deductive model as characterized *by* us "fails to make the crucial logical distinctions between . . . things to be explained, and the description of these things," and he accuses us accordingly of an "incautious amalgamation" of the two[8]; he adds the diagnosis that "the most serious error of all those I believe to be involved in Hempel and Oppenheim's analysis also springs from the very same innocuous-seeming oversimplification: the requirement of deducibility itself, plausible only if we forget that our concern is fundamentally with a phenomenon, not a statement"; and he goes on to say ([24], p. 195), "It may seem unjust to suggest that Hempel and Oppenheim amalgamate the phenomenon and its description . . . when they make clear that the 'conclusion' of the explanation is 'a sentence describing the phenomenon (not that phenomemion itself).'" In view of the passage he quotes here,[9] and in view of the fact that the distinction between empirical phenomena and statements describing them is rigorously observed in all our publications, Scriven's suggestion does not only *seem* unjust: it *is* unjust, and indeed, it is an indefensible misrepresentation.

As for Scriven's remark that "our concern is fundamentally with a phenomenon, not with a statement," it is clear that—no matter how the logical structure of explanation may construed in detail—an explanation will require reference to some explanatory "facts," and that these, as well as the phenomenon to be explained, cannot be dealt with, in the context of scientific explanation, without the use of statements describing them.

(iii) Concerning the status of probabilistic explanation, Scriven ([25], p. 228) speaks of "the mistake [Hempel] originally made in thinking that explanations which fit the deductive models are more scientific than statistical explanations." Again, the attribution of the alleged mistake is not documented, and it could not be; for I have never expressed that opinion. Rather, I have presented deductive and probabilistic explanation as different modes of explanation, each significant in its own right.[10]

(iv) Referring to my analysis of probabilistic explanation, Scriven comments: "Such explanations cannot be subsumed tinder Hempel and Oppenheim's original analysis as it stands, because no *deduction* of a nonprobability statement from them is possible, and hence it is impossible for them to explain any actual occurrence, since actual occurrences have to be described by nonprobability statements" ([25], p. 192). The first part of this sentence is misleading, to say the least; for neither Oppenheim nor I have ever claimed that the deductive model could accommodate probabilistic explanations. And the second part involves a misconception of statistical explanation: In schema (1.2), for example, the explanandum-statement does describe a particular and (if true, as we may assume it to be) an actual event. In an earlier article [23], Scriven states his idea more explicitly: "one cannot *deduce* from any law of the form 'If C then probably E,' combined with the antecedent condition C, that E occurs. One can only deduce that E *probably* occurs, and we are not trying to explain a probability but an event" (p. 457; italics his). But from the premises Scriven mentions, one surely can *not* deduce 'E probably occurs,' for this is not even a significant statement, any more than are such expressions as 'Henry is older' or 'line z is parallel.' For the occurrence of E can be qualified as probable or improbable, or perhaps as having such and such a quantitative probability, only relative to some body of information by reference to which the probability of E's occurrence is appraised. Hence, properly constructed, a statement about the probability of E's occurrence will take some such form as 'On information e, the probability that E occurs is high (or low, or has such and such a numerical value).'

This consideration indicates why, in our earlier discussion, the probabilistic explanation for the occurrence of B in the individual case i is not construed as an argument—deductive or otherwise—with the conclusion 'It is highly probable that Bi,' but as an inductive argument showing, in the manner schematized in (1.2), that 'Bi' is made highly probable by the information adduced in the explanans.

(v) Scriven attributes to me, and expresses disagreement with, the assumption that 'probability statements are statistical in nature" ([25], p. 228). He does not amplify or document this attribution, which is seriously incorrect: I regard it as crucially important—and have said so explicitly in several articles—to distinguish, in the manner of Carnap, between two very different concepts of probability. One of these is the empirical, statistical concept, which figures in sentences of the form '$p_s(G,F) = r$,' or 'the statistical probability of G relative to F is r,' which means, roughly, that the long-run relative frequency with which instances of F are also instances of G is r. Thus, the statistical concept of probability represents a quantitative relation between two kinds, or classes, of events. It is this concept which is used in all probabilistic laws of empirical science, and thus in some of the premises of probabilistic explanations. The second concept of probability, the logical or inductive one, represents a relation, not between kinds of events but between statements, e.g., between the conjunction of the explanans statements and the explanandum statement in a probabilistic explanation. This concept is *not* statistical in character, and it is not invoked in the probabilistic laws of empirical science. Exactly how this concept is to be defined, and to what extent it is capable of precise quantitative characterization is still a matter of debate; a rigorous and systematic quantitative concept of logical probability, applicable to certain types of formalized languages, has been developed by Carnap.[11]

(vi) Another of Scriven's criticisms ([25], p. 209) contends that "Hempel and Oppenheim never deal with the explanation of a particular event instance, but only with events of a certain kind." Since Scriven again does not support this contention, it is hard to offer a pertinent reply other than to point out that it is false. In setting out the deductive model in our article, Oppenheim and I consider first "exclusively the explanation of particular events occurring at a certain time and place"[12] and then go on to examine the explanation of uniformities expressed by general laws.

I might add, moreover, that while both the deductive and the inductive model clearly provide for the explanation of particular event-instances—for example, Bi in a probabilistic explanation of the form (1.2), or a volume increase in a particular body of gas at a particular place and time, as explained by deductive subsumption tinder appropriate gas laws—they do not, strictly speaking, attribute a meaning to the notion of explaining a *kind* of event. For a particular event-instance can be represented by a statement describing it, which may become the explanandum-statement of a covering-law explanation; whereas a certain *kind* of event would be represented by a predicate expression, such as 'is a case of a volume increase' or 'is an instance of drawing a white marble,' which clearly cannot form the explanandum of any covering-law explanation.

4. Different Concepts of Explanation

Since limitations of space preclude any attempt at completeness of coverage in this study of criticisms directed against the covering-law analysis of explanation,[13] I propose to single out for consideration in this final section just one interesting and important group of those criticisms.

As I pointed out at the beginning, the covering-law models are intended to exhibit the logical structure of two basic modes of *scientific explanation*, of two logically different ways in which empirical science answers questions that can typically be put into the form 'Why is it the case that X'?, where the place of

'X' is occupied by some empirical statement. Requests for explanations in this sense are often expressed also by means of other phrasings; but the 'Why?' form, even if not uniformly the simplest or most natural one, is always adequate to indicate—preanalytically, and hence not with the utmost precision—the sense of 'explanation' here under analysis, and to set it apart from the various other senses in which the word 'explain' and its cognates are used. To put forward the covering-law models is not, therefore, to deny that there are many other important uses of those words, and even less is it to claim that all of those other uses conform to one or other of the two models.

For example, an explanation of why every equilateral triangle is equiangular, or why an integer is divisible by 9 whenever the sum of its digits in decimal representation is so divisible requires an argument whose conclusion expresses the proposition in question, and whose premises include general geometrical or arithmetical statements, but not, of course, empirical laws; nor, for that matter, is the explanandum statement an empirical one. This sort of explaining, though rather closely related to the kind with which we are concerned, is not meant to be covered by our models.

Nor, of course, are those models intended to cover the vastly different senses of 'explain' involved when we speak of explaining the rules of a game, or the meaning of a hieroglyphic inscription or of a complex legal clause or of a passage in *Finnegans Wake*, or when we ask someone to explain to us how to repair a leaking faucet. Giving a logical and methodological analysis of scientific explanation is not the same sort of thing as writing an entry on the word 'explain' for the *Oxford English Dictionary*. Hence to complain, as Scriven does, of the "hopelessness" of the deductive model because it does not fit the case of "understanding the rules of Hanoverian succession" ([23], p. 452) is simply to miss the declared intent of the model. And it is the height of irrelevance to point out that "Hempel and Oppenheim's analysis of explanation absolutely presupposes a descriptive language" (which is true), whereas "there are clearly cases where we can explain without language, e.g., when we explain to the mechanic in a Yugoslav garage what has gone wrong with the car" ([25], p. 192). This is like objecting to a definition of 'proof' constructed in metamathematical proof theory on the ground that it does not fit the use of the word 'proof' in 'the proof of the pudding is in the eating,' let alone in '90 proof gin.' I therefore cheerfully concede that wordless gesticulation—however eloquent and successful—which is meant to indicate to a Yugoslav garage mechanic what has gone wrong with the car does not qualify as scientific explanation according to either of the two covering-law models; and I should think that any account of scientific explanation which did admit this case would thereby show itself to be seriously inadequate.

In support of his insistence on encompassing all those different uses of the word 'explain,' Scriven maintains, however, that they have the same "logical function," about which he says: "the request for an explanation presupposes that *something* is understood, and a complete answer is one that relates the object of inquiry to the realm of understanding in some comprehensible and appropriate way. What this way is varies from subject matter to subject matter . . . ; but the *logical function* of explanation, as of evaluation, is the same in each field. And what counts as complete will vary from context to context within a field; but the logical category of complete explanation can still be characterized in the perfectly general way just given" ([25], p. 202; italics his). But while the general observation with which this passage begins may well be true of many kinds of explanation, neither it nor the rest of the statement specifies what could properly be called a *logical* function of explanation; this is reflected in the fact that such terms as 'realm of understanding' and 'comprehensible' do not belong to the vocabulary of logic, but rather to that of psychology. And indeed, the psychological characterization that Scriven offers here of explanation makes excellent sense if one construes explanation as a pragmatic concept. Before considering this construal, I want to indicate briefly why I do not think that explanation in all the different senses envisaged by Scriven can be held, in any useful and enlightening sense, to have the same "logical function."

One of the reasons is the observation that the objects of different kinds of explanation do not even have symbolic representations of the same logical character: Some explanations are meant to indicate the meaning of a word or of a linguistic or nonlinguistic symbol, which will be represented by an expression that is not a statement, but a name ('the integral sign,' 'the swastika') or a definite description ('the first pages of *Finnegans Wake*'); while other explanations are meant to offer reasons, grounds, causes, or the like for something that is properly represented by a statement; for example, a mathematical truth, some particular empirical event, or an empirical uniformity such as that expressed by Galileo's law of free fall. Thus, first of all, the logical character of the explanandum-expression is different in these two classes of explanations.

Secondly, the task of specifying meanings and that of specifying grounds, reasons, causes, etc., surely are not of the same logical character; and still a different kind of task is involved in explaining how to make a Sacher torte or how to program a certain type of digital computer. And while any of these and other kinds of explanation may be said to be capable of enhancing our "understanding" in a very broad sense of this word, it is worth noting that the requisite sense is so inclusive as to be indifferent to the important distinction between knowing (or coming to know) that p, knowing (or coming to know) why p, knowing (or coming to know) the meaning of S, and knowing (or coming to know) how to do Z. To be sure, the application of any concept to two different cases may be said to disregard certain differences between them; but the differences in the tasks to be accomplished by different sorts of explanation reflect, as I have tried to indicate, *differences* precisely *in the logical structure* of the corresponding explanations.

As I suggested a moment ago, Scriven's observations on the essential aspects of explanation are quite appropriate when this concept is understood in a pragmatic sense. Explanation thus understood is always explanation for someone, so that the use of the word 'explain' and its cognates in this pragmatic construal requires reference to someone to whom something is explained, or for whom such and such is an explanation of so and so. One elementary sentence form for the pragmatic concept of explanation is, accordingly, the following:

Person A explains X to person B by means of Y.

Another, simpler, one is

Y is an explanation of X for person B.

Here, Y may be the production of certain spoken or written words, or of gestures; it may be a practical demonstration of some device; or, perhaps, in Zen fashion, a slap or an incongruous utterance.

The pragmatic aspects of explanation have been strongly emphasized by several writers, among them Dray[14] and, as we have seen, Scriven. Indeed, the pragmatic concept may claim psychological and genetic priority over the theoretical nonpragmatic one, which the covering-law models are intended to explicate. For the latter is an abstraction from the former, related to it in a manner quite similar to that in which metamathematical concepts of proof—which might figure in sentences of the form 'String of formulas U is a proof of formula V in system S'—are related to the pragmatic concept of proof, which would typically figure in phrases such as 'Y is a proof of X for person B.' Whether, say, a given argument Y proves (or explains) a certain item X to a given person B will depend not only on X and Y but also quite importantly on B: on his interests, background knowledge, general intelligence, standards of clarity and rigor, state of mental alertness, etc., at the time; and factors of this kind are, of course, amenable to scientific investigation, which might lead to a pragmatic theory of proof, explanation, and understanding. Piaget and his group, for example, have devoted a great deal of effort to the psychological study of what might be called the conception of proof in children of different ages.

But for the characterization of mathematics and logic as objective disciplines, we clearly need a concept of proof which is not subjective in the sense of being relative to, and variable with, individuals; a concept in terms of which it makes sense to say that a string Y of formulas is a proof of a formula X (in such and such a theory), without making any mention of persons who might understand or accept Y; and it is concepts of this nonpraginatic kind which are developed in metamathematical proof theory.

The case of scientific explanation is similar. Scientific research seeks to give an account—both descriptive and explanatory—of empirical phenomena which is objective in the sense that its implications and its evidential support do not depend essentially on the individuals who happen to apply or to test them. This ideal suggests the problem of constructing a nonpragmatic conception of scientific explanation—a conception that requires reference to puzzled individuals no more than does the concept of mathematical proof. And it is this nonpragmatic conception of explanation with which the two covering-law models are concerned.

To propound those models is therefore neither to deny the existence of pragmatic aspects of explanation, nor is it to belittle their significance. It is indeed important to bear in mind that when a particular person seeks an explanation for a given phenomenon, it may suffice to bring to his attention some particular facts he was not aware of; in conjunction with his background knowledge of further relevant facts, this may provide him with all the information he requires for understanding, so that, once the "missing item" has been supplied, everything falls into place for him. In other cases, the search for an explanation may be aimed principally at discovering suitable explanatory laws or theoretical principles; this was Newton's concern, for example, when he sought to account for the refraction of sunlight in a prism. Again at other times, the questioner will be in possession of all the requisite particular data and laws, and what he needs to see is a way of inferring the explanandum from this information. Scriven's writings on explanation suggest some helpful distinctions and illustrations of various kinds of puzzlement that explanations, in this pragmatic sense, may have to resolve in different contexts. But to call attention to this diversity at the pragmatic level is not, of course, to show that nonpragmatic models of scientific explanation cannot be constructed, or that they are bound to be hopelessly inadequate—any more than an analogous argument concerning the notion of proof can establish that theoretically important and illuminating nonpragmatic concepts of proof cannot be constructed. As is well known, the contrary is the case.

On these grounds we can also dismiss time complaint that the covering-law models do not, in general, accord with the manner in which working scientists actually formulate their explanations.[15] Indeed, their formulations are usually chosen with a particular audience—and thus with certain pragmatic requirements—in mind. But so are the formulations which practicing mathematicians give to their proofs in their lectures and writings; and the metamathematical construal of the concept of proof purposely, and reasonably, leaves this aspect out of consideration.

I think it is clear then, from what has been said, that many—though by no means all—of the objections that have been raised against the covering-law models, as well as some of the alternatives to them that have been suggested, miss their aim because they apply to nonpragmatic concepts of explanation certain standards that are proper only for a pragmatic construal.

NOTES

1. For a presentation of the deductive model, with references to earlier proponents of the basic conception, see Hempel and Oppenheim [17]; a shorter statement of the deductive model and a reply to some criticisms of it, as well as a detailed account of the probabilistic model are included in [15]; brief characterizations of both models are also given in [13] and [14].
2. Dray introduced the felicitous terms "covering law" and "covering-law model" in [6]. To avoid a misunderstanding that might arise from some passages in Chapter II, Section 1 of this work, however, it ought to be borne in mind that a covering-law explanation may invoke more than just one law. Also, while Dray used the term "covering-law model" exclusively to refer to the deductive-nomological model, I am using it here to refer to the probabilistic model as well.
3. This point is discussed more fully and precisely in [15], Secs. 10 and 11.

4. This formulation takes into account the remark, made by Scheffler [22], pp. 297–298) and by Scriven ([25], pp. 179–181), that one cannot properly speak of "predicting" the explanandum-phenomenon if the latter is not an individual event but, say, a uniformity as expressed by a general law.
5. *Philosophy of Science*, 15, 138 (1948); p. 322 in reprinted text. For an earlier statement, see [11], Sec. 4.
6. Cf. [22], Sec. 1; a fuller discussion of explanation and prediction in the light of the type-token distinction will be found in Kim [20].
7. This misconstrual is pointed out also by Brodbeck [2], in an essay which includes a critical study of various aspects of Scriven's views on explanation.
8. [25], p. 196 and p. 195.
9. The passage here quoted, which I had called to Professor Scriven's attention by correspondence, is from Hempel and Oppenheim [18], pp. 136–137; p. 321 in reprinted text.
10. The earliest references to the point are in Hempel [12], Sec. 5.3; Hempel and Oppenheim [18], pp. 139–140 (p. 324 in reprinted text); Hempel [13], Sec. 1. The first two of these articles make only brief mention of probabilistic explanation and then concentrate on an analysis of deductive-nomological explanation; the reason being the difficulty of the logical problems posed by statistical explanation, as is explicitly stated at the second of the three places just listed.
11. See especially Carnap, [4]. A very clear general account of the basic ideas will be found in [5].
12. See [18], p. 136; p. 320 in reprinted text.
13. The important discussions and new contributions with which, to my regret, I cannot deal in this essay include, for example, the precise and incisive critique by Eberle et al. [7] of the formal definition, set forth in Sec. 7 of Hempel and Oppenheim [18], of the concept E of potential explanans for languages which have the syntactic structure of the first-order functional calculus without identity. The authors prove that the proposed definition is vastly too liberal to be adequate. An ingenious modification which avoids the difficulty in question has been constructed by one of the three critics, D. Kaplan [19]. An alternative way of remedying the shortcoming in question has been proposed by J. Kim in one section of his doctoral dissertation [20]; his procedure is to be published in a separate article under the title "On the Logical Conditions of Deductive Explanation." Two other illuminating studies both of which raise certain questions about the deductive model of explanation, and which I can only mention here, are Bromberger [3] and Feyerabend [9]. Several earlier critical and constructive comments on the covering-law models are acknowledged and discussed in [15].
14. See, for example, [6], p. 69, where the author says: "as I shall argue further in this, and in succeeding chapters, there is an irreducible pragmatic dimension to explanation."
15. On this point, cf. also the discussion in Bartley [1], Sec. 1, which, among other things, defends Popper's presentation of the deductive model against this charge. For some comments in a similar vein, see Pitt [21], pp. 585, 586.

REFERENCES

1. W. W. Bartley, III, "Achilles, the Tortoise, and Explanation in Science and History," *The British Journal for the Philosophy of Science*, 13, 15–33 (1962).

2. M. Brodbeck, Explanation, Prediction and 'Imperfect' Knowledge,"in Feigl and Maxwell, [8], pp. 231–272.
3. S. Bromberger, "The Concept of Explanation," Ph.D. thesis, Harvard University, 1961.
4. R. Carnap, *Logical Foundations of Probability*, University of Chicago Press, Chicago, 1950.
5. R. Carnap, "Statistical and Inductive Probability." Reprinted, from a pamphlet published in 1955, in E. H. Madden, Ed., *The Structure of Scientific Thought*, Houghton Mufflin, Boston, 1960.
6. W. Dray, *Laws and Explanation in History*, Oxford University Press, London, 1957.
7. R. Eberle, D. Kaplan, and R. Montague, "Hempel and Oppenheim on Explanation," *Philosophy of Science*, 28, 418–428 (1961).
8. H. Feigl and G. Maxwell, Eds., *Minnesota Studies in thc Philosophy of Science*, Vol. III, University of Minnesota Press, Minneapolis, 1962.
9. P. K. Feyerabend, "Explanation, Reduction, and Empiricism," in Feigl and Maxwell [8], 28–97.
10. A. Grünbaum, "Temporally Asymmetric Principles, Parity between Explanation and Prediction, and Mechanism versus Teleology," *Philosophy of Science*, 29, 146–170 (1962).
11. Norwood Russell Hanson, "Retroductive Inference," in B. Baumrin, ed., *Philosophy of Science: The Delaware Seminar, Vol. 1* (1961–62) (New York: Inter-science Publishers, 1963), pp. 21–37.
12. C. G. Hempel, "The Function of General Laws in History," *The Journal of Philosophy*, 39, 35–48 (1942). Reprinted in P. Gardiner, Ed., *Theories of History*, Allen and Unwin, London, and The Free Press, Glencoe, Ill., 1959, pp. 344–356.
13. C. G. Hempel, "The Theoretician's Dilemma," in H. Feigl, M. Scriven, and G. Maxwell, Eds., *Minnesota Studies in the Philosophy of Science*, Vol. II, University of Minnesota Press, Minneapolis, 1958, pp. 37–98.
14. C. G. Hempel, "The Logic of Functional Analysis," in L. Gross, Ed., *Symposium on Sociological Theory*, Row, Peterson and Co., Evanston, Ill., and White Plains, N. Y., 1959, pp. 271–307.
15. C. G. Hempel, "Deductive-Nomological *vs.* Statistical Explanation," in Feigl and Maxwell [8], pp. 98–169.
16. C. G. Hempel, "Rational Action," *Proceedings and Addresses of the American Philosophical Association*, Vol. XXXV, The Antioch Press, Yellow Springs, Ohio, 1962, pp. 5–23.
17. C. G. Hempel, "Explanation in Science and in History," in R. G. Colodny, Ed., *Frontiers of Science and Philosophy*, University of Pittsburgh Press, Pittsburgh, 1962, pp. 7–33.
18. C. G. Hempel and P. Oppenheim, "Studies in the Logic of Explanation," *Philosophy of Science*, 15, 135–175 (1948). Sections 1–7 of this article are reprinted in H. Feigl and M. Brodbeck, Eds., *Readings in the Philosophy of Science*, Appleton-Century-Crofts, New York, 1953, pp. 319–352.
19. D. Kaplan, "Explanation Revisited," *Philosophy of Sciencc*, 28, 429–436 (1961).
20. J. Kim, "Explanation, Prediction, and Rctrodiction: Some Logical and Pragmatic Considerations," Ph.D. thesis, Princeton University, 1962.
21. J. Pitt, "Generalizations in Historical Explanation," *The Journal of Philosophy*, 56, 578–586 (1959).

22. I. Scheffler, "Explanation, Prediction, and Abstraction," *British Journal for the Philosophy of Science*, 7, 293–309 (1957).
23. M. Scriven, "Truisms as the Grounds for Historical Explanations," in P. Gardiner, Ed., *Theories of history*, Allen and Unwin, London, and The Free Press, Glencoe, Ill., 1959, pp. 443–475.
24. M. Scriven, "Explanation and Prediction in Evolutionary Theory," *Science*, 130, 477–482 (1959).
25. M. Scriven, "Explanations, Predictions, and Laws," in Feigl and Maxwell, [8], pp. 170–230.

6

Deductive-Nomological versus Statistical Explanation

1. Introduction

This essay is concerned with the form and function of explanation in the sense in which it is sought, and often achieved, by empirical science. It does not propose to examine all aspects of scientific explanation; in particular, a closer study of historical explanation falls outside the purview of the present investigation. My main object is to propose, and to elaborate to some extent, a distinction of two basic modes of explanation—and similarly of prediction and retrodiction— which will be called the deductive and the inductive mode.[1]

The structure of deductive explanation and prediction conforms to what is now often called the covering-law model: it consists in the deduction of whatever is being explained or predicted from general laws in conjunction with information about particular facts. The logic of this procedure was examined in some earlier articles of mine, and especially in a study carried out in collaboration with P. Oppenheim.[2]

Since then, various critical comments and constructive suggestions concerning those earlier efforts have appeared in print, and these as well as discussions with interested friends and with my students have led me to reconsider the basic issues concerning the deductive model of scientific explanation and prediction. In the first of the two principal parts of this essay, I propose to give a brief survey of those issues, to modify in certain respects the ideas set forth in the earlier articles, and to examine some new questions concerning deductive explanation, deductive prediction, and related procedures.

The second major part of the present study is an attempt to point out, and to shed some light on, certain fundamental problems in the logic of inductive explanation and prediction.

Part I. Deductive-Nomological Systematization

2. The Covering-Law Model of Explanation

The deductive conception of explanation is suggested by cases such as the following: The metal screwtop on a glass jar is tightly stuck; after being placed in warm water for a short while, it can be readily removed. The familiar explanation of this phenomenon is, briefly, to the effect that the metal has a higher coefficient of thermal expansion than glass, so that a given rise in temperature will produce a larger expansion of the lid than of the neck of the glass jar; and that, in addition, though the metal is a good conductor of heat, the temperature of the lid will temporarily be higher than that of the glass—a fact which further increases the difference between the two perimeters. Thus, the loosening of the lid is here explained by showing that it came about, by virtue of certain antecedent circumstances, in accordance with certain physical laws. The explanation may be construed as an argument in which the occurrence of the event in question is inferred from information expressed by statements of two kinds: (a) general laws, such as those concerning the thermal conductivity of metal and the coefficients of expansion for metal and for glass, as well as the law that heat will be transferred from one body to another of lower temperature with which it is in contact; (b) statements describing particular circumstances, such as that the jar is made of glass, the lid of metal; that initially, at room temperature, the lid fitted very tightly on the top of the jar; and that then the top with the lid on it was immersed in hot water. To show that the loosening of the lid occurred "by virtue of" the circumstances in question, and "in accordance with" those laws, is then to show that the statement describing the result can be validly inferred from the specified set of premises.

Thus construed, the explanation at hand is a deductive argument of this form:

$$(2.1) \quad \frac{L_1, L_2, \ldots, L_r}{\underline{C_1, C_2, \ldots, C_k}} \\ E$$

Here, $L_1, L_2, \ldots, L_r$ are general laws and $C_1, C_2, \ldots, C_k$ are statements of particular occurrences, facts, or events; jointly, these premises form the explanans. The conclusion E is the explanandum statement; it describes the phenomenon (or event, etc.) to be explained, which will also be called the explanandum phenomenon (or event, etc.); thus, the word 'explanandum' will be used to refer ambiguously either to the explanandum statement or to the explanandum phenomenon. Inasmuch as the sentence E is assumed to be a logical consequence of the premises, an explanatory argument of form (2.1) deductively subsumes the explanandum under "*covering* laws."[3] I will say, therefore, that (2.1) represents the *covering-law model of explanation*. More specifically, I will refer to explanatory arguments of the form (2.1) as *deductive-nomological*, or briefly as deductive, explanations: as will be shown later, there are other explanations invoking general laws that will have to be construed as inductive rather than as deductive arguments.

In my illustration, the explanandum is a particular event, the loosening of a certain lid, which occurs at a definite place and time. But deductive subsumption under general laws can serve also to explain general uniformities, such as those asserted by laws of nature. For example, the uniformity expressed by Galileo's law for free fall can be explained by deduction from the general laws of mechanics and Newton's law of gravitation, in conjunction with statements specifying the mass and radius of the earth. Similarly, the uniformities expressed by the law of geometrical optics can be explained by deductive subsumption under the principles of the wave theory of light.[4]

3. Truth and Confirmation of Deductive Explanations

In SLE (Section 3) two basic requirements are imposed upon a scientific explanation of the deductive-nomological variety:[5] (i) It must be a deductively valid argument of the form (2.1), whose premises include at least one general law essentially, i.e., in such a way that if the law were deleted, the argument would no longer be valid. Intuitively, this means that reliance on general laws is essential to this type of explanation; a given phenomenon is here explained, or accounted for, by showing that it conforms to a general nomic pattern. (ii) The sentences constituting the explanans must be true, and hence so must the explanandum sentence. This second requirement was defended by the following consideration: suppose we required instead that the explanans be highly confirmed by all the relevant evidence available, though it need not necessarily be true. Now it might happen that the explanans of a given argument of the form (2.1) was well confirmed at a certain earlier stage of scientific research, but strongly disconfirmed by the more comprehensive evidence available at a later time, say, the present. In this event, we would have to say that the explanandum was correctly explained by the given argument at the earlier stage, but not at the later one. And this seemed counterintuitive, for common usage appeared to construe the correctness of a given explanation as no more time dependent than, say, the truth of a given statement. But this justification, with its reliance on a notion of correctness that does not appear in the proposed definition of explanation, is surely of questionable merit. For in reference to explanations as well as in reference to statements, the vague idea of correctness can be construed in two different ways, both of which are of interest and importance for the logical analysis of science: namely, as truth in the semantical sense, which is independent of any reference to time or to evidence; or as confirmation by the available relevant evidence—a concept which is clearly time dependent. We will therefore distinguish between *true explanations*, which meet the requirement of truth for their explanans, and *explanations that* are more or *less well confirmed* by a given body of evidence (e.g., by the total evidence available). These two concepts can be introduced as follows:

First, we define a *potential explanation* (of deductive-nomological form)[6] as an argument of the form (2.1) which meets all the requirements indicated earlier, except that the statements forming its explanans and explanandum need not be true. But the explanans must still contain a set of sentences, $L_1, L_2, \ldots, L_r$, which are lawlike, i.e., which are like laws except for possibly being false.[7] Sentences of this kind will also be called *nomic*, or *nomological, statements*. (It is this notion of potential explanation which is involved, for example, when we ask whether a tentatively proposed but as yet untried theory would be able to explain certain puzzling empirical findings.)

Next, we say that a given potential explanation is more or less highly confirmed by a given body of evidence according as its explanans is more or less highly confirmed by the evidence in question. If the explanation is formulated in a formalized language for which an adequate quantitative concept of degree of confirmation or of inductive probability is available, we might identify the probability of the explanation relative to e with the probability of the explanans relative to e.

Finally, by a *true explanation* we understand a potential explanation with true explanans—and hence also with true explanandum.

4. Causal Explanation and the Covering-Law Model

One of the various modes of explanation to which the covering-law model is relevant is the familiar procedure of accounting for an event by pointing out its "cause." In our first illustration, for example, the expansion of the lid might be said to have been caused by its immersion in hot water. Causal attributions of this sort presuppose appropriate laws, such as that whenever metal is heated under constant pressure, it expands. It is by reason of this implicit presupposition of laws that the covering-law model is relevant to the analysis of causal explanation. Let us consider this point more closely.

We will first examine general statements of causal connections, i.e., statements to the effect that an event of a given kind A—for example, motion of a magnet near a closed wire loop—will cause an event of some specified kind B—for example, flow of a current in the wire. Thereafter, we will consider statements concerning causal relations among individual events.

In the simplest case, a general statement asserting a causal connection between two kinds of events, A and B, is tantamount to the statement of the general law that whenever and wherever an instance of A occurs, it is accompanied by an instance of B. This analysis fits, for example, the statement that motion of a magnet causes a current in a neighboring wire loop. Many general statements of causal connection call for a more complex analysis, however. Thus, the statement that in a mammal, stoppage of the heart will cause death presupposes that certain "normal" conditions prevail, which are not explicitly stated, but which are surely meant to preclude, for example, the use of a heart-lung machine. "To say that X causes Y is to say that under proper conditions, an X will be followed by a Y," as Scriven[8] puts it. But unless the "proper conditions" can be specified, at least to some extent, this analysis tells us nothing about the meaning of 'X causes Y.' Now, when this kind of causal locution is used in a given context, there usually is at least some general understanding of the kind of background conditions that have to be assumed; but still, to the extent that those conditions remain indeterminate, a general statement of causal connection falls short of snaking a definite assertion and has at best the character of a promissory note to the effect that there are further background factors whose proper recognition would yield a truly general connection between the "cause" and "effect" under consideration.

Sentences concerning causal connections among individual events show similar characteristics. For example, the statement that the death of a certain person was caused by an overdose of phenobarbital surely presupposes a generalization, namely, a statement of a general causal connection between one kind of event, a person's taking an overdose of phenobarbital, and another, the death of that person.

Here again, the range of application for the general causal statement is not precisely stated, but a sharper specification can be given by indicating what constitutes an overdose of phenobarbital for a person—this will depend, among other things, on his weight and on his habituation to the drug—and by adding the proviso that death will result from taking such an overdose if the organism is left to itself, which implies, in particular, that no countermeasures are taken. To explain the death in question as having been caused by the antecedent taking of phenobarbital is therefore to claim that the explanandum event followed according to law upon certain antecedent circumstances. And this argument, when stated explicitly, conforms to the covering-law model.

Generally, the assertion of a causal connection between individual events seems to me unintelligible unless it is taken to make, at least implicitly, a nomological claim to the effect that there are laws which provide the basis for the causal connection asserted. 'When an individual event, say b, is said to have been caused by a certain antecedent event, or configuration of events, a, then surely the claim is intended that whenever "the same cause" is realized, "the same effect" will recur. This claim cannot be taken to mean that whenever a recurs then so does b; for a and b are individual events at particular spatiotemporal locations and thus occur only once. Rather, a and b are, in this context, viewed as particular events of certain *kinds*—e.g., the expansion of a piece of metal or the death of a person—of which there may be many further instances. And the law tacitly implied by the assertion that b, as an event of kind B, was caused by a, as an event of kind A, is a general statement of causal connection to the effect that, under suitable circumstances, an instance of A is invariably accompanied by an instance of B. In most causal explanations offered in other than advanced scientific contexts, the requisite circumstances are not fully stated; for these cases, the import of the claim that b, as an instance of B, was caused by a may be suggested by the following approximate formulation: event b was in fact preceded by an event a of kind A, and by certain further circumstances which, though not fully specified or specifiable, were of such a kind that an occurrence of an event of kind A under such circumstances is universally followed by an event of kind B. For example, the statement that the burning (event of kind B) of a particular haystack was caused by a lighted cigarette carelessly dropped into the hay (particular event of kind A) asserts, first of all, that the latter event did take place; but a burning cigarette will set a haystack on fire only if certain further conditions are satisfied, which cannot at present be fully stated; and thus, the causal attribution at hand implies, second, that further conditions of a not fully specifiable kind were realized, under which an event of kind A will invariably be followed by an event of kind B.

To the extent that a statement of individual causation leaves the relevant antecedent conditions—and thus also the requisite explanatory laws—indefinite, it is like a note saying that there is a treasure hidden somewhere. Its significance and utility will increase as the location of the treasure is narrowed down, as the revelant conditions and the corresponding covering laws are made increasingly explicit. In some cases, such as that of the barbiturate poisoning, this can be done quite satisfactorily; the covering-law structure then emerges, and the statement of individual causal connection becomes amenable to test. When, on the other hand, the relevant conditions or laws remain largely indefinite, a statement of causal connection is rather in the nature of a program, or of a sketch, for an explanation in terms of causal laws; it might also be viewed as a "working hypothesis" which may prove its worth by giving new, and fruitful, direction to further research.

I would hike to add here a brief comment on Scriven's observation that "when one asserts that X causes Y one is certainly committed to the generalization that an identical cause would produce an identical effect, but this in no way commits one to any necessity for producing laws not involving the term 'identical,' which justify this claim. Producing laws is one way, not necessarily more conclusive, and usually less easy than other ways of supporting the causal statement."[9] I think we have to distinguish here two questions, namely (i) what is being claimed by the statement that X causes Y, and in particular, whether asserting it commits one to a generalization, and (ii) what kind of evidence would support the causal statement, and in particular, whether such support can be provided only by producing generalizations in the form of laws.

As for the first question, I think the causal statement does imply the claim that an appropriate law or set of laws holds by virtue of which X causes Y; but, for reasons suggested above, the law or laws in question cannot be expressed by saying that an identical cause would produce an identical effect. Rather, the general claim implied by the causal statement is to the effect that there are certain "relevant" conditions of such a kind that whenever they occur in conjunction with an event of kind X, they are invariably followed by an event of kind Y.

In certain cases, some of the laws that are claimed to connect X and Y may be explicitly statable—as, for example, in our first illustration, the law that metals expand upon heating; and then, it will be possible to provide evidential support (or else disconfirmation) for them by the examination of particular instances; thus, while laws are implicitly claimed to underlie the causal connection in question, the claim can be supported by producing appropriate empirical evidence consisting of particular cases rather than of general laws. When, on the other hand, a nomological claim made by a causal statement has merely the character of an existential statement to the effect that there are relevant factors and suitable laws connecting X and Y, then it may be possible to lend some credibility to this claim by showing that under certain conditions an event of kind X is at least very frequently accompanied by an event of kind Y. This might justify the working hypothesis that the background conditions could be further narrowed down in a way that would eventually yield a strictly causal connection. It is this kind of statistical evidence, for example, that is adduced in support of such claims as that cigarette smoking is "a cause of" or "a causative factor in" cancer of the lung. In this case, the supposed causal laws cannot at present be explicitly stated. Thus, the nomological claim implied by this causal conjecture is of the existential type; it has the character of a working hypothesis that gives direction to further research. The statistical evidence adduced lends support to the hypothesis and justifies the program, which clearly is the aim of further research, of determining more precisely the conditions under which smoking will lead to cancer of the lung.

The most perfect examples of explanations conforming to the covering-law model are those provided by physical theories of deterministic character. A theory of this kind deals with certain specified kinds of physical systems, and limits itself to certain aspects of these, which it represents by means of suitable parameters; the values of these parameters at a given time specify the state of the system at that time; and a deterministic theory provides a system of laws which, given the state of an isolated system at one time, determine its state at any other time. In the classical mechanics of systems of mass points, for example, the state of a system at a given time is specified by the positions and momenta of the component particles at that time; and the principles of the theory—essentially the Newtonian laws of motion and of gravitation—determine the state of an isolated system of mass points at any time provided that its state at some one moment is given; in particular, the state at a specified moment may be fully explained, with the help of the theoretical principles in question, by reference to its state at some earlier time. In this theoretical scheme, the notion of a cause as a more or less narrowly circumscribed antecedent event has been replaced by that of some antecedent state of the total system, which provides the "initial conditions" for the computation, by means of the theory, of the later state that is to be explained; if the system is not isolated, i.e., if relevant outside influences act upon the system during the period of time from the initial state invoked to the state to be explained, then the particular circumstances that must be stated in the explanans include also those "outside influences"; and it is these "boundary conditions" in conjunction with the "initial" conditions which replace the everyday notion of cause, and which have to be thought of as being specified by the statements $C_1, C_2, \ldots, C_k$ in the schematic representation (2.1) of the covering-law model.

Causal explanation in its various degrees of explicitness and precision is not the only type of explanation, however, to which the covering-law model is relevant. For example, as was noted earlier, certain empirical regularities, such as that represented by Galileo's law, can be explained by deductive subsumption under more comprehensive laws or theoretical principles; frequently, as in the ease of the explanation of Kepler's laws by means of the law of gravitation and the laws of mechanics, the deduction yields a conclusion of which the generalization to be explained is only an approximation. Then the explanatory principles not only show why the presumptive general law holds, at least in approximation, but also provide an explanation for the deviations.

Another noncausal species of explanation by covering laws is illustrated by the explanation of the period of swing of a given pendulum by reference to its length and to the law that the period of a mathematical pendulum is proportional to the square root of its length. This law expresses a mathematical relation between the length and the period (a dispositional characteristic) of a pendulum *at the same time;* laws of this kind are sometimes referred to as *laws of coexistence,* in contradistinction to *laws of succession,* which concern the changes that certain systems undergo in the course of time. Boyle's, Charles's, and Van der Waals's laws for gases, which concern concurrent values of pressure, volume, and temperature of a gas; Ohm's law; and the law of Wiedemann and Franz (according to which, in metals, electric conductivity is proportional to thermal conductivity) are examples of laws of coexistence. Causal explanation in terms of antecedent events clearly calls for laws of succession in the explanans; in the case of the pendulum, where only a law of coexistence is invoked, we would not say that the pendulum's having such and such a length at a given time caused it to have such and such a period.[10]

It is of interest to note that in the example at hand, a statement of the length of a given pendulum in conjunction with the law just referred to will much more readily be accepted as explaining the pendulum's period, than a statement of the period in conjunction with the same law would be considered as explaining the length of the pendulum; and this is true even though the second argument has the same logical structure as the first: both are cases of deductive subsumption, in accordance with the schema (2.1), under a law of coexistence. The distinction made here seems to me to result from the consideration that we might change the length of the pendulum at will and thus control its period as a "dependent variable," whereas the reverse procedure does not seem possible. This idea is open to serious objections, however; for clearly, we can also change the period of a given pendulum at will, namely, by changing its length; and in doing so, we will change its length. It is not possible to retort that in the first case we have a change of length independently of a change of the period; for if the location of the pendulum, and thus the gravitational force acting on the pendulum bob, remains unchanged, then the length cannot be changed without also changing the period. In cases such as this, the common-sense conception of explanation appears to provide no clear and reasonably defensible grounds on which to decide whether a given argument that deductively subsumes an occurrence under laws is to qualify as an explanation.

The point that an argument of the form (2.1), even if its premises are assumed to be true, would not always be considered as constituting an explanation is illustrated even more clearly by the following example, which I owe to my colleague Mr. S. Bromberger. Suppose that a flagpole stands vertically on level ground and subtends an angle of 45 degrees when viewed from the ground level at a distance of 80 feet. This information, in conjunction with some elementary theorems of geometry, implies deductively that the pole is 80 feet high. The theorems in question must here be understood as belonging to physical geometry and thus as having the status of general laws, or, better, general theoretical principles, of physics. Hence, the deductive argument is of the type (2.1). And yet, we would not say that its premises *explained* the fact that the pole is 80 feet high, in the sense of showing why it is that the pole has a height of 80 feet. Depending on the context in which it is raised, the request for an explanation might call here for some kind of causal account of how it came about that the pole was given this height, or perhaps for a statement of the purpose for which this height was chosen. An account of the latter kind would again be a special ease of causal explanation, invoking among the antecedent conditions certain dispositions (roughly speaking, intentions, preferences, and beliefs) on the part of the agents involved in erecting the flagpole.

The geometrical argument under consideration is not of a causal kind; in fact, it might be held that if the particular facts and the geometrical laws here invoked can be put into an explanatory connection at all, then at best we might say that the height of the pole—in conjunction with the other particulars and the laws—explains the size of the substended angle, rather than vice versa. The consideration underlying this view would be similar to that mentioned in the case of the pendulum: It might be said that by changing the height of the pole, a change in the angle can be effected, but not vice versa. But here as in the previous case, this contention is highly questionable. Suppose that the other factors involved, especially the distance from which the pole is viewed, are kept constant; then the angle can be changed, namely by changing the length of the pole; and thus, if the angle is made to change, then, trivially, the length of the pole changes. The notion that somehow we can "independently" control the length and thus make the angle a dependent variable, but not conversely, does not seem to stand up under closer scrutiny.

In sum then, we have seen that among those arguments of the form (2.1) which are not causal in character there are some which would not ordinarily be considered as even potential explanations; but ordinary usage appears to provide no clear general criterion for those arguments which are to be qualified as explanatory. This is not surprising, for our everyday conception of explanation is strongly influenced by preanalytic causal and teleological ideas; and these can hardly be expected to provide unequivocal guidance for a more general and precise analysis of scientific explanation and prediction.

5. Covering Laws: Premises or Rules?

Even if it be granted that causal explanations presuppose general laws, it might still be argued that many explanations of particular occurrences as formulated in everyday contexts or even in scientific discourse limit themselves to adducing certain particular facts as the presumptive causes of the explanandum event, and that therefore a formal model should construe these explanations as accounting for the explanandum by means of suitable statements of particular fact, C_1, C_2, ..., C_k, alone. Laws would have to be cited, not in the context by *giving* such an explanation, but in the context of *justifying* it; they would serve to show that the antecedent circumstances specified in the explanans are indeed connected by causal laws with the explanandum event. Explanation would thus be comparable to proof by logical deduction, where explicit reference to the rules or laws of logic is called for, not in stating the successive steps of the proof, but only in justifying them, i.e., in showing that they conform to the principles of deductive inference. This conception would construe general laws and theoretical principles, not as scientific statements, but rather as extralogical rules of scientific inference. These rules, in conjunction with those of formal logic, would govern inferences—explanatory, predictive, retrodictive, etc.—that lead from given statements of particular fact to other statements of particular fact.

The conception of scientific laws and theories as rules of inference has been advocated by various writers in the philosophy of science.[11] In particular, it may be preferred by those who hesitate, on philosophic grounds, to accord the status of bona fide statements, which are either true or false, to sentences which purport to express either laws covering an infinity of potential instances or theoretical principles about unobservable "hypothetical" entities and processes.[12]

On the other hand, it is well known that in rigorous scientific studies in which laws or theories are employed to explain or predict empirical phenomena, the formulas expressing laws and theoretical principles are used, not as rules of inference, but as statements—especially as premises—quite on a par with those sentences which presumably describe particular empirical facts or events. Similarly, the formulas expressing laws also occur as conclusions in deductive arguments; for example, when the laws governing the motion of the components of a double star about their common center of gravity are derived from broader laws of mechanics and of gravitation.

It might also be noted here that a certain arbitrariness is involved in any method of drawing a line between those formulations of empirical science which are to count as statements of particular fact and those which purport to express general laws, and which accordingly are to be construed as rules of inference. For any term representing an empirical characteristic can be construed as dispositional, in which case a sentence containing it acquires the status of a generalization. Take, for example, sentences which state the boiling point of helium at atmospheric pressure, or the electric conductivity of copper: are these to be construed as empirical statements or rather as rules? The latter status could be urged on the grounds that (i) terms such as 'helium' and 'copper' are dispositional, so that their application even to one particular object involves a universal assertion, and that (ii) each of the two statements attributes a specific disposition to *any* body of helium or of copper at *any* spatio-temporal location, which again gives them the character of general statements.

The two conceptions of laws and theories—as statements or as rules of inference—correspond to two different formal reconstructions, or models, of the language of empirical science; and a model incorporating laws and theoretical principles as rules can always be replaced by one which includes them instead as scientific statements.[13] And what matters for our present purposes is simply that in either mode of representation, explanations of the kind here considered "presuppose" general theoretical principles essentially: either as indispensable premises or as indispensable rules of inference.

Of the two alternative construals of laws and theories, the one which gives them the status of statements seems to me simpler and more perspicuous for the analysis of the issues under investigation here; I will therefore continue to construe deductive-nomological explanations as having the form (2.1).

6. Explanation, Prediction, Retrodiction, and Deductive Systematization—a Puzzle about "About"

In a deductive-nomological explanation of a particular past event, the explanans logically implies the occurrence of the explanandum event; hence, we may say of the explanatory argument that it could also have served as a predictive one in the sense that it could have been used to predict the explanandum event if the laws and particular circumstances adduced in its explanans had been taken into account at a suitable earlier time.[14] Predictive arguments of the form (2.1) will be called *deductive-nomological predictions*, and will be said to conform to the covering-law model of prediction. There are other important types of scientific prediction; among these, statistical prediction, along with statistical explanation, will be considered later.

Deductive-nomological explanation in its relation to prediction is instructively illustrated in the fourth part of the *Dialogues Concerning Two New Sciences*. Here, Galileo develops his laws for the motion of projectiles and deduces from them the corollary that if projectiles are fired from the same point with equal initial velocity, but different elevations, the maximum range will be attained when the elevation is 45°. Then, Galileo has Sagredo remark: "From accounts given by gunners, I was already aware of the fact that in the use of cannon and mortars, the maximum range ... is obtained when the elevation is 45° ... but to understand why this happens far outweighs the mere information obtained by the testimony of others or even by repeated experiment."[15] The reasoning that affords such understanding can readily be put into the form (2.1); it amounts to a deduction, by logical and mathematical means, of the corollary from a set of premises which contains (i) the fundamental laws of Galileo's theory for the motion of projectiles and (ii) particular statements specifying that all the missiles considered are fired from the same place with the same initial velocity. Clearly then, the phenomenon previously noted by the gunners is here *explained*, and thus *understood*, by showing that its occurrence was to be expected, under the specified circumstances, in view of certain general laws set forth in Galileo's theory. And Galileo himself points with obvious pride to the predictions that may in like fashion be obtained by deduction from his laws; for the latter imply "what has perhaps never been observed in experience, namely, that of other shots those which exceed or fall short of 45° by equal amounts have equal ranges." Thus, the explanation afforded by Galileo's theory "prepares the mind to understand and ascertain other facts without need of recourse to experiment,"[16] namely, by deductive subsumption under the laws on which the explanation is based.

We noted above that if a deductive argument of the form (2.1) explains a past event, then it could have served to predict it if the information provided by the explanans had been available earlier. This remark makes a purely logical point; it does not depend on any empirical assumptions. Yet it has been argued, by Rescher, that the thesis in question "rests upon a tacit but unwarranted assumption as to the nature of the physical universe."[17]

The basic reason adduced for this contention is that "the explanation of events is oriented (in the main) towards the past, while prediction is oriented towards the future,"[18] and that, therefore, before we can decide whether (deductive-nomological) explanation and prediction have the same logical structure, we have to ascertain whether the natural laws of our world do in fact permit inferences from the present to the future as well as from the present to the past. Rescher stresses that a given system might well be governed by laws which permit deductive inferences concerning the future, but not concerning the past, or conversely; and on this point he is quite right. As a schematic illustration, consider a model "world" which consists simply of a sequence of colors, namely, Blue (B), Green (G), Red (R), and Yellow (Y), which appear on a screen during successive one-second intervals $i_1, i_2, i_3, \ldots$ Let the succession of colors be governed by three laws:

(L_1) B is always followed by G.
(L_2) G and R are always followed by Y.
(L_3) Y is always followed by R.

Then, given the color of the screen for a certain interval, say i_3, these laws unequivocally determine the "state of the world," i.e., the screen color, for all later intervals, but not for all earlier ones. For example, given the information that during i_3 the screen is Y, the laws predict the colors for the subsequent intervals uniquely as RYRYRY ...; but for the preceding states i_1 and i_2, they yield no unique information, since they allow here two possibilities: BG and YR.

Thus, it is possible that a set of laws governing a given system should permit unique deductive *predictions* of later states from a given one, and yet not yield unique deductive *retrodictions* concerning earlier states; conversely, a set of laws may permit unique retrodiction, but no unique prediction. But—and here lies the flaw in Rescher's argument—this is by no means the same thing as to say that such laws, while permitting deductive prediction of later states from a given one, do not permit explanation; or, in the converse case, that while permitting explanation, they do not permit prediction. To illustrate by reference to our simple model world: Suppose that during i_3 we find the screen to be Y, and that we seek to explain this fact. This can be done if we can ascertain, for example, that the color for i_1 had been B; for from the statement of this particular antecedent fact we can infer, by means of L_1, that the color for i_2 must have been G and hence, by L_2, that the color for i_3 had to be Y. Evidently, the same argument, used before i_3, could serve to predict uniquely the color for i_3 on the basis of that for i_1. Indeed, quite generally, any predictive argument made possible by the laws for our model world can also be used for explanatory purposes and vice versa. And this is so although those laws, while permitting unique predictions, do not always permit unique retrodictions. Thus, the objection under consideration misses its point because it tacitly confounds explanation with retrodiction.[19]

The notion of scientific retrodiction, however, is of interest in its own right; and, as in the case of explanation and prediction, one important variety of it is the deductive-nomological one. It has the form (2.1), but with the statements $C_1, C_2, \ldots, C_k$ referring to circumstances which occur later than the event specified in the conclusion E. In astronomy, an inference leading, by means of the laws of celestial mechanics, from data concerning the present positions and movements of the sun, the earth, and Mars to a statement of the distance between earth and Mars a year later or a year earlier illustrates deductive-nomological prediction and retrodiction, respectively; in this case, the same laws can be used for both purposes because the processes involved are reversible.

It is of interest to observe here that in their predictive and retrodictive as well as in their explanatory use, the laws of classical mechanics, or other sets of deterministic laws for physical systems, require among the premises not only a specification of the state of the system for some time, t_0, earlier or later than the time, say t_1, for which the state of the system is to be inferred, but also a statement of the boundary conditions prevailing between t_0 and t_1; these specify the external influences acting upon the system during the time in question. For certain purposes in astronomy, for example, the disturbing influence of celestial objects other than those explicitly considered may be neglected as insignificant, and the system under consideration may then be treated as "isolated"; but this should not lead us to overlook the fact that even those laws and theories of the physical sciences which provide the exemplars of deductively nomological prediction do not enable us to forecast certain future events strictly on the basis of information about the present: the predictive argument also requires certain premises concerning the future—e.g., absence of disturbing influences, such as a collision of Mars with an unexpected comet—and the temporal scope of these boundary conditions must extend up to the very time at which the predicted event is to occur. The assertion therefore that laws and theories of deterministic form enable us to predict certain aspects of the future from information about the present has to be taken with a considerable grain of salt. Analogous remarks apply to deductive-nomological retrodiction and explanation.

I will use the term 'deductive-nomological systematization' to refer to any argument of the type (2.1), irrespective of the temporal relations between the particular facts specified by $C_1, C_2, \ldots, C_k$ and the particular events, if any, described by E. And, in obvious extension of the concepts introduced in Section 3 above, I will speak of *potential* (deductive-nomological) *systematizations*, of *true systematizations*, and of *systematizations* whose joint premises are more or less well confirmed by a given body of evidence.

To return now to the characterization of an explanation as a potential prediction: Scriven[20] bases one of his objections to this view on the observation that in the causal explanation of a given event (e.g., the collapse of a bridge) by reference to certain antecedent circumstances (e.g., excessive metal fatigue in one of the beams) it may well happen that the only good reasons we have for assuming that the specified circumstances were actually present lie in our knowledge that the explanandum event did take place. In this situation, we surely could not have used the explanans predictively since it was not available to us before the occurrence of the event to be predicted. This is an interesting and important point in its own right; but in regard to our conditional thesis that an explanation could have served as a prediction *if* its explanans had been taken account of in time, the argument shows only that the thesis is sometimes counterfactual (i.e., has a false antecedent), but not that it is false.

In a recent article, Scheffler[21] has subjected the idea of the structural equality of explanation and prediction to a critical scrutiny; and I would like to comment here briefly on at least some of his illuminating observations.

Scheffler points out that a prediction is usually understood to be an assertion rather than an argument. This is certainly the case; and we might add that, similarly, an explanation is often formulated, not as an argument, but as a statement, which will typically take the form 'q because p.' But predictive statements in empirical science are normally established by inferential procedures (which may be deductive or inductive in character) on the basis of available evidence; thus, there arises the question as to the logic of predictive arguments in analogy to the problem of the logic of explanatory arguments; and the idea of structural equality should be understood as pertaining to explanatory, predictive, retrodictive, and related arguments in science.

Scheffler also notes that a scientific prediction statement may be false, whereas, under the requirement of truth for explanations as laid down in Section 3 of *SLE*, no explanation can be false. This remark is quite correct; however, I consider it to indicate, not that there is a basic discrepancy between explanation and prediction, but that the requirement of truth for scientific explanations is unduly restrictive. The restriction is avoided by the approach that was proposed above in Section 3, and again in the present section in connection with the general characterization of scientific systematization; this approach enables us to speak of explanations no less than of predictions as being possibly false, and as being more or less well confirmed by the empirical evidence at hand.

Another critical observation Scheffler puts forth concerns the view, presented in *SLE*, that the difference between an explanatory and a predictive argument does not lie in its logical structure, but is "of a pragmatic character. If . . . we know that the phenomenon described by E has occurred, and a suitable set of statements C_1, $C_2, \ldots, C_k, L_1, L_2, \ldots, L_r$ is provided afterward, we speak of an explanation of the phenomenon in question. If the latter statements are given and E is derived prior to the occurrence of the phenomenon it describes, we speak of a prediction."[22] This characterization would make explanation and prediction mutually exclusive procedures, and Scheffler rightly suggests that they may sometimes coincide, since, for example, one may reasonably be said to be both predicting and explaining the sun's rising when, in reply to the question 'Why will the sun rise tomorrow?' one offers the appropriate astronomical information.[23]

I would be inclined to say, therefore, that in an explanation of the deductive-nomological variety, the explanandum event—which may be past, present, or future—is taken to be "given," and a set of laws and particular statements is then adduced which provides premises in an appropriate argument of type (2.1); whereas in the case of prediction, it is the premises which are taken to be "given," and the argument then yields a conclusion about an event to occur after the presentation of the predictive inference. Retrodiction may be construed analogously. The argument referred to by Scheffler about tomorrow's sunrise may thus be regarded, first of all, as predicting the event on the basis of suitable laws and presently available information about antecedent circumstances; then, taking the predicted event as "given," the premises of the same argument constitute an explanans for it.

Thus far, I have dealt with the view that an explanatory argument is also a (potentially) predictive one. Can it be held equally that a predictive argument always offers a potential explanation? In the case of deductive-nomological predictions, an affirmative answer might be defended, though as was illustrated at the end of Section 4, there are some deductive systematizations which one would readily accept as predictions while one would find it at least awkward to qualify them as explanations. Construing the question at hand more broadly, Scheffler, and similarly Scriven,[24] have rightly pointed out, in effect, that certain sound predictive arguments of the nondeductive type cannot be regarded as affording potential explanations. For example, from suitable statistical data on past occurrences, it may be possible to "infer" quite soundly certain predictions concerning the number of male births, marriages, or traffic deaths in the United States during the next month; but none of these arguments would be regarded as affording even a low-level explanation of the occurrences they serve to predict. Now, the inferences here involved are inductive rather than deductive in character; they lead from information about observed finite samples to predictions concerning as yet unobserved samples of a given population. However, what bars them from the role of potential explanations is not their inductive character (later I will deal with certain explanatory arguments of inductive form) but the fact that they do not invoke any general laws either of strictly universal or of statistical form: it appears to be characteristic of an explanation, though not necessarily of a prediction, that it present the inferred phenomena as occurring in conformity with general laws.

In concluding this section, I would like briefly to call attention to a puzzle concerning a concept that was taken for granted in the preceding discussion, for example, in distinguishing between prediction and retrodiction. In drawing that distinction, I referred to whether a particular given statement, the conclusion of an argument of form (2.1), was "about" occurrences at a time earlier or later than some specified time, such as the time of presentation of that argument. The meaning of this latter criterion appears at first to be reasonably clear and unproblematic. If pressed for further elucidation, one might be inclined to say, by way of a partial analysis, that if a sentence explicitly mentions a certain moment or period of time then the sentence is about something occurring at that time. It seems reasonable, therefore, to say that the sentence 'The sun rises on July 17, 1958,' says something about July 17, 1958, and that, therefore, an utterance of this sentence on July 16, 1958, constitutes a prediction.

Now the puzzle in question, which might be called the puzzle of 'about,' shows that this criterion does not even offer a partially satisfactory explication of the idea of what time a given statement is about. For example, the statement just considered can be equivalently restated in such a way that, by the proposed criterion, it is about July 15 and thus, if uttered on July 16, is about the past rather than about the future. The following rephrasing will do: 'The sun plus-two-rises on July 15,' where plus-two-rising on a given date is understood to be the same thing as rising two days after that date. By means of linguistic devices of this sort, statements about the future could be reformulated as statements about the past, or conversely; we could even replace all statements with temporal reference by statements which are, all of them, ostensibly "about" one and the same time.

The puzzle is not limited to temporal reference, but arises for spatial reference as well. For example, a statement giving the mean temperature at the North Pole can readily be restated in a form in which it speaks ostensibly about the South Pole; one way of doing this is to attribute to the South Pole the property of having, in such and such a spatial relation to it, a place where the mean temperature is such and such; another device would be to use a functor, say 'm,' which, for the South Pole, takes as its value the mean temperature at the North Pole. Even more generally there is a method which, given any particular object o, will reformulate any statement in such a way that it is ostensibly about o. If, for example, the given statement is 'The moon is spherical,' we introduce a property term, 'moon-spherical,' with the understanding that it is to apply to o just in case the moon is spherical; the given statement then is equivalent to 'o is moon-spherical.'

The puzzle is mentioned here in order to call attention to the difficulties that face an attempt to explicate the idea of what a statement is "about," and in particular, what time it refers to; and that idea seems essential for the characterization of prediction, retrodiction, and similar concepts.[25]

Part II. Statistical Systematization

7. Laws of Strictly General and Statistical Form

The nomological statements adduced in the explanans of a deductive-nomological explanation are all of a strictly general form: they purport to express strictly unexceptionable laws or theoretical principles interconnecting certain characteristics (i.e., qualitative or quantitative properties or relations) of things or events. One of the simplest forms a statement of this kind can take is that of a universal conditional: 'All (instances of) F are (instances of) G.' When the attributes in question are quantities, their interconnections are usually expressed in terms of mathematical functions, as is illustrated by many of the laws and theoretical principles of the physical sciences and of mathematical economics.

On the other hand, there are important scientific hypotheses and theoretical principles which assert that certain characters are associated, not unexceptionally or universally, but with a specified long-range frequency; we will call them statistical generalizations, or laws (or theoretical principles) of statistical form, or (statistical) probability statements. The laws of radioactive decay, the fundamental principles of quantum mechanics, and the basic laws of genetics are examples of such probability statements. These statistical generalizations, too, are used in science for the systematization of various empirical phenomena. This is illustrated, for example, by the explanatory and predictive applications of quantum theory and of the basic laws of genetics, as well as by the postdictive use of the laws of radioactive decay in dating archeological relics by means of the radiocarbon method.

The rest of this essay deals with some basic problems in time logic of statistical systematizations, i.e., of explanatory, predictive, or similar arguments which make essential use of statistical generalizations.

Just as in the case of deductive-nomological systematization, arguments of this kind may be used to account not only for particular facts or events, but also for general regularities, which, in this case, will be of a statistical character. For example, from statistical generalizations stating that the six different results obtainable by rolling a given die are equiprobable and statistically independent of each other, it is possible to deduce the statistical generalization that the probability of rolling two aces in succession is $1/36$; thus the latter statistical regularity is accounted for by subsumption (in this case purely deductive) under broader statistical hypotheses.

But the peculiar logical problems concerning statistical systematization concern the role of probability statements in the explanation, prediction, and postdiction of individual events or finite sets of such events. In preparation for a study of these problems, I shall now consider briefly the form and function of statistical generalizations.

Statistical probability hypotheses, or statistical generalizations, as understood here, bear an important resemblance to nomic statements of strictly general form: they make a universal claim, as is suggested by the term 'statistical law,' or 'law of statistical form.' Snell's law of refraction, which is of strictly general form, is not simply a descriptive report to the effect that a certain quantitative relationship has so far been found to hold, in all cases of optical refraction, between the angle of incidence and that of refraction: it asserts that that functional relationship obtains universally, in all cases of refraction, no matter when and where they occur.[26] Analogously, the statistical generalizations of genetic theory or the probability statements specifying the half lives of various radioactive substances are not just reports on the frequencies with which certain phenomena have been found to occur in some set of past instances; rather, they serve to assert certain peculiar but universal modes of connection between certain attributes of things or events.

A statistical generalization of the simplest kind asserts that the probability for an instance of F to be an instance of G is r, or briefly that p(G,F) = r; this is intended to express, roughly speaking, that the proportion of those instances of F which are also instances of G is r. This idea requires clarification, however, for the notion of the proportion of the (instances of) G among the (instances of) F has no clear meaning when the instances of F do not form a finite class. And it is characteristic of probability hypotheses with their universal character, as distinguished from statements of relative frequencies in some finite set, that the reference class—F in this case—is not assumed to be finite; in fact, we might underscore their peculiar character by saying that the probability r does not refer to the class of all actual instances of F but, so to speak, to the class of all its potential instances.

Suppose, for example, that we are given a homogeneous regular tetrahedron whose faces are marked 'I,' 'II,' 'III,' 'IV.' We might then be willing to assert that the probability of obtaining a III, i.e., of the tetrahedron's coming to rest on that face, upon tossing it out of a dice box is ¼; but while this assertion would be meant to say something about the frequency with which a III is obtained as a result of rolling the tetrahedron, it could not be construed as simply specifying that frequency for the class of all tosses which are in fact ever performed with the tetrahedron. For we might well maintain our probability hypothesis even if the given tetrahedron were tossed only a few times throughout its existence, and in this case, our probability statement would certainly not be meant to imply that exactly or even nearly, one-fourth of those tosses yielded the result III. In fact, we might clearly maintain the probability statement even if the tetrahedron happened to be destroyed without ever having been tossed at all. We might say, then, that the probability hypothesis ascribes to the tetrahedron a certain disposition, namely, that of yielding a III in about one out of four cases in the long run. That disposition may also be described by a subjunctive or counterfactual statement: If the tetrahedron were to be tossed (or had been tossed) a large number of times, it would yield (would have yielded) the result III in about one-fourth of the cases.[27]

Let us recall here in passing that nomological statements of strictly general form, too, are closely related to corresponding subjunctive and counterfactual statements. For example, the lawlike statement 'All pieces of copper expand when heated' implies the subjunctive conditional 'If this copper key were heated, it would expand' and the counterfactual statement, referring to a copper key that was kept at constant temperature during the past hour, 'If this copper key had been heated half an hour ago, it would have expanded.'[28]

To obtain a more precise account of the form and function of probability statements, I will examine briefly the elaboration of the concept of statistical probability in contemporary mathematical theory. This examination will lead to the conclusion that the logic of statistical systematization differs fundamentally from that of deductive-nomological systematization. One striking symptom of the difference is what will be called here *the ambiguity of statistical systematization*.

In Section 8, I will describe and illustrate this ambiguity in a general manner that presupposes no special theory of probability; then in Section 9, I will show how it reappears in the explanatory and predictive use of probability hypotheses as characterized by the mathematical theory of statistical probability.

8. The Ambiguity of Statistical Systematization

Consider the following argument which represents, in a nutshell, an attempt at a statistical explanation of a particular event: "John Jones was almost certain to recover quickly from his streptococcus infection, for he was given penicillin, and almost all cases of streptococcus infection clear up quickly upon administration of penicillin." The second statement in the explanans is evidently a statistical generalization, and while the probability value is not specified numerically, the words 'almost all cases' indicate that it is very high.

At first glance, this argument appears to bear a close resemblance to deductive-nomological explanations of the simplest form, such as the following: 'This crystal of rock salt, when put into a Bunsen flame, turns the flame yellow, for it is a sodium salt, and all sodium salts impart a yellow color to a Bunsen flame. This argument is basically of the form:

(8.1) All F are G.
      x is F.
      ―――――
      x is G.

The form of the statistical explanation, on the other hand, appears to be expressible as follows:

(8.2) Almost all F are G.
      x is F.
      ――――――――――――――――
      x is almost certain to be G.

Despite this appearance of similarity, however, there is a fundamental difference between these two kinds of argument: A nomological explanation of the type (8.1) accounts for the fact that x is G by stating that x has another character, F, which is uniformly accompanied by G, in virtue of a general law. If in a given case these explanatory assumptions are in fact true, then it follows logically that x must be G; hence x cannot possibly possess a character, say H, in whose presence G is uniformly absent; for otherwise, x would have to be both G and non-G. In the argument (8.2), on the other hand, x is said to be almost certain to have G because it has a character, F, which is accompanied by G in almost all instances. But even if in a given case the explanatory statements are both true, x may possess, in addition to F, some other attribute, say H, which is almost always accompanied by non-G. But by the very logic underlying (8.2), this attribute would make it almost certain that x is not G.

Suppose, for example, that almost all, but not quite all, penicillin-treated, streptococcal infections result in quick recovery, or briefly, that almost all P are R; and suppose also that the particular case of illness of patient John Jones which is under discussion—let us call it j—is an instance of P. Our original statistical explanation may then be expressed in the following manner, which exhibits the form (8.2):

(8.3a) Almost all P are R.
        j is P.
        ——————————————————
        j is almost certain to be R.

Next, let us say that an event has the property P* if it is either the event j itself or one of those infrequent cases of penicillin-treated streptococcal infection which do not result in quick recovery. Then clearly j is P*, whether or not j is one of the cases resulting in recovery, i.e., whether or not j is R. Furthermore, almost every instance of P* is an instance of non-R (the only possible exception being j itself). Hence, the argument (8.3a) in which, on our assumption, the premises are true can be matched with another one whose premises are equally true, but which by the very logic underlying (8.3a), leads to a conclusion that appears to contradict that of (8.3a):

(8.3b) Almost all P* are non-R.
        j is P*.
        ——————————————————
        j is almost certain to be non-R.

If it should be objected that the property P* is a highly artificial property and that, in particular, an explanatory statistical law should not involve essential reference to particular individuals (such as j in our case), then another illustration can be given which leads to the same result and meets the contemplated requirement. For this purpose, consider a number of characteristics of John Jones at the onset of his illness, such as his age, height, weight, blood pressure, temperature, basal metabolic rate, and IQ. These can be specified in terms of numbers; let $n_1, n_2, n_3, \ldots$ be the specific numerical values in question. We will say that an event has the property S if it is a case of streptococcal infection in a patient who at the onset of his illness has the height n_1, age n_2, weight n_3, blood pressure n_4, and so forth. Clearly, this definition of S in terms of numerical characteristics no longer makes reference to j. Finally, let us say that an event has the property P** if it is either an instance of S or one of those infrequent cases of streptococcal infection treated with penicillin which do not result in quick recovery. Then evidently j is P** because j is S; and furthermore, since S is a very rare characteristic, almost every instance of P** is an instance of non-R. Hence, (8.3a) can be matched with the following argument, in which the explanatory probability hypothesis involves no essential reference to particular cases:

(8.3c) Almost all P** are non-R.
        j is P**.
        ——————————————————
        j is almost certain to be non-R.

The premises of this argument are true if those of (8.3a) are, and the conclusion again appears to be incompatible with that of (8.3a).

The peculiar phenomenon here illustrated will be called the *ambiguity of statistical explanation*. Briefly, it consists in the fact that if the explanatory use of a statistical generalization is construed in the manner of (8.2), then a statistical explanation of a particular event can, in general, be matched by another one, equally of the form (8.2), with equally true premises, which statistically explains the nonoccurrence of the same event. The same difficulty arises, of course, when statistical arguments of the type (8.2) are used for predictive purposes. Thus, in the case of our illustration, we might use either of the two arguments (8.3a) and (8.3c) in an attempt to predict the effect of penicillin treatment in a fresh case, j, of streptococcal infection; and even though both followed the same logical pattern—that exhibited in (82)—and both had true premises, one argument would yield a favorable, the other an unfavorable forecast. We will, therefore, also speak of the *ambiguity of statistical prediction* and, more inclusively, of the *ambiguity of statistical systematization*.

This difficulty is entirely absent in nomological systematization, as we noted above; and it evidently throws into doubt the explanatory and predictive relevance of statistical generalizations for particular occurrences. Yet there can be no question that statistical generalizations are widely invoked for explanatory and predictive purposes in such diverse fields as physics, genetics, and sociology. It will be necessary, therefore, to examine more carefully the logic of the arguments involved and, in particular, to reconsider the adequacy of the analysis suggested in (8.2). And while for a general characterization of the ambiguity of statistical explanation it was sufficient to use an illustration of statistical generalization of the vague form 'Almost all F are G,' we must now consider the explanatory and predictive use of statistical generalizations in the precise form of quantitative probability statements: 'The probability for an F to be a G is r.' This brings us to the question of the theoretical status of the statistical concept of probability.

9. The Theoretical Concept of Statistical Probability and the Problem of Ambiguity

The mathematical theory of statistical probability[29] seeks to give a theoretical systematization of the statistical aspects of random experiments. Roughly speaking, a random experiment is a repeatable process which yields in each case a particular finite or infinite set of "results," in such a way that while the results vary from repetition to repetition in an irregular and practically unpredictable manner, the relative frequencies with which the different results occur tend to become more or less constant for large numbers of repetitions. The theory of probability is intended to provide a "mathematical model," in the form of a deductive system, for the properties and interrelations of such long-run frequencies, the latter being represented in the model by probabilities.

In the mathematical theory of probability, each of the different outcomes of a random experiment which have probabilities assigned to them is represented by a set of what might be called elementary possibilities. For example, if the experiment is that of rolling a die, then getting an ace, a deuce, and so forth, would normally be chosen as elementary possibilities; let us refer to them briefly as I, II, ..., VI, and let F be the set of these six elements. Then any of those results of rolling a die to which probabilities are usually assigned can be represented by a subset of F: getting an even number, by the set (II, IV, VI); getting a prime number, by the set (II, III, V); rolling an ace, by the unit set (I); and so forth. Generally, a random experiment is represented in the theory by a set F and a certain set, F*, of its subsets, which represent the possible outcomes that have definite probabilities assigned to them. F* will sometimes, but not always, contain all the subsets of F. The mathematical theory also requires F* to contain, for each of its member sets, its complement in F; and also for any two of its member sets, say G_1 and G_2, their sum, $G_1 \vee G_2$, and their products, $G_1 \cdot G_2$. As a consequence, F* contains F as a member set.[30] The probabilities associated with the different outcomes of a random experiment then are represented by a real-valued function $p_F(G)$ which ranges over the sets in F*.

The postulates of the theory specify that p_F is a nonnegative additive set function such that $p_F(F) = 1$; i.e., for all G in F*, $p_F(G) \geq 0$; if G_1 and G_2 are mutually exclusive sets in F* then $p_F(G1 \vee G_2) = p_F(G_1) + p_F(G_2)$. These stipulations permit the proof of the theorems of elementary probability theory; to deal with experiments that permit infinitely many different outcomes, the requirement of additivity is suitably extended to infinite sequences of mutually exclusive member sets of F*.

The abstract theory is made applicable to empirical subject matter by means of an interpretation which connects probability statements with sentences about long-run relative frequencies associated with random experiments. I will state the interpretation in a form which is essentially that given by Cramér,[31] whose book *Mathematical Methods of Statistics* includes a detailed discussion of the foundations of mathematical probability theory and its applications. For convenience, the notation '$p_F(G)$' for the probability of G relative to F will now be replaced by 'p (G, F).'

> (9.1) *Frequency interpretation of statistical probability*: Let F be a given kind of random experiment and G a possible result of it; then the statement that p(G, F) = r means that in a long series of repetitions of F, it is practically certain that the relative frequency of the result G will be approximately equal to r.

Evidently, this interpretation does not offer a precise definition of probability in statistical terms: the vague phrases 'a long series,' 'practically certain,' and 'approximately equal' preclude that. But those phrases are chosen deliberately to enable formulas stating precisely fixed numerical probability values to function as theoretical representations of near-constant relative frequencies of certain results in extended repetitions of a random experiment.

Cramér also formulates two corollaries of the above rule of interpretation; they refer to those cases where r differs very little from 0 or from 1. These corollaries will be of special interest for an examination of the question of ambiguity in the explanatory and predictive use of probability statements, and I will therefore note them here (in a form very similar to that chosen by Cramér):

(9.2a) If $0 \leq p(G, F) < \epsilon$, where ϵ is some very small number, then, if a random experiment of kind F is performed one single time, it can be considered as practically certain that the result G will not occur.[32]

(9.2b) If $1 - \epsilon < p(G, F) \leq 1$, where ϵ is some very small number, then if a random experiment of kind F is performed one single time, it can be considered as practically certain that the result G will occur.[33]

I now turn to the explanatory use of probability statements. Consider the experiment, D, of drawing, with subsequent replacement and thorough mixing, a ball from an urn containing one white ball and 99 black ones of the same size and material. Let us suppose that the probability, $p(W, D)$, of obtaining a white ball as a result of a performance of D is .99. According to the statistical interpretation, this is an empirical hypothesis susceptible of test by reference to finite statistical samples, but for the moment, we need not enter into the question how the given hypothesis might be established. Now, rule (9.2b) would seem to indicate that this hypothesis might be used in statistically explaining or predicting the results of certain individual drawings from the urn. Suppose, for example, that a particular drawing, d, produces a white ball. Since $p(W, D)$ differs from 1 by less than, say, .015, which is a rather small number, (9.2b) yields the following argument, which we might be inclined to consider as a statistical explanation of the fact that d is W:

(9.3) $1 - .015 < p(W, D) \leq 1$; and .015 is a very small number.
d is an instance of D.
―――――――――――――――――
It is practically certain that d is W.

This type of reasoning is closely reminiscent of our earlier argument (8.3a), and it leads into a similar difficulty, as will now be shown. Suppose that besides the urn just referred to, which we will assume to be marked '1,' there are 999 additional urns of the same kind, each containing 100 balls, all of which are black. Let these urns be marked '2,' '3' ... '1000.' Consider now the experiment E which consists in first drawing a ticket from a bag containing 1000 tickets of equal size, shape, etc., bearing the numerals '1,' '2' ... '1000,' and then drawing a ball from the urn marked with the same numeral as the ticket drawn. In accordance with standard theoretical considerations, we will assume that $p(W, E)$ = .00099. (This hypothesis again is capable of confirmation by statistical test in view of the interpretation (9.1).) Now, let e be a particular performance of E in which the first step happens to yield the ticket numbered 1. Then, since e is an instance of E, the interpretative rule (9.2a) permits the following argument:

(9.4a) $0 \leq p(W, E) < .001$; and .001 is a very small number.
e is an instance of E.

It is practically certain that e is not W.

But on our assumption, the event e also happens to be an instance of the experiment D of drawing a ball from the first urn; we may therefore apply to it the following argument:

(9.4b) $1 - .015 < p(W, D) \leq 1$; and .015 is a very small number.
e is an instance of D.

It is practically certain that e is W.

Thus, in certain cases the interpretative rules (9.2a) and (9.2b) yield arguments which again exhibit what was called above the ambiguity of statistical systematization.

This ambiguity clearly springs from the fact that (a) the probability of obtaining an occurrence of some specified kind G depends on the random experiment whose result G is being considered, and that (b) a particular instance of G can normally be construed as an outcome of different kinds of random experiment, with different probabilities for the outcome in question; as a result, under the frequency interpretation given in (9.2a) and (9.2b), an occurrence of G in a particular given case may be shown to be both practically certain and practically impossible. This ambiguity does not represent a flaw in the formal theory of probability: it arises only when the empirical interpretation of that theory is brought into play.

It might be suspected that the trouble arises only when an attempt is made to apply probability statements to individual events, such as one particular drawing in our illustration: statistical probabilities, it might be held, have significance only for reasonably large samples. But surely this is unconvincing since there is only a difference in degree between a sample consisting of just one case and a sample consisting of many cases. And indeed, the problem of ambiguity recurs when probability statements are used to account for the frequency with which a specified kind G of result occurs in finite samples, no matter how large.

For example, let the probability of obtaining recovery (R) as the result of the "random experiment" P of treating cases of streptococcus infection with penicillin be $p(R, P) = .75$. Then, assuming statistical independence of the individual cases, the frequency interpretation yields the following consequence, which refers to more or less extensive samples: For any positive deviation d, however small, there exists a specifiable sample size n_d such that it is practically certain that in one single series of n_d repetitions of the experiment P, the proportion of cases of R will deviate from .75 by less than d.[34] It would seem therefore that a recovery rate of close to 75 percent in a sufficiently large number of instances of P could be statistically explained or predicted by means of the probability statement that $p(R, P) = .75$. But any such series of instances can also be construed as a set of cases of another random experiment for which it is practically certain that almost all the cases in the sample recover; alternatively, the given cases can be construed as a set of instances of yet another random experiment for which it is practically certain that none of the cases in a sample of the given size will recover. The arguments leading to this conclusion are basically similar to those presented in connection with the preceding illustrations of ambiguity; the details will therefore be omitted.

In its essentials, the ambiguity of statistical systematization can be characterized as follows: If a given object or set of objects has an attribute A which with high statistical probability is associated with another attribute C, then the same object or set of objects will, in general, also have an attribute B which, with high statistical probability, is associated with non-C. Hence, if the occurrence of A in the particular given case, together with the probability statement which links A with C, is regarded as constituting adequate grounds for the predictive or explanatory conclusion that C will almost certainly occur in the given case, then there exists, apart from trivial exceptions, always a competing argument which in the same manner, from equally true premises, leads to the predictive or explanatory conclusion that C will not occur in that same case. This peculiarity has no counterpart in nomological explanation: If an object or set of objects has a character A which is invariably accompanied by C then it cannot have a character B which is invariably accompanied by non-C.[35]

The ambiguity of statistical explanation should not, of course, be taken to indicate that statistical probability hypotheses have no explanatory or predictive significance, but rather that the above analysis of the logic of statistical systematization is inadequate. That analysis was suggested by a seemingly plausible analogy between the systematizing use of statistical generalizations and that of nomic ones—an analogy which seems to receive strong support from the interpretation of statistical generalizations which is offered in current statistical theory. Nevertheless, that analogy is deceptive, as will now be shown.

10. The Inductive Character of Statistical Systematization and the Requirement of Total Evidence

It is typical of the statistical systematizations considered in this study that their "conclusion" begins with the phrase 'It is almost certain that,' which never occurs in the conclusion of a nomological explanation or prediction. The two schemata (8.1) and (8.2) above exhibit this difference in its simplest form. A nomological systematization of the form (8.1) is a deductively valid argument: if its premises are true then so is its conclusion. For arguments of the form (8.2), this is evidently not the case. Could the two types of argument be assimilated more closely to each other by giving the conclusion of (8.1) the form 'It is certain that x is G'? This suggestion involves a misconception which is one of the roots of the puzzle presented by the ambiguity of statistical systematization. For what the statement 'It is certain that x is G' expresses here can be restated by saying that the conclusion of an argument of form (8.1) cannot be false if the premises are true, i.e., that the conclusion is a logical consequence of the premises. Hence, the certainty here in question represents not a property of the conclusion that x is G, but rather a relation which that conclusion bears to the premises of (8.1). Generally, a sentence is certain, in this sense, relative to some class of sentences just in case it is a logical consequence of the latter. The contemplated reformulation of the conclusion of (8.1) would therefore be an elliptic way of saying that

(10.1) 'x is G' is certain relative to, i.e., is a logical consequence of, the two sentences 'All F are G' and 'x is F.'[36]

But clearly this is not equivalent to the original conclusion of (8.1); rather, it is another way of stating that the entire schema (8.1) is a deductively valid form of inference.

Now, the basic error in the formulation of (8.2) is clear: near certainty, like certainty, must be understood here not as a property but as a relation; thus, the "conclusion" of (8.2) is not a complete statement but an elliptical formulation of what might be more adequately expressed as follows:

(10.2) 'x is G' is almost certain relative to the two sentences 'Almost all F are G' and 'x is F.'

The near certainty here invoked is sometimes referred to as (high) probability; the conclusion of arguments like (8.2) is then expressed by such phrases as '(very) probably, x is G,' or 'it is (highly) probable that x is G'; a nonelliptic restatement would then be given by saying that the sentences 'Almost all F are G' and 'x is F' taken jointly lend strong support to, or confer a high probability or a high degree of rational credibility upon, 'x is G.' The probabilities referred to here are logical or inductive probabilities, in contradistinction to the statistical probabilities mentioned in the premises of the statistical systematization under examination. The notion of logical probability will be discussed more fully a little later in the present section.

As soon as it is realized that the ostensible "conclusions" of arguments such as (8.2) and their quantitative counterparts, such as (9.3), are elliptic formulations of relational statements, one puzzling aspect of the ambiguity of statistical systematization vanishes: the apparently conflicting claims of matched argument pairs such as (8.3a) and (8.3b) or (9.4a) and (9.4b) do not conflict at all. For what the matched arguments in a pair claim is only that each of two contradictory sentences, such as 'j is R' and 'j is not R' in the pair (8.3), is strongly supported by certain other statements, which, however, are quite different for the first and for the second sentence in question. Thus far then, no more of a "conflict" is established by a pair of matched statistical systematizations than, say, by the following pair of deductive arguments, which show that each of two contradictory sentences is even conclusively supported, or made certain, by other suitable statements which, however, are quite different for the first and for the second sentence in question:

(10.3a)   All F are G.
          a is F.
          ―――――――
          a is G.

(10.3b) No H is G.
         a is H.
         ―――――――
         a is not G.

The misconception thus dispelled arises from a misguided attempt to construe arguments containing probability statements among their premises in analogy to deductive arguments such as (8.1)—an attempt which prompts the construal of formulations such as 'j is almost certain to be R' or 'probably, j is R' as self-contained complete statements rather than as elliptically formulated statements of a relational character.[37]

The idea, repeatedly invoked in the preceding discussion, of a statement or set of statements e (the evidence) providing strong grounds for asserting a certain statement h (the hypothesis), or of e lending strong support to h, or making h nearly certain is, of course, the central concept of the theory of inductive inference. It might be conceived in purely qualitative fashion as a relation S which h bears to e if e lends strong support to h; or it may be construed in quantitative terms, as a relation capable of gradations which represents the extent to which h is supported by e. Some recent theories of inductive inference have aimed at developing rigorous quantitative conceptions of inductive support: this is true especially of the systems of inductive logic constructed by Keynes and others and recently, in a particularly impressive form, by Carnap.[38] If—as in Carnap's system—the concept is construed so as to possess the formal characteristics of a probability, it will be referred to as the logical (or inductive) probability, or as the degree of confirmation, $c(h, e)$, of h relative to e. (This inductive probability, which is a function of statements, must be sharply distinguished from statistical probability, which is a function of classes of events.) As a general phrase referring to a quantitative notion of inductive support, but not tied to any one particular theory of inductive support or confirmation, let us use the expression '(degree of) inductive support of h relative to e.'[39]

An explanation, prediction, or retrodiction of a particular event or set of events by means of principles which include statistical generalizations has then to be conceived as an inductive argument. I will accordingly speak of *inductive systematization* (in contradistinction to *deductive systematization*, where whatever is explained, predicted, or retrodicted is a deductive consequence of the premises adduced in the argument).

When it is understood that a statistical systematization is an inductive argument, and that the high probability or near certainty mentioned in the conclusions of such arguments as (8.3a) and (8.3b) is relative to the premises, then, as shown, one puzzle raised by the ambiguity of statistical explanation is resolved, namely, the impression of a conflict, indeed a near incompatibility, of the claims of two equally sound inductive systematizations.

But the same ambiguity raises another, more serious, problem, which now calls for consideration. It is very well to point out that in (8.3a) and (8.3b) the contradictory statements 'j is R' and 'j is not R' are shown to be almost certain by referring to different sets of "premises": it still remains the case that both of these sets are true. Here, the analogy to (10.3a) and (10.3b) breaks down: in these deductive arguments with contradictory conclusions the two sets of premises cannot both be true. Thus, it would seem that by statistical systematizations based on suitably chosen bodies of true information, we may lend equally strong support to two assertions which are incompatible with each other. But then—and this is the new problem—which of such alternative bodies of evidence is to be relied on for the purposes of statistical explanation or prediction?

An answer is suggested by a principle which Carnap calls *the requirement of total evidence*. It lays down a general maxim for all applications of inductive reasoning, as follows: "in the application of inductive logic to a given knowledge situation, the total evidence available must be taken as basis for determining the degree of confirmation."[40] Instead of the total evidence, a smaller body, e_1, of evidence may be used on condition that the remaining part, e_2, of the total evidence is inductively irrelevant to the hypothesis h whose confirmation is to be determined. If, as in Carnap's system, the degree of confirmation is construed as an inductive probability, the irrelevance of e_2 for h relative to e_1 can be expressed by the condition that $c(h, e_1 \cdot e_2) = c(h, e_1)$.[41]

The general consideration underlying the requirement of total evidence is obviously this: If an investigator wishes to decide what credence to give to an empirical hypothesis or to what extent to rely on it in planning his actions, then rationality demands that he take into account all the relevant evidence available to him; if he were to consider only part of that evidence, he might arrive at a much more favorable, or a much less favorable, appraisal, but it would surely not be rational for him to base his decision on evidence he knew to be selectively biased. In terms of the concept of degree of confirmation, the point might be stated by saying that the degree of confirmation assigned to a hypothesis by the principles of inductive logic will represent the rational credibility of the hypothesis for a given investigator only if the argument takes into account all the relevant evidence available to the investigator.

The requirement of total evidence is not a principle of inductive logic, which is concerned with relations of potential evidential support among statements, i.e., with whether, or to what degree, a given set of statements supports a given hypothesis. Rather, the requirement is a maxim for the *application* of inductive logic; it might be said to state a necessary condition of rationality in forming beliefs and making decisions on the basis of available evidence. The requirement is not limited to arguments of the particular form of statistical systematizations, where the evidence, represented by the "premises," includes statistical generalizations: it is a necessary condition of rationality in the application of any mode of inductive reasoning, including, for example, those cases in which the evidence contains no generalizations, statistical or universal, but only data on particular occurrences.

Let me note here that in the case of deductive systematization, the requirement is automatically satisfied and thus presents no special problem.[42] For in a deductively valid argument whose premises constitute only part of the total evidence available at the time, that part provides conclusive grounds for asserting the conclusion; and the balance of the total evidence is irrelevant to the conclusion in the strict sense that if it were added to the premises, the resulting premises would still constitute conclusive grounds for the conclusion. To state this in the language of inductive logic: the logical probability of the conclusion relative to the premises of a deductive systematization is 1, and it remains 1 no matter what other parts of the total evidence maybe added to the premises.

The residual problem raised by the ambiguity of probabilistic explanation can now be resolved by requiring that if a statistical systematization is to qualify as a rationally acceptable explanation or prediction (and not just as a formally sound *potential* explanation or prediction), it must satisfy the requirement of total evidence. For under this requirement, the "premises" of an acceptable statistical systematization whose "conclusion" is a hypothesis h must consist either of the total evidence e or of some subset of it which confers on h the same inductive probability as e; and the same condition applies to an acceptable systematization which has the negation of h as its "conclusion." But one and the same evidence, e, cannot—if it is logically self-consistent—confer a high probability on h as well as on its negation, since the sum of the two probabilities is unity. Hence, of two statistical systematizations whose premises confer high probabilities on h and on the negation of h, respectively, at least one violates the requirement of total evidence and is thus ruled out as unacceptable.

The preceding considerations suggest that a statistical systematization may be construed generally as an inductive argument showing that a certain statement or finite set of statements, e, which includes at least one statistical law, gives strong but not logically conclusive support to a statement h, which expresses whatever is being explained, predicted, retrodicted, etc. And if an argument of this kind is to be acceptable in science as an empirically sound explanation, prediction, or the like—rather than only a formally adequate, or potential one—then it will also have to meet the requirement of total evidence.

But an attempt to apply the requirement of total evidence to statistical systematizations of the simple kind considered so far encounters a serious obstacle. This was noted, among others, by S. Barker with special reference to "statistical syllogisms," which are inductive arguments with two premises, very similar in character to the arguments (9.4a) and (9.4b) above. Barker points out, in effect, that the statistical syllogism is subject to what has been called here the ambiguity of statistical systematization, and he goes on to argue that the principle of total evidence will be of no avail as a way to circumvent this shortcoming because generally our total evidence will consist of far more than just two statements, which would moreover have to be of the particular form required for the premises of a statistical syllogism.[43] This observation would not raise a serious difficulty, at least theoretically speaking, if an appropriate general system of inductive logic were available: the rules of this system might enable us to show that that part of our total evidence which goes beyond the premises of our simple statistical argument is inductively irrelevant to the conclusion in the sense specified earlier in this section. Since no inductive logic of the requisite scope is presently at hand, however, it is a question of great interest whether some more manageable substitute for the requirement of total evidence might not be formulated which would not presuppose a full system of inductive logic and would be applicable to simple statistical systematizations. This question will be examined in the next section on the basis of a closer analysis of simple statistical systematizations offered by empirical science.

11. The Logical Form of Simple Statistical Systematizations: A Rough Criterion of Evidential Adequacy

Let us note, first of all, that empirical science offers many statistical systematizations which accord quite well with the general characterization to which we were led in the preceding section.

For example, by means of Mendelian genetic principles it can be shown that in a random sample taken from a population of pea plants each of whose parent plants represents a cross of a pure white-flowered and a pure red-flowered strain, approximately 75 percent of the plants will have red flowers and the rest white ones. This argument, which may be used for explanatory or for predictive purposes, is a statistical systematization; what it explains or predicts are the approximate percentages of red- and white-flowered plants in the sample; the "premises" by reference to which the specified percentages are shown to be highly probable include (1) the pertinent laws of genetics, some of which are statistical generalizations, whereas others are of strictly universal form; and (2) particular information of the kind mentioned above about the genetic make-up of the parent generation of the plants from which the sample is taken. (The genetic principles of strictly universal form include the laws that the colors in question are tied to specific genes; that the red gene is dominant over the white one; and various other general laws concerning the transmission, by genes, of the colors in question—or, perhaps, of a broader set of gene-linked traits. Among the statistical generalizations invoked is the hypothesis that the four possible combinations of color-determining genes—WW, WR, RW, RR—are statistically equiprobable in their occurrence in the offspring of two plants of the hybrid generation.) These premises may fairly be regarded as exhausting that part of the total available evidence that is relevant to the hypothesis about the composition of the sample. Similar considerations apply to the kind of argument that serves retrodictively to establish the time of manufacture of a wooden implement found at an archeological site when the estimate is based on the amount of radioactive carbon the implement contains. Again, in addition to statements of particular fact, the argument invokes hypotheses of strictly universal form as well as a statement, crucial to the argument at hand, concerning the rate of decay of radioactive carbon; this statement has the form of a statistical probability hypothesis.

Let us now examine one further example somewhat more closely. The statistical law that the half life of radon is 3.82 days may be invoked for a statistical explanation of the fact that within 7.64 days, a particular sample consisting of

10 milligrams of radon was reduced, by radioactive decay, to a residual amount falling somewhere within the interval from 2 to 3 milligrams; it could similarly be used for predicting a particular outcome of this kind. The gist of the explanatory and predictive argument is, briefly, this: The statement giving the half life of radon conveys two statistical laws, (i) that the statistical probability for an atom of radon to undergo radioactive decay within a period of 3.82 days is ½, and (ii) that the decaying of different radon atoms constitutes statistically independent events. One further premise needed is the statement that the number of atoms in 10 milligrams of radon is enormously large (in excess of 10^{19}). As mathematical probability theory shows, the two laws in conjunction with this latter statement imply deductively that the statistical probability is exceedingly high that the mass of the radon atoms surviving after 7.64 days will not deviate from 2.5 milligrams by more than .5, i.e., that it will fall within the specified interval. More explicitly, the consequence deducible from the two statistical laws in conjunction with the information on the large number of atoms involved is another statistical law to this effect: The statistical probability is very high that the random experiment F of letting 10 milligrams of radon decay for 7.64 days will yield an outcome of kind G, namely, a residual amount of radon whose mass falls within time interval from 2 to 3 milligrams. Indeed, the probability is so high that, according to the interpretation (9.2b), if the experiment F is performed just once, it is "practically certain" that the outcome will be of kind G. In this sense, it is rational on the basis of the given information to expect the outcome G to occur as the result of a single performance of F; and also in this sense, the information concerning the half life of radon and the large number of atoms involved in an experiment of kind F affords a statistical explanation or prediction of the occurrence of G in a particular performance of the experiment.[44]

In the statistical systematization here outlined, the requirement of total evidence is satisfied at least in the broad sense that according to the total body of present scientific knowledge, the rate of radioactive decay of an element is independent of all other factors, such as temperature and pressure, ordinary magnetic and electric influences, and chemical interactions; so none of these need be taken into consideration in appraising the probability of the specified outcome.

Other statistical explanations offered in science for particular phenomena follow the same general pattern: To account for the occurrence of a certain kind of event under specified (e.g., experimental) conditions, certain laws or theories of statistical form are adduced, and it is shown that as a consequence of these, the statistical probability for an outcome of the specified kind under circumstances of the specified kind is extremely high, so that that outcome may be expected with practical certainty in any one case where the specified conditions occur. (For example, the probabilistic explanation provided by wave mechanics for the diffraction of an electron beam by a narrow slit is essentially of this type.)

Let us examine the logic of the argument by reference to a simple model case: Suppose that a statistical explanation is to be given of the fact that a specified particular sequence S of 10 successive flippings of a given coin yielded heads at least once—let this fact be expressed by the sentence h; and suppose furthermore we are given the statements that the statistical probabilities of heads and of tails for a flipping of the given coin both equal ½, and that the results of different flippings are statistically independent of each other. These statements might then be invoked to achieve the desired explanation; for jointly, they imply that the probability for a set of 10 successive flippings of the given coin to yield heads at least once is $1 - (½)^{10}$, which is greater than .999. But this probability is still statistical in character; it applies to a certain kind of event (heads at least once) relative to a certain other kind of event (10 flippings of the given coin), but not to any individual event, such as the appearance of heads at least once in the particular unique set S of 10 flippings. If the statistical probability statement is to be used in explaining this latter event, then an additional principle is needed which makes statistical probabilities relevant to rational expectations concerning the occurrence of particular events.

One such principle is provided by the interpretation of a very high statistical probability as making it practically certain that the kind of outcome in question will occur in any one particular case (see (9.2b) above). This idea can be expressed in the following rule:

(11.1) On the information that the statistical probability p(G, F) exceeds $1 - \epsilon$ (where ϵ is some very small positive number) and that b is a particular instance of F, it is practically certain that b is an instance of F.

Another way of giving statistical probability statements relevance for rational expectations concerning individual events would be to develop a system of inductive logic for languages in which statistical probability statements can be expressed. Such a system would assign, to any "hypothesis" h expressible in the language, a logical probability c(h, e) with respect to any logically consistent evidence sentence e in that language. Choosing as evidence the sentence e_1, 'The statistical probability of obtaining heads at least once in a set of 10 flippings of this coin is $1 - (½)^{10}$, and S is a particular set of such flippings,' and as hypothesis the sentence h_1, 'S yields heads at least once,' we would then obtain the logical probability conferred by e_1 on h_1. Now the systems of inductive logic presently available—by far the most advanced of which is Carnap's—do not cover languages rich enough to permit the formulation of statistical probability statements.[45] However, for the simple kind of argument under consideration here, it is clear that the value of the logical probability should equal that of the corresponding statistical probability, i.e., that we should have $c(h_1, e_1) = 1 - (½)^{10}$. Somewhat more generally, the idea may be expressed in the following rule:

(11.2) If e is the statement '(p(G, F) = r) · Fb' and h is 'Gb,' then c(h, e) = r.

This rule is in keeping with the conception, set forth by Carnap, of logical probability as a fair betting quotient for a bet on h on the basis of e; and it accords equally with Carnap's view that the logical probability on evidence e of the hypothesis that a particular case b will have a specified property M may be regarded as an estimate, based on e, of the relative frequency of M in any class K of cases on which the evidence e does not report. Indeed, Carnap adds that the logical probability of 'Mb' on e may in certain cases be considered as an estimate of the statistical probability of M.[46] If, therefore, e actually contains the information that the statistical probability of M is r, then it seems clear that the estimate, on e, of that statistical probability, and thus the logical probability of 'Mb' on e, should be r as well.

The rules (11.1) and (11.2) may be regarded as schematizing at least simple kinds of statistical systematization. But, arguments conforming to those rules will constitute acceptable explanations or predictions only if they satisfy the principle of total evidence. For example, suppose that the total evidence e contains the information e_1 that F_1b and $p(G, F_1) = .9999$; then e_1 makes it practically certain that Gb; and yet it would not be acceptable as the premise of a statistical explanation or prediction of 'Gb' if e also contained the information, e_2, and F_2b and $p(G, F_2) = .0001$. By itself, e_2 makes it practically certain that b is not G; and if e consists of just e_1 and e_2, then the simple rule (11.2) does not enable us to assign a logical probability to 'Gb.' But suppose that, besides e_1 and e_2, e also contains e_3: '$p_3(G, F_1 \cdot F_2) = .9997$, and nothing else (i.e., nothing that is not logically implied by e_1, e_2, and e_3 in conjunction). Then it seems reasonable to say that the probability of 'Gb' on e should be equal, or at least close, to .9997. Similarly, if e contains just the further information that F_3b and $p(G, F_1 \cdot F_2 \cdot F_3) = .00002$ then the probability of 'Gb' on e should be close to .00002, and so on.

This consideration suggests the possibility of meeting the desideratum expressed at the end of Section 10 by the following rough substitute for the requirement of total evidence:

(11.3) *Rough criterion of evidential adequacy for simple statistical systematizations*: A statistical systematization of the simple type indicated in rules (11.1) and (11.2) may be regarded as satisfying the requirement of total evidence if it is based on the statistical probability of G within the narrowest class, if there is one, for which the total evidence e available provides the requisite statistical probahility.[47] More explicitly, a statistical systematization with the premises 'Fb' and '$p(G, F) = r$' may be regarded as roughly satisfying the requirement of total evidence if the following conditions are met: (i) the total evidence e contains (i.e., explicitly states or deductively implies) those two premises; (ii) e implies[48] that F is a subclass of any class F* for which e contains the statement that F*b and in addition a statistical law (which must not be simply a theorem of formal probability theory)[49] stating the value of the probability $p(G, F^*)$. The classes F, F*, etc., are of course understood here simply as the classes of those elements which have the characteristics F, F*, etc.

Condition (ii) might be liberalized by the following qualification: F need not be the narrowest class of the kind just specified; it suffices if e implies that within any subclass of F to which e assigns b, the statistical probability of G is the same as in F. For example, in the prediction, considered above, of the residual mass of radon, the total information available may well include data on temperature, pressure, and other characteristics of the given sample s: In this case, e assigns the particular event under study to a considerably narrower class than the class F of cases where a 10 milligram sample of radon is allowed to decay for 7.64 days. But the theory of radioactivity, likewise included in e, implies that those other characteristics do not affect the probability invoked in the prediction; in other words, the statistical probability of decay in the corresponding subclasses of F is the same as in F itself.

The working rule suggested here would also avoid an embarrassment which the general requirement of total evidence creates for the explanatory use of statistical systematizations. Suppose, for example, that an individual case b has been found to have the characteristic G (or to belong to the class G); and consider a proposed explanation of Gb by reference to the statements 'Fb' and 'p(G, F) = .9999.' Even assuming that nothing else is known, the total evidence then includes, in addition to these latter two statements, the sentence 'Gb.' Hence, if we were strictly to enforce the requirement of total evidence, then the explanans, by virtue of containing the explanandum, would trivially imply the latter without benefit of any statistical law, and would confer upon it the logical probability 1. Thus, no nontrivial inductive explanation would be possible for any facts or events that are known (reported by e) to have occurred. This consequence cannot be avoided by the convention that e with the explanandum statement omitted is to count as total evidence for the statistical explanation of an event known to have occurred; for despite its apparent clarity, the notion of omitting the explanandum statement from e does not admit of a precise logical explication. It is surely not a matter of just deleting the explanandum sentence from e, for the total evidence can always be so formulated as not to contain that sentence explicitly; for example, 'Gb' may be replaced by the two sentences 'Gb $\vee$ Fb' and 'Gb $\vee$ $-$ Fb.'

On the other hand, the working rule would circumvent the difficulty. For even though, in the illustration, e contains 'Gb,' the rule qualifies the statistical explanation of 'Gb' by means of 'Fb' and 'p(G, F) = .9999' alone as satisfying the requirement of total evidence. For the statistical law invoked here specifies the probability of C for the narrowest reference class to which e assigns b, namely, the class F. (To be sure, e also assigns b to the narrower reference class F $\cdot$ G, for which clearly p(G, F $\cdot$ G) = 1. It will be reasonable to say that e (trivially) contains this latter statement since it is simply a logical consequence of the measure-theoretical postulates for statistical probability. But precisely for this reason, the statement 'p(G, F $\cdot$ G) = 1' is not an empirical law; hence, under the working rule, this part of the content of e need not be taken into consideration.)[50]

But while a rule such as (11.3) does seem in accord with the rationale of scientific arguments intended to explain or to predict individual occurrences by means of statistical laws, it offers no more than a rough working principle, which must be used with caution and discretion. Suppose, for example, that the total evidence e consists of the statements 'Fb,' 'Hb,' 'p(G, F) = .9999,' and a report on 10,000 individual cases other than b, to the effect that all of them were H and non-G. Then the statistical argument with 'Fb' and 'p(G, F) = .9999' as its premises and 'Gb' as its conclusion would qualify, under the rule, as meeting the requirement of total evidence; but even though e does not state the statistical probability of G relative to H, its sample statistics on 10,000 cases of H, in conjunction with the statement that b is H, must surely cast serious doubt upon the acceptability of the proposed statistical argument as an explanation or prediction of Gb. Hence, the information relevant to 'Gb' that is provided by e cannot generally and strictly be identified with the information provided by e concerning the statistical probability of G in the narrowest available reference class; nor, of course, can the logical probability of 'Gb' on e be strictly equated with the statistical probability of G in that narrowest reference class. Thus, as a general condition for a statistical systematization that is to be not only a formally correct argument (a potential systematization) but also a scientifically acceptable one, the requirement of total evidence remains indispensable.

12. On Criteria of Rational Credibility

Besides the requirement of total evidence, there is a further condition which it might seem any statistical systematization ought to satisfy if it is to qualify as an adequate explanation, prediction, or retrodiction; namely, that the information contained in its "premises" e should provide so strong a presumption in favor of the "conclusion" h as to make it rational, for someone whose total evidence is e, to believe h to be true, or, as I will also say, to include h in the set of statements accepted by him as presumably true. In a deductive-nomological systematization, the premises afford such presumption in an extreme form: they logically imply the conclusion; hence, someone whose system of accepted statements includes those premises has the strongest possible reason to accept the conclusion as well.

Thus, the study of inductive generalization gives rise to the question whether it is possible to formulate criteria for the rational acceptability of hypotheses on the basis of information that provides strong, but not conclusive, evidence for them.

I will first construe this question in a quite general fashion without limiting it specifically to the case where the supporting information provides the premises of a statistical systematization. Toward the end of this section I will return to this latter, special case.

Let us assume that the total body of scientific knowledge at a given time t can be represented by a set K_t, or briefly K, whose elements are all the statements accepted as presumably true by the scientists at time t. The class K will contain statements describing particular events as well as assertions of statistical and universal law and in addition various theoretical statements. The membership of K will change in the course of time; for as a result of continuing research, additional statements come to be established, and thus accepted into K; while others, formerly included in K, may come to be disconfirmed and then eliminated from the system.

We can distinguish two major ways in which a statement may be accepted into K: *direct acceptance*, on the basis of suitable experiences or observations, and *inferential acceptance*, by reference to previously accepted statements. An observer who records the color of a bird or notes the reading of an instrument accepts the corresponding statements directly, as reporting what he immediately observes, rather than as hypotheses whose acceptability is warranted by the fact that they can be inferred from other statements, which have been antecedently accepted and thus are already contained in K. Inferential acceptance may be either deductive or (strictly) inductive, depending on whether the statement in question is logically implied or only more or less highly supported by the previously accepted statements.

This schematic model does not require, then, that the statements representing scientific knowledge at a given time be true; rather, it construes scientific knowledge as the totality of beliefs that are accepted at a given time as warranted by appropriate scientific procedures. I will refer to this schematization as the *accepted-information model of scientific knowledge*.

Now, we have to consider the rules of acceptance or rejection which regulate membership in K. In its full generality, this question calls for a comprehensive set of principles for the formulation, test, and validation of scientific hypotheses and theories. In the context of our investigation, however, it will suffice to concentrate on some general rules for indirect acceptance; the question of criteria for direct acceptance, which would bear on standards for observational and experimental procedures, is not relevant to the central topic of this essay.

The rules to be discussed here may be considered as stating certain necessary conditions of rationality in the formation of beliefs. One very obvious condition of this kind is the following:

(CR1)   Any logical consequence of a set of accepted statements is likewise an accepted statement; or, K contains all logical consequences of any of its subclasses.

The reason for this requirement is clear: If an investigator believes a certain set of statements, and thus accepts them as presumably true, then, to be rational, he has to accept also their logical consequences because any logical consequence of a set of true statements is true.

Note that (CR1) does not express a rule or principle of logic but rather a maxim for the rational *application* of the rules of deductive logic. These rules, such as *modus ponens* or the rules of the syllogism, simply indicate that if sentences of a specified kind are true, then so is a certain other sentence; but they say nothing at all about what it is rational to believe. Another rule is the following:

(CR2)   The set K of accepted statements is logically consistent.

Otherwise, by reason of (CR1), K would also contain, for every one of its statements, its contradictory. This would defeat the objective of science of arriving at a set of presumably true beliefs (if a statement is presumably true, its contradictory is not); and K could provide no guidance for expectations about empirical phenomena since whatever K asserted to be the case it would also assert not to be the case.

(CR3) The inferential acceptance of any statement h into K is decided on by reference to the total system K (or by reference to a subset K' of it whose complement is irrelevant to h relative to K').

This is simply a restatement of the requirement of total evidence. As noted earlier, it is automatically satisfied in the case of deductive acceptance.

Now we must look for more specific rules of inferential acceptance. The case of deductive acceptance is completely settled by (CR1), which makes it obligatory for rational procedure to accept all statements that are deductively implied by those already accepted. Can analogous rules be specified for rational inductive acceptance? Recent developments in the theory of inductive procedures suggest that this question might best be considered as a special case of the general problem of establishing criteria of rationality for choices between several alternatives; in time case at hand, the choice would be that of accepting a proposed new statement h into K, rejecting it (in the strong sense of accepting its contradictory), or leaving the decision in suspense (i.e., accepting neither h nor its contradictory).

I will consider the problem first on the assumption that a system of inductive logic is available which, for any hypothesis h and for any logically consistent "evidence" sentence e, determines the logical probability, or the degree of confirmation, $c(h, e)$, of h relative to e.

The problem of specifying rational rules of decision may now be construed in the following schematic fashion: An agent X has to choose one from among n courses of action, $A_1, A_2, \ldots, A_n$, which, on the total evidence e available to him, are mutually exclusive and jointly exhaust all the possibilities open to him. Each of these may eventuate, with certain probabilities (some of which may be zero), in any one of m outcomes, $O1, O2, \ldots, O_m$, which, on the evidence e, are mutually exclusive and exhaustive. The agent's decision to choose a particular course of action, say A_k, will be rational only if it is based on a comparison of its probable consequences with those of the alternative choices that are open to him. For such a comparison, inductive logic would provide one important tool. Let $a_1, a_2, \ldots, a_n$ be statements to the effect that X follows course of action $A_1, A_2 \ldots, A_n$, respectively; and let $o_1, o_2, \ldots, o_m$ be statements asserting the occurrence of $O_1, O_2, \ldots, O_m$, respectively. Then the probability, relative to e, for a proposed course of action, say A_j, to yield a specified outcome, say O_k, is given by $c(o_k, e \cdot a_j)$. The principles of the given system of inductive logic would determine all these probabilities, but they would not be sufficient to determine a rational course of action for X. Indeed, rationality is a relative concept; a certain decision or procedure can be qualified as rational only relative to some objective, namely by showing, generally speaking, that the given decision or procedure offers the optimal prospect of attaining the stated objective.

One theoretically attractive way of specifying such objectives is to assume that for X each of the outcomes $O_1, O_2, \ldots, O_m$ has a definite value or disvalue, which is capable of being represented in quantitative terms by a function assigning to any given outcome, say O_k, a real number u_k, the utility of O_k for X at the time in question. The idea of such a utility function raises a variety of problems which cannot be dealt with here, but which have been the object of intensive discussion and of much theoretical as well as experimental research.[51] The utility function, together with the probabilities just mentioned, determines the expectation value, or the probability estimate, based on e, of the utility attached to A_j for X:

(12.1) $\quad u'(A_j, e) = c(o_1, c \cdot a_j) \cdot u_1 + \ldots + c(o_m, e \cdot a_j) \cdot u_m.$

In the context of our schematization, the conception of rationality of decision as relative to some objective can now be taken into account in a more precise form; this is done, for example, in the following rule for rational choice, which was proposed by Carnap:

> *Rule of maximizing* the *estimated utility*: In the specified circumstances, X acts rationally if he chooses a course of action, A_j, for which the expectation value of the utility is maximized, i.e., is not exceeded by that associated with any of the alternative courses of action.[52]

I will now attempt to apply these considerations to the problem of establishing criteria of rational inductive acceptance. The decision to accept, or to reject, a given hypothesis, or to leave it in suspense might be considered as a special kind of choice required of the scientific investigator. This conception invites an attempt to obtain criteria of rational inductive belief by applying the rule of maximizing the expected utility to this purely scientific kind of choice with its three possible "outcomes": K enlarged by the contemplated hypothesis h; K enlarged by the contradictory of h; K unchanged. But what could determine the utilities of such outcomes?

The pursuit of knowledge as exemplified by pure scientific inquiry, by "basic research" not directly aimed at any practical applications with corresponding utilities, is often said to be concerned with the discovery of truth. This suggests that the acceptance of a hypothesis might be considered a choice as a result of which either a truth or a falsehood is added to the previously established system of knowledge. The problem then is to find a measure of the purely scientific utility, or, as I will say, the *epistemic utility*, of such an addition.

It seems reasonable to say that the epistemic utility of adding h to K depends not only on whether h is true or false but also on how much of what h asserts is new, i.e., goes beyond the information already contained in K. Let k be a sentence which is logically implied by K, and which in turn implies every sentence in K, just as the conjunction of the postulates in a finite axiomatization of geometry implies all the postulates and theorems of geometry. Then k has the same informational content as K. Now, the common content of two statements is expressed by their disjunction, which is the strongest statement logically implied by each of them. Hence, the common content of h and K is given by h ∨ k. But h is equivalent to (h ∨ k) · (h ∨ −k), where the two component sentences in parentheses have no common, content: their disjunction is a logical truth. Hence, that part of the content of h which goes beyond the information contained in K is expressed by (h ∨ −k). To indicate *how much* is being asserted by this statement, we make use of the concept of a content measure for sentences in a (formalized) language L. By a *content measure function* for a language L we will understand a function m which assigns, to every sentence s of L, a number m(s) in such a way that (i) $0 \leq m(s) \leq 1$; (ii) m(s) = o if and only if s is a logical truth of L; (iii) if the contents of s_1 and s_2 are mutually exclusive—i.e., if the disjunction $s_1 \vee s_2$ is a logical truth of L—then $m(s_1 \cdot s_2) = m(s_1) + m(s_2)$.[53] (If these requirements are met, then m can readily be seen to satisfy also the following conditions: (iv) m(s) = 1 − m(−s); (v) if s_1 logically implies s_2, then $m(s_1) \geq m(s_2)$; (vi) logically equivalent sentences have equal measures.)

Let m be a content measure function for an appropriately formalized language suited to the purposes of empirical science. Then, in accordance with the idea suggested above, it might seem plausible to accept the following:

(12.2) *Tentative measure of epistemic utility*: The epistemic utility of accepting a hypothesis h into the set K of previously accepted scientific statements is m(h ∨ −k) if h is true, and −m(h ∨ −k) if h is false; the utility of leaving h in suspense, and thus leaving K unchanged, is 0.

The rule of maximizing the estimated utility now qualifies the decision to accept a proposed hypothesis as epistemically rational if time probability estimate of the corresponding utility is at least as great as the estimates attached to the alternative choices. The three estimates can readily be computed. The probability, on the basis of K, that the proposed hypothesis h is true is c(h, k), and that it is false, 1 − c(h, k). Denoting the three alternative actions of accepting h, rejecting h, and leaving h in suspense by 'A,' 'R,' 'S,' respectively, we obtain the following formulas for the estimated utilities attached to these three courses of action:

(12.3a) u'(A, k) = c(h, k) · m(h ∨ −k) − (1 − c(h, k)) · m(h ∨ −k) = m(h ∨ −k) · (2c(h, k) − 1).

Analogously, considering that rejecting h is tantamount to accepting −h, which goes beyond K by the assertion −h ∨ −k, we find

(12.3b) u'(R, k) = m(− h ∨ −k) · (1 − 2c(h, k)).

Finally, we have

(12.3c) u'(S, k) = 0.

Now the following can be readily verified:[54]

(i) If $c(h, k) = \frac{1}{2}$, then all three estimates are zero;
(ii) If $c(h, k) > \frac{1}{2}$, then $u'(A, k)$ exceeds the other two estimates;
(iii) If $c(h, k) < \frac{1}{2}$, then $u'(R, k)$ exceeds the other two estimates.

Hence, the principle of maximizing the estimated utility leads to the following rule:

(12.4) *Tentative rule for inductive acceptance*: Accept or reject h, given K, according as $c(h, k) > \frac{1}{2}$ or $c(h, k) < \frac{1}{2}$; when $c(h, k) = \frac{1}{2}$, h may be accepted, rejected, or left in suspense.

It is of interest to note that the principle of maximizing the estimated utility, in conjunction with the measure of epistemic utility specified in (12.2), implies this rule of acceptance quite independently of whatever particular inductive probability function c and whatever particular measure function m might be adopted.

Unfortunately, time criteria specified by this rule are far too liberal to be acceptable as general standards governing the acceptance of hypotheses in pure science. But this does not necessarily mean that the kind of approach attempted here is basically inadequate: the fault may well lie with the oversimplified construal of epistemic utility. It would therefore seem a problem definitely worth further investigation whether a modified version of the concept of epistemic utility cannot be construed which, via the principle of maximizing estimated utility, will yield a more satisfactory rule for the inductive acceptance or rejection of hypotheses in pure science. Such an improved measure of epistemic utility might plausibly be expected to depend, not only on the change of informational content, but also on other changes in the total system of accepted statements which the inductive acceptance of a proposed hypothesis h would bring about. These would presumably include the change in the simplicity of the total system, or, what may be a closely related characteristic, the change in the extent to which the theoretical statements of the system would account for, or systematize, the other statements in the system, in particular those which have been directly accepted as reports of previous observational or experimental findings. As yet, no fully satisfactory general explications of these concepts are available, although certain partial results have been obtained.[55] And even assuming that the concepts of simplicity and degree of systematization can be made explicit and precise, it is yet another question whether the notion of epistemic utility permits a satisfactory explication, which can serve as a basis for the construction of rules of inductive acceptance.

We will now consider briefly an alternative construal of scientific knowledge, which would avoid the difficulties just outlined: it will be called the *pragmatist* or *instrumentalist model*. Let us note, first of all, that the epistemic utilities associated with the decision inductively to accept (or to reject, or to leave in suspense) a certain hypothesis would have to represent "gains" or "losses" as judged by reference to the objectives of "pure" or "basic" scientific research; in contradistinction to what will be called here *pragmatic utilities*, which would represent gains or losses in income, prestige, intellectual or moral satisfaction, security, and so forth, that may accrue to an individual or to a group as a result of "accepting" a proposed hypothesis in the practical sense of basing some course of action on it. Theories of rational decision making have usually been illustrated by, and applied to, problems in which the utilities are of this pragmatic kind, as for example, in the context of quality control. The hypotheses that have to be considered in that case concern the items produced by a certain technological process during a specified time; e.g., vitamin capsules which must meet certain standards, or tablets containing a closely specified amount of a certain toxic ingredient, or ball bearings for whose diameter a certain maximum tolerance has been fixed, or light bulbs which must meet various specifications. The hypothesis under test will assert, in the simplest case, that the members of the population (e.g., the output produced by a given industrial plant in a week) meet certain specified standards (e.g., that certain of their quantitative characteristics fall within specified numerical intervals). The hypothesis is tested by selecting a random sample from the total population and examining its members in the relevant respects. The problem then arises of formulating a general decision rule which will indicate, for every possible outcome of the test, whether on the evidence afforded by that outcome the hypothesis is to be accepted or rejected. But what is here referred to as acceptance or rejection of a hypothesis clearly amounts to adopting or rejecting a certain practical course of action (e.g., to ship the ball bearings to the distributors, or to reprocess them). In this kind of situation, we may distinguish four possible "outcomes": the hypothesis may be accepted and in fact true, rejected though actually true, accepted though actually false, or rejected and in fact false. To each of these outcomes there will be attached a certain positive or negative utility, which in cases of the kind considered might be represented, at least approximately, in monetary terms. Once such utilities have been specified, it is possible to formulate decision rules which will indicate for every possible outcome of the proposed testing procedure whether, on the evidence provided by the outcome and in consideration of the utilities involved, the hypothesis is to be accepted or to be rejected. For example, the principle of maximizing estimated utilities affords such a rule, which presupposes, however, that a suitable inductive logic is available which assigns to any proposed hypothesis h, relative to any consistent "evidence" statement e, a definite logical probability, $c(h, e)$.

Alternatively, there have been developed, in mathematical statistics and in the theory of games, certain methods of arriving at decision rules which do not require any such general concept of inductive or logical probability. These methods are limited to certain special types of hypotheses and evidence sentences; normally, their application is to hypotheses in the form of probability statements (statistical generalizations), and to evidence sentences in the form of reports on statistical findings in finite samples. One of the best known methods of this kind is based on the minimax principle. This method uses the concept of probability only in its statistical form. It is intended to select the most rational from among various possible rules that might be followed in deciding on the acceptance or rejection of a proposed hypothesis h in consideration of (i) the results of a specified kind of test and (ii) the utilities assigned to the possible "outcomes" of accepting or rejecting the hypothesis. Briefly, the minimax principle directs that we adopt, from among the various possible decision rules, one that minimizes the maximum risk, i.e., one for which the largest of the (statistically defined) probability estimates of the losses that might be incurred in the given context as a result of following this rule is no greater than the largest of the corresponding risks (loss estimates) attached to any of the alternative decision rules.[56]

Clearly, the minimax principle is not itself a decision rule, but rather a metarule specifying a standard of adequacy, or of rationality, for decision rules pertaining to a suitably characterized set of alternative hypotheses, plus testing procedure, and a given set of utilities.[57]

But whatever decision rules, or whatever general standards for the choice of decision rules, may be adopted in situations of the kind referred to here, the crucial point remains that the pragmatic utilities involved, and thus the decision dictated by the rule once the test results are given, will depend on, and normally vary with, the kind of action that is to be based upon the hypothesis. Consider, for example, the hypothesis that all of the vials of vaccine produced during a given period of time by a pharmaceutical firm meet certain standards of purity; and suppose that a test has been performed by analyzing the vials in a random sample. Then the gains or losses to be expected from correct or incorrect assumptions as to the truth of the hypothesis will depend on the action that is intended, for example, on whether the vaccine is to be administered to humans or to animals. By reason of the different utilities involved, a given decision rule—be it the rule of maximizing estimated utility or a rule selected in accordance with the minimax, or a similar, standard—may then well specify, on one and the same evidence, that the hypothesis is to be rejected in the case of application to human subjects, but accepted if the application is to be to animals.

Clearly then, in cases of this kind we cannot properly speak of a decision to accept or reject a hypothesis per se; the decision is rather to adopt one of two (or more) alternative courses of action. Moreover, it is not even clear on what grounds the acceptance or rejection of this hypothesis per se, on the given evidence, could be justified—unless it is possible to specify a satisfactory concept of epistemic utility, whose role for the decisions of pure science would be analogous to that of pragmatic utility in decisions concerning actions based on scientific hypotheses.

Some writers on the problems of rational decision have therefore argued that one cannot strictly speak of a decision to accept a scientific hypothesis, and that the decisions in question have to be construed as concerning choices of certain courses of action.[58] A lucid presentation and defense of this point of view has been given by R. C. Jeffrey, who accordingly arrives at the conclusion that the scientist's proper role is to provide the rational agents of his society with probabilities for hypotheses which, on the more customary account, he would be described as simply accepting or rejecting.[59]

This view, then, implies a rejection of the accepted-information model of scientific knowledge and suggests an alternative which might be called a tool-for-optimal-action model, or, as I said earlier, an instrumentalist model of scientific knowledge. This label is meant to suggest the idea that whether a hypothesis is to be accepted or not will depend upon time sort of action to be based on it, and on the rewards and penalties attached to the possible outcomes of such action. An instrumentalist model might be formulated in different degrees of refinement. A very simple version would represent the state of scientific knowledge at a given time t by a set D, or more explicitly, D_t, of directly accepted statements, plus a theory of inductive support which assigns to each proposed hypothesis, or to at least some of them, a certain degree of support relative to D_t. Like K in the accepted-information model, D_t would be assumed to be logically consistent and to contain any statement logically implied by any of its own subsets. But no statement other than those in D_t, however strongly confirmed by D_t, would count as accepted, or as belonging to the scientific knowledge at the given time. Rather, acceptance would be understood pragmatically in the context of some contemplated action, and a decision would then depend on the utilities involved.

If, in particular, the theory of inductive support assumed here is an inductive logic in Carnap's sense, then it will assign a degree of confirmation c(h, e) to any statement h relative to any logically consistent statement e in the language of science, which we assume to be suitably chosen and formalized. In this case, scientific knowledge at a given time t might be represented by a functional k_t assigning to every sentence S that is expressible in the language of science a real-number value, $k_t(S)$, which lies between 0 and 1 inclusive. The value $k_t(S)$ would simply be the logical probability of S relative to D_t; in particular, for any S included in or logically implied by D_t, $k_t(S)$ would be 1; for any S logically incompatible with D_t, $k_t(S)$ would be 0. Temporal changes of scientific knowledge would be reflected by changes of k_t, and thus by changes in the numbers assigned to some of the sentences in the language of science.

But clearly, this version of a pragmatist model is inadequate: It construes scientific knowledge as consisting essentially of reports on what has been directly observed, for the formal theory of inductive probability which it presupposes for the appraisal of other statements would presumably be a branch of logic rather than of empirical science. This account of science disregards the central importance of theoretical concepts and principles for organizing empirical data into patterns that permit explanation and prediction. So important is this aspect of science that theoretical considerations will often strongly influence the decision as to whether a proposed report on some directly observed phenomenon is to be accepted: What is a fact is to some extent determined by theory. In this respect the notion of a system D_t of statements which are accepted directly and independently of theoretical considerations, and by reference to which the rational credibility of all other scientific statements is adjudged, is a decided oversimplification. And it is an oversimplification in yet another respect: In theoretically advanced disciplines, many of the terms that the experimenter would use to record his observations, and thus to formulate his directly accepted statements, belong to time theoretical vocabulary rather than to that of everyday observation and description; and the appropriate theoretical framework has to be presupposed if those statements are to make sense.

Another inadequacy of the model lies in the assumption that any individual hypothesis that may be proposed in the language of science can be assigned a reasonable degree of confirmation by checking it against the total set D_t of statements that have been directly accepted, for the test of any even moderately advanced scientific hypothesis will require the assumption of other hypotheses in addition to observational findings. As Duhem emphasized so strongly, what can be tested experimentally is never a single theoretical statement, but always a comprehensive and complexly interconnected body of statements.

If we were to try to construct a pragmatist model on the basis of statistical decision theory, the difficulties would become even greater; for this theory, as noted earlier, eschews the assignment of degrees of support to statements relative to other statements. Hence, here, the scientist would have to assume the role of a consultant who, in a limited class of experimental contexts, provides decisions concerning the acceptance or rejection of certain statistical hypotheses for the guidance of action, provided that the pertinent utilities have been furnished to him.

At present, I do not know of a satisfactory *general* way of resolving the issue between the two conceptions of science which are schematized by our two models. But the preceding discussion of these models does seem to suggest an answer to the question raised at the beginning of this section, namely, whether it should be required of a statistical explanation, prediction, etc. in science that its premises make its conclusion rationally acceptable.

The preceding considerations seem to indicate that it would be pointless to formulate criteria of acceptability by reference to pragmatic utilities; for we are concerned here with purely theoretical (in contrast to applied) explanatory and predictive statistical arguments. We might just add the remark that criteria of rational acceptability based on pragmatic utilities might direct us to accept a certain predictive hypothesis, even though it was exceedingly improbable on the available evidence, on the ground that, if it were true, time utility associated with its adoption would be exceedingly large. In other words, if a decision rule of this kind, which is based on statistical probabilities and on an assignment of utilities, singles out, on the basis of evidence e, a certain hypothesis h from among several alternatives, then what is qualified as rational is, properly speaking, not the decision to believe h to be true, but the decision to act in the given context *as if* one believed h to be true, even though e may offer very little support for that belief.

The rational credibility of the conclusion, in a sense appropriate to the purely theoretical, rather than the applied, use of statistical systematizations will thus have to be thought of as represented by a suitable concept of inductive support (perhaps in conjunction with a concept of epistemic utility). And at least for the types of statistical systemization covered by rule (11.1) or (11.2), the statistical or logical probability specified in the argument itself may serve as an indicator of inductive support; the requirement of high credibility for the conclusion can then be met by requiring, in the case of (11.1), that ϵ be sufficiently small, and in the case of (11.2), that r be sufficiently large.

But the notions "sufficiently small" and "sufficiently large" invoked here cannot well be construed as implying the existence of some fixed probability value, say r*, such that a statistical systematization will meet the requirement of rational credibility just in case the probability associated with it exceeds r*: The standards of rational credibility will vary with the context in which a statistical systematization is used.[60] It will therefore be more satisfactory, for an explication of the logic of statistical explanation, prediction, and similar arguments, explicitly to construe *statistical systematization as admitting of degrees*: The evidence e adduced in an argument of this kind may then be said to explain, or predict, or retrodict, or generally to systematize its "conclusion" h to degree r, where r is the inductive support that e gives to h. In this respect, statistical systematization differs fundamentally from its deductive-nomological counterpart: In a deductive-nomological systematization, the explanandum follows logically from the explanans and thus is certain relative to the latter; no higher degree of rational credibility (relative to the information provided by the premises) is possible, and anything less than it would vitiate the claim of a proposed argument to constitute a deductive systematization.

13. The Nonconjunctiveness of Statistical Systematization

Another fundamental difference between deductive and statistical systematization is this: Whenever a given explanans e deductively explains each of n different explananda, say $h_1, h_2, \ldots, h_n$, then e also deductively explains their conjunction; but if an explanans e statistically explains each of n explananda, $h_1, h_2, \ldots, h_n$, to a positive degree, however high, it may still attribute a probability of zero to their conjunction. Thus, e may statistically explain (or analogously, predict, retrodict, etc.) very strongly whatever is asserted by each of n hypotheses, but not at all what is asserted by them conjointly: statistical systematization is, in this sense, nonadditive, or nonconjunctive (whereas deductive systematization is additive, or conjunctive). This point can be stated more precisely as follows:

(13.1) *Nonconjunctiveness of statistical systematization*: For any probability value p*, however close to 1, there exists a set of statistical systematizations which have the same "premise" e, but different "conclusions," $h_1, h_2, \ldots, h_n$, such that e confers a probability of at least p* on every one of these conclusions but the probability zero on their conjunction.

The proof can readily be outlined by reference to a specific example. Let us assume that p* has been chosen as .999 (and similarly, that ϵ, for use of the rule (11.1), has been chosen as .001). Then consider the case, mentioned earlier, of ten successive flippings of a given coin. Choose as "premise" the statement e: 'The statistical probabilities of getting heads and of getting tails by flipping this coin are both ½; the results of different flippings are statistically independent; and S is a particular sequence of 10 flippings of this coin'; furthermore, let h_1 be 'S does not yield tails 10 times in succession'; h_2: 'S does not yield 9 tails followed by 1 head'; h_3: 'S does not yield 8 tails followed by 2 heads'; and so on to h_{1024}: 'S does not yield heads 10 times in succession.' Each of these hypotheses h_j ascribes to S a certain kind of outcome O_j; and as is readily seen, the probability statements included in e imply logically that for each of these 2^{10} different possible outcomes, the statistical probability of obtaining it as a result of 10 successive flippings of the given coin is $1 - (½)^{10}$. But according to rule (11.1), this makes it practically certain, for any one of the O_j that the particular sequence S will have O_j as its outcome; in other words, this makes it practically certain, for each one of our hypotheses h_j, that h_j is true. Rule (11.2) more specifically ascribes the logical probability $1 - (½)^{10}$ to each of the h_j on the basis of the statistical probability for O_j which is implied by e.[61]

On the other hand, the conjunction of the h_j is tantamount to the assertion that none of the 10 particular flippings that constitute the individual sequence S will yield either heads or tails—a kind of outcome, say O^*, for which e implies the statistical probability zero. This, under rule (11.1), makes it practically certain that this outcome will not occur in S, i.e., that the conjunction of the h_j is false—even though each of the conjoined hypotheses is practically certain to be true. And if (11.2) is invoked, then the statement that the statistical probability of O^* is zero confers upon the conjunction of the h_j the logical probability zero even though, on the basis of statistical information also provided by e, each of the h_j has a logical probability exceeding .999.[62]

A similar argument can be presented for the case, considered earlier, of the radioactive decay of a particular sample S of 10 milligrams of radon over a period of 7.64 days. For the interval from 2 to 3 milligrams referred to in our previous discussion can be exhaustively divided into mutually exclusive subintervals $i_1, i_2, \ldots, i_n$, which are so small that for each i_j there is a statistical probability exceeding .999999, let us say, that the residual mass of radon left of an initial 10 milligrams after 7.64 days will not lie within i_j. Hence, given the information that the half life of radon is 3.82 days, it will be practically certain, according to rule (11.1) that if the experiment is performed just with the one particular sample S, the residual mass of radon will not lie within the interval i_1; it will also be practically certain that the residual mass will not lie within i_2; and so forth. But conjointly these hypotheses, each of which is qualified as practically certain, assert that the residual mass will not lie within the interval from 2 to 3 milligrams; and as was noted earlier, the law stating the half life of radon makes it practically certain that precisely the contradictory of this assertion is true! Thus, the statistical information about the half life of radon statistically explains (or predicts, etc., depending on the context) to a very high degree each of the individual hypotheses referring to the sub-intervals; but it does not thus explain (or predict, etc.) their conjunction.

Though superficially reminiscent of the ambiguity of statistical systematization, which was examined earlier, this nonadditivity is a logically quite different characteristic of statistical systematization. In reference to statistical systematizations of the simple kind suggested by rule (11.1), ambiguity can be characterized as follows: If the fact that b is G can be statistically explained (predicted) by a true explanans stating that b is F and that $p(G, F) > 1 - \epsilon$, then there is in general another true statement to the effect that b is F' and that $p(-G, F') > 1 - \epsilon$, which in the same sense statistically explains (predicts) that b is non-G. This ambiguity can be prevented by requiring that a statistical systematization, to be scientifically acceptable, must satisfy the principle of total evidence; for one and the same body of evidence cannot highly confirm both 'Gb' and '−Gb.'

But the principle of total evidence does not affect at all the nonconjunctiveness of statistical systematization, which lies precisely in the fact that one and the same set of inductive "premises" (one and the same body of evidence) e may confirm to within $1 - \epsilon$ each of n alternative "conclusions" (hypotheses), while confirming with equal strength also the negation of their conjunction. This fact is rooted in the general multiplication theorem of elementary probability theory, which implies that the probability of the conjunction of two items (characteristics or statements, according as statistical or logical probabilities are concerned) is, in general, less than the probability of either of the items taken alone. Hence, once the connection between "premises" and "conclusion" in a statistical systematization is viewed as probabilistic in character, nonconjunctiveness appears as inevitable, and as one of time fundamental characteristics that distinguish statistical systematization from its deductive-nomological counterpart.

14. Concluding Remarks

Commenting on the changes that the notion of causality has undergone as a result of the transition from deterministic to statistical forms of physical theory, R. von Mises holds that "people will gradually come to be satisfied by causal statements of this kind: It is *because* the die was loaded that the 'six' shows more frequently (but we do not know what the next number will be); or, *Because* the vacuum was heightened and the voltage increased, the radiation became more intense (but we do not know the precise number of scintillations that will occur in the next minute)."[63] This passage clearly refers to statistical explanation in the sense considered in the present essay; it sets forth what might be called a statistical-probabilistic concept of "because," in contradistinction to a strictly deterministic one, which would correspond to deductive-nomological explanation. Each of the two concepts refers to a certain kind of subsumption under laws—statistical in one case, strictly universal in the other; but, as has been argued in the second part of this study, they differ in a number of fundamental logical characteristics: The deterministic "because" is deductive in character, the statistical one is inductive; the deterministic "because" is an either-or relation, the statistical one permits degrees; the deterministic "because" is unambiguous, while the statistical one exhibits an ambiguity which calls for relativization with respect to the total evidence available; and finally, the deterministic "because" is conjunctive whereas the statistical one is not.

The establishment of these fundamental logical differences is at best just a small contribution toward a general analytic theory of statistical modes of explanation and prediction. The fuller development of such a theory raises a variety of other important issues, some of which have been touched upon in these pages; it is hoped that those issues will be further clarified by future investigations.

NOTES

1. This distinction was developed briefly in Hempel [25], Sec. 2.
2. See Hempel [24], especially Secs. 1–4; Hempel [25]; and Hempel and Oppenheim [26]. This latter article will henceforth be referred to as *SLE*. The point of these discussions was to give a more precise and explicit statement of the deductive model of scientific explanation and to exhibit and analyze some of the logical and methodological problems to which it gives rise: the general conception of explanation as deductive subsumption under more general principles had been set forth much earlier by a variety of authors, some of whom are listed in *SLE*, fn. 4. In fact, in 1934 that conception was explicitly presented in the following passage of an introductory textbook: "Scientific explanation consists in subsuming under some rule or law which expresses an invariant character of a group of events, the particular event it is said to explain. Laws themselves may be explained, and in the same manner, by showing that they are consequences of more comprehensive theories." (Cohen and Nagel [10], p. 397.) The conception of the explanation of laws by deduction from theories was developed in great detail by N. R. Campbell; for an elementary account see his book [4], which was first published in 1921. K. R. Popper, too, has set forth this deductive conception of explanation in several of his publications (cf. fn. 4 in *SLE*); his earliest statement appears in Sec. 12 of his book [38], which has at long last been published in a considerably expanded English version [40].
3. The suggestive terms 'covering law' and 'covering-law model' are borrowed from Dray, who, in his book [13], presents a lucid and stimulating critical discussion of the question whether, or to what extent, historical explanation conforms to the deductive pattern here considered. To counter a misunderstanding that might be suggested by some passages in Ch. II, Sec. 1 of Dray's book, I would like to emphasize that the covering-law model must be understood as permitting reference to *any number of laws* in the explanation of a given phenomenon: there should be no restriction to just one "covering law" in each case.
4. More accurately, the explanation of a general law by means of a theory will usually show (1) that the law holds only within a certain range of application, which may not have been made explicit in its standard formulation; (2) that even within that range, the law holds only in close approximation, but not strictly. This point is well illustrated by Duhem's emphatic reminder that Newton's law of gravitation, far from being an inductive generalization of Kepler's laws, is actually incompatible with them, and that the credentials of Newton's theory lie rather in its enabling us to compute the perturbations of the planets, and thus their deviations from the orbits assigned to them by Kepler. (See Duhem [14], pp. 312ff, and especially p. 317. The passages referred to here are included in the excerpts from P. P. Wiener's translation of

Duhem's work that are reprinted in Feigl and Brodbeck [15], under the title "Physical Theory and Experiment.")

Analogously, Newtonian theory implies that the acceleration of a body falling freely in a vacuum toward the earth will increase steadily, though over short distances it will be very nearly constant. Thus, strictly speaking, the theory contradicts Galileo's law, but shows the latter to hold true in very close approximation within a certainrange of application. A similar relation obtains between the principles of wave optics and those of geometrical optics.

5. No claim was made that this is the only kind of scientific explanation; on the contrary, at the end of Sec. 3, it was emphasized that "Certain cases of scientific explanation involve 'subsumption' of the explanandum under a set of laws of which at least some are statistical in character. Analysis of the peculiar logical structure of that type of subsumption involves difficult special problems. The present essay will be restricted to an examination of the causal type of explanation . . ." A similar explicit statement is included in the final paragraph of Sec. 7 and in Sec. 5.3 of the earlier article, Hempel [24]. These passages seem to have been overlooked by some critics of the covering-law model.
6. This was done already in *SLE*, Sec. 7.
7. The term 'lawlike sentence' and the general characterization given here of its intended meaning are from Goodman [20]. The difficult problem of giving an adequate general characterization of those sentences which if true would constitute laws will not be dealt with in the present essay. For a discussion of the issues involved, see, for example, *SLE*, Secs. 6–7; Braithwaite [3], Ch. IX, where the central question is described as concerning "the nature of the difference, if any, between 'nomic laws' and 'mere generalizations'"; and the new inquiry into the subject by Goodman [20, 21]. All the sentences occurring in a potential explanation are assumed, of course, to be empirical in the broad sense of belonging to some language adequate to the purposes of empirical science. On the problem of characterizing such systems more explicitly, see especially Scheffler's stimulating essay [46].
8. [49], p. 185.
9. Ibid., p. 194.
10. Note, however, that from a law of coexistence connecting certain parameters it is possible to derive laws of succession concerning the rates of change of those parameters. For example, the law expressing the period of a mathematical pendulum as a function of its length permits the derivation, by means of the calculus, of a further law to the effect that if the length of the pendulum changes in the course of time, then tile rate of change of its period at any moment is proportional to the rate of change of its length, divided by the square root of its length, at that moment.
11. Among these is Schlick [48], who gives credit to Wittgenstein for the idea that a law of nature does not have the character of a statement, but rather that of an instruction for the formation of statements. Schlick's position in this article is prompted largely by the view that a genuine statement must be definitively verifiable—a condition obviously not met by general laws. But this severe verifiability condition cannot be considered as an acceptable standard for scientific statements.

More recently, Ryle—see, for example, [44], pp. 121–123—has described law statements as statements which are true or false, but one of whose jobs is to serve as inference tickets: they license their possessors to move from the assertion of some factual statements to the assertion of others.

Toulmin [53], has taken the view, more closely akin to Schlick's, that laws of nature and physical theories do not function as premises in inferences leading to observational statements, but serve as modes of representation and as rules of inference according to which statements of empirical fact may be inferred from other such statements. An illuminating discussion of this view will be found in E. Nagel's review of Toulmin's book, in *Mind*, 63:403-412 (1954); it is reprinted in Nagel [35], pp. 303-315.

Carnap [5], par. 51, makes explicit provision for the construction of languages with extralogical rules of inferences. He calls the latter physical rules, or P-rules, and emphasizes that whether, or to what extent, P-rules are to be countenanced in constructing a language is a question of expedience. For example, adoption of P-rules may oblige us to alter the rules—and thus the entire formal structure—of the language of science in order to account for some new empirical findings which, in a language without P-rules, would prompt only modification or rejection of certain statements previously accepted in scientific theory.

The admission of material rules of inference has been advocated by W. Sellars in connection with his analysis of subjunctive conditionals; see [51, 52]. A lucid general account and critical appraisal of various reasons that have been adduced in support of construing general laws as inference rules will be found in Alexander [1].

12. For detailed discussions of these issues, see Barker [2], especially Ch. 7; Scheffler [46], especially Secs. 13-18; Hempel [25], especially Sec. 10.
13. On this point, see the review by Nagel mentioned in fn. 11.
14. This remark does not hold, however, when all the laws invoked in the explanans are laws of coexistence (see Sec. 4) and all the particular statements adduced in the explanans pertain to events that are simultaneous with the explanandum event. I am indebted to Mr. S. Bromberger for having pointed out to me this oversight in my formulation.
15. [18], p. 265.
16. Ibid.
17. [42], p. 282.
18. Ibid., p. 286.
19. In Sec. 3 of *SLE*, to which Rescher refers in his critique, an explanation of a past event is explicitly construed as a deductive argument inferring the occurrence of the event from "antecedent conditions" and laws; so that the temporal direction of the inference underlying explanation is the same as that of a predictive nomological argument, namely, from statements concerning certain initial (and boundary) conditions to a statement concerning the *subsequent* occurrence of the explanandum event.

    I should add, however, that although all this is said unequivocally in *SLE*, there is a footnote in *SLE*, Sec. 3, which is certainly confusing, and which, though not referred to by Rescher, might have encouraged him in his misunderstanding. The footnote, numbered 2a, reads: "The logical similarity of explanation and prediction, and the fact that one is directed towards past occurrences, the other towards future ones, is well expressed in the terms 'postdictability' and 'predictability' used by Reichenbach [in *Philosophic Foundations of Quantum Mechanics*, p. 13]." To reemphasize the point at issue: postdiction, or retrodiction, is not the same thing as explanation.
20. [50].
21. [47].
22. *SLE*, Sec. 3
23. Scheffler [47], p. 300.

24. See ibid., p. 296; Scriven [49].
25. Professor Nelson Goodman, to whom I had mentioned my difficulties with the notion of a statement being "about" a certain subject, showed me a draft of an article entitled "About," which has now appeared in Mind, 70: 1–24 (1961); in it, he proposes an analysis of the notion of aboutness which will no doubt prove helpful in dealing with the puzzle outlined here, and which may even entirely resolve it.
26. It is sometimes argued that a statement asserting such a universal connection rests, after all, only on a finite, and necessarily incomplete, body of evidence; that, therefore, it may well have exceptions which have so far gone undiscovered, and that, consequently, it should be qualified as probabilistic, too. But this argument fails to distinguish between the claim made by a given statement and the strength of its evidential support. On the latter score, all empirical statements have to count as only more or less well supported by the available evidence; but the distinction between laws of strictly universal form and those of statistical form refers to the claim made by the statements in question: roughly speaking, the former attribute a certain character to all members of a specified class; the latter, to a fixed proportion of its members.
27. The characterization given here of the concept of statistical probability seems to me to be in agreement with the general tenor of the "propensity interpretation" advocated by Popper in recent years. This interpretation "differs from the purely statistical or frequency interpretation only in this—that it considers the probability as a characteristic property of the experimental arrangement rather than as a property of a sequence"; the property in question is explicitly construed as *dispositional*. (Popper [39], pp. 67–68. See also time discussion of this paper at the Ninth Symposium of the Colston Research Society, in Körner [30], pp. 78–89 passim.) However, the currently available statements of the propensity interpretation are all rather brief (for further references, see Popper [40]); a fuller presentation is to be given in a forthcoming book by Popper.
28. In fact, Goodman [20], has argued very plausibly that one symptomatic difference between lawlike and nonlawlike generalizations is precisely that the former are able to lend support to corresponding subjunctive or counterfactual conditionals; thus the statement 'If this copper key were to be heated it would expand' can be supported by the law mentioned above. By contrast, the general statement 'All objects ever placed on this table weigh less than one pound' is nonlawlike, i.e., even if true, it does not count as a law. And indeed, even if we knew it to be true, we would not adduce it in support of corresponding counterfactuals; we would not say, for example, that if a volume of *Merriam-Webster's Unabridged Dictionary* had been placed on the table, it would have weighed less than a pound. Similarly, it might be added, general statements of this latter kind possess no explanatory power: this is why the sentences $L_1, L_2, \ldots, L_n$ in the explanans of any deductive-nomological explanation are required to be lawlike.

The preceding considerations suggest the question whether there is a category of statistical probability statements whose status is comparable to that of accidental generalizations. It would seem clear, however, that insofar as statistical probability statements are construed as dispositional in the sense suggested above, they have to be considered as being analogous to lawlike statements.
29. The mathematical theory of statistical probability has been developed in two major forms. One of these is based on an explicit definition of probabilities as limits of relative frequencies in certain infinite reference sequences. The use of this limit definition is

an ingenious attempt to permit the development of a simple and elegant theory of probability by means of the apparatus of mathematical analysis, and to reflect at the same time the intended statistical application of the abstract theory. The second approach, which offers certain theoretical advantages and is now almost generally adopted, develops the formal theory of probability as an abstract theory of certain set-functions and then specifies rules for its application to empirical subject matter. The brief characterization of the theory of statistical probability given in this section follows the second approach. However, the problem posed by the ambiguity of statistical systematization arises as well when the limit definition of probability is adopted.

30. See, for example, Kolmogoroff [31], Sec 2.
31. [11], pp 148–149. Similar formulations have been given by other representatives of this measure theoretical conception of statistical probability, for example, by Kolmogoroff [31], p. 4.
32. Cf. Cramér [11], p. 149; see also the very similar formulation in Kolmogoroff [31], p. 4.
33. Cf. Cramér [11],p. 150.
34. Ibid., pp. 197–198.
35. My manuscript here originally contained the phrase 'is invariably (or even in some cases) accompanied by non-C.' By reading the critique of this passage as given in the manuscript of Professor Scriven's contribution to the present volume, I became aware that the chains made in parentheses is indeed incorrect. Since the point is entirely inessential to my argument, I deleted the parenthetical remark after having secured Professor Scriven's concurrence. However, Professor Scriven informed me that he would not have time to remove whatever references to this lapse his manuscript might contain: I therefore add this note for clarification.
36. A sentence of the form 'It is certain that x is C' ostensibly attributes the modality of certainty to the proposition expressed by the conclusion in relation to the propositions expressed by the premises. For the purposes of the present study, involvement with propositions can be avoided by construing the given modal sentence as expressing a logical relation that the conclusion, taken as a sentence, bears to the premise sentences. Concepts such as near certainty and probability can, and will here, equally be treated as applying to pairs of sentences rather than to pairs of propositions.
37. These remarks seem to me to be relevant, for example, to C. I. Lewis's notion of categorical, as contradistinguished from hypothetical, probability statements. For in [32], p. 319, Lewis argues as follows: "Just as 'If D then (certainly) P, and D is the fact,' leads to the categorical consequence, "Therefore (certainly) P'; so too, 'If D then probably P, and D is the fact,' leads to a categorical consequence expressed by 'It is probable that P'. And this conclusion is not merely the statement over again of the probability relation between 'P' and 'D'; any more than 'Therefore (certainly) P' is the statement over again of 'If D then (certainly) P'. 'If the barometer is high, tomorrow will probably be fair; and the barometer *is* high,' categorically assures something expressed by 'Tomorrow will probably be fair'. This probability is still relative to the grounds of judgment; but if these grounds are actual, and contain all the available evidence which is pertinent, then it is not only categorical but may fairly be called *the* probability of the event in question."

'This position seems to me to be open to just those objections which have been suggested in the main text. If 'P' is a statement, then the expressions 'certainly P' and 'probably P' as envisaged in the quoted passage are not statements: if we ask how one would go about trying to ascertain whether they were true, we realize that we are entirely at a loss unless and until a reference set of statements or assumptions is specified relative to which P may then be found to be certain, or to be highly probable, or neither. The expressions in question, then, are essentially incomplete; they are elliptic formulations of relational statements; neither of them can be the conclusion of an inference. However plausible Lewis's suggestion may seem, there is no analogue in inductive logic to *modus ponens*, or the "rule of detachment" of deductive logic, which, given the information that 'D,' and also 'if D then P,' are true statements, authorizes us to detach the consequent 'P' in the conditional premise and to assert it as a self-contained statement which must then be true as well.

At the end of the quoted passage, Lewis suggests the important idea that 'probably P' might be taken to mean that the total relevant evidence available at the time confers high probability upon P; but even this statement is relational in that it tacitly refers to some unspecified time; and besides, his general notion of a categorical probability statement as a conclusion of an argument is not made dependent on the assumption that the premises of the argument include all the relevant evidence available.

It must be stressed, however, that elsewhere in his discussion, Lewis emphasizes the relativity of (logical) probability, and thus the very characteristic which rules out the conception of categorical probability statements.

38. See especially [7, 8], and, for a very useful survey [6].
39. In a recent study, Kemeny and Oppenheim [29], have proposed, and theoretically developed, an interesting concept of "degree of factual support" (of a hypothesis by given evidence), which differs from Carnap's concept of degree of confirmation, or inductive probability, in important respects; for example, it does not have the formal character of a probability function. For a suggestive distinction and comparison of different concepts of evidence, sec Reacher [43].
40. Carnap [7], p. 211. In his comments, pp. 211–213, Carnap points out that in less explicit form, the requirement of total evidence has been recognized by various authors at least since Bernoulli. The idea also is suggested in the passage from Lewis [32], quoted in fn. 36. Similarly, Williams, whose book *The Ground of Induction* centers about various arguments that have the character of statistical systematizations, speaks of "the most fundamental of all rules of probability logic, that 'the' probability of any proposition is its probability in relation to the known premises and them only." (Williams [55], p. 72.)

I wish to acknowledge here my indebtedness to Professor Carnap, to whom I turned in 1945, when I first noticed the ambiguity of statistical explanation, and who promptly pointed out to me in a letter that this was but one of several apparent paradoxes of inductive logic which result from violations of the requirement of total evidence.

In his recent book, Barker [2], pp. 70–78, concisely and lucidly presents the gist of the puzzle under consideration here and examines the relevance to it of the principle of total evidence.
41. Cf. Carnap [7], pp. 211, 494.
42. Carnap [7], p. 211, says "There is no analogue to this requirement [of total evidence] in deductive logic"; but it seems more accurate to say that the requirement is automatically met here.
43. See Barker [2], pp. 76–78. The point is made in a more general form by Carnap [7], p. 404.
44. By reference to a physical theory that makes essential use of statistical systematization, Hanson [23], has recently advanced an interesting argument against the view

that any explanation constitutes a potential prediction. According to Hanson, that view fits the character of the explanations and predictions made possible by the laws of Newtonian classical mechanics, which are deterministic in character; but it is entirely inappropriate for quantum theory, which is fundamentally nondeterministic. More specifically, Hanson holds that the laws of quantum theory do not permit the *prediction* of any individual quantum phenomenon P, such as the emission of a beta particle from a radioactive substance, but that "P can be completely *explained* ex post facto; one can understand fully just what kind of event occurred, in terms of the well-established laws of . . . quantum theory . . . These laws give the *meaning* of 'explaining single microevents'." (Hanson [23], p. 354; the italics are the quoted author's.) I quite agree that by reason of their statistical character, the laws of quantum theory permit the prediction of events such as the emission of beta particles by a radioactive substance only statistically and not with deductive-nomological certainty for an individual case. But for the same reason it is quite puzzling in what sense those laws could be held to permit a complete explanation ex post facto of the single event P. For if the explanans contains the statement that P has occurred, then the explanation is unilluminatingly circular; it might be said, at best, to provide a description of what in fact took place, but surely not an understanding of why it did; and to answer the question 'why?' is an essential task of explanation in the characteristic sense with which we have been, and will be, concerned throughout this essay. If, on the other hand, the explanans does not contain the statement that P has occurred, but only statements referring to antecedent facts plus the laws of quantum theory, then the information thus provided can at best show that an event of the kind illustrated by P—namely, emission of a beta particle—was highly probable under the circumstances; this might then be construed, in the sense outlined in the text, as constituting a probabilistic explanation for the occurrence of the particular event P. Thus, it still seems correct to say that an explanation in terms of statistical laws is also a potential prediction, and that both the explanation and the prediction are statistical-probabilistic in character, and provide no complete accounts of individual events in the manner in which deductive-nomological systematization permits a complete account of individual occurrences.

45. I learned from Professor Carnap, however, that in as yet unpublished work, his system of inductive logic has been extended to cover also statistical probability statements.
46. See Carnap [7] pp. 168–175.
47. This idea is closely related to one used by Reichenbach (see [41], Sec. 72) in an attempt to show that it is possible to assign probabilities to individual events within the framework of a strictly statistical conception of probability. Reichenbach proposed that the probability of a single event, such as the safe and successful completion of a particular scheduled flight of a given commercial plane, be construed as the statistical probability which the *kind* of event considered (safe and successful completion of a flight) possesses within the narrowest reference class to which the given case (the specified flight of the given plane) belongs, and for which reliable statistical information is available (this might be, for example, the class of scheduled flights undertaken so far by planes of the line to which the given plane belongs, and under weather conditions similar to those prevailing at the time of the flight in question). Our working rule, however, assigns a probability to (a statement describing) a single event only if the total evidence specifies the value of the pertinent statistical probability; whereas Reichenbach's interpretation refers to the case where the total evidence provides a statistical report on a finite sample from the

specified reference class (in our illustration, a report on the frequencies of safe completion in the finite class of similar flights undertaken so far); note that such a sample report is by no means equivalent to a statistical probability statement, though it may well suggest such a statement and may serve as supporting evidence for it. (On this point. cf. also Sec. 7 of the present essay.)
48. This requirement of implication serves to express the idea that F is the narrowest class of which b is known (namely, as a consequence of the total evidence) to be an element.
49. Statistical probability statements which are theorems of mathematical probability theory cannot properly he regarded as affording an explanation of empirical subject matter. The condition will prove significant in a context to be discussed a little later in this section.
50. While here our rule permits us to disregard, as it were, the occurrence of the explanandum in the total evidence, this is not so in all cases. Suppose, for example, that the total evidence e consists of the following sentences: e_1: 'p(G, F) = .4'; e_2: 'p(G,(G ∨ H) · F) = .9999'; e_3: 'Fb'; e_4: 'Gb.' Here again, e assigns b to the class F · G, for which p(G, F · G) = 1; as before, we may disregard this narrowest reference class. But e implies as well that b belongs to the class (G ∨ H) · F, which is the narrowest reference class relative to which e also specifies an empirical probability for G. Hence under our rule the statistical systematization with the premises e_2 and '(Gb ∨ Hb) · Fb' and with the conclusion 'Gb' satisfies the requirement of total evidence (whereas the argument with e_1 and e_3 as premises and 'Gb' as conclusion does *not*). Thus, we have here an argument that statistically explains b's being G by reference to b's being (G ∨ H) · F, though to establish this latter fact, we made use of the sentence 'Gb.' In this case, then, our rule does not allow us to disregard time occurrence of the explanandum in the total evidence.

The logical situation illustrated here seems to be analogous to that described by Scriven [50], in reference to causal explanation. Scriven points out that when we causally explain a certain event by reference to certain antecedent circumstances, it may happen that practically the only ground we have for assuming that those explanatory antecedents were in fact present is the information that the explanandum event did occur. Similarly, in our illustration, the information that h is G provides the ground for the assertion that b has the explanatory characteristic (G ∨ H) · F.
51. For details and further bibliographic references, see, for example, Neumann and Morgenstern [36]; Savage [45]; Luce and Raiffa [33]; Carnap [7], par. 51.
52. Cf. Carnap [7], p. 269; the formulation given there is "Among the possible actions choose that one for which the estimate of the resulting utility is a maximum." Carnap proposes this rule after a critical examination, by reference to instructive illustrations, of several other rules for rational decision that might seem plausible (ibid., Secs. 50, 51).
53. Content measures satisfying the specified conditions can readily be constructed for various kinds of formalized languages. For a specific measure function applicable to any first-order functional calculus with a finite number of predicates of any degrees, and a finite universe of discourse, see *SLE*, par. 9, or Carnap and Bar-Hillel [9], Sec. 6.
54. We have

$$u'(A, k) - u'(R, k) = (2c(h, k) - 1) \cdot (m(h \vee -k) + m(-h \vee -k)).$$

Since m is nonnegative, the second factor on the right could be 0 only if both of its terms were 0. But this would require h ∨ -k as well as -h ∨ -k to be logically true, in which case k would logically imply both h and –h; and this is precluded by the consistency requirement, (CR2), for K. Hence, u'(A, k) exceeds u'(R, k) or is excluded by it according as c(h, k) is greater or less than ½; and whichever of the two estimates is the greater will also be positive and thus greater than u'(S, k).

55. For an illuminating discussion of the concept of simplicity of a total system of statements, see Barker [2] (especially Chs. 5 and 9); also see the critical survey by Goodman [22]. One definition (applicable only to formalized languages of rather simple structure) of the systematizing power of a given theory with respect to a given class of data has been proposed in *SLE*, Secs. 8 and 9.

56. The minimax principle was proposed and theoretically developed by A. Wald; see especially his book [54]. A lucid and stimulating less technical account and appraisal of the minimax method, of special interest from a philosophical point of view, is given in Braithwaite [3], Ch. VII. Recent very clear presentations of the fundamentals of minimax theory, plus critical comments and further developments, may be found, for example, in Savage [45], and in Luce and Raiffa [33]. Carnap [7], par. 98, gives an instructive brief comparison of those methods of estimation which are based on inductive logic with those which, like the minimax method, have been developed within the framework of statistical probability theory, without reliance on a general inductive logic.

57. The standard set up by the minimax principle is by no means the only possible standard of rationality that can be proposed for decision rules in problem situations of the kind referred to here; and indeed, the minimax standard has been criticized in certain respects, and alternatives to it have been suggested by recent investigators. For details, see, for example, Savage [45], Ch. 13; Luce and Raiffa [33], Ch. 13. In an article which includes a lucid examination of the basic ideas of the minimax principle, R. C. Jeffrey points out that in applying this principle the experimenter acts on the assumption that this is the worst of all possible worlds for him; thus "the minimax criterion is at the pessimistic end of a continuum of criteria. At the other end of this continuum is the 'minimin' criterion, which advises each experimenter to minimize his minimum risk. Here each experimenter acts as if this were the best of all possible worlds for him." (Jeffrey [28], p. 244.)

58. See, for example, De Finetti [12], p. 219; Neyman [37], pp. 259–260. Savage [45], Ch. 9, Sec. 2, strongly advocates a "behavioralistic" as opposed to a "verbalistic" outlook on statistical decision problems; he argues that these problems are concerned with acts rather than with "assertions" (i.e., of scientific hypotheses). However, he grants the possibility of considering an "assertion" as a special kind of behavioral act and thus does not rule out explicitly the possibility of speaking of the acceptance— as presumably the same thing as "assertion"—of hypotheses in science. Savage here also makes some suggestive though all too brief remarks on the subtle practical consequences resulting from the assertion of a hypothesis in pure science (such as that the velocity of light is between 2.99×10^{10} and 3.01×10^{10} cm/sec); those consequences would presumably have to be taken into account, from his behavioralistic point of view, in appraising the utilities attached to the acceptance or rejection of purely scientific hypotheses. But Savage stresses that "many problems described according to the verbalistic outlook as calling for decisions between assertions really call only for decisions between much more down-to-earth acts, such as whether to issue single— or double—edged razors to an army." [45], p. 161. A distinction similar to that drawn by Savage is considered by Luce and Raiffa [33], who contrast a classical statistical inference" with "modern statistical decision theory" (Ch. 13, Sec. 10). In this con-

text, the authors briefly consider the question of how to appraise the losses from falsely rejecting or accepting a scientific research hypothesis. They suggest that no such evaluation appears possible, but conclude with a remark that seems to hint at what I have called the concept of epistemic utility: "if information is what is desired, then this requirement should be formalized and attempts should be made to introduce the appropriate information measures as a part of the loss structure. This hardly ends the controversy, however, for decision theorists are only too aware that such a program is easier suggested than executed!" ([33] p. 324.)

59. Jeffrey [28], p. 245.
60. Even decision rules of the kind discussed earlier, which are formulated by reference to certain probabilities and utilities, provide only a comparative, not an absolute (classificatory) concept of rationality, i.e., they permit, basically, a comparison of any two in the proposed set of alternative choices and determine which of them, if any, is more rational than the other; thus, they make it possible to single out a most rational choice from among a set of available alternatives. But they do not yield a classificatory criterion which would characterize any one of the alternatives, either as rational or as nonrational in the given context.
61. Thus the basis for the assignment, under rule (11.1), of the probability $1 - (½)^{10}$ to each h_j, is, strictly speaking, not e, but the sentence e_j: 'The statistical probability of obtaining O_j, as a result of 10 successive flippings of the coin is $1 - (½)^{10}$, and S is a particular set of n such successive flippings'; this e_j is a logical consequence of, but not equivalent to, e. Now, in general, if $c(h^*, e^*) = q$ and e^{**} is a logical consequence of e^*, then $c(h^*, e^{**})$ need not equal q at all; but in our case, it is extremely plausible to assume that whatever information e contains beyond e_j is inductively irrelevant to h_j; and on this assumption, we then have $c(h_j, e) = 1 - (½)^{10}$ for each $_j$. The requisite assumption may also be expressed more generally in the following rule, which is a plausible extension of (11.2):

Let e be a sentence which (i) specifies, for various outcomes G_k of a random experiment F, their statistical probabilities $p(G_k, F) = r_k$, (ii) states that the outcomes of different performances of F are statistically independent of each other, and (iii) asserts that a certain particular event b is a case of n successive performances of F; and let e assert nothing else. Next, let h be a statement ascribing to each of the particular performances of F that constitute b some particular one of the various outcomes G_k. Then c(h, e) equals the product of the statistical probabilities of those n outcomes. (For example, if b consists of three performances of F and h asserts that the first and third of these yield G_2, and the third G_4, then $c(h, e) = r_2 \cdot r_4 \cdot r_2$.)
62. The observation made in the preceding note applies here in an analogous manner.
63. Mises [34], p. 188; italics in original text.

REFERENCES

1. Alexander, H. Gavin. "General Statements as Rules of Inference?" in *Minnesota Studies in the Philosophy of Science*, Vol. II, H. Feigl, M. Scriven, and G. Maxwell, eds. Minneapolis: University of Minnesota Press, 1958. Pp. 309–329.
2. Barker, S. F. *Induction and Hypothesis*. Ithaca: Cornell University Press, 1957.
3. Braithwaite, R. B. *Scientific Explanation*. Cambridge: Cambridge University Press, 1953.
4. Campbell, Norman. *What Is Science?* New York: Dover Press, 1952.
5. Carnap, R. *The Logical Syntax of Language*. New York: Harcourt, Brace, and Co., 1937.

6. Carnap, R. "On Inductive Logic," *Philosophy of Science*, 12:72–97 (1945).
7. Carnap, R. *Logical Foundations of Probability*. Chicago: University of Chicago Press, 1950.
8. Carnap, R. *The Continuum of Inductive Methods*. Chicago: University of Chicago Press, 1952.
9. Carnap, R., and Y. Bar-Hillel. *An Outline of a Theory of Semantic Information*. Massachusetts Institute of Technology, Research Laboratory of Electronics. Technical Report No. 247, 1952.
10. Cohen, M. R., and E. Nagel. *An Introduction to Logic and Scientific Method*. New York: Harcourt, Brace, and Co., 1934.
11. Cramér, H. *Mathematical Methods of Statistics*. Princeton: Princeton University Press, 1946.
12. De Finetti, Bruno. "Recent Suggestions for the Reconciliations of Theories of Probability," in *Proceedings of the Second Berkeley Symposium on Mathematical Statistics and Probability*, J. Neyman, ed. Berkeley: University of California Press, 1951. Pp. 217–226.
13. Dray, W. *Laws and Explanation in History*. London: Oxford University Press, 1957.
14. Duhem, Pierre. *La Théorie physique, son objet et sa structure*. Paris: Chevalier et Rivière, 1906.
15. Feigl, H., and May Brodbeck, eds. *Readings in the Philosophy of Science*. New York: Appleton-Century-Crofts, 1953.
16. Feigl, H., and W. Sellars, eds. *Readings in Philosophical Analysis*. New York: Appleton-Century-Crofts, 1949.
17. Feigl, H., M. Scriven, and G. Maxwell, eds. *Minnesota Studies in the Philosophy of Science*, Vol. II. Minneapolis: University of Minnesota Press, 1958.
18. Galilei, Galileo. *Dialogues Concerning Two New Sciences*. Transl. by H. Crew and A. de Salvio. Evanston, Ill.: Northwestern University, 1946.
19. Gardiner, Patrick, ed. *Theories of History*. Glencoe, Ill.: Free Press, 1959.
20. Goodman, Nelson. "The Problem of Counterfactual Conditionals," *Journal of Philosophy*, 44:113–128 (1947). Reprinted, with minor changes, as the first chapter of Goodman [21].
21. Goodman, Nelson. *Fact, Fiction, and Forecast*. Cambridge, Mass.: Harvard University Press, 1955.
22. Goodman, Nelson. "Recent Developments in the Theory of Simplicity," *Philosophy and Phenomenological Research*, 19:429–446 (1959).
23. Hanson, N. R. "On the Symmetry between Explanation and Prediction," *Philosophical Review*, 68:349–358 (1959).
24. Hempel, C. G. "The Function of General Laws in History," *Journal of Philosophy*, 39:35–48 (1942). Reprinted in Feigl and Sellars [16], and in Jarrett and McMurrin [27].
25. Hempel, C. G. The Theoretician's Dilemma," in *Minnesota Studies in the Philosophy of Science*, Vol. II, H. Feigl, M. Scriven, and G. Maxwell, eds. Minneapolis: University of Minnesota Press, 1958. Pp. 37–98.
26. Hempel, C. G., and P. Oppenheim. "Studies in the Logic of Explanation," *Philosophy of Science*, 15:135–175 (1948). Secs. 1–7 of this article are reprinted in Feigl and Brodbeck [15].
27. Jarrett, J. L., and S. M. McMurrin, eds. *Contemporary Philosophy*. New York: Henry Holt, 1954.
28. Jeffrey, R. C. "Valuation and Acceptance of Scientific Hypotheses," *Philosophy of Science*, 23:237–246 (1956).

29. Kemeny, J. G., and P. Oppenheim. "Degree of Factual Support," *Philosophy of Science*, 19:307–324 (1952).
30. Körner, S., ed. *Observation and Interpretation.* Proceedings of the Ninth Symposium of the Colston Research Society. New York: Academic Press Inc., 1957. London: Butterworth, 1957.
31. Kolmogoroff, A. *Grundbegriffe der Wahrscheinlichkeitrechnung.* Berlin: Springer, 1933.
32. Lewis, C. I. *An Analysis of Knowledge and Valuation.* La Salle, Ill.: Open Court Publishing Co., 1946.
33. Luce, R. Duncan, and Howard Raiffa. *Games and Decisions.* New York: Wiley, 1957.
34. Mises, Richard von. *Positivism. A Study in Human Understanding.* Cambridge, Mass.: Harvard University Press, 1951.
35. Nagel, E. *Logic without Metaphysics.* Glencoe, Ill.: The Free Press, 1956.
36. Neumann, John von, and Oskar Morgenstern. *Theory of Games and Economic Behavior.* Princeton: Princeton University Press, 2d ed., 1947.
37. Neyman, J. *First Course in Probability and Statistics.* New York: Henry Holt, 1950.
38. Popper, K. R. *Logik der Forschung.* Vienna: Springer, 1935.
39. Popper, K. R. "The Propensity Interpretation of the Calculus of Probability, and the Quantum Theory," in *Observation and Interpretation*, S. Körner, ed. Proceedings of the Ninth Symposium of the Colston Research Society. New York: Academic Press Inc., 1957. London: Butterworth, 1957. Pp. 65–70.
40. Popper, K. R. *The Logic of Scientific Discovery.* London: Hutchinson, 1959.
41. Reichenbach, H. *The Theory of Probability.* Berkeley and Los Angeles: University of California Press, 1949.
42. Rescher, N. "On Prediction and Explanation," *British Journal for the Philosophy of Science*, 8:281–290 (1958).
43. Rescher, N. "A Theory of Evidence," *Philosophy of Science*, 25:83–94 (1958).
44. Ryle, G. *The Concept of Mind.* London: Hutchinson, 1949.
45. Savage, L. J. *The Foundations of Statistics.* New York: Wiley, 1954.
46. Scheffler, I. "Prospects of a Modest Empiricism," *Review of Metaphysics*, 10:383–400, 602–625 (1957).
47. Scheffler, I. "Explanation, Prediction, and Abstraction," *British Journal for the Philosophy of Science*, 7:293–309 (1957).
48. Schlick, M. "Die Kausalitaf in der gegenwärtigen Physik," *Die Naturwissenschaften*, 19:145–162 (1931).
49. Scriven, M. "Definitions, Explanations, and Theories," in *Minnesota Studies in the Philosophy of Science*, Vol. II, H. Feigl, M. Scriven, and G. Maxwell, eds. Minneapolis: University of Minnesota Press, 1958. Pp. 99–195.
50. Scriven, M. "Explanations, Predictions, and Laws," in *Minnesota Studies in the Philosophy of Science*, Vol. III, H. Feigl and G. Maxwell, eds. Minneapolis: University of Minnesota Press, 1962. Pp. 170–230.
51. Sellars, W. "Inference and Meaning," *Mind*, 62:313–338 (1953).
52. Sellars W. "Counterfactuals, Dispositions, and the Causal Modalities," in *Minnesota Studies in the Philosophy of Science*, Vol. II, H. Feigl, M. Scriven, and G. Maxwell, eds. Minneapolis: University of Minnesota Press, 1958. Pp. 225–308.
53. Toulmin, S. *The Philosophy of Science.* London: Hutchinson, 1953.
54. Wald, A. *Statistical Decision Functions.* New York: Wiley, 1950.
55. Williams, D. C. *The Ground of Induction.* Cambridge, Mass.: Harvard University Press, 1947.

7

Maximal Specificity and Lawlikeness in Probabilistic Explanation

1. The Rationale of the Requirement of Maximal Specificity (RMS)

In this article, I propose to reconsider certain basic issues in the logic of probabilistic-statistical explanation and to respond to some criticisms and constructive suggestions concerning my previous writings on the subject.

In my articles [4] and [5], I contrasted probabilistic, or inductive-statistical (I-S), explanation with deductive-nomological (D-N) explanation. A D-N explanation is an argument in which the explanandum sentence, which describes whatever is being explained, is deduced from a set of explanans sentences which include one or more laws or theoretical principles and usually, though not necessarily, also certain statements of particulars, such as initial or boundary conditions. An argument of this kind explains the explanandum phenomenon by showing that it was to be expected in view of the general laws adduced, given the particular circumstances specified. Such an account might, therefore, be said to exhibit the nomic expectability of the explanandum phenomenon.

A statistical explanation, too, relies on laws; but at least one of these is of a probabilistic-statistical character. The simplest laws of this kind have the form: 'the statistical probability for an F to be a G is r', or '$p(G, F) = r$' for short; they are the probabilistic counterparts of strictly general laws of the form 'All F are G'. But while a law of the latter kind, combined with the particular statement 'i is F' deductively implies 'i is G' and thus affords a corresponding D-N explanation, the statistical law '$p(G, F) = r$' combined with 'i is F' can be said to explain i's being G only inductively, i.e., in the sense that it lends more or less strong inductive support to the explanandum sentence 'i is G'. For reasons indicated in [5], pp. 377–378, I took this inductive support numerically to equal to r, and I schematized the resulting I-S explanation thus:

$$(1) \quad \frac{\begin{array}{c} p(G, F) = r \\ Fi \end{array}}{Gi} \quad [r]$$

Explanatory arguments having this structure I called I-S explanations of basic form; and, as in my previous articles, I will limit my discussion here to the logic of this simplest type of probabilistic explanation. The number indicated in brackets is "the probability associated with the explanation"; it is not a statistical probability, but an inductive one in Carnap's sense, namely, the probability of the explanandum relative to the explanans. The argument *explains* i's being G by showing that this is to be expected, with probability r, in view of the general statistical law and the statement of particular fact included in the explanans. The argument will be considered as explanatory only if r is sufficiently close to 1; but no specific common lower bound for r can reasonably be imposed on all probabilistic explanations.

In offering an explanation of either kind for a given phenomenon, we claim of course not only that the argument in question is "valid"—that its "conclusion" bears the requisite logical relation to the "premisses'—but we also affirm the premisses: just as in making an assertion of the form 'B because A' we implicitly claim that A is the case.

Explanations based on probabilistic-statistical laws may exhibit what I have called statistical ambiguity. Given an argument of type (1) in which both premisses are *true* and the associated probability r is close to 1, there may exist a "conflicting" I-S argument

(2)
$$\frac{p(-G, H) = s}{-Gi} \quad [s]$$
$$Hi$$

which also has *true* premisses and an associated probability close to 1, and which therefore explains i's not being G just as the first account explains i's being G.

For example, let j be some particular individual, Jones; let 'Ix' stand for 'x has been infected with malaria plasmodium'; 'Mx' for 'x contracts malaria'; and 'Sx' for 'x is a heterozygote in regard to the sickle hemoglobin gene', which means that x has acquired that gene from one, but not both, of his parents. This characteristic S has been found to afford strong protection against malaria.[1] Let us assume, to be specific, that $p(-M, S) = .95$ and $p(M, I) = .9$. Suppose now that Jones has been infected with malaria plasmodium, but has the protective characteristic S. Then the following two arguments have true premisses and thus form conflicting inductive-statistical accounts:

(3a)
$$\frac{p(M, I) = .9}{Mj} \quad [.9]$$
$$Ij$$

(3b)
$$\frac{p(-M, S) = .95}{-Mj} \quad [.95]$$
$$Sj$$

This possibility of "explaining" by true statements both the occurrence of a phenomenon and its nonoccurrence throws serious doubt on any claim to explanatory power that might be made for such inductive-statistical arguments. (D-N explanations are not subject to any such ambiguity: the existence of a true D-N explanans for the occurrence of a given phenomenon logically precludes the existence of a true D-N explanans for its nonoccurrence.)

As a way of bypassing (though not resolving) this ambiguity, I suggested that I-S explanation, in contrast to D-N explanation, be construed as an epistemological concept explicitly relativized with respect to a given "knowledge situation." The latter would be formally represented by a class K containing all those sentences—whether actually true or false—which are accepted as presumably true by the person or persons in the given knowledge situation. By way of idealization, the class K will be assumed to be logically consistent and closed under the relation of logical consequence, and to contain the theories of the statistical and the logico-inductive concepts of probability.

We are thus led to consider the concept of "I-S explanation relative to K (or, in K)." If an argument of the form (1) is to qualify as such an explanation, its premisses will have to be in K.[2] But this condition alone cannot prevent explanatory ambiguity from recurring in a new variant: a class K—for example, the class of sentences accepted in contemporary science—may evidently contain the explanans sentences for two conflicting explanatory arguments, such as (3a) and (3b). This variant might be referred to as *epistemic ambiguity*, in contrast to the kind described first, which could be called *ontic ambiguity* of I-S explanation. The latter results from the existence of nomic and particular facts, or of corresponding true sentences—no matter whether known or believed—which give rise to I-S arguments with true premisses and logically incompatible conclusions; epistemic ambiguity results from the fact that the class K of sentences believed or accepted in a given knowledge situation—no matter whether they are true or not—may similarly contain premiss-sets for incompatible conclusions. (D-N explanation, let it be noted in passing, cannot be epistemically ambiguous, any more than it can be ontically ambiguous.)

A simple and plausible way out suggests itself here. An explanation is normally asked for only when the explanandum phenomenon is taken to have occurred, i.e., when the explanandum sentence belongs to K. Suppose we require accordingly of any I-S explanation in K that its explanandum sentence belong to K: then the consistency of K precludes conflicting explanations like (3a) and (3b). I find the requirement a very reasonable one for a concept of explanation that refers explicitly to what is taken to be the case; and I will therefore adopt it. But though it never grants explanatory status to two arguments with logically incompatible premisses, it does not eliminate what seems to me the objectionable aspect of explanatory ambiguity. For in a case where K contains the premisses and the conclusion of (3a), as well as the premisses of (3b), we would still be able to say: "i is G, and we can explain that by (3a); but if i had turned out not to be G, our total knowledge would just as readily have afforded an explanation for that namely, (3b)." If, as I think, an explanation exhibits the strong nomic expectability of a phenomenon then surely that claim cannot hold good.

Should we, then, stipulate instead that an argument like (3a) qualifies as an I-S explanation relative to K only if K does not contain the premises of any "conflicting" I-S argument, such as (3b)? This requirement would indeed bar explanatory ambiguities, but it is too restrictive. For suppose that K contains the premises of (3a) and (3b) and also those of the argument

(3c)
$$\frac{p(-M, S \cdot I) = .95}{Sj \cdot Ij} \quad [.95]$$
$$-Mj$$

Assuming that K contains no further statements—i.e., none that are not logically implied by those just specified—we would presumably say that (3c) afforded an I-S explanation, relative to K, of why Jones did not catch malaria. And we would grant (3c) this status even though our knowledge K also contains the premises for the conflicting argument (3a). For while K does inform us that Jones belongs to the class I, and that within I, the feature M has the high statistical probability .9, K contains the further information that Jones also belongs to another class S, and hence to the class $S \cdot I$; and that, among the members of that subclass of S, M has the very low probability .05. Finally—and this is decisive—K does not assign Jones to a still narrower reference class; hence, the I-S argument (3c) is based on the most specific information we have about Jones in the given knowledge situation; and that would seem to confer on (3c), but not on (3a), the status of an explanation relative to K. What of (3b)? Since, according to the first premise of (3c), the statistical probability of not contracting malaria is the same in $S \cdot I$ as in S, the factor I probabilistically irrelevant to $-M$ relative to S. For this reason, (3b), too, may count as an explanation of Jones's not contracting malaria.[3]

Considerations of this kind led me to propose, in [5], a "requirement of maximal specificity" intended to preclude explanatory ambiguity. In substance it provided that an argument

(4)
$$\frac{p(G, F) = r}{Fi} \quad [r]$$
$$Gi$$

where r is close to 1 and both premises are contained in K, constitutes a probabilistic explanation relative to K only if it meets the following condition:

(RMS)   For any class F_1 for which K contains statements to the effect that F_1 is a subclass of F and that F_1i, K also contains a probabilistic-statistical law to the effect that $p(G, F_1) = r_1$, where $r_1 = r$ unless the law is a theorem of probability theory.[4]

The 'unless'-clause is meant to allow K to contain, without prejudice to the explanatory status of (4), pairs of sentences such as '$Fi \cdot Gi$' and '$p(G, F \cdot G) = 1$';

the latter, being a theorem of the probability calculus, is thus not reckoned as an explanatory empirical law—just as '$(x)[(Fx \cdot Gx) \supset Gx]$' does not qualify as an explanatory law on which a D-N explanation can be based. Similar remarks apply to probability statements such as '$p(G, F \cdot -G) = 0$'.

In our example concerning Jones's malaria, the condition RMS is met by (3b) and (3c), but not by (3a), which is just the result we want.

2. A Clarification: Maximal Specificity versus Total Evidence

A requirement to essentially the same effect as RMS was proposed already in my essay [4] (pp. 146–148), but on grounds which, I now think, misconstrued the relationship of RMS to the requirement of total evidence. I argued there that explanatory ambiguities like those arising for a class K containing the premisses of both (3a) and (3b) are properly and readily avoided by heeding the requirement of total evidence, i.e., by assigning inductive probabilities to the conflicting explanandum sentences on the basis of the total evidence available in the given knowledge situation, that is, on the basis of the entire class K. This, after all, is a principle which, as stressed by Carnap and others, must be observed in all rational applications of probabilistic inference.[5] Now, whatever the class K may be, it cannot confer high inductive probabilities on both of two contradictory explanandum sentences since their probabilities must add up to unity. Thus, I concluded, adherence to the requirement of total evidence is the way to avoid explanatory ambiguity.

But, I noted further, even if we assume that an adequate general definition of logical probability can be given—perhaps in the general manner proposed by Carnap—it would be a hopelessly complex task actually to compute the probabilities of two conflicting explanandum sentences with respect to the vast set K representing our total putative knowledge. It would be desirable, therefore, to have a practicable method of assigning to those sentences at least approximations of their probabilities relative to K. And I suggested, in effect, that if K contains the premisses of an argument of type (4) and satisfies the maximal specificity condition with respect to it, then the logical probability r of 'Gi' relative to the premisses of (4) may be considered as an approximation of the probability of 'Gi' with respect to the total evidence K ([4], pp. 146–147). In support of this suggestion, I offered some plausibility considerations for special cases; but I also noted, by reference to a specific example, that under certain conditions, the value of r may be quite different from the probability of 'Gi' on K ([4], pp. 148–149). I therefore characterized my rule as only a "rough substitute for the requirement of total evidence" ([4], p. 146) and concluded that "the requirement of total evidence remains indispensable" for the assignment of probabilities to the conclusions of I-S explanations ([4], p. 149).

But this reasoning confounds two quite different questions. One of these concerns the strength of the evidence for the *assertion that* the explanandum event did occur; the other, the probability associated with an I-S *explanation of why* the explanandum event occurred. In reference to our simple schema (4), the first question might be put thus:

(5a) What degree of belief, or what probability, is it rational to assign to the statement 'Gi' in a given knowledge situation?

Here, the requirement of total evidence applies. It directs that the probability should be determined by reference to the total evidence available, i.e., by reference to the entire class K.

The second question does not concern the grounds on which, and the degree to which, it is rational to believe that i is G, but the grounds on which, and the strength with which, i's being G can be explained, or shown to be nomically expectable, in a given knowledge situation:

(5b) Does K contain sentences that can serve as the explanans of an I-S explanation of i's being G; and if so, what is the associated probability which the explanans confers on the explanandum sentence 'Gi'?

The inductive probabilities referred to in the two questions are largely independent of each other. For example, as noted earlier, when an explanation of i's being G is sought, the sentence 'Gi' is normally included in K. In that case, the probability of 'Gi' on K is 1; yet if K contains sentences like the premises of (4), which can serve to explain i's being G, these sentences will confer upon 'Gi' a probability that is less than 1. And that is quite reasonable; for *the point of an explanation is not to provide evidence for the occurrence of the explanandum phenomenon, but to exhibit it as nomically expectable.* And the probability attached to an I-S explanation is the probability of the conclusion relative to the explanatory premises, not relative to the total class K. Thus, the requirement of total evidence simply does not apply to the determination of the probability associated with an I-S explanation, and the requirement of maximal specificity is *not* "a rough substitute for the requirement of total evidence."

As noted in my earlier articles on the subject, my conception of the maximal specificity condition was influenced by Reichenbach's rule of the narrowest reference class. I will therefore briefly indicate how I see the relation between that rule and the requirements of total evidence and of maximal specificity. Reichenbach proposed his rule as a method of assigning a probability, or a "weight," to what he called "a single case," such as recovery of an individual patient from his illness, or getting heads as the result of the next flipping of a given coin. He held that "there exists only one legitimate concept of probability," namely, the statistical one, "which refers to classes," not to individual events; and that, therefore "the pseudoconcept of a probability of a single case must be replaced by a substitute constructed in terms of class probabilities" ([8], p. 375). This substitute notion of the weight to be assigned to the occurrence of a certain kind of event, say G, in a particular single case, say i, Reichenbach construed as the estimated statistical probability of G in the "*narrowest*" reference class containing *i* "*for which reliable statistics can be compiled,*"[6] i.e., in our parlance: for which K includes reliable statistical information.

Reichenbach's rule then is intended to answer questions of type (5a). (Indeed, as far as I am aware, he never explicitly examined the logic of explanations based on probabilistic laws.) The rule may, in fact, be viewed as Reichenbach's version of the requirement of total evidence: it requires consideration of the total evidence and specifies what parts of it count as relevant to the weight of a single case, and how they are to be used in computing that weight.

RMS, on the other hand, pertains to questions of type (5b). Its function is not to assign a probability to 'Gi', but to specify conditions under which two sentences in K—a law, '$p(G, F) = r$' and a singular sentence 'Fi'—can serve to explain, relative to K, i's being G. The necessary condition laid down by RMS is that for every subclass of F to which K assigns i—*and hence also for the narrowest of these, their intersection*, say, S—the class K must contain a statement to the effect that within that subclass, the probability of G equals r (except when the probability in question is determined by the calculus of probability alone). The link to Reichenbach's principle lies in the fact that RMS has an implication concerning "the narrowest reference class." But this link, as it now appears, is substantively rather tenuous. For the narrowest reference class in the sense of Reichenbach's principle is by no means always the intersection S just characterized; and while RMS normally requires, among other things, that K contain a probability statement concerning S, Reichenbach's rule imposes no such condition on K concerning S or concerning the narrowest reference class in his own sense.

Just like Reichenbach's principle, RMS requires reference to the total evidence K, though for a different purpose, namely, in order to ensure that K does not contain premises suitable for an explanation conflicting with (4). The requirement of maximal specificity was intended to guarantee fulfillment of this latter condition by barring explanatory ambiguity. Recently, however, several writers have argued that in the form I have given it, the requirement falls short of its objective. I will now examine their reasons.

3. The Lawlikeness of Explanatory Probability Statements and Its Significance for RMS

W. C. Humphrey's [6] has argued that RMS is so restrictive as to deprive virtually all the usual probabilistic-statistical laws of an explanatory role. In place of his illustration, I will give a strictly analogous one by reference to our earlier examples.

Let the content of K amount to just what follows from the premises of (3a) and (3c) and the further information that Jones is a member of a certain subclass S' of $S \cdot I$ which has exactly four members. Then, Humphreys argues, by reason of this latter bit of information, of an almost always available and quite trivial sort, the argument (3c) violates RMS and is thus barred from explanatory status in K. He reasons as follows: The value of $p(-M, S')$, even though not explicitly specified by K, obviously must be one of the numbers 0, ¼, ½, ¾, or 1 because according to K, the class S' has just four elements. Now, according to RMS, (3c) qualifies as an I-S explanation relative to K only if either (a) K implies that the probability of $-M$ in S' is the same as in S · I, namely, .95, or (b) the probability $p(-M, S')$ is determined by the mathematical theory of probability alone. Humphreys does not explicitly consider the latter possibility, although his five "obvious" probability values are presumably determined by purely mathematical-combinatorial considerations. But he rightly points out that condition (a) surely is not met, and he concludes that therefore (3c) is ruled out as an I-S explanation. Similarly, he reasons, virtually any argument of the form (4) that would normally count as an I-S explanation can be disqualified on the basis of RMS by showing, and noting in K, that the individual case i belongs to some small finite subclass of the reference class F mentioned in (4).

But Humphreys' counterexamples can be barred by reference to the proviso, mentioned above, that the statistical probability statements on which probabilistic explanations can be based must be lawlike, must have the character of potential laws.[7] I will argue that the predicates occurring in lawlike probability statements—let us call them nomic predicates—must meet certain conditions, which are violated in the examples constructed by Humphreys.

The first condition, which in one form or another has been suggested by several writers, was adumbrated in my observation that laws of the universal conditional form

(6a) $\qquad\qquad (x)(Fx \supset Gx)$

and statistical laws of the form

(6b) $\qquad\qquad p(G, F) = r$

"share an important feature, which is symptomatic of their nomological character: both make general claims concerning a class of cases that might be said to be potentially infinite" ([5], p. 377). This is vague, however. For a class is either finite or infinite, leaving no room for "potentially infinite" classes. The quoted remark should be understood as ruling out any reference class that is finite on purely logical grounds, i.e., as a consequence of the way in which it is characterized by the "reference predicate expression" that occupies the place of 'F'. Accordingly, the expression 'Fx' in (6a) may not be logically equivalent to '$x = a \lor x = b \lor x = c$'; for then (6a) would be logically equivalent to '$Ga \cdot Gb \cdot Gc$', and such a finite conjunction of singular sentences lacks the force—explanatory and otherwise—of a law.

The two predicate expressions in a lawlike probability statement of form (6b) are subject to the same requirement:

(N1) In a lawlike sentence of form (6b), neither the reference predicate expression, which takes the place of 'F', nor the outcome predicate expression, which takes the place of 'G', must have an extension that is finite on purely logical grounds.[8]

Indeed, the two predicates either stand for properties of things that can have "indefinitely many" instances (such as having blue-eyed parents and being blue-eyed); or they stand for *kinds* of events that are conceived as "indefinitely repeatable," such as flipping a penny, and the penny landing heads up; or a ten-second irradiation of a phosphorescent screen with alpha-particles from a given source, and the occurrence of from six to eight scintillations on the screen; or infection of a person with malaria plasmodium, and the person's catching malaria.[9] The probability statement then asserts, briefly, that in increasingly long series of cases instantiating the reference predicate, the proportion of those having the specified outcome tends to come, and remain, as close as we please to r. This claim presupposes that the predicate expressions stand for kinds of objects or events that can be conceived of as having infinitely many instances—at least "in principle," i.e., without logical inconsistency. And this is what N1 requires.

Now, predicates like that characterizing Humphreys' four-membered reference class S' clearly violate the requirement N1 and are therefore barred by it.

The probability statements adduced by Humphreys are disqualified for yet another reason. Explanatory laws, whether of universal or of probabilistic-statistical form, must be empirical in character; and the statement that the probability of $-M$ in S' must have one of the values 0, ¼, ½, ¾, 1 is not: it simply expresses the logical truth that these are all logically possible proportions of those among four individuals who may have the property $-M$. This shows, moreover, that the statement is not a statistical probability statement at all; it does not concern the long-run relative frequency of the outcome $-M$ as the result of some repeatable kind of event.

I now turn to some criticisms and constructive suggestions made by G. J. Massey[10] concerning the adequacy of my formulation of RMS and concerning my earlier construal of the requirement as a "rough substitute for the requirement of total evidence."

In regard to that construal, Massey argues that I have furnished "no unobjectionable total evidence requirement which [the] rough criterion might subserve as a rule of thumb." Presumably his point comes to this: the total evidence requirement as formulated by Carnap is not even roughly subserved by RMS, for reasons stated in section 2 above; and that indeed I have offered no other version of the total evidence requirement for whose satisfaction RMS might play the role of a rule of thumb. I entirely agree.

Massey then suggests that I should have treated RMS "as a *bona fide* substitute for the defunct total evidence requirement." Here I resist the implication that the total evidence requirement is defunct. In its proper place, namely, in determining the credence rationally assignable to a statement, I think it does ideally apply: the rational credibility in question depends on *all* the evidence available in the given situation. Practically, of course, we rely, explicitly or tacitly, on various judgments of irrelevance to trim the evidence down to manageable size.[11] On the (other hand, I agree with Massey in viewing RMS as a *bona fide* rule in its own right; but for the reasons given in section 2, I consider that rule as pertaining strictly to the probabilistic explanation of empirical phenomena, not to appraisals of the credibility of their occurrence: and quite possibly this is what Massey has in mind, too.

Massey then questions the adequacy of my formulation of RMS on two grounds. His first objection is based on counterexamples very similar to those constructed independently by Humphreys. But he goes on to suggest, correctly, that I would presumably reject the counterexamples on the ground that they use nonlawlike probability statements.

Massey's second objection is to the effect that if an argument like (4) is to qualify as an explanation in K, RMS requires K to contain "a wholly unreasonable number of statistical laws": for normally, many predicates besides 'F' will be known—i.e., will be said in K—to apply to i; the conjunction of any one of these with 'F' determines a subclass of F; and K is required to contain laws stating the probability of G within each of these subclasses (with the exception stated in RMS). This complaint will be met in section 4 below by a modification of RMS which is less stringent in this respect.

Another objection here to be considered has been raised by R. Wójcicki,[12] who illustrates his point, in effect, by one of my earlier examples for the ambiguity of statistical explanation ([4], p. 126). He argues that while RMS may eliminate such ambiguity, it qualifies as explanatory certain arguments that no scientist would regard as such. He reasons as follows:[13] Let 'Px' stand for 'x is a person with a streptococcal infection who has been treated with penicillin'; 'Rx' for 'x recovers quickly'; 'Qx' for 'x has the property P, but he also has a certain physiological characteristic whose presence makes quick recovery very unlikely'; finally, let 'j' again be short for 'Jones'. Now suppose that K contains just the following sentences and their consequences:

$$p(R, P) = .95$$
$$p(-R, Q) = .96$$
$$(x)(Qx \supset Px)$$
$$Pj$$

as well as the definition

$$Q^+x =_{df} Qx \lor (x = j)$$

and consequently also the sentence

$$(x)(Q^+x \supset Px)$$

Since the class determined by 'Q^+' has at most one element more than that determined by 'Q', nonrecovery will have the same probability in Q^+ as in Q, so that K also contains the sentence

$$p(-R, Q^+) = .96$$

Hence, K contains the premises of the following two rival arguments:

(7a)
$$\begin{array}{c} p(R, P) = .95 \\ Pj \\ \hline Rj \end{array} \quad [.95]$$

(7b) $$\frac{\begin{array}{c} p(-R, Q^+) = .96 \\ Q^+j \end{array}}{-Rj} \quad [.96]$$

Of these, Wójcicki rightly points out, RMS qualifies the second rather than the first as explanatory since K contains the information that Q^+ is a subclass of P. Undeniably, this is a counterintuitive and unreasonable verdict. The objection can be met, however, by arguing that 'Q^+' does not qualify as a nomic predicate. For 'Q^+' applies to a certain individual, namely j, on purely logical grounds: 'Q^+j' is short for '$Qj \vee (j = j)$', which is a logical truth. And this violates a second condition which, I would suggest, must be met by nomic predicates in probabilistic laws: no such predicate must *demonstrably* apply to any particular individual, or:

(N2) No full sentence formed from a predicate in a lawlike probability statement and an (undefined) individual name is a logical truth.

In its application to probabilistic-statistical laws, N2 can be supported by this consideration: The predicate 'P' in (7a) stands for an indefinitely repeatable kind of event: streptococcal infection of a person, followed by penicillin treatment of that person. Similarly for 'Q': suppose, for example, that the antirecovery factor is a grave allergy to penicillin; then the repeatable kind of event is a streptococcal infection, followed by penicillin treatment, of a person suffering from that kind of allergy. But there is no analogous way of construing 'Q^+'; for what kind of event would '$x = j$' stand for? If this consideration is correct, then it just makes no sense to assign a statistical probability to nonrecovery with respect to the reference class characterized by 'Q^+'.

This consideration supports N2 specifically for the reference predicates of probabilistic laws. But I think it may properly be extended to all nomic predicates, in lawlike sentences of probabilistic and of universal character; for it reflects the conception that it is not individual events or objects as such, but individuals as bearers of general characteristics that can enter into nomic connections.

While in earlier articles on I-S explanation, I had stressed that the relevant probabilistic-statistical statements must be lawlike, I had not explicitly stated conditions like N1 and N2 for the predicates permissible in such statements. I am indebted to the critics I have mentioned for obliging me to consider this point more closely.

In introducing N1, we noted in passing that the predicate expression in the place of 'F' in a universal law of form (6a) must not be logically equivalent to an expression of the form '$x = a_1 \vee x = a_2 \vee \ldots \vee x = a_n$', where '$a_1$', '$a_2$', etc. are individual names. Nevertheless, the extension of the reference predicate in question may well be finite in fact and may even be the null class. For example, Newton's law of gravitation logically implies various more specific laws, most of them never explicitly formulated. One of these concerns the case where the two bodies in question are solid copper spheres having masses of 10^7 and

10^8 grams, respectively; and it expresses the mutual attractive force they exert upon each other as a function of the distance between the centers of the spheres. Now the reference class thus characterized, the class of all pairs of such spheres, may well, as a matter of fact, be finite—quite possibly null. But this does not follow logically from its characterization, and the consequence in question has the character of a law: it can serve, for example, to support subjunctive conditional statements such as predictions as to what forces *would* come into play *if* two such spheres *were* to be produced. Thus, in a lawlike sentence of the form (6a), the reference predicate may in fact, but not on logical grounds alone, be coextensive with an expression of the form '$x = a_1 \lor x = a_2 \lor \ldots \lor x = a_n$'. Replacement of the reference predicate by that expression would then turn the lawlike sentence into another that has the same truth value but is no longer lawlike.[14] Similarly, while the reference predicate in a lawlike sentence of form (6a) must satisfy N2, it may in fact be coextensive with another predicate expression which violates that requirement; replacement of the former by the latter again yields a nonlawlike sentence of the same truth value.

Epitomizing these observations we might say that a lawlike sentence of universal, nonprobabilistic character is not about classes or about the extensions of the predicate expressions it contains, but about those classes or extensions *under certain descriptions*.

An analogous remark applies to lawlike sentences of probabilistic-statistical form. Take, for example, the sentence

(8) $\qquad p(H, C) = 1 - (\tfrac{1}{2})^{100,000}$

where 'Cx' and 'Hx' are short for 'x is a succession of 100,000 flippings of a regular coin', and 'x is a succession of coin flippings at least one of which yields heads'. The reference class, i.e., the extension of 'Cx', then contains all and only those events each of which consists of 100,000 successive flippings of a regular coin; and the total number of such events ever to occur may well be small or even zero. Suppose there are exactly three such events, e_1, e_2, and e_3. Then 'Cx' is coextensive with '$x = e_1 \lor x = e_2 \lor x = e_3$', which I will abbreviate by 'C^*x'. Replacement of 'C' by 'C^*' in the lawlike probability statement (8) yields an expression that violates N1 as well as N2 and that is, therefore, not lawlike. Indeed, it might be added that because of these violations, the expression is not a significant probability statement at all since its reference predicate does not characterize an indefinitely repeatable kind of event. Accordingly, we note:

> Statistical probability statements, and in particular those which are lawlike, concern the long-run frequency of a specified characteristic within a *reference class under some particular description*, and the predicate expressions that can serve as such descriptions must satisfy conditions N1 and N2.

It follows that, properly, the requirement of maximal specificity should not be formulated as a condition on reference classes and their subclasses, as is RMS above, but as a condition on certain predicate expressions. A version which meets this condition will be developed in the next section.

It should be noted that conditions N1 and N2, though presumably necessary, are not sufficient to ensure lawlikeness. It is possible, for example, to construct probability statements which, though satisfying the two conditions, appear to be analogues to Goodman's examples of generalizations that receive no confirmation from their instances (cf. [3], pp. 72–83). Statements of the form '$p(G, F) = r$' do not, to be sure, have individual instances in the sense in which Goodman speaks of instances of universal conditional sentences; but they can receive support or disconfirmation from pertinent statistical evidence, i.e., from findings concerning the frequency of G in finite sets of events of the kind characterized by 'F'. This consideration suggests that statistical probability statements analogous to Goodman's nonlawlike universal conditionals might now be constructed in the manner of the following example: Let us say that an event is a fleagaroo jump, or a J for short, if either it is examined before January 1, 2000 and is a jump made by a flea or it is not so examined and is a jump made by a kangaroo. And let us say that a jump is short, or S, if the distance it spans is less than one foot. Consider now the sentence:

(9) $$p(S, J) = .9$$

It meets the conditions N1 and N2. Suppose now that we gather relevant statistical data, which would concern exclusively the distances covered by flea jumps; and suppose further that in the vast, and steadily growing, set of observed jumps, the proportion of short ones is, and remains, very close to .9. This would not tend to support the general claim made by (9) because of its implications for J's examined after the twentieth century.

The expression (9), then, is a rather close analogue, for statistical probability statements, to Goodman's nonlawlike universal conditionals. Doubtless, we would not qualify (9) as lawlike. But we might even question whether it constitutes a significant probability statement, albeit a presumably false one. A profitable assessment of this issue would require a much more thorough analysis than has here been suggested of the "meaning" of statistical probability statements, and I will not pursue this question further in the present context.

4. A Revision of RMS

The requirement of maximal specificity was meant to preclude the possibility of conflicting I-S explanations, and in [5] (p. 401) I offered an argument purporting to prove that, stated in a form essentially tantamount to RMS above, it does have the desired effect. But since then, my colleague, Dr. Richard Grandy, has pointed out to me[15] that my argument is fallacious, and has illustrated this by the following counterexample:

Let K contain the premises of the following two arguments (and, of course, their logical consequences)

Maximal Specificity and Lawlikeness in Probabilistic Explanation   159

(10a)
$$\frac{p(G, F \vee G) = .9}{Gb} \quad [.9]$$
$$Fb \vee Gb$$

(10b)
$$\frac{p(-G, -F \vee G) = .9}{-Gb} \quad [.9]$$
$$Fb \vee Gb$$

The first of these arguments satisfies RMS. For the only subclasses of $F \vee G$ which K tells us contain b are $(F \vee G)(-F \vee G)$, which is G; and $(F \vee G)G$, which again is G. But $p(G, G) = 1$ by virtue of the theory of probability alone. Thus, (10a) fulfills RMS. But an analogous argument shows that RMS is satisfied by (10b) as well. Thus, both of the two conflicting arguments qualify, under RMS, as I-S explanations relative to K.

Grandy's example raises a difficulty also for the alternative to *RMS* proposed by Humphreys [6] under the name "The rule of complete evidence." Humphreys' own formulation is somewhat vague because, among other things, it fails to observe the distinction between what is the case and what is known or believed, i.e., what is asserted by K to be the case. But the rule he intended seems to come to this: Let K contain statements to the effect that the individual case n belongs to the classes $C_1, C_2, \ldots, C_m$, and that the probabilities of W relative to these are $p(W, C_i) = r_i$ ($i = 1, 2, \ldots, m$); and let the C_i be *all* the classes for which K provides this twofold information. Then a probabilistic explanation of n being W (or of n being $- W$) is possible in K if and only if K contains a law specifying the probability of W with respect to the intersection, C, of all the C_i; and it is on this law, '$p(W, C) = r$', that the explanation must be based.

This rule has one clear advantage over RMS: it makes much less stringent demands concerning the probabilistic-statistical laws that K is required to contain, and it thus goes a long way toward meeting the objection, mentioned above, that Massey has raised against RMS on this score. But when applied to Grandy's example, Humphreys' rule implies that with respect to the given class K, the following is a proper explanatory argument:

$$\frac{p(G, G) = 1}{Gb} \quad [1]$$
$$Gb$$

And this surely is unacceptable.

I will now suggest a modified version, RMS*, of the requirement of maximal specificity, which avoids the pitfalls we have considered. First, some auxiliary concepts and observations.

Let 'F_1' and 'F_2' be short for two one-place predicate expressions. Then the first will be said to *entail* the second if '$(x)(F_1x \supset F_2x)$' is logically true; and it will be called *stronger* than the second if it entails, but is not entailed by, the second. Two predicate expressions that entail each other will be called *logically equivalent*. If 'F' entails 'G' or '$- G$', then, by the theory of probability alone, $p(G, F)$ equals 1 or 0, respectively; and, as noted at the end of section 1, the 'unless'-clause in RMS is meant to refer to just those probability statements in which the reference predicate thus entails the "outcome predicate" or its negate.

'F_1' will be called an *i-predicate in K* if K contains 'F_1i'; and 'F_1' will be said to be *statistically relevant* to 'Gi' in K if (1) 'F_1' is an i-predicate that entails neither 'G' nor '$- G$' and (2) K contains a lawlike sentence '$p(G, F_1) = r$' specifying the probability of 'G' in the reference class characterized by 'F_1'.

Now, one essential feature of RMS* will be this: RMS imposes conditions on all classes to which K assigns i; or, more accurately, on all i-predicates (save those entailing 'G or '$- G$') by which those classes are characterized in K. In RMS*, only those i-predicates which are statistically relevant to 'Gi' will be subject to similar conditions. In this respect, RMS* is analogous to Humphreys' rule; and like the latter, it is much less demanding than RMS in regard to the probabilistic laws that K is required to contain.

Another modification of RMS is intended to avoid the difficulty noted by Grandy. Let us call a predicate expression, say 'M', *a maximally specific predicate related to* 'Gi' in K if (1) 'M' is logically equivalent to a conjunction of predicates that are statistically relevant to 'Gi' in K; (2) 'M' entails neither 'G' nor '$- G$'; (3) no predicate expression stronger than 'M' satisfies (1) and (2); i.e., if 'M' is conjoined with a predicate that is statistically relevant to 'Gi' in K, the resulting expression entails 'G' or '$-G$', or else it is just equivalent to 'M'. Every such most specific predicate is evidently an i-predicate in K.

The proposed modification of the requirement of maximal specificity can now be stated as follows: An argument

(11)
$$\frac{p(G, F) = r \\ Fi}{Gi} \quad [r]$$

where r is close to 1 and all constituent statements are contained in K, qualifies as an I-S explanation relative to K only if the following condition is met:

> (RMS*) For any predicate, say 'M', which either (a) is a maximally specific predicate related to 'Gi' in K or (b) is stronger than 'F' and statistically relevant to 'Gi' in K, the class K contains a corresponding probability statement, '$p(G, M) = r$', where, as in (11), $r = p(G, F)$.

To illustrate: let 'F', 'G', 'H', 'J', 'N' be logically independent predicate constants, and let K contain just the following sentences and their consequences: 'Fi', 'Gi', 'Hi', 'Ji', 'Ni'; '$p(G, F) = .95$', '$p(G, F \cdot H \cdot J) = .95$'. Then the i-predicates in K are the five predicate constants just mentioned and all the predicate expressions that they singly or jointly entail. The predicates statistically relevant to 'Gi' in K are 'F' and '$F \cdot H \cdot J$'; and, apart from logically equivalent versions, the last of these is the only maximally specific i-predicate related to 'Gi'. Hence, if the argument (11), with $r = .95$, is to qualify as an explanation in K, RMS* requires K to contain the sentence '$p(G, F \cdot H \cdot J) = .95$': and this condition is satisfied. RMS* is satisfied also by an alternative argument with 'Gi' as the explanandum: its explanans consists of the sentences '$Fi \cdot Hi \cdot Ji$' and '$p(G, F \cdot H \cdot J) = .95$'.

Now let K' be the class obtained by adjoining to K the sentence '$p(G, F \cdot H) = .1$' (and by closing this set under the relation of logical consequence). Then there are the same i-predicates in K' as in K, and just one more of them is statistically relevant to 'Gi', namely, '$F \cdot H$'; again, '$F \cdot H' \cdot J$' is the only maximally specific i-predicate related to 'Gi'. Thus, condition (a) of RMS* is satisfied for (11) in K'. But (11) should not have explanatory status in K'[16]; for the information it adduces, that i is F and that the probability for an F to be a G is high, cannot count as explaining i's being G since K' tells us further that i belongs to a subclass, $F \cdot H$, of F for whose members the probability of being G is very small. But by reason of containing this latter probability statement, K' violates clause (b) of RMS*, and (11) is thus denied explanatory status in K'. Relative to K', there is essentially just one explanatory argument with 'Gi' as its conclusion: its explanans sentences are '$p(G, F \cdot H \cdot J) = .95$' and '$Fi \cdot Hi \cdot Ji$.'

Note that RMS, in contrast to RMS*, would have barred (11) from explanatory status in the class K of our first illustration; for it requires K to contain a number of additional statements, assigning the probability .95 to G with respect to the reference classes determined by '$F \cdot H$,' '$F \cdot J$,' '$F \cdot N$,' '$F \cdot H \cdot J \cdot N$,' '$F \cdot (H \lor N)$,' and so forth.

But while RMS* is less demanding than RMS in this respect, it is more exacting in another; for its clause (a) imposes a condition on *all* maximally specific predicates related to 'Gi', and not only on those that entail 'F'.

It is by reason of this stricter condition that RMS* escapes the pitfall noted by Grandy. In the class K of Grandy's illustration, the b-predicates are '$F \lor G$' and '$-F \lor G$' and all the predicate expressions they entail. The two expressions just cited are the only ones that are statistically relevant to 'Gb' in K; and each of them is also a maximally specific predicate related to 'Gb'. Thus, if (10a) is to be an explanation, RMS* requires that K contain the statement '$p(G, -F \lor G) = .9$'; and this condition is not satisfied.

RMS* quite generally precludes the possibility of conflicting explanations. For suppose that K contains the premises of the arguments

$$\begin{array}{c} p(G, F_1) = r_1 \\ F_1 i \\ \hline Gi \end{array} \quad [r_1]$$

and

$$\begin{array}{c} p(G, F_2) = r_2 \\ F_2 i \\ \hline Gi \end{array} \quad [r_2]$$

Let 'F' be short for one of the maximally specific predicates related to 'Gi' in K. Then *both* of the arguments qualify as explanations in K only if K contains statements to the effect that $p(G, F) = r_1$ and $p(G, F) = r_2$; but then, $r_1 = r_2$, and there is no conflict. Thus, it appears that in the version RMS*, the requirement of maximal specificity serves the purpose for which it was intended.

In conclusion, I summarize the construal here proposed for the concept of probabilistic, or I-S, explanation: An argument of the form

$$\frac{p(G, F) = r}{Gi} \quad [r]$$

is a probabilistic explanation of basic form relative to a class K (of the general kind characterized in section 1) if and only if

(1) K contains the explanans and the explanandum sentences of the argument.
(2) r is close to 1.
(3) The probability statement in the explanans is lawlike.
(4) The requirement of maximal specificity as expressed in RMS* is satisfied.

For the predicate expressions permissible in lawlike probability statements, the conditions N1 and N2 were proposed as necessary, but presumably not sufficient.

NOTES

1. See, for example, Glass [2], pp. 57–58.
2. In order also to accommodate explanations which are merely proposed or contemplated rather than asserted—i.e. whose explanans sentences are not, at least as yet, included in K—my earlier treatment of the subject made slightly more complicated provisions (cf. [5], p. 400, note 20); but the logically crucial points can be stated more simply if we require, as I do here, that the explanans sentences of any I-S explanation relative to K must belong to K.
3. For convenience of formulation, I permit myself here to speak of one-place predicates as standing for properties, features, or characteristics of objects or events, and alternatively as standing for the corresponding classes. An important caveat concerning this usage is noted in section 3 below.
4. For the reasons referred to in note 2, my formulation of the requirement in [5], p. 400, is slightly more involved; but the substance of the earlier version is essentially the same as that of RMS above.
5. See [1], pp. 211–213.
6. [8], p. 374 (emphasis in the original). Reichenbach acknowledged (p. 375) that his rule does not determine the weight in question univocally. It might be noted also that, even if the narrowest relevant reference class were always uniquely specifiable, the concept characterized by the rule would not have all the formal properties of a probability. For example, when two predicates, 'G_1' and 'G_2', logically exclude

each other, the "probability," or weight, of '$G_1 i \vee G_2 i$' is not always the sum of the weights of '$G_1 i$' and '$G_2 i$'; for the narrowest reference classes available for determining the three weights may well not be identical, and this may result in a violation of the addition principle for probabilities. For a sympathetically critical discussion and revised statement of the rationale of Reichenbach's rule, see Salmon [9], pp. 90–94.
7. See the discussion of this point in [4], pp. 121–124 and [5], pp. 376–380.
8. More explicitly: Let us say that an open sentence S in one free predicate variable U is a logical finiteness condition if (i) S contains no constants other than those of logic and set theory, and (ii) S is satisfied only by predicate expressions with finite extensions. Condition N1 is meant to require that if the reference predicate or the outcome predicate is expanded in primitive terms, and then substituted for U in any logical finiteness condition S, the result is not a truth of logic or set theory.
9. The predicates here said to stand for kinds of events need not be construed, however, as applying to entities of a special kind, namely, individual events (like those envisaged by Donald Davidson in "Causal Relations," *The Journal of Philosophy*, vol. 64, 1967, pp. 691–703); they may be treated instead as two-place predicates that apply to individual objects at certain times or during certain time intervals. Thus, the repeatable kind of event, flipping a penny, need not be represented by a predicate that is true of certain individual events, namely, those that are penny-flippings: it can be symbolized instead by a two-place predicate that applies to an object x at time t just in case x is a penny that undergoes a flipping at t.
10. See [7]. My discussion is based on a draft of [7] which Professor Massey sent me in January, 1966.
11. An example is discussed in [4], pp. 142–143. See also Carnap's remarks on this point: [1], p. 494.
12. In his review article [10], which was published in 1966. Dr. Wójcicki had presented the idea to me already in January 1965. His article also contains interesting critical comments on my construal of deductive-nomological explanation; but these cannot be discussed here.
13. I slightly tighten Wójcicki's formulation so as to make explicit the requisite relativization with respect to K; and, for clarity of statement, I specify definite quantitative probabilities instead of speaking of high probabilities or near-certainties, as does Wójcicki, and as I had done in [4], p. 126.
14. On this point, my conception of lawlikeness differs from that advanced by Goodman in his pioneering work on the subject. Goodman would presumably assign lawlike status to any sentence obtained by replacing a predicate in a lawlike sentence by a coextensive one. For he characterizes the predicates that occur in lawlike sentences in terms of their entrenchment and stresses that entrenchment carries over from one predicate expression to any coextensive one; so that, in effect, "not the word itself but the class it selects is what becomes entrenched, and to speak of the entrenchment of a predicate is to speak elliptically of the entrenchment of the extension of that predicate." ([3], p. 95; cf. also my remarks on this point in [5], p. 343.)
15. In an unpublished note he wrote in February 1966, as a graduate student at Princeton.
16. I am much indebted to Dr. Richard Grandy, who pointed out to me the need to meet situations of the kind here under discussion, and who suggested that this might be done by means of clause (b) in RMS*.

REFERENCES

1. Carnap, Rudolf, *Logical Foundations of Probability*, University of Chicago Press, 1950.
2. Glass, Bentley, *Science and Ethical Values*, The University of North Carolina Press, Chapel Hill, 1965.
3. Goodman, Nelson, *Fact, Fiction, and Forecast* (Second edition), The Bobbs-Merrill Company, Inc., Indianapolis, 1965.
4. Hempel, C. G., "Deductive-Nomological vs. Statistical Explanation," in *Minnesota Studies in the Philosophy of Science*, vol. III (eds. H. Feigl and G. Maxwell), University of Minnesota Press, Minneapolis, 1962, pp. 98–169.
5. Hempel, C. G., "Aspects of Scientific Explanation," in *Aspects of Scientific Explanation and other Essays in the Philosophy of Science*, The Free Press, New York, 1965, pp. 331–496.
6. Humphreys, W. C., "Statistical Ambiguity and Maximal Specificity," *Philosophy of Science*, vol. 35, No. 2, 1968. pp. 112–115.
7. Massey, Gerald J., "Hempel's Criterion of Maximal Specificity," *Philosophical Studies*, vol. 19, No. 3, 1968, pp. 43–47.
8. Reichenbach, Hans, *The Theory of Probability*, The University of California Press, Berkeley and Los Angeles, 1949.
9. Salmon, Wesley C., *The Foundations of Scientific Inference*, The University of Pittsburgh Press, Pittsburgh, 1967.
10. Wójcicki, Ryszard. "Filozofia Nauki w Minnesota Studies," *Studia Filozoficzne* for 1966, pp. 143–154. This is a review article on *Minnesota Studies in the Philosophy of Science*, vol. III (eds. H. Feigl and G. Maxwell), University of Minnesota Press, Minneapolis, 1962.

8

Postscript 1976

More Recent Ideas on the Problem of Statistical Explanation

In this postscript I would like to take issue with the ideas on the problem of statistical explanation which recently have been developed by R. C. Jeffrey,[1] W. C. Salmon,[2] and W. Stegmüller,[3] and which deviate significantly from that in the view stated above. In short, Jeffrey and Salmon reject my statement of statistical explanation in the form of "argument," my demand of a high assigned probability value, and the epistemological relativization of statistical explanation to given knowledge situations; however, they grant statistical laws an explanatory role (Jeffrey, certainly, only to a restricted extent) and characterize this in a new kind of way.

Stegmüller, on the other hand, is of the opinion that Salmon's and my statements on a theory of statistical explanation are concerned with two entirely different concepts of what has to be explained: that what I designate as statistical arguments are not *explanations* presenting the bases of existence, but rather, at best, statistical substantiation of rational arguments; whereas Salmon (and indeed also Jeffrey) strive for the clarification of an important scientific method that, to be sure, cannot also be regarded as explanation, but rather as "*statistical deep-analysis*" leading to a "*statistical situation understanding.*" With reference to the central concept which Salmon's and my inquiries seek to explain, Stegmüller comes to the conclusion that there is in general no such thing as statistical explanation.

3.7.2 Jeffrey's Critique of the Concept of I-S Explanation

Statistical explanations were represented above as arguments that in the simplest case have the form

(30)
$$\frac{p(G,F) = r}{F(b)} \quad [r]$$

whereby the explanans satisfies the requirement of maximal specificity (relative to the given knowledge situation K) and r lies near 1, so that, in view of the explanans, the occurrence of the explanandum-event is to be expected with high probability.

Jeffrey now rejects the idea that an event is the better explained by means of a statistical law, the greater the probability is, which the law attributes to the event.[4] Consider, for example, a probability experiment E, repeatable at will, whose single completion consists of throwing a fair coin 10 times in succession. The probability that E yields the result of "heads at least once"—or in short R—is then $p(R,E) = 1-(½)^{10}$. This probability lies near 1; as Jeffrey remarks, this justifies us, on a single completion of E, to wager with a high betting ratio on the result R, which is not the case for not-R, *namely, ten tails*.

Hence, the numerical value of the probability is, according to Jeffrey, "a good measurement of the correct strength of our expectation" that the result in question will occur ([1971], p. 23); it has, however, nothing to do with the worth of the explanation of the occurrence of the result. It is all the same whether one single completion of E yields the probable result R or the improbable result not-R; our statistical law explains the one result no better than the other. For our understanding of the result rests here on knowledge of the underlying stochastic process and the explanation is finally the same for both cases; it consists in the description of this process and the statistical laws which specify the probabilities of the various possible results. According to Jeffrey one could say that the question *how* a particular event took place is answered by means of the information that it was the result of a stochastic process for which this particular statistical law holds good; to the question *why* this one and not another of the different possible results occurred, one can only answer: by means of chance, i.e., as a result of that chance kind of process. One can further add that the event in question had, on the basis of the law, such and such a probability; that is then, however, an aside and no substantial part of the explanation ([1971], pp. 24–25).

I think that Jeffrey in his fundamental ideas is right: the understanding that statistical laws give us regarding particular events is not the deeper or more complete, the greater is the probability which the laws attribute to the event. The explanation consists of the characterization of the stochastic process that produced the event and of what the statistical law specifies for the probabilities of the various results of this process. These statements explain then the occurrence of an improbable event no less well than the occurrence of a very probable result. My demand that a statistical explanation must show that the explanandum event was to be expected with high probability is thus to be given up.

Jeffrey draws from his considerations the further conclusion that statistical explanations in general do not have the character of inferences or arguments. In this respect he places statistical explanations in sharp contrast to causal explanations. The latter show, as he expresses it, *why* an event occurred in that they, with the help of universal laws, prove deductively *that* it took place. Statistical laws make possible no such proofs and the explanations dependent upon them are in general not inferences. Jeffrey is willing, however, to grant to a certain narrow class of statistical explanations the character of inferences; these are the "beautiful" cases as they occur for instance in statistical mechanics and in the theory of radioactive decay; here the statistical laws confer upon the explanandum-event a probability so near to 1 that its occurrence must, for all practical purposes, be considered as certain ([1971], pp. 20–23, 27). Jeffrey adds that he considers it to be misleading to consider statistical arguments—with the exception of the "beautiful cases"—in general as explanations.

Jeffrey's distinguishing of the "beautiful" cases seems to me, however, untenable.[5] For there is of course no logical difference between a "beautiful" statistical explanation and one that attributes to its explanandum no very great probability; the size of the probability value can certainly not be decisive as to whether a statistical explanation possesses the character of an argument. The further question of whether statistical arguments or analyses can be thought to be explanations not at all, or perhaps only in cases of very high probability, will be discussed in more detail in the following section. Here I will only say that in my opinion there are good, although not decisive, grounds given for so widening the explanatory concept that even statistical arguments with small assigned probabilities also have explanatory validity. Salmon's theory expresses this view clearly.

I would further assert that the representation in statistical argument form—but without the demand for high probability—remains completely admissible but is not necessary. The latter may, however, also be said apropos the representation of D-N explanations in argument form. One could indeed designate instead a class of statements as a D-N explanans for a given explanandum-statement, if the above-mentioned conditions are satisfied; in this case that the corresponding argument is deductively valid could also be spoken of as an "aside."

In the representation of statistical inferences in argument form, there is, however, to be noticed a fundamental distinction between deductive and inductive inferences, already set forth at the beginning of section 3.3 of the essay "Aspects of Scientific Exploration" in the book by the same name [Hempel 1965]: the conclusion of a deductive argument may be asserted independently as true, if the truth of the premises is assured; for inductive arguments there is no analogous separation rule which allows a "conclusion" to be detached from the "premises" and asserted independently. This holds naturally also for "beautiful" statistical explanations: the conclusion is practically certain only relative to the respective premises and it may be false even when the premises are true.

Jeffrey says nothing about the ambiguity of statistical explanation emphasized in section 3.4, to avoid which I proposed an epistemologically relativized conception of this kind of explanation. His explanations express an absolute concept without, however, making explicit his attitude toward the problematic nature of this concept. Since Jeffrey in general attributes no argument form to statistical explanation, one could say that for him the ambiguity problem in the form presented in section 3.4 of "Aspects" in Hempel [1965] does not arise at all. It does, however, exist by the very nature of the matter. In my opinion, in the case of an ambiguity the explanandum can be represented as the result of various stochastic processes which, however, yield this result with entirely different probabilities: Are these representations all to be regarded equally as explanations? An affirmative answer would deprive the absolute concept of statistical explanation of its scientific theoretical interest; this is indeed evident from the examination of the ambiguity in section 3.4 and may become more clear in the following discussion of Salmon's statements.

3.7.3 Salmon on Statistical Explanation and Homogeneity

Salmon agrees with me that statistical explanations rest essentially on statistical laws. He rejects completely, however, my representation of such explanations in argument form, and indeed apparently chiefly because there is for statistical inferences no rule of detachment of the kind characteristic of deductive arguments ([1971b], p. 77). As just stated I nowhere assume such rules; I here employ the word 'argument' rather in a broader sense which is illustrated by means of schema (3o). Salmon says further that, for the statistical explanation of an event, we need to represent the explanandum not as the conclusion of a statistical argument, but rather must only give the "weights" or rational expectation values which are to be attributed to the explanandum in connection with wagers and other practical decisions ([1971b], p. 78). But the probability r which is attributed to the explanation in schema (3o) is indeed to be understood as the expectation value which the explanans confers on the explanandum. Thus these considerations do not testify against my representation of statistical explanations in the form of arguments.

Salmon rejects, just as Jeffrey does, my demand that a statistical explanation must show the occurrence of the explanandum-event as very probable. As already stated I agree with this criticism. Salmon next points out correctly ([1971b], pp. 80–81) that the abandonment of the demand of high probability eliminates the nonconjunctivity of statistical explanations: in the example given in section 3.6, the given statistical law now supplies also an explanation of $E_{500}(w)$, attributing to this event, however, a very small expectation value.

Salmon's conception of statistical explanation is *absolute* in the sense that it is not relativized to actual total knowledge, to a class K of propositions which are at any given time accepted. In order to avoid certain potential difficulties of this absolute conception, Salmon demands of each statistically explanatory-powerful law that it must appeal to a *homogeneous reference class*. This demand rests, in short, on the following consideration: the information that a certain single event b is of kind F (in short, that $F(b)$) and that, further, $p(G,F) = r$ can serve as an explanation that b is of kind G (in short, that $G(b)$), only if F is for G purely of a chance kind in the sense that no possibility exists to select from the particular cases of F a certain subclass—for instance the class of all those cases which possess a certain further characteristic C—such that, in the subclass so selected, the probability $p(G,F.C) = r'$ is different from r. For otherwise, if for instance b has the characteristic C, the explanation of $G(b)$ can be made sharper in the sense that, on the basis of the specific law '$p(G,F.C) = r'$', there is to be attributed to the occurrence of $G(b)$ an expectation value r' different from (but not necessarily greater than) r. Therefore, Salmon demands of each explanatory-powerful statistical law $p(G,F) = r$ that in principle it is capable of no such sharpening that thus for each relevant characteristic C it has the value: $p(G,F.C) = p(G,F)$.

Not each characteristic C is "relevant" in the sense here intended. Let U be the class of drawings from an urn which contains 1000 otherwise identical balls of which 800 are red (R), such that $p(R,U) = .8$. If now the characteristic R itself is chosen for C, then obviously $p(R,U \cdot R) = 1 \neq .8$ is valid. But this probability statement makes possible no sharpened empirical-statistical explanation of the drawing of a red ball in a particular case, since it is a nonempirical theorem of the pure probability theory. The characteristic R is therefore not "relevant"; it is not significant for the question whether U in reference to R is a stochastic process. Salmon's general characterization of the intended relevancy may be then expressed in short as: a characteristic C is not relevant for a subdivision of the reference class F in '$p(G,F) = r$'. if, even in principle, we cannot know whether a particular individual possesses the characteristic C *without already knowing* whether the particular individual has the characteristic G. Of each explanatory-powerful statistical law '$p(G,F) = r$' Salmon demands that for each relevant characteristic C, it has the value: $p(G,F \cdot C) = p(G,F)$. If this condition is satisfied, Salmon terms the reference class F *homogeneous* with respect to G.[6]

A law '$p(G,F) = r$' (where $r \neq 1$) that satisfies this condition can be called a fundamental statistical law: it is capable of no further sharpening. If b is a particular instance of F, so that $G(b)$, then b can possess especially no characteristic C which is connected with G by a strictly universal law '$(x) (C(x) \supset G(x))$'; because otherwise it would also be valid that $(x)[(F(x) \cdot C(x)) \supset G(x)]$ and consequently $p(G,F \cdot C) = 1 \neq r$. In this sense a statistically explainable event would thus, according to the homogeneity condition, be capable of no deterministic explanation.[7]

According to Salmon's view, a statistical explanation gives to the question 'Why does the particular event b have the attribute G?' an answer of the following kind: b belongs to the reference class F which is homogeneous with respect to G, and it has the value $p(G,F) = r$.[8] Epistemological relativization is here thus avoided just as it was by Jeffrey.

Salmon's view is certainly intuitively very plausible; various laws can also be cited which, according to current knowledge, satisfy the homogeneity condition, for example, the laws of radioactive decay.[9]

Nevertheless, important problems, some of which will be discussed in what follows, stand in the way of a rigorous characterization of the homogeneity concept.

Salmon's formulation limits the homogeneity condition '$p(G,F \cdot C) = p(G,F)$' to those criterion-attributes C for which one in principle *cannot know* whether a particular case has the attribute C without already knowing whether it has the attribute G ([1971b], pp. 43, 50). This formulation suggests the supposition that an epistemological relativization upon the actual complete knowledge K here lies hidden; but the proviso 'in principle' should seemingly make such a relativization unnecessary and furnish *absolute* concepts of homogeneity and statistical regularity, in that it to some extent refers to the class K^* everything which "in principle" is accessible to human knowledge. This is, however, a very unclear notion.

A part of this notion can certainly be explained more clearly. As our example of the drawing of a red ball from the urn illustrated, one can in general say that, for the probability statement '$p(G,F) = r$', the attribute G itself determines no relevant subdivision of the reference class: the fact that $p(G,F \cdot G) = 1 \neq r$ does not prove F as inhomogeneous: and in fact we can only then know that b has the attribute G if we know that b has the attribute G. Likewise $G \cdot H \cdot J$ yields no relevant subdivision of F with respect to G, etc. The reference to that which we can know is here, however, not essential; it is satisfactory—and is clearer—in place of that simply to say that C is, in any case, then not a relevant attribute, if $p(G,F \cdot C) = s'$ is a theorem of the pure probability calculus; for such a theorem yields indeed no empirical explanation at all.[10]

Salmon brings up a further example, however, which shows that his conception of the nonrelevant selective attributes is more comprehensive than the one just cited. Let us consider once again the drawings U from the urn, for which the probability of getting a red ball is $p(R,U) = .8$. Let E be the attribute of a drawing which yields a ball of that color which lies at the opposite end of the visible spectrum from the color violet. Then a drawing has the attribute R iff it has the property E. Thus $p(R,U \cdot E) = 1$ is valid. Now, Salmon is of the opinion that this fact does not violate the homogeneity condition, and indeed because we cannot know that a given drawing has the attribute E without already being acquainted with "the result" of the drawing, thus without knowing that it has the attribute R ([1971b], pp. 50,43). This argument is, however, not conclusive. That a drawing has the attribute E iff it has the attribute R is an empirical law. Let us assume that the drawings are carried out by an observer who does not know this law, but who knows that the color of a physical object has the attribute E iff the object reflects electromagnetic radiance in a determined wavelength interval w. Now there could be connected to the urn an apparatus which produced a click iff an object of the above-named kind is picked from the urn. If our observer is aware of this arrangement, he can then know whether the drawing just executed by him has the attribute E without knowing that it also has the attribute R. "In principle" a situation of this kind is completely possible; the homogeneity condition is thus in this case violated.

In general it seems clear that, according to Salmon's view, the homogeneity condition for a statistical statement '$p(G,F) = r$' (with $r \neq 1$) is certainly then not satisfied when for a certain C the statement '$(x)[(F(x) \cdot C(x)) \supset Gx]$' is a natural law; because in this case there does exist a subclass of F in which the probability of G is not equal to r, and for whose members, the fact that they possess the attribute G, constitutes a deterministic explanation.

Stegmüller, who enlists, for the definition of his concept of a statistical deepanalysis, a homogeneity demand, does not exclude cases of the just mentioned kind, but rather explicitly admits them.[11] This seems to me, however, not to agree with his intention. For the homogeneity condition should, as he said, make precise the thought "that there is in principle no subdivision of the reference class Y statistically relevant for B."[12] In the case just mentioned the strict universal law would indeed supply a statistically relevant subdivision of the reference class, since it implies that $p(G, F \cdot C) = 1$; this situation would make possible a corresponding sharpened deepanalysis as well.

The difficulties, indicated above, of the homogeneity concept were already foreshadowed by Mises's definition of a collective; they have, in my opinion, found as yet no satisfactory solution. Salmon himself emphasizes the necessity of further clarification.[13]

3.7.4 On Salmon's Demand of Maximum Reference Classes

My requirement of maximal specificity for statistical explanations has a consequence which Salmon considers unacceptable and which he excludes in his definition by means of an additional condition.

Let us consider a case which is adapted from one of his examples ([1971b], p. 33). Why was Hans cured of his headcold within a week? He took vitamin C and the probability of someone who catches a head cold [S] and takes vitamin C [V] to be cured after a week [G] is $p(G, S \cdot V) = .9$. Let us now assume that the taking of vitamin C is irrelevant to the duration of a headcold, i.e., that the probability of getting rid of a headcold within a week is in general equal to .9, ($p(G,S) = .9$) and that S moreover is a homogeneous reference class for G. Then, according to Salmon's view, the explanation which mentions the taking of the vitamin C should not be admitted, since it adduces information which is irrelevant to the explanandum: the reference class $S \cdot V$ is, to be sure, also homogeneous with respect to G, but is narrowed down in a statistically irrelevant manner.

In order to exclude such undesirable cases Salmon introduces a condition that a statistical law '$p(G,F) = r$' may only then be drawn upon for an explanation when F is a *maximum* homogeneous reference class in respect to G, i.e., when F is not a proper subclass of a homogeneous class F' for which $p(G,F') = p(G,F)$.[14] This condition is obviously violated in our example.

Two comments on this: (i) reference to a nonmaximum reference class seems to me for explanatory purposes not in general inadmissible; (ii) Salmon's alternative leads to a dual counterpart of the just considered *admission of undesirably narrow reference classes*: one could designate it as the *demand of undesirably broad reference classes*.

Comment (i) is illustrated by the following consideration:

Let us assume that the class K includes all of the explanans-statements occurring in both of the following schemas ('h' stands here for 'Hans'), and that these schemas satisfy the condition of maximum specification relative to K:

(3p)
$$\frac{p(G,S) = .9}{G(h)} \quad [.9]$$

(3q)
$$\frac{p(G,S \cdot V) = .9}{G(h)} \quad [.9]$$

Then a statistical explanation of Hans's recovery is represented according to my characterization of *each* of the two schemas; the unnecessarily narrow class $S \cdot V$ is thus permitted, but not *demanded*.

The impression that (3q) cannot be valid as an explanation rests, at least in part, on the following misleading way of reformulating the argument in words: it was almost certain that Hans would recover within a week from his headcold *because he took vitamin C* and because almost all headcold patients who take vitamin C recover in a week.[15] The paraphrase has the appearance of attributing the explanatory role to the attribute V, although V is statistically irrelevant in the class S in respect to G.

Certainly it is desirable in a statistical explanation to avoid the mention of particulars which are systematically irrelevant and psychologically misleading. But—and this brings me to comment (ii)—it is by no means clear, how inadmissible limitations of the reference class are actually to be defined precisely.

Salmon's demand of maximum homogeneous reference classes supplies no satisfactory answer because it leads to consequences which are intuitively no less unacceptable than those just discussed.

Let us consider for instance the statement: This sample i of a crystalline substance dissolved in water ($W(i)$) because i was common salt ($K(i)$) and $p(W,K) = 1$. Salmon's condition would not allow this as an explanation; for i belongs, in respect to water solubility, to a homogeneous reference class V which is much broader than K and which includes, among others, all substances consisting of sodium chloride, sodium bromide, sodium carbonate, potassium chlorate, sugar, or . . . Salmon's maximum-condition demands that a statistical explanation must classify the sample i in this maximum class V, for which $p(W,V) = 1$ is valid. However, this demand is thoroughly incomprehensible.[16] The effort to exclude "unnecessarily narrow" reference classes, which my analysis permits but does not demand, leads thus, in the demand of maximum reference classes, to a condition which not only permits "unnecessarily broad" reference classes, but also rather demands them.

The reason, therefore, that we would not consider the class V, in spite of its statistical homogeneity for water solubility, as a "natural" reference class appropriate for explanatory purposes, lies indeed in our knowledge that the way table salt dissolves in water differs in important aspects from the way each of the other water-soluble substances dissolves, such as in the degree of solubility and its temperature dependence, in the molecular processes occurring in the solution, etc. Sometimes such continuing theoretical reflections can also serve to show the reference classes invoked in certain statistical explanations as being unnecessarily *narrow*.

I do not believe, however, that considerations of this kind, mentioned also by Salmon (1973) and especially by Koertge (1975), can lead to a clear characterization of a "correct" reference class for a statistical explanation. To demand of each statistical explanation that it must rest on a maximum homogeneous reference class seems to me in principle just as unjustified as it would be to demand of each deductive-nomological explanation that it must rest on laws which are maximally far-reaching in the sense that they do not represent special cases of still more general laws.

According to Salmon a statistical explanation is an answer to a question of the form: "Why does this b, which is an element of F, have the attribute G?" where the explanatory answer supplies the following information: (i) the reference class F determined by the question is divided into n maximum homogeneous subclasses $F \cdot C_1$, for $F \cdot C_2, \ldots, F \cdot C_n$ for G; (ii) in these classes the attribute G has the probabilities $p(G, F \cdot C_i) = r_i (i = 1, 2, \ldots, n)$; (iii) the particular case b belongs to such and such a subclass—e.g. $F \cdot C_k$ ([1971b], pp. 76–77). The thus formulated explanation is therefore no argument; it lays down, however, a statistical law, '$p(G, F \cdot C) = r_k$', that is applicable to the given case and thereby also determines a rational expectation value for $G(b)$. This answer is then valid also as an explanation, if the expectation value r_k is small, so that therefore a statistical explanation in Salmon's sense does not in general show that the explanandum-event was to be expected.

The form of the explanation-demanding question required in this definition restricts, however, the admissible explanatory answer in an implausible and undesirable manner. Thus, for example, an admissible answer to the question "Why does Hans Schmidt, the dentist, have blue eyes?" must establish a subdivision of dentists into subclasses which are maximum and homogeneous for blue eyecolor, it must specify the probability of blue-eyedness for each of these subclasses, and finally must classify Hans Schmidt into one of these subclasses. This demand is certainly not in accord with the intuition which led Salmon to his demand of maximum reference classes. Furthermore, it does not comply with scientific procedure. And finally, it would make it dependent on the contingencies of the question-formulation whether certain statements have explanatory value. To answer the question "Why does body K, which is spherical, sink in water?", laws about the sinking of spherical bodies would be necessary; as answer to the question "Why does body b sink in water?" such laws would be inadmissible because of the demand of maximum reference classes.

In his postscript (1971c) Salmon acknowledges that it was an error to attribute that narrow form to questions seeking statistical explanations; he does not say, however, in which way he would modify his definition. For statistical explanations of the relatively simple kind, to which Salmon's definition refers, it seems best to me to ascribe to the corresponding questions the form: "Why does b have the attribute G?" An answer in Salmon's sense could, moreoever, then have the form laid down by him, where now, however, the reference class F is not dictated by the question, but rather, let us say, is suggested by the current knowledge. An explanation of the simplest kind would say that $F(b)$ and $p(G,F) = r$, where F is homogeneous in respect to G. This form would, it seems to me, preserve all the essential features of the Salmonian explanation-conception.

3.7.5 On the Epistemological Relativization of the Homogeneity Concept

Salmon also expresses the conjecture that my requirement of maximal specificity too requires a further clarification of the homogeneity concept ([1971b], p. 50). It seems to me, however, that the epistemological relativization of statistical explanations set forth in section 3.4.2 of "Aspects" in Hempel [1965] avoids the difficulties discussed above of the absolute homogeneity concept—and does so even if the demand of high probability is abandoned.

My definition, to be sure, makes no explicit use of the concept of homogeneity, but it imposes in effect a homogeneity condition on the reference class of an explanatory statistical law, that is by means of the requirement of maximal specificity. While Salmon's homogeneity concept, however, is to be based on simply all the relevant sub-classes $F \cdot C$ of the reference class F, the requirement of maximal specificity represents a relativized condition as regards the knowledge situation K: it concerns only those classes F_i for which K logically implies that they are subclasses of F. The requirement of maximal specificity seems to me entirely to avoid especially those conceptual difficulties which result, for the definition of the absolute homogeneity concept, from the reference to that "which one in principle can know."

My earlier formulation of the maximum specification with reference to (3o) mentions, however, explicitly the particular case b with which the explanation is concerned. This leads, however, to undesirable consequences. Let us assume that the requirement of maximal specificity is satisfied in (3o); that, however, K furthermore includes the statements '$(x) (F_1(x) \supset F(x))$' and '$p(G,F_1) = r_1 \neq r$,' and the probability statement referring to G in F is not a theorem of the probability calculus. Although K does not include the statement '$F_1(b)$', we would certainly not wish to allow (3o) as an explanation of $G(b)$ in K since '$p(G,F) = r$' would then not be the most specific law included in K that concerns the occurrence of G in F. Hence, I would replace my earlier formulation of the requirement of maximal specificity with the more general condition (suggested by Salmon's formulation) that, in an explanation of the form (3o), the statistical law must satisfy the following condition of the *homogeneity relative to K*: for each pair of statements included in K of the form '$(x)(F_1(x) \supset F(x))$' and '$p(G,F_1)$

$= r_1'$, it holds that $r_1 = r$, except when the second statement is a theorem of the pure probability calculus.

While the relativization of the concepts of homogeneity and statistical explanation avoids the previously discussed problems of the corresponding absolute concepts, it is not, however, to be denied that an explanation, epistemologically relativized in my sense, is at least intuitively but a faint counterpart to a statistical explanation of the absolute kind intended by Salmon and Jeffrey. If one explains an event deductive nomologically—for instance, the behavior of Dewey's soap bubbles dicussed in section 2.1—, then one advances empirical assertions, namely, the statements of the explanans. These describe previous events and boundary conditions which led to the event to be explained, and do so on the basis of natural laws, which likewise are cited in the explanans. Formulated clearly in a simplified way, a deductive-nomological explanation thus says this: the explanandurn-event occurred because such and such other events had occurred, which, with nomological necessity, had the given event as their consequence.

It is certainly a reasonable and attractive idea to conceive statistical explanations in an analogous way as being absolute. They would then explain the explanandum by means of bases of existence in a weaker sense, namely, by reference to a stochastic process which, in accordance with an explicitly invoked statistical natural law, entailed the given event with a corresponding probability. Jeffrey's, as well as Salmon's, conceptions of statistical explanation, are absolute in this sense. I would certainly prefer an absolute characterization of statistical explanation to my relativized one, if the problem of the ontological ambiguity of such absolutely understood explanations can be satisfactorily resolved—be it through an adequate explication of the demand of homogeneous reference classes for explanatorily powerful statistical laws or in a different way. In the next section we intend to consider briefly some newer tendencies in this direction.

3.7.6 Newer Tendencies Toward the Nonrelativized Explication of Homogeneity and Statistical Explanation

Coffa has sketchily expressed the view that a probability statement '$p(G,F) = r$' can only have explanatory value if F is "nomologically relevant" for G, where "a predicate is nomologically relevant to a second if a natural law determines that alterations in the first bring about alterations in the second."[17] This characterization of explanatorily powerful laws, and Coffa's characterization of statistical explanations grounded thereupon ([1974], p. 162), is, however, only programmatic so far and needs, as Coffa himself remarks, sharper formulation and further development.

Independently of Coffa, Krüger expressed the fundamentally similar view that explanatorily powerful probability laws have the character of indeterministic causal laws; thereby, to put it briefly, '$p(G,F) = r$' would possess this characteristic only if the production or removal of F "influences" the occurrence of G.[18]

This idea too is, however, only sketched and needs tighter working out. In particular, the question here arises whether a statistical-causal "influence" of G by the production or elimination of F would not require that the probability of the occurrence of G must be the same in the presence of F as in the presence of $F \cdot C$ where C is any additional condition. This would then lead back again to the central question of the homogeneity concept, namely, which conditions or attributes are to be considered "relevant" for this demand.

Stegmüller restricts the attributes C, relevant for homogeneity, by means of an important proviso, whose fundamental idea on closer examination exhibits an interesting kinship with Coffa's and Krüger's ideas. His definition is in short the following: let '$p(G,F) = r$' be a statistical law. Then let "F be called an . . . absolutely homogeneous reference class for G . . . [In the German text this is given as 'G gdw'] for each nomological class or for each nomological attribute C, whichever of these two logically or nomologically implies neither G nor ¬G, having the value: $p(G,F \cdot C) = p(G,F)$."[19]

This definition therefore takes into account for the homogeneity condition only those subclasses of F which are determined by means of "nomological attributes." Although Stegmüller does not, to the best of my knowledge, define the concept of a nomological attribute, it seems perfectly clear that he has in mind Goodman's concept of an entrenched predicate. In a different context—in his definition of a statistical proof—Stegmüller demands, namely, that the relevant statistical laws must be formulated exclusively by means of nomological predicates, and remarks that this demand should serve to avoid the paradoxes resting on Goodman's "pathological" predicates.[20]

As is mentioned in section 2.1, Goodman suggested that natural laws differ from merely "accidentally true" generalizations in that, inter alia, the former can serve as support for subjunctive conditional statements, and developed the view that lawlike statements are characterized by the entrenchment, in a sense which he then thoroughly develops, of the predicates occurring in them.

Considered in this light, Stegmüller's demand of nomological attributes expresses the view also emphasized above in section 3.1, namely, that, for the purposes of statistical explanation, only *lawlike* statistical statements are to be taken into consideration. His definition of absolute homogeneity assumes thus that the statement in question, '$p(G,F) = r$', is a statistical law and demands then that there is no statistical law of the type '$p(G,F \cdot C) = r'$', for which $r' \neq r$. Coffa's demand of nomological relevance and Krüger's idea of the "influencing" of G by F obviously require connections through statistical *laws* Stegmüller's demand of nomological attributes can thus be considered as a step toward the clarification of these ideas.

Restriction to lawlike statistical statements would also make it possible to avoid an old objection to the homogeneity concept. If, for instance, the reference class of a statistical law is conceived of, in the sense of the relative frequency concept of probability, as an infinite sequence of events (repetitions of a stochastic process), then it is a purely mathematical fact that each such sequence F, in which the probability (the limit of the relative frequency) of G is equal to r, includes infinitely many subsequences F' in which the limit frequency of G possesses any imaginable value at all r', distinguishable from r. How would the sequences relevant for the homogeneity condition have to be distinguished here?

Moreover, the number of all infinite subsequences of F has the cardinality of the continuum. Since, however, a scientific language S contains only an enumerable set of expressions, most of those subsequences cannot in general be named or characterized in S. To which of these subclasses should the homogeneity condition refer?

Restriction to reference classes, which are characterized by means of entrenched predicates, makes the number of the classes affected by the homogeneity condition enumerably infinite. Furthermore, it can now not have significance as more than a purely mathematical fact that each reference class possesses certain subclasses with varying probability values for G. Let us consider this point somewhat more closely.

A sequence M of throws of a given coin, in which the limiting frequency is ½ for the result "heads" (K) always includes, as said, subsequences in which the result K possesses a different limiting frequency, for instance ¾. Let us think of the throws belonging to one such subsequence as designated by the place numbers which they possess in the sequence M, and let us understand by $M \cdot C$ the subsequence of M determined by these place numbers. The purely mathematically provable fact that there is such a deviant subclass in no way justifies our assuming that the statement '$p(K, M \cdot C) = ¾$' is a statistical *law*. This becomes especially evident when we view statistical laws in the manner which was illustrated in section 3.1 in the example of the tetrahedron game. According to this view, the law '$p(K,M) = ½$' attributes to the coin the disposition to yield the result K in approximately one-half of the cases in long sequences of throws. The statement '$p(K, M \cdot C) = ¾$' permits, however, no analogous interpretation: we have no grounds for assuming that the coin has the disposition to yield result sequences in which at any given time the subsequence selected by the place number–criterion for C includes the result K with a relative frequency of approximately ¾. The existence of subsequences of the deviant kind illustrated by $M \cdot C$ represents therefore no violation of the homogeneity condition.

In view of these considerations it seems to me an important task to analyze more thoroughly and explicate more sharply the concept of a statistical law as it underlies the assessments of Stegmüller, Krüger, and Coffa.

Referring to the characterization of lawlike statements by means of the entrenchment of the predicates occurring in them is to think, as Goodman does, that the entrenchment of a predicate is in short dependent on the history of its hitherto existing use in inductive projections (Goodman (1955), chap. IV), and hence, strictly speaking, represents a linguistically and historically relativized concept.

Now, however, Goodman's concepts of the inductive projectability and the entrenchment of a predicate are closely akin to the idea of a predicate which characterizes a "natural kind."[21] This suggests the question whether the classes concerned with the homogeneity condition can be characterized perhaps as such classes as form natural kinds. This would, however, require an absolute view of the concept of a natural kind: and as yet there seems to be no satisfactory explication of this idea available.[22]

An entirely different idea for the absolute characterization of the concepts here discussed was proposed by Coffa (1974, pp. 151–153): a statistical explanation in the absolute sense was to be simply definable as a relativized (in my sense) explanation in the special case where the class K of the accepted statements is identical with the class T of all true statements. In this case the requirement of maximal specificity would not of course exclude only those homogeneity conditions which, according to our current knowledge, exist, but would rather guarantee that there are no subclasses at all of F which violate the homogeneity condition in reference to G; for the existence of such classes would be expressed in each case by the corresponding true statements, thus statements in T. Homogeneity relative to the class T would thus be absolute homogeneity. As Coffa correctly remarks, this argument assumes that the class T is one of the admissible values for 'K', that T thus represents an at least logically possible knowledge situation. Coffa considers it, however, as improbable that this assumption is false.

Actually, however, the expression 'the class of all true statements' has no clear sense. For a row of symbols can be denoted as a true—or false—statement always only with reference to a specified language. A row of symbols may well form a grammatically correct statement in two different languages, true in one of the languages, but false in the other. The expression 'the class of all true statements' characterizes thus no class of statements at all and hence especially none of the classes K, which represent the possible knowledge situations.

Can Coffa's thought be upheld in such a way that we apply, instead of T, the class T_S of all true statements of a certain very extensive scientific language S, in which especially all theories which are currently accepted or are coming up for discussion, can be formulated? If, as it would be natural to assume, S contains an enumerably infinite vocabulary and only statements of any length, which is, however, finite, then the number of all statements from S—and hence also the number T_S—is enumerably infinite. In this case it is not logically impossible that the class of the statements accepted at a given time coincides with the class of all true statements in S. It would, however, not be the case that T_S includes for each fact at least one statement describing it. Let us consider, for instance, the facts which are that the absolute temperature at a given fixed point in space and at time t has such and such a value, where t passes through all time points of a determined time interval. The totality of the time points in question is then uncountable and it is thus plausible to say that they determine likewise an uncountable class of facts, whereas the class T_S of all true statements in S is only countably infinite.[23]

3.7.7 On Stegmüller's Critique of the Idea of a Statistical Explanation

The ideas concerning statistical explanation, developed by Jeffrey, Salmon, and myself, were very thoroughly and critically illuminated, constructively modified, and taken further by Stegmüller. The limited scope of this "Postscript" does not allow his widely ramified investigations to be taken into account in detail; yet in addition to those ideas of Stegmüller previously touched on, let us briefly discuss at least his fundamental ideas on our topic.

Stegmüller thinks that the analyses undertaken by Jeffrey, Salmon, and myself contribute to the clarification of two explanatory concepts of importance in the theory of science, which, however, are to be sharply distinguished from one another and none of which is to be understood as statistical explanation; Stegmüller characterizes them as statistical proof and statistical deepanalysis.[24] With reference to the concept which the ostensible explanandum forms especially from my and Salmon's investigations, Stegmüller comes to the conclusion that *"there are in general no such things as statistical explanations."*[25]

In his critique of my explication Stegmüller refers inter alia to a requirement which he calls the Leibniz-condition: an explanation must, with reference to the explanandum E, show "cur potius sit quam non sit"; this requires that it *"gives a proof why E has occurred and why it has not failed to occur."*[26]

If proof here is understood in the strictly deductive sense, then this demand would obviously exclude entirely the possibility of statistical explanations. Stegmüller demands, however, less: the law used for the statistical proof of $G(b)$ must attribute to events of kind G at least a probability of more than ½.[27]

Even this liberal conception of the Leibniz-condition, however, still disqualifies for instance Salmon's view, which indeed lays down no lower limit for the relevant probability values.

The arguments characterized by me in section 3.4.2 as explanations obviously satisfy the Leibniz-condition, as there it was demanded that the pertinent probability value lies near 1. For all that, Stegmüller denies to them, however, the character of explanations, and indeed for a reason which he formulates in reference to a nonrelativized conception of the explanation concept in question. Let us assume that the premises are true in the following argument and that the statistical law satisfies the conditions regarded as relevant:

$$\frac{p(G,F) = .99}{G(b)} \quad [.99]$$

Whether this argument, which has the form (3o), is an explanation is then dependent upon whether $G(b)$ is actually the case: if so, then the argument is an explanation of $G(b)$; if not, then there is no fact which is explained by the argument: the argument is then not an explanation at all. Whether the given true premise-statements explain something thus depends on whether $G(b)$ or rather $\neg G(b)$ is the case. If one accepts this view of statistical explanation, "then one allows it to depend on chance whether anything at all explains anything else. This might conflict with all of our intuitions" ([1973 II], p. 284). This "paradox of the explanation of the improbable" ([1973 II], p. 281), is Stegmüller's major reason for completely abandoning the concept of statistical explanation ([1973 II], p. 285).

Stegmüller is, however, of the opinion that statistical laws play a substantial role in a different connection, namely, in the formulation of statistical proofs. Speaking generally, such a proof shows that the occurrence of a certain event (in the past, present, or future), which is not accepted at the time as fact, "is rationally to be expected" ([1973 II], p. 325).

One could approximately say that an argument which satisfies the conditions formulated in section 3.4.2 for statistical explanations, yields, according to Stegmüller, at best a statistical proof for the hypothesis '$G(b)$' in the knowledge situation K. Stegmüller gives a more general definition, likewise explicitly relativized to a knowledge situation, of statistical proof in which he avoids the argument form for proofs in a similar way to Salmon in his definition of statistical explanations; thereby he avoids at the same time the difficulty of the supposition that the supporting information confers a definite inductive probability on the supported hypothesis. In Stegmüller's view a statistical proof yields no bases of existence, no explanation of the event described in the proved hypothesis; instead it gives rational proofs, i.e., pieces of information in view of which it is rational, in the given knowledge situation, to count on the occurrence and not rather on the non-occurrence of the events in question.[28]

The second of the procedures distinguished by Stegmüller, in which statistical laws play a central role, is statistical deepanalysis. Just like the quest for an explanation, such an analysis can be motivated by the need to gain insight into the occurrence of a particular event. This event could for example be that a certain particular case b of a stochastic process of kind F yields the result G. According to Stegmüller a deepanalysis can, it is true, answer no explanation-demanding why-question concerning the occurrence of $G(b)$, but perhaps "*a question having in view the situation understanding,* which could then for instance be formulated: '*How is it to be understood that b, which is an F, is also a G?*'"[29] Leaving out certain refinements, the answer provided by Stegmüller to this question has the same form and the same information content as the statement which Salmon would characterize as a statistical explanation of $G(b)$. That is to say a deepanalysis-answer determines a subdivision of the class F into maximum homogeneous subclasses $F\,C_i$ ($i = 1, 2, \ldots n$), specifies the values of the probabilities $p(G, F \cdot C_1)$, and classifies b into a specified subclass of F among those named. As laid down in section 3.7.6, Stegmüller restricts the homogeneity condition to reference classes which are characterized by means of nomological predicates. In addition to the concept of deepanalysis here characterized as absolute, Stegmüller also introduces an epistemologically relativized counterpart.[30]

Stegmüller's idea of a statistical situation understanding made possible through a deepanalysis is also akin to Jeffrey's view that the question, *how* did the event $G(b)$ occur, is to be answered by the statement that it was the result of a stochastic process for which the statistical law cited in the analysis is valid; whereas the question, why exactly $G(b)$ and not rather $\neg G(b)$ occurred, permits only the answer: through chance, as the result of the named stochastic process.

One of the major distinctions between Stegmüller's and Jeffrey's views is that Jeffrey, as remarked above, permits statistical explanations in argument form at least for certain "beautiful" cases. A second distinction is that Jeffrey advocates an absolute view of statistical analyses and laws, without taking an explicit position on the questions of ambiguity, homogeneity, and statistical fundamental laws, or characterizing more closely the peculiarity of the probability or "stochastic" processes to which the statistical laws apply.

Stegmüller's deepanalysis adopts from Salmon's analysis of statistical explanation (i) the assumption that the class F subdivided in the analysis is determined through an asserted question and (ii) the demand that the subclasses must be *maximum* homogeneous reference classes for G in F. From the reasons specified in section 3.7.4, I would think that these conditions should also be eliminated from Stegmüller's explication of the statistical deepanalysis. A deepanalysis would then answer a question of the type: "How is it to be understood that this b is a G?" And irrespective of whether one speaks in this context of deepanalyses or of explanations, it is, in my opinion, to be required of an adequate answer only that it classify the particular case b into a reference class homogeneous for G and invoke a statistical law which establishes the probability of G in this class; the statement of laws for the probability of G in different reference classes seems to me unnecessary. Thus, for instance, a statistical explanation or analysis of the emission of alpha-particles through a sample of a particular element in a given time interval certainly does not also need to cite statistical laws for the emission of alpha-particles through other elements.

From this it seems to me to follow that the information to be supplied by a statistical explanation or a statistical analysis of the occurrence of $G(b)$ consists precisely of statements of the kind which already occur as premises in my schema (3o), namely, '$F(b)$' and '$p(G,F) = r$'.

Is the explanation or deepanalysis thus set up to be understood more precisely as an "inductive argument" of the type (3o)? This question is answered negatively with certain exceptions by Jeffrey and by Salmon and Stegmüller without exceptions. Avoidance of the argument form certainly has the advantage of avoiding the problems of a theory of inductive probability; yet certain analogous questions arise for the concept, invoked by Salmon in place of this, of the rational expectation value which the explanatory statistical information attributes to the explanandurn, and which Salmon numerically identifies with the probability occurring in the relevant statistical law.[31] A Salmonian explanation in the simplified form just suggested may thus be expressed as follows: "b belongs to the reference class F homogeneous for G for which the law '$p(G,F) = r$' holds; this information confers the expectation value r to the supposition that $G(b)$ is the case."

Certainly this is not an argument that yields a separable conclusion; however, this holds also for the representation of an explanation in the form of the argument (3o). On this point my view seems to me to be thus in fundamental accord with that of Salmon and Jeffrey.

Can, however, information of the just-named kind be valid as an explanation of $G(b)$, that is, as it complies with Salmon's view, irrespective of whether the numerical value of r is large or small? The Leibniz-condition would demand a negative answer, since the given information could otherwise serve as the explanation of the occurrence of $G(b)$ just as well as of the nonoccurrence of $G(b)$. And the Leibniz-condition does certainly have a very plausible ring; I myself have borrowed and used it—although not under this name—in earlier works as an adequacy condition for every rationally acceptable explanation (e.g., above in section 2.4).

Is the fulfilment of this condition, however, really to be viewed as an "essential," irrevocable attribute of all empirical explanations? Within the limits of a theory with fundamental laws of a statistical form, one could plausibly say that it can give no deeper or more complete understanding of the occurrence of the relevant events—whether they are very probable or very improbable—i.e., that the theoretical fundamental laws provide the fullest possible understanding of the events in question. This suggests the thought of expanding correspondingly the concept of explanation.[32]

A statistical explanation certainly does not answer the question, why exactly has $G(b)$ occurred, although, according to the theory, $\neg G(b)$ too was possible. But perhaps one could say that it yields the most complete nomological insight into the occurrence of the events in question that is objectively possible. In contrast to a deductive-nomological explanation, it does not show that the explanandum-event was to be expected with nomological necessity,[33] but perhaps it does show that it was to be expected with a certain nomological probability.

Important similarities as well as profound distinctions exist between the understanding of empirical phenomena that is established through laws of strong universal form and causal theories, and the understanding made possible through statistical laws. The more clearly this is grasped—an undertaking which was promoted through all the works here discussed—the less important it seems whether, with Salmon, we consider items of statistical information of the relevant kind as explanations, or characterize them, with Stegmüller, as statistical deepanalyses which indeed yield no explanation, but rather a statistical situation understanding.

The classification of such pieces of information as explanations would indeed gain in plausibility if one could compare with the absolute concept of a causal natural law a clearly explained, likewise absolute, statistical counterpart. The problem with the absolute concept which cropped up for example in the question of ambiguity, induced me to give a general epistemological relativization of the concept of statistical explanation. From the reasons given in this "Postscript" I would like, however, to modify in a certain way the view, formulated in section 3.4.2 of "Aspects" of my relativized concept: the demand of high probability values is to be given up as erroneous; the representation in argument form I consider, furthermore, as admissible, but not as necessary. The requirement of maximal specificity I would generalize in the manner indicated in section 3.7.5. I would retain the term "explanation" in the correspondingly expanded statistical sense. And while, toward the end of section 3.4.2 of "Aspects", I characterized the concept of statistical explanation as "essentially" relative to a knowledge situation, I would like once more to emphasize that the difficulties mentioned by me certainly do not prove that a clear explication of an absolute concept is impossible, and that I regard it as very desirable.

NOTES

This essay was translated by Hazel Maxian, University of Melbourne (1979).

1. Jeffrey (1971).
2. Cf. especially Salmon (1971b), but also (1971a) and (1971c). For some interesting remarks on Salmon's view, cf. Niiniluoto and Tuomela (1973), chap. IV.
3. See Stegmüller (1969), chap. IX, and especially (1973 II), part IV, where also the ideas of Jeffrey and Salmon are dealt with thoroughly.
4. I had indeed not asserted this and had also, to my knowledge, nowhere characterized the probability in question as the "strength" of the explanation as Jeffrey says [1971], p. 23). Jeffrey's important main thoughts are, however, largely independent of these particulars, as will be shown forthwith.
5. Stegmüller criticizes this view in detail in (1973 II), pp. 315–316.
6. Cf. Salmon (1971b), p. 43. I use the word 'relevant' here for the sake of brevity and simplicity. Salmon characterizes the kinds relevant for the homogeneity condition to determine subclasses of the reference class, with the expression 'place selection', which goes back to R. von Mises; see (1971b), pp. 42–43, 49–50.
7. Cf. on this Salmon (1971b), pp. 45–46, 64. Jeffrey does not deal with the question touched on here in detail, but he does remark that certain explanation-demanding questions (as, for example, "Why was my first child a boy?") allow a correct statistical, as well as a correct causal answer; cf. (1971), p. 25.
8. This is, in my opinion, the core idea of the Salmonian view. The definition of statistical explanation which Salmon formulates in (1971b), pp. 76–77, is more complex. It is discussed in section 3.7.4, where I attempt to show that certain further conditions included in the definition should be abandoned.
9. Cf. the example of the decay of radon handled in section 3.4.2 of "Aspects" in ASPECTS in each subclass which may be chosen by means of limited conditions such as pressure, temperature, surrounding electrical and magnetic fields, etc., the decay probability is exactly the same as in the whole class of the decay events of radon-atoms. Salmon gives further examples (1971b), pp. 45–46.
10. This criterion has an important consequence. Whether a given subclass of F satisfies the homogeneity condition for

$$(S) \quad p(G,F) = r$$

depends not only on the probability of G in this subclass, but also on the "manner of description" of the subclass, i.e., on the predicate expression by means of which it is characterized. Let us assume, by way of example, that

$$(S') \quad (x)\,[G(x) \equiv (F(x) \cdot H(x))]$$

is an empirical law. Then it is "in principle" (viz., if one does not know the law) possible to know of a particular case b of which $F(b) \cdot H(b)$ holds, without knowing that $G(b)$ holds. Now S' implies, however, the statement

$$(S'') \quad p(G, F \cdot H) = 1 \neq r$$

so that the homogeneity condition is violated for S.

On the other hand, the class characterized by '$F \cdot H$' is, because of S', identical with the one which is characterized by 'FGH'. Yet the probability statement

$$(S''') \quad p(G, F \cdot G \cdot H) = 1 \neq r$$

expresses no violation of the homogeneity condition since it is a theorem of the probability calculus.

The homogeneity condition concerns thus not classes pure and simple, but rather classes in this or that manner of description. Without question this remark would also hold for the homogeneity condition in the more general view, which Salmon seeks to characterize in reference to what one in principle can know.

11. Stegmüller (1973 II), pp. 344–345 (Stegmüller's definition is cited below in section 3.7.5). For the proof of this idea compare (1973 II), pp. 298–299.
12. Stegmüller (1973 II) p. 3145. The symbols 'B' and 'Y' correspond to 'G' and 'F' in our version.
13. Cf. Mises (1928), chap. I and III, as well as Salmon's remarks (1971b), p. 50, and (1971c), p. 106; and the representation in Stegmüller (1973 II), pp. 342–344, where also the relationship between Salmon's and von Mises' ideas is clearly worked out.
14. Cf. Salmon (1971b), pp. 43–45, as well as the alteration sketched in (1971c).
15. Salmon formulates this and similar examples in this way, which does sound plausible, but which does not fully do justice to the content of the explanation; cf. e.g. (1971b), pp. 33–34.
16. This objection was already essentially raised by Lehman (1972). Cf. also Salmon's reply (1973) and the discussion by Koertge (1975), which seeks to defend Salmon's ideas against various objections.

    In the demand for maximum reference classes, Koertge sees the reasonable investigation maxim, viz., to strive for the deepest and most comprehensive explanations possible.
17. Coffa (1974), p. 161. Cf. also the remarks on Coffa's and Salmon's ideas in Niiniluoto (1976).
18. Krüger (1976), p. 144. This essay also includes very stimulating reflections on other questions touched on in this Postscript.
19. Stegmüller (1973 II) p. 344. For the sake of uniformity I have replaced some of the symbols used by Stegmüller by those appropriate here and used in what goes before.

    As suggested toward the end of section 3.7.3, I would think that in the proviso 'which ever of these two logically or nomologically implies neither G nor $\neg G$' the words 'or nomologically' should be struck out.

    The then remaining condition affecting the case of logical bonds is, however, too weak; because if perchance G is not indeed logically implied by F alone, but rather by $F \cdot C$, with the consequence that $p(G, F \cdot C) = 1$, then we would certainly not view this as a violation of the homogeneity condition since the misleading probability statement can yield no empirical explanation. As already said in section 3.7.3, the still more general demand seems to me to be indicated: if $p(G, F \cdot C) = r' \neq r$, then the statement that $p(G, F \cdot C) = r'$ must be a theorem of the pure probability calculus.
20. Stegmüller (1973 II), pp. 324–325. For Goodman's paradoxes cf. Goodman (1955), chap. III, and Stegmüller (1969), pp. 276–282.
21. See the short remarks in Goodman (1955), p. 119, and the thorough expositions in Quine (1969) and Schlesinger (1968).

22. Goodman would certainly be of the opinion that the concept of natural kinds is, by his theory of the inductive projectability and the underlying entrenchment of predicates—therefore by the linguistic expressions—to be clarified, and not vice versa.
23. This consideration certainly employs the term 'fact' only intuitively; especially in that it pays no attention to the question whether a sharp criterion of identity can, after all, be formulated for facts; still it seems to me to be adequate for the present purpose.
24. This view is presented in Part IV of Stegmüller (1973 II); for a short summary, see pp. 350–351.
25. Stegmüller (1973 II), p. 356 (emphasized in the original); the same thesis was expressed by Stegmüller already in (1969), pp. 701–702.
26. (1973 II), p. 313 (emphasized in the original); cf. also the similar formulation in (1969), pp. 220, 684.
27. Stegmüller gives no general explication of the Leibniz-condition; the minimum demand mentioned here is clearly expressed, however, in (1973 II), pp. 324–325.
28. For the major points, encompassing the definition of the proof concept, cf. Stegmüller (1973 II), pp. 324–325.
29. Stegmüller (1973 II), p. 346 (emphasis added). For the sake of uniformity, the individual constant 'a' in Stegmüller's text was replaced here by our constant 'b'.
30. Cf. Stegmüller (1973 II), pp. 330–357; with reference to the points mentioned here, see especially pp. 344–346. Stegmüller characterizes analyses of this kind also as "statistical-*causal* deepanalyses" (pp. 345, 347) in order to indicate that the cited statistical laws cannot be tightened up with regard to the homogeneity condition, that they are thus fundamental laws (in the terminology used by me) which come as near to causal laws as is possible.
31. Cf. Salmon (1971b), pp. 78–79. Stegmüller assumes in principle the same idea in his characterization of the statistical proof (cf. for instance [1973 II], p. 325, paragraph 4), referring to it, however, not explicitly in his definition of the statistical deepanalysis. Also the Leibniz-condition requires naturally a concept of rational expectation.
32. Cf. once more the remarks by R. von Mises cited at the end of section 3.3. of "Aspects."
33. I.e., that the explanandum-statement is logically deducible from the description of the preceding events given in the explanans and the natural laws cited in the explanans.

REFERENCES

Coffa, J. A. "Hempel's Ambiguity." *Synthese* 28 (1974), pp. 141–163.
Goodman, N., *Fact, Fiction, and Forecast* (Cambridge, MA: Harvard University Press, 1955).
Hempel, C. G., *Aspects of Scientific Exploration*. New York: The Free Press, 1965.
Hempel, C. G., "Aspects of Scientific Exploration," in Hempel [1965], pp. 331–496.
Jeffrey, R. C. "Statistical Explanation." In Rescher et al. (1969), pp. 104–113. Printed in Salmon (1971), pp. 19–28. Page references in the present text refer to the second-named publication; here cited as Jeffrey (1971).
Kim, J. "Events and Their Descriptions." In Rescher *et al.* (1969), pp. 198–215.
Koertge, N. "An Exploration of Salmon's S-R Model of Statistical Explanation." *Philosophy of Science* 42 (1975), pp. 270–274.

Krüger, L. "Are Statistical Explanations Possible?" *Philosophy of Science* 43 (1976), pp. 129–146.

Lehman, H. "Statistical Explanation." *Philosophy of Science* 39 (1972), pp. 500–506.

Mises, R. v., *Probability, Statistics, and Truth* (London, UK: William Hodge and Company, 1939).

Niiniluoto, I. "Inductive Explanation, Propensity, and Action." In Manninen and Tuomela (ed.), *Essays on Explanation and Understanding.* Dordrecht: D. Reidel Publishing Co., 1976 pp. 335–368.

Niiniluoto, I. and Tuomela, R. *Theoretical Concepts and Hypothetico-Inductive Inference.* Dordrecht: D. Reidel Publishing Co., 1973.

Quine, W. V. "Natural Kinds." In Rescher et al. (1969), pp. 5–23, and in Quine, W. V. *Ontological Relativity and Other Essays.* New York/London: Columbia University Press, 1969; pp. 114–138. References in the present text refer to the last-named publication.

Rescher, N. et al. (ed.). *Essays in Honor of Carl G. Hempel.* Dordrecht: D. Reidel Publishing Co., 1969.

Salmon, W.C. "Reply to Lehman." *Philosophy of Science* 40 (1973), pp. 397–402.

Salmon, W. C., "Explanation and Relevance: Comments on James G. Greeno's 'Theoretical Entities in Statistical Explanation,'" in R. C. Buck and R. S. Cohen, eds., *PSA 1970* (Dordrecht, Holland: D. Reidel, 1971), pp. 27–37.

Salmon, W. C., "Statistical Explanation," in W. C. Salmon, ed., *Statistical Explanation and Statistical Relevance* (Pittsburgh, PA: University of Pittsburgh Press, 1971), pp. 29–87.

Salmon, W. C., "Introduction," in W. C. Salmon, ed., *Statistical Explanation and Statistical Relevance* (Pittsburgh, PA: University of Pittsburgh Press, 1971), pp. 3–17.

Schlesinger, C. "Natural Kinds." In Cohen, R. S. and Wartofsky, M. W. (ed.), *Boston Studies in the Philosophy of Science*, vol. 3. Dordrecht: D. Reidel and New York: The Humanities Press, 1968; pp. 108–122.

Stegmüller, W. *Personelle und Statistische Wahrscheinlichkeit*, Zweiter Halbband: *Statistisches Schliessen—Statistische Begründung—Statistische Analyse.* Berlin/Heidelberg/New York: Springer-Verlag, 1973 [Referred to in the text as (1973 II)].

Stegmuller, W., *Wissenschaftliche Erklarung and Begrundung* (Berlin, Heidelberg, and New York: Springer-Verlag).

III

SCIENTIFIC THEORIES

9

Reduction

Ontological and Linguistic Facets

1. Introduction

Among the many issues in the logic and methodology of science to whose clarification Ernest Nagel has contributed by his thorough and illuminating analyses, the subject of theoretical reduction holds a place of special philosophical interest. Reduction affords an important mode of scientific explanation; it raises intriguing questions concerning the relation between the concepts of the reducing and of the reduced theory; it is a procedure that reaches across the boundary lines of scientific disciplines; and its potential scope encompasses all the major branches of empirical science, since problems of reducibility arise in fields as diverse as physics and chemistry, biology, psychology, and the social and historical disciplines. All these aspects lend philosophical interest and importance to the topic of reduction. But one further reason for the fascination the subject has held for philosophers lies, I think, in the ontological roots of many questions concerning reduction—questions such as these: Are mental states nothing else than brain states? Are social phenomena simply compounds of individual modes of behavior? Are living organisms no more than complex physicochemical systems? Are the objects of our everyday experience nothing else than swarms of electrons and other subatomic particles? Or is it the case, as the doctrine of emergence would have it, that as we move from subatomic particles to atoms and molecules, to macroscopic objects, to living organisms, to individual human minds, and to social and cultural phenomena, we encounter at each stage various novel phenomena which are irreducible, which cannot be accounted for in terms of anything that is to be found on the preceding levels?

These questions appear to concern ontological issues, namely, the basic identity or difference of various kinds of empirical states and processes which, prima facie, exhibit striking differences. Yet in philosophical studies in the past several decades, the problems of reduction have generally been given a decidedly linguistic turn. Thus, that chapter in Ernest Nagel's book *The Structure of Science* which deals with the concept of reduction characteristically bears the title "The Reduction of Theories" and examines the relations that obtain between the terms and the laws of two theories when one of them is reduced to the other. The linguistic construal of reduction goes back at least to early logical empiricist studies, especially to the work of Carnap and Neurath on unitary languages for empirical science and to Carnap's subsequent emphasis on the possibility and importance of stating all problems of the philosophy of science in the "formal mode" rather than in the "material mode of speech" or in "pseudo-object-sentences."[1] The view then held by Carnap, that all significant issues in the philosophy of science can be restated so as to concern exclusively the *syntax* of the language of science, is much too restrictive, as Carnap himself has long since noted;[2] but a construal that is linguistic in a wider sense has proved remarkably fruitful—so much so, in fact, that many recent and current studies of reduction characterize the subject from the very beginning as one concerning the language of science rather than the "nature" and the interrelationships of various kinds of entities, states, and processes. This linguistic turn is indicative of philosophical misgivings about the ontological construal of the issues in question, and it is an attempt to explicate the latter by restating them in a clearer and philosophically more satisfactory fashion. But the reasons for rejecting the ontological versions and for construing problems of reduction as concerning the language of science are not, as a rule, made very explicit. It may be of some interest, therefore, to reflect upon the rationale of this linguistic turn and to explore some of its implications for current problems in the theory of reduction. This I will attempt to do in the present essay.

2. On Mechanism in Biology

As an example of an originally ontological view that can be construed as a reductionist thesis, consider the mechanistic conception of biological phenomena, according to which all biological systems, states, and processes are basically nothing else than complex physicochemical systems, states, and processes, and thus are governed by purely physicochemical laws and explainable in terms of these.

Much like the statement that the morning star is the same thing as the evening star, this thesis presupposes the possibility of making at least a conceptual distinction between the two classes of phenomena that it declares to be basically of the same kind. How, then, are biological systems, states, and processes to be conceptually distinguished from physical, chemical, psychological, or sociological systems and occurrences? Let us note first that objects, states, and events cannot be unambiguously divided into mutually exclusive classes of "physical entities," "chemical entities," "biological entities," and so forth; for any individual object or event—the sun, the Atlantic Ocean, the periods of the Black Death in London, the assassination of President Kennedy—can become a subject of investigation for many different disciplines of empirical science. These disciplines, however, are concerned with different aspects of the phenomena in question; and the aspects of concern to a particular discipline will be characterized by means

3. Ontological and Linguistic Construals of Psychophysical Association

The reductionist thesis I wish to consider has been said to express the empirical component of the psychophysical identity theory. As Shaffer puts it:

> Identity Theory rests on an empirical hypothesis. It hypothesizes that for each particular mental event there is some particular physical event which always occurs and is such that whenever that physical event occurs then the mental event occurs. (The theory proposes to explain this by the further hypothesis that the physical event and the mental event turn out to be one and the same event.)[3]

I shall refer to the "empirical hypothesis" noted in this passage as the *hypothesis of universal psychophysical association*, or briefly, as the hypothesis U. The mention made in U of "particular events" evidently is not meant to refer to *individual* physical and mental events, for these have definite temporal locations and therefore cannot recur in the specified manner; the hypothesis must rather be taken to refer to *kinds* or *types* of mental and physical occurrences. Moreover, the physical events are understood to occur in the body of the individual who experiences the mental events. The hypothesis may therefore be restated as follows:

(U) For every kind M of mental state or event there exists a kind B of bodily state or event such that a person experiences a state or event of kind M if and only if a state or event of kind B occurs in his body.

But it seems that this statement can be shown to be true without benefit of empirical evidence. For let us adopt the following definition:

(D) The body of a person will be said to be in state M' if and only if the person is in mental state M. (Similarly for events; I will henceforth refrain from always spelling this out.)

If bodily states of this kind are taken into account, hypothesis U becomes trivially true. It may seem natural to say that states defined by D do not properly qualify as bodily states; but if the notion of mental state is taken to be intelligible, as U plainly does, then definition D must surely be said to specify various states that can intelligibly be attributed to a human body.

It will be necessary, therefore, to supplement U by specifying the kinds of bodily (or of physical) states and events to which the hypothesis is meant to refer. And just as in the case of the mechanistic thesis, this will have to be done by indicating the conceptual apparatus, and the associated vocabulary, in terms of which bodily states and events are to be characterized. In the context of the mind-body problem, the conceptual framework that serves to specify bodily states or brain states will normally be meant to comprise not only physics but chemistry and biology as well.

The reference made in U to "every kind of mental state" requires analogous qualifications. This need is made clear by the arguments already presented; but it might be further illustrated by this example: Suppose that a physico-chemical-biological theory has been specified in terms of whose concepts the bodily states mentioned in U are to be characterized. And suppose further that P is a certain physicochemical feature (such as, perhaps, the occurrence of rapid small oscillations of the average brain temperature) which is describable in the specified theory and which, as a matter of empirical fact, satisfies the following condition: A person's brain has the feature P if and only if the person "is in a pain-state," i.e., feels some pain or other. In this case, it would be correct to say that every mental state which is a pain-state has a corresponding kind of brain-state associated with it in the manner envisaged by the hypothesis U. However, the associated brain-state would be of the same kind, namely, the kind characterized by the presence of feature P, for all pain-states, irrespective of the qualities and intensities of the pain; whereas the hypothesis of universal psychophysical association is surely meant to imply that mental states of different kinds are associated with bodily states of different kinds. And indeed, for an opponent of a reductive identification of the mental with the physical, any difference between mental states that was not matched by a difference between the associated brain states would constitute evidence that mental phenomena could not be completely reduced to brain states. But when do two mental states count as different in a sense relevant to the hypothesis U? Do my toothache in the morning and my headache in the evening count as mental states of the same kind, since both are pain-states, or do they count as different kinds because of the phenomenally different locations of the pain? Or take my toothache in the morning and my toothache in the evening: suppose that they are indistinguishable for me in terms of location, quality, and intensity, yet the later one is accompanied by a feeling of fatigue—does this qualify them as mental states of different kinds in the context of the claim made by U? What kinds of occurrences count as mental states or events, and what features count as establishing differences among them, evidently depends on our changing conceptions of mental phenomena. Being possessed by a demon would not count today as a kind of mental state, whereas being under hypnosis might. And if extrasensory perception came to be well established, then certain kinds of, and differences between, mental states might be theoretically countenanced which currently are not, such, perhaps, as types and degrees of extrasensory receptivity.

If U is to be given a clear meaning, therefore, the scope of the notions of mental event and mental state must be indicated. Ideally, this would have to be done by specifying a psychological theory whose conceptual framework would then settle these matters: different mental states would be those distinguishable in terms of the concepts of the theory.

As a result, the distinction between bodily and mental states which underlies the hypothesis U and related views comes to be construed as the distinction between states characterized in terms of the concepts of physico-chemical-biological theories, and states characterized with essential use of psychological terms. This, I think, provides strong grounds for giving the psychophysical hypothesis in question a linguistic turn.

For those who hold the identity theory, there is a further reason to espouse such a construal. For if every mental state is identical with a physical state, then there can be no empirical characteristics that differentiate mental states from physical states: the distinction will rather pertain to alternative ways of characterizing "the same" state—in terms of physical, chemical, and biological concepts, or in terms of psychological (specifically, mentalistic) ones; thus, the distinction will concern states-under-a-theoretical-characterization.

On this understanding, U would have to be construed as asserting, roughly, that for every characterization of a state or event that makes essential use of psychological terms, there is a coextensive one expressed in physical, chemical, and biological terms alone. Considerations analogous to those presented earlier suggest that since U refers to all possible mental states, a reformulation would have to assert this: For every mental-state characterization expressible in a true psychological theory, there exists a coextensive bodily–state characterization in some true physico-chemical-biological theory. But while this reformulation comes close to the intent of the ontological view it is to paraphrase, it clearly is problematic in essentially the same respects as the latter.

4. The Linguistic-Ontological Dilemma for Universal Determinism

It may be of interest to note, by way of a brief digression, that the thesis of universal determinism is subject to a similar difficulty. To see this, let us recall that a theory is called deterministic with respect to systems of some particular kind if, for any such system, it specifies (a) a set of magnitudes called variables of state, whose values at a given time are said fully to characterize the state of the system at that time, and (b) a set of laws which, given the state of such a system at some time t_0, mathematically determine the state of that system at any other time. Newton's theory of gravitation and of motion is a deterministic theory with respect to any isolated system of point masses exerting only gravitational forces upon each other—provided that the notion of state for such a system is limited to the positions and momenta of the constituent point masses. In astronomical applications of the theory, the point masses are represented by astronomical bodies that are small by comparison with their distances; and the theory has been used with great success to compute earlier or later states on the basis of a given one. Astronomical objects may also change, of course, in many respects other than position and momentum: in mass, temperature, surface texture, internal structure, magnetic characteristics, and so forth. On these, however, the Newtonian theory has no bearing, for they cannot be expressed in terms of its variables of state; and Newton's theory is deterministic only with respect to its specified, quite narrow, concept of state. Systems covered by a deterministic theory may then be called deterministic systems *with respect to the relevant notion of state*.[4]

Universal determinism may then be viewed as an ontological thesis to the effect that the entire universe is deterministic in all respects, that it forms a deterministic system with respect to a concept of state sufficiently comprehensive to permit a full characterization of every empirical feature of the universe at a given time. More explicitly, this would come to the assertion that there is a true theory which is deterministic and whose concept of state is so rich that all aspects of the universe at any given time—physicochemical, biological, psychological, sociocultural, etc., and also aspects of kinds that we may be entirely unaware of—can be fully specified, down to the last detail, by a suitable assignment of numerical values to the variables of state of that theory.

Under this boundlessly inclusive construal of the total state of the universe, however, determinism turns into a trivial truth. For it is surely one of the characteristics of the state of the universe at time t_0 that at a certain other time t_1—i.e., at a specified interval before or after t_0—it displays such and such features. Accordingly, a complete specification of the state of the universe at t_0 would have to include a full characterization of the universe at all times; and thus, the total state of the universe at one time would trivially determine its total state at all other times—all those states being identical. Any significant statement of universal determinism requires, therefore, a suitable limitation of those features that are to count as constituting the state of the universe at any given time.

Here, then, is another example of the contrast between a conception explicitly concerned with systems and states characterizable by means of the conceptual apparatus of a specified theory and a related, self-defeatingly comprehensive ontological idea. And as in the cases of mechanism and the thesis of universal psychophysical association, a linguistic turn of determinism, involving relativization with respect to some theory, however rich, cannot preserve the full intuitive intent of its ontological conception.

5. Reduction of Theoretical Principles

Although the ontological theses of mechanism in biology and of universal psychophysical association have proved to be obscure in important respects, they are by no means totally unintelligible; in fact, they might be regarded as overextensions of much more restricted and intelligible assertions concerning the reducibility of the concepts and principles of some particular scientific theory to those of another. How the relevant notion of reduction might be construed is suggested in part by the preceding discussion. Let us consider a few of the details.

The reduction of the principles, or statements, of one theory, for example, a biological theory B, to those of another, such as some physicochemical theory P, would consist in the derivation of the statements of B from those of P. But a statement of B contains terms of the characteristic vocabulary, V_B, of B; whereas the statements of P do not. The derivation of the former from the latter will therefore normally require additional premises in the form of statements that contain terms from the vocabularies of both B and P. It is possible, to be sure, to deduce from any sentence containing only physicochemical terms various sentences which contain biological terms essentially (i.e., which are not logically equivalent to sentences not containing those biological terms). For example, the sentence

(S_1) When sodium chloride crystals are put into water, they dissolve

logically implies

(S_2) When sodium chloride crystals are put into water they dissolve or turn into bacteria;

and the statement

(S_3) Any body falling freely near the earth moves with constant acceleration

implies

(S_4) Any living body falling freely near the earth moves with constant acceleration.

But these reductive derivations are not specifically biological, even though each of the derived sentences S_2 and S_4 contains an essential occurrence of a biological term. For if that term is replaced by any other term of the same logical type—whether from the vocabulary of biology or from that of some other discipline—the resulting sentence will equally follow from the given physicochemical premise. Thus, if in S_2 the word "bacteria" is replaced by "fish," by "nonbacteria," or by "iron filings," the resulting sentence still follows from S_1; and similarly, S_3 also implies the sentences obtainable from S_4 by replacing "living" by "nonliving," "iron," "spherical," etc.

It is the absence of this peculiarity which characterizes what might be called reductions specific to biology. In these, the derivation of a biological principle from physicochemical ones must make use of additional premises which contain biological as well as physicochemical terms essentially, and which thus establish connections between the two theories. Following Ernest Nagel's similar usage,[5] will refer to them as connecting principles. For example, the chemical analysis of photosynthesis and cellular respiration in plants permits a reduction of some uniform features of these phenomena to physicochemical laws. Such reduction makes use of connecting principles broadly to the effect that the plant cells involved in photosynthesis contain certain characteristic chemicals, including chlorophyll; that the latter is a mixture of two substances of such and such molecular structures; and so forth. To this information, principles of physics and chemistry can then be applied to account for certain physicochemical processes that take place in the cells and for the production of energy by oxidation, which is the principal aspect of cellular respiration. In its barest logical essentials, the structure of the resulting reductive accounts may be schematically characterized as follows: A law or theoretical principle couched in terms of physicochemical theory

$$(x)\,(Q_1 x \supset Q_2 x)$$

is combined with connecting principles

$$(x)\,(R_1 x \supset Q_1 x)$$
$$(x)\,(Q_2 x \supset R_2 x)$$

to derive a biological law

$$(x)\,(R_1 x \supset R_2 x)$$

The reduction of a theoretical principle of B to principles of P thus does not require a definitional or quasidefinitional reduction of the relevant biological terms; the connecting principles need not provide full definitions nor necessary and sufficient conditions of application for the biological terms in physicochemical language; statements expressing necessary or sufficient physicochemical conditions for biological concepts may well suffice.

The discovery of such principles is a matter of scientific research; and the connecting principles that are presently available do not, of course, even remotely suffice to reduce all the laws and theoretical principles of current biological theory to those of current physicochemical theory.

6. Reduction of Theoretical Terms

Even a set of connective principles powerful enough to permit a reduction of all the principles of a biological theory B to those of a physicochemical theory P need not permit a corresponding reduction of concepts; i.e., it need not provide a characterization of every concept of B by means of concepts of P. And only if this, too, had been accomplished could the biological theory be said to have been reduced to a chapter of the physicochemical one. For short of this, the reduction of B to P would amount to the establishment of a set of connecting principles which, by our earlier criterion, would constitute additional biological laws and would thus simply expand the theory B, but would not make its conceptual apparatus dispensable.

Full reduction of concepts in this strict sense would require, for every term of B, a connective law of biconditional form, specifying a necessary and sufficient condition for its applicability in terms of concepts of P alone. Such a law could then be used to "define" the biological term and thus, theoretically, to avoid it. The "definition" of the temperature of a gas as the mean kinetic energy of its molecules illustrates this kind of connective principle. Examples pertaining to biological concepts are provided by the connective principles characterizing such substances are chlorophyll and various hormones in terms of their molecular structures. These principles do not, of course, express synonymies ascertained by meaning analysis, but empirical connections ascertained by chemical analysis. In fact, the biological terms and the corresponding chemical characterizations are not even strictly coextensive if the former are taken to refer to substances found in certain kinds of cell or produced by certain kinds of gland: the chemical "definition" of the biological terms then involves a broadening of their extensions, with the effect of covering also certain substances synthesized in the laboratory.

As was noted in Section 2, the possibility of such reduction of biological concepts is of crucial importance to mechanism; for the claim that all—or even certain specified—biological characteristics "are" basically physicochemical implies at least that they can be extensionally characterized in physicochemical terms. But as in the case of the theoretical principles, the reduction of all concepts of current biological theory (or of the terms standing for them) to those of current physics and chemistry is not remotely in sight. This fact does not, of course, lend support to such specific antimechanistic conceptions of life as neovitalism, which fails to meet the most basic requirements concerning the formulation and testability of scientific theories, or holism and emergentism, which involve various logical and methodological confusions—including an implicitly ontological construal of reduction—as Ernest Nagel has convincingly shown.[6]

The reasoning outlined in this section and in the preceding one can be applied analogously also to the questions concerning the reducibility of the concepts and laws of psychological theories to the concepts and laws of physicochemical and biological theories. Both kinds of reduction require the establishment of appropriate connective principles, and whatever principles of that kind may be currently available are very far indeed from providing a basis for a wholesale reduction.

7. Reduction and Scientific Change

Interest in the concept of theoretical reduction has recently increased as a result of the debate among philosophers and historians of science about the logical aspects of the process of scientific change in which one theory comes to be replaced by another. According to a familiar conception, the old theory will often be linked to the new one by a reductive relationship. Thus, classical thermodynamics is said to have been reduced to statistical mechanics; geometrical optics to wave optics; Galileo's and Kepler's laws to Newton's theory of motion and of gravitation, and so forth.

But our earlier characterization of reduction does not strictly fit these cases; for, as has been stressed by several writers, in none of these standard examples does the supposedly reducing theory imply the principles of the supposedly reduced theory; on the contrary, it contradicts them. Newton's theory implies that the acceleration of a body in free fall is not constant, as asserted by Galileo's law, but increases throughout the period of fall; and that when several planets move about the sun, their orbits will show perturbational deviations from an elliptic shape. Wave optics implies that light does not always travel in straight lines; and the kinetic theory is incompatible with the second law of thermodynamics in its strict classical form.[7] Thus, the notion of deductive reducibility of laws does not fit the logical relationship between such successive theories.

Indeed, several philosophers and historians of science have argued that the idea of deductive reducibility reflects a fundamentally mistaken conception which views scientific change as a cumulative process in which each new theory incorporates, and thus deductively implies, the knowledge embodied in its predecessor; whereas in fact, according to this view, the transition to a new theory involves the replacement of one conceptual scheme and its characteristic set of theoretical principles by another one that is incompatible and indeed incomparable with it, and whose concepts, even if represented by terms from the vocabulary of the preceding theory, have new and different meanings.[8]

This is not the place for a detailed examination of this conception of scientific change; but I do wish to comment here on some of Feyerabend's ideas on the subject, which are directly relevant to our topic. Feyerabend notes the logical incompatibility of many scientific theories with their predecessors and then comments on the consequences of this fact for the meanings of the key terms which are retained in the transition:

> After all, the meaning of every term we use depends upon the theoretical context in which it occurs. Words ... obtain their meanings by being part of a theoretical system. Hence if we consider two contexts with basic principles that either contradict each other or lead to inconsistent consequences in certain domains, it is to be expected that some terms of the first context will not occur in the second with exactly the same meaning.[9]

Feyerabend goes on to say that the conceptual systems of two mutually inconsistent theories must be "*mutually irreducible* (their primitives cannot be connected by bridge laws that are meaningful *and* factually correct)."[10]

But two theories using the same terms with different meanings may be perfectly compatible even if their formulations are contradictories of each other.[11] For example, the two formally contradictory expressions '$(x) (Fx \supset Gx)$' and '$(\exists x) (Fx . \sim Gx)$' need not be incompatible if the meanings of 'F' and 'G' in the first formula differ from those in the second.

Feyerabend's puzzling thesis that successive theories are incompatible and that their key terms differ in meaning may have its root in an overly narrow construal of the idea that terms "obtain their meanings by being part of a theoretical system." If one construes this idea as implying that the meanings of the key terms of a theory must be such as to make the theory true, then the terms are, as it were, "implicitly defined" by the theoretical principles, and the latter are true simply in virtue of the meanings thus assigned to the terms—they are analytic. And a formal incompatibility between two theories with the same key terms must then be taken to indicate simply that those terms have different meanings in the two systems.

But while this consideration may lend plausibility to Feyerabend's contention, its initial assumption hardly fits the character of scientific theorizing. For if the principles asserted by a scientific theory are implicit definitions of its key terms and hence analytic, the role of experiment and observation and the need for empirical evidence are thrown into question. If—to construct a schematic example—Galileo's law and Kepler's laws were taken to be definitive of "free fall" and of "planetary motion," then there would be no need for experimental or observational test. Moreover, empirical data on the actual fall of bodies near the earth or on the actual motion of planets about the sun would be irrelevant to those laws. If the findings did not conform to the law, this would show only that actual fall is not free fall as implicitly defined by Galileo's law or that the actual motion of the planets is not planetary motion as implicitly defined by Kepler's laws. The laws would be analytic; in order to make them applicable to their usual empirical subject matter and thus to restore the relevance of empirical testing, they would have to be supplemented by laws to this effect: the fall of a body in a vacuum near the surface of the earth is free fall as characterized by Galileo's formula; the motion of the planets about the sun is planetary motion as characterized by Kepler's laws. But the "theories" obtained by such supplementation clearly are no longer analytic; their terms are not implicitly defined by them. It does not follow, therefore, that the transition to a new theory using some or all of the key terms of its predecessor must involve a total change in the meanings of those terms, including even those that serve to describe observational findings.[12] Indeed, it is quite unclear how the two theories could be held to present different and even conflicting conceptions of some class of empirical phenomena if the sameness of the subject matter and the details of the theoretical conflict could not be characterized, at least in part, by means of terms whose meanings remain constant in the transition.

But if we grant that, in the development of science, a new theory will normally be incompatible with the one it replaces, and if on the other hand we reject the view that the two theories in question are conceptually "incommensurable" in the sense that "the meanings of their main descriptive terms depend on mutually inconsistent principles"[13]—then how are we to construe the logical relationship between such theories, and how is the concept of reduction to be modified to fit situations of this kind?

Kemeny and Oppenheim have proposed a construal of reduction[14] which, with slight stretching, might accommodate cases of this sort, though it was avowedly not intended for that purpose.[15] According to this construal, reducibility is not a relation just between two theories but a relation in which one theory, T_2, may stand to another, T_1, with respect to some set of observational data O. Briefly, T_2 is said to be reduced to T_1 relative to O if (a) T_1 explains any part of O that is explained by T_2, (b) T_1 is at least as well systematized as T_2 (this notion is made clear only in an intuitive fashion), and (c) the theoretical vocabulary of T_2 contains terms that are not contained in the theoretical vocabulary of T_1. If a more or less close, but not exact, accord of the data with the theoretical principles is taken to suffice for purposes of explanation, then this characterization would seem to fit the relationship of successive theories which, strictly speaking, are incompatible, provided that O is taken to be a set of data that T_2 has in fact served to explain. Thus, the Newtonian theory is much more highly systematized (in an intuitively plausible sense) than the combination of Galileo's and Kepler's laws; its theoretical vocabulary does not contain the terms "free fall" and "the sun," for example; and it accounts quantitatively for the empirical findings that bear out, and are explained by, the more limited set of laws.

But this construal of reduction also has some awkward features, which throw its adequacy into doubt. Suppose that two theories are equally well systematized, that they have no theoretical terms (and hence no theoretical principles) in common, and that each of them explains a certain set of data, O. Then either of the theories would count as reduced to the other (relative to O), even though they do not share a single theoretical term or principle. Reducibility of one theory to another requires more than that the latter account for a set of data that the former can explain: The point of the reduced theory is, after all, to offer general principles of which the observed cases are just particular instances; and the reducing theory will somehow have to account for these principles. In fact, the arguments that are actually used in particular instances to show that a given theory is reducible to another normally make little if any specific mention of the data explained by the reduced theory; they rather aim to show that the general laws asserted by the old theory are—within a limited domain, to which its supporting data were restricted—approximations of what the new theory implies for that domain. For example, the Newtonian theory implies that within small altitudes above the surface of the earth, free fall will have nearly constant acceleration, and that under specific conditions that are in fact met by the planets of our solar system, the planets will move in very close accordance with Kepler's laws.

Several writers have therefore suggested a liberalized construal of the reductive relationship between theories, or between a theory and a set of previously established empirical laws that the theory is said to explain. Under this construal, the new or reducing theory would be required to imply that, within a certain domain, close approximations of the principles of the old theory hold good. This view has recently been urged by Smart and by Putnam[16] in a discussion of Feyerabend's position. Feyerabend's reply is that in this kind of reasoning, "*real theories, theories which have been discussed in the scientific literature, are replaced by emasculated caricatures*"[17]—namely, by the approximations of them that are derivable from the new theory.

But this reply, it seems to me, is a verbal quibble. I think that both sides in the dispute have contributed important insights, as the following analogy—which in some respects is more than an analogy—might serve to show.

Consider a group of investigators who live in a small area on the surface of a large sphere and who try to ascertain the geometrical structure of that surface by suitable measurements. Initially, when their limited technology restricts their movements to relatively short distances and permits measurements of only moderate precision, they obtain a set of geometrical findings which are very satisfactorily explained by the theory E, that their surface is a Euclidean plane. Later, with an advanced technology that increases the region accessible to them and improves the accuracy of their measurements, he investigators obtain results that deviate more and more markedly from what E would lead them to expect; and a new theory S gains acceptance, which declares the surface to be spherical.

This new theory strictly contradicts its predecessor. But in a perfectly reasonable sense, we may say that S does not show E to have been totally mistaken; for S implies that within regions as small as that in which the original measurements were made, the propositions of E hold in very close approximation: the angle sum of a triangle equals almost exactly two right angles; the ratio of the circumference to the diameter of a circle is almost exactly equal to π; two "straight lines" which intersect a third one at equal angles do not intersect each other, and indeed their distance is practically constant; and so forth. In this sense, the new theory incorporates the old one as a close approximation applicable in a limited domain. To be sure, he new theory shows the old one in a new light; and as a rule, in the development of science, it is only in the light of the new theory that the nature and the extent of the limitations of the old one become identifiable and explainable. But within the limits thus indicated, the old theory may well continue to be used (as is illustrated by the cases of geometrical optics versus wave optics, classical mechanics versus relativistic mechanics, etc.).

It is certainly important to note that the new theory contradicts the old one, and that the resulting change in science is not cumulative in the strict sense of preserving the content of the old theory and simply adding on to it. But neither is the change entirely discontinuous; as the preceding considerations show, it does not, in general, amount to a total rejection of the earlier ideas and their replacement by new and "incommensurable" ones. Indeed, the two theories are concerned with the same subject matter in a sense that is intuitively clear and that can be made more explicit by examining the relations between the key concepts of the theories E and S.

The vocabularies of these theories overlap to a large extent; both of them contain the terms "point," "line," "incident on," "intersects," "angle," "triangle," "circle," and many others. Do these terms have the same meanings in the two systems?

The standards of a very narrow "operationism" might suggest an affirmative answer on the ground that the basic operational criteria of application for each of these terms are the same in both cases: in either theory, a straight-line segment would be operationally represented by a straightedge or by a taut string; points might be interpreted as small marks on the surface under study or as scratches on straightedges or as knots in strings, congruence of line segments by the coincidence of boundary markers. Also, the basic measurement of distances and of angles would follow the same procedure in both cases, and so would the construction of triangles or circles. There is good reason, therefore, for denying that the two theories are incomparable, and for maintaining that they concern the geometrical structure of the same surface, and that the concepts in terms of which they characterize that structure agree to a considerable extent in their empirical reference as characterized by their operational criteria of application. Indeed, otherwise it is not clear how they could be held to contradict each other. The contradiction manifests itself, for example, in the fact that according to S, the ratio of the circumference of a circle to its diameter is less than π, whereas according to E, it equals π; and these conflicting claims lend themselves to test by means of the common operational procedures specified for the relevant measurements in both theories.

But the agreement in basic empirical interpretation assuredly cannot be taken to show that the meanings of the terms in question remain entirely unchanged in the transition from E to S; in a plausible sense, they are indeed affected by the difference of the theoretical principles in which they function. This must be admitted even from a more liberal operationist point of view; for the principles of a scientific theory normally give rise to what might be called derivative operational criteria of application for some of its terms. Thus, the principles of E imply that two lines which intersect a third one at right angles are parallel; and this consequence may be used as one operational criterion for parallelism. But this criterion is not transferable to the theory S, whose basic principles do not yield the requisite implication; indeed, there are no parallels at all in spherical geometry.

There remains the question whether, apart from the matter of derivative operational criteria, the meanings of scientific terms are determined by the theoretical principles in which they occur.[18] Judging the issue intuitively, it seems plausible to say that to some extent this is indeed the case. Thus, the principle that mass is additive—that the mass of a physical system equals the sum of the masses of its components—might be regarded as expressing part of what the term "mass" means in classical mechanics; and the same view might be taken of the principle of conservation of mass. Similarly, the law that all electric charges are integral multiples of the charge of the electron might be regarded as expressing part of what is meant by the term "electron."

But, plausible though they are, these construals may well be questioned. Suppose, for example, that the existence of particles with charges smaller than that of the electron were to be established. It is quite likely that the term "electron" would then continue to be used, but without the claim that the electronic charge is the smallest possible one. In this event, would we have to say that there had been a change in the meaning of the word "electron," that we were no longer speaking of the same kind of thing as before—or could the change be described instead as one in our empirical beliefs about electrons, as a change showing that we had learned something new about electrons? I think the process could be described in either way; neither of them is unequivocally correct nor even decisively more illuminating than the other.[19]

The heated debate over the question whether theoretical terms change their meanings as a result of changes in theory has largely been carried on without benefit of a clearly delineated notion of meaning by reference to which claims and counterclaims might be adjudged. What would be needed is a concept, and an associated criterion, that would determine a division of all sentences asserted by science at any given time into two classes: those which reflect the meanings of their constituent terms and which are true solely by virtue of those meanings; and those expressing empirical claims. As for this desideratum, however, I share the doubts expressed by Quine[20] and others concerning the adequacy of the various philosophical attempts at explicating the analytic-synthetic distinction—and even concerning the significance of that distinction. What is being claimed for a statement by saying that it holds true by virtue of the meanings of its constitutive terms? One familiar conception of analyticity would suggest that the sentence is then regarded as inaccessible to refutation by empirical evidence. But, as we noted, the statement that electrons carry the smallest possible charges, even if viewed as reflecting part of the meaning of "electron," may well be given up in the face of certain new empirical findings. The reply that the change in this case is one in meaning rather than in beliefs concerning empirical facts is hardly very helpful for an attempt to clarify the distinction between truth in virtue of meanings and truth in virtue of facts.

In any particular scientific investigation of a given subject matter—such as electrons and their charges—it will surely be necessary to agree on some identifying characteristics of the subject matter and of those of its features that are to be examined; and the statements specifying those characteristics may then be said to indicate the meanings of the relevant terms. But the class of statements serving this purpose for a given term or set of terms may change, to some extent, from one investigation to another. Statements might perhaps be said, with Quine, to be more or less central to the meanings of the terms they contain, greater centrality indicating greater reluctance to abandon the statement for the purpose of accommodating new evidence, and also a stronger inclination to view such abandonment as effecting a conceptual change. For the term "electron," for example, each of the following statements would seem to have greater centrality than the one following it: electrons are subatomic particles; electrons carry uniform negative charges; any electric charge is an integral multiple of the charge of the electron; the charge of an electron equals 4.802×10^{-10} electrostatic units. Laws or theoretical principles that may plausibly be construed as relevant to the logical form of a term would be very central to its meaning; thus, the transition from classical mechanics to the special theory of relativity is often said to have effected a change in the concept of mass, which in the classical theory represents an intrinsic characteristic of a physical body, in relativity theory a relational feature.[21]

Our discussion of conceptual change had as its point of departure a comparison of the two theories E and S in our example. The relations we noted between these theories seem to me to obtain also between a set of empirical laws (such as Galileo's and Kepler's) and a theory that is said to explain them (e.g., Newton's theory of gravitation and of motion); and similarly between two theories of which the first is said to be reducible to the second (as in the case of classical mechanics and the mechanics of special relativity, or geometrical optics and wave optics, or classical thermodynamics and statistical mechanics, or Bohr's theory of atomic structure and Sommerfeld's refinement of it).

But the analogy to the relation between E and S does not extend, it seems, to all pairs of scientific theories one of which supersedes the other. The phlogiston theory and the oxygenation theory of combustion are not related in such a way that approximations of the principles of the former are derivable from the latter; according to the oxygenation theory, phlogiston does not participate sometimes or to some degree in chemical reactions: there just is no such thing as phlogiston, and combustion is a process quite different from dephlogistication. Cases of this kind seem to accord better with the conception, proposed by Kemeny and Oppenheim, that one theory is reduced to another if the second accounts in a more systematic and economical fashion for the empirical findings explainable by the first.

In sum, then, the construal of theoretical reduction as a strictly deductive relation between the principles of two theories, based on general laws that connect the theoretical terms, is indeed an untenable oversimplification which has no strict application in science and which, moreover, conceals some highly important aspects of the relationship to be analyzed. But the characterization of that relationship in terms of incompatibility and incommensurability overemphasizes certain significant differences and neglects those affinities by virtue of which the reducing theory may be said to offer a more adequate account of the subject matter with which the reduced theory is concerned.

NOTES

1. R. Carnap, *The Logical Syntax of Language*; New York: Harcourt, Brace and Co., 1937, Part V. A less technical statement of the basic idea is given in Carnap, *Philosophy and Logical Syntax*; London: Kegan Paul, Trench, Trubner and Co., 1935.
2. See, for example, pp. 249–250 of R. Carnap, *Introduction to Semantics*; Cambridge, Mass.: Harvard University Press, 1942.
3. J. A. Shaffer, "Recent Work on the Mind-Body Problem," *American Philosophical Quarterly* 2, 81–104 (1965); p. 93. (The identity theory is not, however, the only one that would account for the empirical association: A dualistic parallelism would be another possibility; but this issue will not concern us here.)
4. For a thorough discussion of the concepts of deterministic theory and deterministic system, see Nagel, *The Structure of Science*; New York: Harcourt, Brace and World, 1961; chapter 10.
5. Nagel refers to the requirements that there be such principles as the "condition of connectability" (see ibid. p. 354, p. 433). In accordance with the convention introduced in Section 2, such connecting principles would count as additional biological laws.
6. See Nagel, ibid., Chapter 11, Parts IV and V, and Chapter 12. On pp. 364–366, Nagel comments critically on what is, in effect, an implicitly ontological construal of reduction.
7. For a fuller discussion of these and other examples, see P. K. Feyerabend, "Problems of Empiricism," in R. G. Colodny (ed.), *Beyond the Edge of Certainty*, Englewood Cliffs, N. J.: Prentice-Hall, Inc., 1965, pp. 145–260; especially pp. 166–177.
8. This brief summary, and the additional remarks in the following discussion, cannot, of course, do justice to the many facets of this challenging thesis, nor do they differentiate between the views and the supporting arguments of different proponents of this idea. Fuller accounts by leading advocates of this view will be found, for example, in the following publications: P. K. Feyerabend, "Explanation, Reduction, and Empiricism," in H. Feigl and G. Maxwell (eds.) *Minnesota Studies in the Philosophy of Science*, Volume III, Minneapolis: University of Minnesota Press, 1962, pp. 28–97; "Problems of Empiricism"; "Reply to Criticism," in R. S. Cohen and M. W. Wartofsky (eds.), *Boston Studies in the Philosophy of Science*, Volume II: Humanities Press, 1965, pp. 223–261. N. R. Hanson, *Patterns of Dis-*

covery, Cambridge: Cambridge University Press, 1958. T. S. Kuhn, *The Structure of Scientific Revolutions*, Chicago: University of Chicago Press, 1962. The central ideas of these writers, and especially of Feyerabend, on scientific change are lucidly surveyed and critically examined in D. Shapere, "Meaning and Scientific Change," in R. G. Colodny (ed.), *Mind and Cosmos*, Pittsburgh: University of Pittsburgh Press, 1966, pp. 41–85.

9. Feyerabend, "Problems of Empiricism," p. 180.
10. Ibid. (italics quoted).
11. This has been emphasized also by Shapere (see, for example, "Meaning and Scientific Change," p. 57). Feyerabend has tried to meet this objection (see his "Reply to Criticism," pp. 231–233), but I have been unable to understand the possibilities he suggests of escaping the difficulty.
12. Feyerabend deals with the meanings of observational terms and observational reports in "Problems of Empiricism," pp. 180–181, 202–218.
13. Feyerabend, "Problems of Empiricism, p. 227.
14. J. G. Kemeny and P. Oppenheim, "On Reduction," *Philosophical Studies* VII, 6–19 (1956).
15. Ibid. p. 13.
16. J. J. C. Smart, "Conflicting Views about Explanation," in R. S. Cohen and M. W. Wartofsky (eds.), *Boston Studies in the Philosophy of Science*, Volume II: Humanities Press, 1965, pp. 157–169; see pp. 160–161. H. Putnam, "How Not to Talk about Meaning," in the same volume, pp. 205–222; see p. 207.
17. Feyerabend, "Reply to Criticism," p. 229 (italics quoted).
18. For an illuminating critical discussion of this issue, supplemented by apt examples, see also P. Achinstein, "On the Meaning of Scientific Terms," *The Journal of Philosophy* 61, 497–510 (1964).
19. I find myself in essential agreement on this point with Achinstein's remark: 'The fact that a 'changed in meaning' label may be unwarranted in a given case does not necessarily imply that an 'unchanged in meaning' label is better. Both car be misleading . . . (ibid., p. 504) and with Shapere's conclusion: "Both the thesis of theory-dependence of meanings . . . and its opponent, the condition of meaning invariance, rest on the same kind of mistake (or excess). This does not mean that there is not considerable truth . . . in both theses." ("Meaning and Scientific Change," p. 70.)
20. As set forth, for example, in "Two Dogmas of Empiricism," reprinted in W. V. O. Quine, *From a Logical Point of View*; Cambridge, Mass.: Harvard University Press, 2d ed. 1961; and "Carnap and Logical Truth" in P. A. Schilpp (ed.) *The Philosophy of Rudolf Carnap*; La Salle, Illinois: Open Court, 1963; pp. 385–406. Carnap's reply on pp. 915–922 of the same volume is of great interest for the problem here under discussion, and so is H. Putnam's essay, "The Analytic and the Synthetic," in H. Feigl and G. Maxwell (eds.), *Minnesota Studies in the Philosophy of Science*, Vol. III; Minneapolis: University of Minnesota Press, 1962, pp. 358–397.
21. On this point, see the fuller discussions by Nagel, in *The Structure of Science*, p. 111 and by Feyerabend, in "Problems of Empiricism," pp. 168–169.

10

The Meaning of Theoretical Terms

A Critique of the Standard Empiricist Construal

1. The Standard Empiricist Construal of Scientific Theories

This essay is concerned with a general characterization of scientific theories that has been developed, with certain individual differences, by various thinkers sharing a broadly empiricist outlook and a precise logico-analytical approach to problems in the philosophy of science; among them, F. P. Ramsey, N. R. Campbell, R. B. Braithwaite, Hans Reichenbach, Rudolf Carnap, Herbert Feigl, and Ernest Nagel. This characterization is often referred to as the standard construal, or the standard analysis, of scientific theories.

Let me indicate first what was perhaps the basic problem this analysis was meant to clarify. A scientific theory usually accounts for a class of empirical phenomena by positing particular kinds of entities and processes, which are taken to be governed by specified laws of their own, and which are, intuitively speaking, farther removed from the realm of our everyday experience than are the phenomena the theory is to explain. In characterizing those entities and processes, a theory typically employs a set of new terms, which are said to form its theoretical vocabulary. They are new in the sense that they are not among those already in use in the given discipline; in particular, they do not occur in the vocabulary used to describe the phenomena to be explained.

Now, it seems obvious—and logical empiricism certainly took this view—that the sentences of such a theory can have objective empirical significance and can explain empirical phenomena only if the theoretical terms they contain have clearly specifiable meanings. Thus, there arose the problem of characterizing those meanings and indicating how they are assigned to the theoretical terms. This was one of the principal questions to which the standard analysis was addressed; let us call it the meaning problem for theoretical expressions.

In tackling this issue, the thinkers associated with the standard construal appear to have been committed to a fundamental assumption which was generally left tacit, and which imposed a significant constraint on what would count as a permissible solution. This was the idea that it should be possible to characterize the meanings of the theoretical terms of a given theory by explicitly linguistic means, namely, by specifying a set of sentences which interpret those terms in an empirical vocabulary that is fully and clearly understood. An adequate formulation of the theory would then contain those sentences, and it would be the task of philosophical analysis to set them apart from other sentences of the theory by indicating their distinctive characteristics. Thus, the solutions to the meaning problem for theoretical expressions were tacitly subjected to what I shall call the *requirement of explicit linguistic specification* of the meanings in question.

I now turn to a brief characterization of those features of the standard construal that are relevant for the following discussion. According to the standard conception, a theory can be analytically divided into two constituent classes of sentences. Roughly, and in realistic parlance, the first of these might be said to contain the internal principles of the theory, which specify its 'theoretical scenario' by describing the underlying entities and processes postulated by the theory and stating the laws or theoretical principles assumed to govern them. The other set might be said to contain the bridge principles, which indicate the ways in which occurrences at the level of the scenario are held to be linked to the phenomena the theory is to explain.

The standard construal characterizes the two classes in formal rather than in realistic parlance and refers to them by different names. The sentences of the first class are viewed simply as a set of sentential formulas containing certain uninterpreted extralogical constants, namely, the theoretical terms. Let us assume these formulas to be axiomatized; the result will be an uninterpreted axiomatized formal system. The standard construal conceives the first set as such a system; it is sometimes called the theoretical calculus, C. The sentences of the second class are viewed as affording interpretations of theoretical expressions in a vocabulary whose terms have fully determinate and clearly understood empirical meanings; they are said to constitute the set R of rules of correspondence, or of interpretative sentences, for the calculus C. According to the standard conception, then, a scientific theory can be characterized by means of two constituent classes of sentences, C and R, which jointly determine its content.[1]

2. Implicit Definition by Postulates

Let us now consider the kinds of answer that have been offered, within the framework of the standard construal, to the meaning problem for theoretical expressions.

According to all versions of the standard conception, the characteristic new terms of a theory are given an empirical interpretation, at least in part, by the rules of correspondence; but some versions hold further that the meanings of those terms are partly determined also by the postulates of the theoretical calculus, which are said to constitute 'implicit definitions' for its primitive terms.

This latter idea, which I will discuss first, has its roots in a remark made by Hilbert, in *Die Grundlagen der Geometrie*, to the effect that in his axiomatization of Euclidean geometry, the axioms of betweenness *define* the concept 'between', and the axioms of congruence *define* the concept of congruence. Hilbert, incidentally, did not use the term 'implicit definition' in this context: that was done by Schlick, who gave Hilbert's idea a construal not to be found, I think, in Hilbert's writings. According to Schlick, Hilbert had sought "to construct geometry on a foundation whose absolute certainty is nowhere endangered by appeals to intuition," and he had achieved this end "simply by stipulating: the primitive concepts are to be *defined* by their satisfying the axioms." This, Schlick says, 'is the so-called definition by axioms . . . or implicit definition'.[2]

Later, Reichenbach referred to Schick's conception in setting forth his well-known distinction between purely mathematical geometry and physical geometry. In geometry as an axiomatized mathematical theory, he held, the geometrical terms have no meaning other than that determined by the axioms; whereas in geometry as part of physical theory, the geometrical terms are assigned physical interpretations by means of what he called coordinative definitions: these turn the sentences of a purely mathematical system of geometry into a system of physical hypotheses. Reichenbach's characterization of physical geometry was thus an early—and very influential—example of the standard construal of a scientific theory. The idea that the postulates of a theoretical calculus implicitly define its theoretical terms was subsequently endorsed by several other empiricist writers.

But what exactly is being asserted by this doctrine of implicit definition for theoretical terms? The term 'definition' suggests terminological convention or legislation; and indeed, according to Schlick, the postulates of a mathematical theory acquire a definitory role by the *stipulation* that the primitives *are to be understood* so as to ensure the truth of the postulates. But whatever may be the merits of this view in regard to mathematical theories, which alone Schlick refers to in his discussion, it does not apply to empirical science. Indeed, if the truth of the theoretical postulates were enforced by terminological *fiat* then the entire theory would be made true a priori; it could be known to be true independently of any empirical evidence—and regardless, moreover, of what interpretations the correspondence rules may assign to the empirical terms.

But the theories of empirical science are in fact subjected to empirical test, and their sentences are held open to modification in response to new test results. The doctrine that the meanings of theoretical terms are implicitly specified, at least in part, by the theoretical calculus must therefore be rejected.

The reason just given for this rejection does not hinge at all on the recently much debated obscurities of the intensional notion of meaning, which have led Quine and other analytic philosophers to reject that notion, along with kindred ones like synonymy and analyticity, as hopelessly unclear. The considerations I have adduced remain in force also for a purely extensional construal of the doctrine of implicit definition. For such a construal would presuppose the stipulation that the theoretical terms are to be understood as having extensions of which the postulates are true, and that stipulation would, again, render the postulates true a priori.

Henceforth, I shall use the term 'interpretation' to refer ambiguously to the specification of meanings or intensions, and to the specification of extensions.

It is clear that the doctrine of implicit definition by postulates cannot be saved even in a weaker version, according to which not all postulates, but only a certain subset of them, referred to as meaning postulates, provide implicit definitions.[3] For there is nothing in science to show for the distinction between meaning postulates and empirical postulates of a theory. As Quine, in particular, has argued forcefully, no sentences in a theory are governed by a convention that protects them from rejection no matter what adverse evidence the theory may come to face.

The preceding considerations have a bearing on a provocative and suggestive thesis that has been eloquently propounded by Feyerabend and others, namely that, as Feyerabend puts it, "the meaning of every term we use depends on the theoretical context in which it occurs" (1965, p. 180), and that therefore a change of the theoretical principles in which a term is used effects a change in its meaning. The only plausible construal I can envisage for the basic thesis is that the meanings—the intensions or the extensions—of theoretical terms are or must be understood to be such as to make the corresponding theory true. But this assertion, as I have argued, is untenable. As a consequence, the thesis that every theoretical change entails a change in the meanings of the terms involved does not seem to me sufficiently clear to permit of a satisfactory appraisal.

3. Correspondence Rules: Basic Vocabulary

I now turn to the idea of the correspondence rules of a theory. As mentioned earlier, these are assumed, in the standard construal, to specify the intensions or the extensions of the theoretical terms by means of a vocabulary whose terms have definite and fully understood empirical meanings.

That vocabulary, the interpretation base, has usually been conceived as a set of observational predicates, each standing for some property or relation of physical objects that is directly observable in the sense that, under suitable conditions, a normal human observer is able to ascertain its presence or absence in a particular case by means of immediate observation, without reliance on instruments or inferences. The assumption of such an observational interpretation base made it possible to present theoretical knowledge in empirical science as grounded on the data of direct observation in ways made explicit by the correspondence rules. More specifically, the scheme provided an objective and public evidential basis for empirical science by construing the basic evidence for scientific theories as expressed in the form of 'observation sentences' asserting the presence or absence of a directly observable attribute in a particular instance:

and about such sentences, requiring for their formulation only an observational vocabulary, different normal observers would be in agreement.

Yet recent discussion has shown this conception untenable. Let me mention just one reason. What kinds of terms a so-called normal observer is able to apply to particular instances on the basis of unaided observation depends not only on his biological and psychological endowment as a member of the species *Homo sapiens*, but quite essentially also on his prior conditioning, especially on his linguistic and scientific training. As has been pointed out by several writers, among them Feyerabend and Putnam, suitably trained scientific observers will be able to apply, "on the basis of direct observation," and with high interpersonal agreement, a variety of terms which the man in the street cannot thus apply, and which surely were not meant, under the standard construal, to count as observational.

The expression 'observational predicate', therefore, cannot be taken to refer to a reasonably determinate class of predicates; it must rather be treated as a relational term, properly used in contexts of the form 'term t is an observational predicate for person p'. As a consequence, the public, intersubjective character of the evidence by which scientific theories are tested cannot be taken to be secured by the exclusive use of observational predicates in the description of the evidence.

Besides, the requirement of an observational interpretation base for scientific theories is unnecessarily artificial. The phenomena which a theory is to explain as well as those by reference to which it is tested are usually described in terms which are by no means observational in a narrow intuitive sense, but which have a well-established use in science and are employed by investigators in the field with high intersubjective agreement. I shall say that such terms belong to the *antecedently available vocabulary*. Often, such terms will have been introduced into the language of science in the context of an earlier theory. For example, Bohr's and Sommerfeld's theories of atomic structure were developed to account for certain characteristic features of the spectra of chemical elements. Those features were described in terms of wavelengths and intensities of radiation emitted or absorbed, and thus by means of a vocabulary that clearly is not observational in an intuitive sense; yet this vocabulary was used by physicists with high accuracy and interpersonal uniformity; the principles for their use, e.g., for the measurement of wavelengths, having been provided by earlier theories, including wave optics. It seems reasonable, therefore, to construe the interpretation base of a theory as consisting, not of observational predicates, but of antecedently available ones.

The concept of antecedent availability is, again, relational: a predicate, say 'electrically charged' or 'introverted', cannot be said to be antecedently available *tout court*, but only with respect to the introduction of a given theory. The concept is thus of a historic-pragmatic character. But although antecedent availability is a relative notion, the conception of the interpretation base as consisting of antecedently available terms affords a plausible construal of the public and intersubjective character of the evidential base for scientific theories by linking it to the uniformity with which the antecedent vocabulary is used by scientists trained in the field.

4. Correspondence Rules: Logical Form

The interpretative sentences of a theory have been variously construed as formulated within the language of science or as sentences effecting a semantical interpretation and thus belonging to a suitable metalanguage. My remarks will be limited to the first, more common, construal; my basic objections can readily be transferred, however, to the alternative one.

What is the logical form of interpretative sentences? For a philosophical analyst with strict standards of intelligibility, the most desirable form—indeed, perhaps the only acceptable one—would be that of a definitional biconditional stating, in terms of the antecedent vocabulary, a necessary and sufficient condition of applicability for the theoretical term to be interpreted. But it is now rather generally agreed that theoretical terms do not usually admit of such definitions. In the nature of the case, there can be no conclusive proof of this view: strict proofs concerning definability can be given at best for expressions of a precisely formalized language.[4] But what appeared to be plausible ways of defining theoretical terms in antecedent ones have repeatedly been found inadequate.

Thus, e.g., in his essay "Testability and Meaning," which was not even specifically concerned with theoretical terms, Carnap showed that in extensional languages, dispositional traits cannot be defined by reference to their characteristic manifestations, and he therefore construed dispositional predicates as introduced by means of "reduction sentences" of certain specific logical forms, which were said to provide only a partial or incomplete specification of meaning.

Interpretative sentences of quite a different form were envisaged by Ramsey and later by Carnap and by Braithwaite, namely, sentences that explicitly define some antecedently available term by means of theoretical ones; thus, terms like 'water' or 'chlorophyll' might be definable in terms of theoretically characterized molecular structures.

A more general conception was embodied in N. R. Campbell's idea of a 'dictionary' relating theoretical expressions to experimental ones; its entries were assumed to take the form of statements to the effect that a certain theoretical sentence is true if and only if a corresponding empirical sentence, couched in antecedently available experimental terms, is true.

But in scientific theories, theoretical expressions, i.e., terms or sentences, may be linked to expressions in antecedent terms in still various other ways. With a view to maximal generality and flexibility, the correspondence rules of a theory might be characterized as a finite set R of sentences such that R contains essential occurrences of at least some theoretical and some antecedent terms, but of no other extralogical terms; R is logically compatible with the calculus C; and C, as interpreted by R, has empirical implications, i.e., the conjunction of R and

C formally implies a set of sentences not implied by R alone, which contains essential occurrences only of antecedent terms.[5] This conception differs considerably from the earlier idea of coordinative definitions for individual theoretical terms: an interpretative system is not required to specify necessary or sufficient conditions for each theoretical term in C or even for some of them or at least for certain logical compounds of theoretical terms or for certain theoretical sentences; the notion of interpretation here characterized is global, it applies to a theoretical system as a whole.

5. Correspondence Rules: Force

But this liberalization leaves quite unaffected a basic difficulty which presents itself when we ask what specific claim is being made for a sentence by calling it an interpretative sentence of a theory.

The fact that interpretative sentences are often referred to as *rules* of correspondence, as operational *definitions,* or as coordinative *definitions* suggests that they are viewed as signaling terminological conventions. They would then be governed by a stipulation to the effect that the intensions or the extensions of the theoretical terms they contain are to be understood to be such as to make the interpretative sentences true. But this conception faces the same difficulty as the idea of implicit definition by theoretical postulates. The crucial point is again that there are no sentences in a scientific theory whose truth is vouchsafed by convention.

Let us note in particular that even when a sentence is introduced as a criterion of application for a theoretical term, it is not for that reason true by convention and thus true *a priori*. Suppose, for example, that in an early stage of the study of heat, the term 'temperature' is introduced by means of a sentence which numerically identifies the temperature of a body with the reading of a mercury thermometer brought into contact with it. Subsequent research employing this 'criterion sentence' may well lead to a theory of heat transfer which implies that unless the thermometer happens initially to have the same temperature as the body, there will be an exchange of heat between them, with the result that the temperature of the body is changed and thus not correctly recorded by the thermometer. This consequence implies that the original criterion sentence is false. Here, then, a sentence that was originally accepted by convention comes to be rejected in response to empirical findings, namely, those that lent support to, and prompted the acceptance of, the laws of heat transfer.

It may seem plausible to reply that the initial acceptance of the criterion sentence represents, after all, nothing more than the decision to use the word 'temperature' for the purpose of conveniently formulating certain sentences about thermometer readings, and that therefore empirical evidence cannot conflict with it, compelling its abandonment in the way in which the discovery of black swans compels the abandonment of the statement 'All swans are white'. At best—so the reply might continue—the evidence about heat transfer might be said to show that if the criterion sentence were preserved, then certain empirical laws, such as those of heat transfer, would have to be given a vastly more complicated form than if the criterion were abandoned in favor of a suitably refined version; thus, the evidence would *suggest* the latter course in the interest of theoretical economy. But this reply is not telling. For, as has often been noted, adverse evidence confronting a theory never compels the abandonment of a specific theoretical sentence: it can be accommodated by a variety of alternative changes in the theory: and the choice among these is again a matter of decision, informed by considerations of overall theoretical simplicity and fit.

The mode of introduction of the criterion sentence in our example illustrates, I believe, what Quine has called legislative postulation. This process, he remarks, "affords truth by convention unalloyed" (Quine, 1963, p. 395); but such truth, he adds, attaches only to a process of adoption and is not "a significant lingering trait of the legislatively postulated sentence" (1963, p. 396). I would argue, however, that what legislative postulation confers upon a sentence like the temperature criterion is not even initially and temporarily the semantic trait of truth, but the epistemic one of acceptance, of membership in the class of sentences accepted-as-true by science or by a scientist at the time; and such acceptance, of course, by no means entails truth. Once it is realized that legislative postulation confers acceptance rather than truth, it is not surprising that a sentence introduced by such convention may come to be abandoned later on: for acceptance, no matter what the supporting considerations may be, is conferred in science not permanently, but only until further notice.

Truth by convention, then, is not a trait that could serve to single out the interpretative sentences of a theory. But neither, we must add, is acceptance by convention: this is illustrated by our example.

6. Conclusion

We have considered several attempts to single out, from among the sentences of a theory, a subclass indicating what meanings or what extensions the theoretical terms are intended to have. All of those attempts proved to be unavailing.

Nor can a satisfactory solution of the meaning problem for theoretical terms be expected from a refinement of the kinds of approach here considered. For the root of the difficulties we have noted lies in a misconception inherent in the construal of the problem itself, namely, in the requirement of explicit linguistic specification, according to which a philosophically adequate answer to the meaning problem must produce a set of sentences which specify the meanings of the theoretical terms with the help of an antecedently available empirical vocabulary. This idea calls for the singling out, from among all the sentences asserted by a theory, of certain sentences which interpret the theoretical terms, and which hold true by virtue of linguistic legislation: and this general notion cannot be made good.

It might seem that a theory lacking an interpretation in terms that are clearly understood must be deemed, by strict analytic standards, not to be objectively intelligible, to lack objective cognitive significance. But the standard here invoked, which has been very influential in logical empiricism and, indeed, in much of analytic philosophy, is much too restrictive. New concepts can become intelligible, the use of new expressions can be learned, by means other than explicit linguistic interpretation; and, as the history of scientific theorizing illustrates, the new linguistic apparatus thus introduced can come to be employed with high interpersonal agreement.

Such uniformity of usage is achieved in part, to be sure, by the explicit formulation of a body of theoretical principles linking the new theoretical terms to each other and to antecedently available terms: but no partition of those sentences into interpretative and descriptive ones is required in this context. In addition, however, precision and uniformity in the use of theoretical terms are further secured through various kinds of conditioning by not explicitly linguistic means which the scientists receive in the course of their professional training.[6]

To conclude: the introduction of a new scientific theory normally extends the language of science in a manner that is not purely definitional. This much was, in fact, acknowledged within the tradition of logical empiricism ever since Carnap's characterization of reduction sentences as incomplete definitions; and as for the putative method of implicit definition, its proponents never claimed that it assigned unique extensions to the theoretical primitives. I have argued that it is unnecessary and indeed unwarranted to think of theoretical terms as introduced or governed at all by sentences with a special interpretative function characterized by a distinct logical or methodological status. There are no such sentences, and there is no need, therefore, in an analytic account of scientific theories, to make provision for them. Hence, at least one of the major problems to which the standard conception was addressed, the problem of meaning specification for theoretical terms, rests on a mistaken presupposition and thus requires no solution.

NOTES

1. A thorough and substantial discussion of the subject will be found in Nagel (1961, Chapter 5); briefer accounts are given, for example, in Feigl (1970) and Hempel (1970); an important original variant is that set forth by Carnap (1956). I have to omit from consideration in this essay a third constituent which some versions of the standard construal attribute to scientific theories, namely, a so-called analogy or model; on this subject, see Hesse (1961, 1963); Nagel (1961, pp. 107–117); and, for brief comments, Hempel (1970, Sections 2 and 5).
2. Translated from Schlick (1925, p. 31). Hilbert, I think, was concerned, not with the "certainty" of the postulates, but only with their consistency.
3. This conception, which goes back to ideas of Carnap and Kemeny, was recently advocated with specific reference to theories in Kyburg (1968).
4. The issue is discussed more fully in Hempel (1958, Section 7).
5. This is a somewhat modified version of the characterization of the "interpretative system" of a theory as proposed in Section 8 of my essay (1958). The changes made

here are intended to meet certain objections to my earlier formulation that have been raised in Chapter 5, Section 5 of Stegmüller's (1970) comprehensive and searching work.
6. For suggestive observations that bear on this point, see Kuhn (1970, pp. 187–198).

REFERENCES

Carnap, R., 1956, *The methodological character of theoretical concepts*, in: Minnesota Studies in the Philosophy of Science, vol. 1, eds. H. Feigl and M. Scriven (University of Minnesota Press, Minneapolis), pp. 38–76.
Feigl, H., 1970, *The "orthodox" view of theories: remarks in defense as well as critique*, in: Minnesota Studies in the Philosophy of Science, vol. 4, eds. M. Radner and S. Winokur (University of Minnesota Press, Minneapolis), pp. 3–16.
Feyerabend, P. K., 1965, *Problems of empiricism*, in: Beyond the Edge of Certainty, ed. R. G. Colodny (Prentice-Hall, Englewood Cliffs, New Jersey), pp. 145–260.
Hempel, C. G., 1958, *The theoretician's dilemma*, in: Minnesota Studies in the Philosophy of Science, vol. 2, eds. H. Feigl, M. Scriven and G. Maxwell (University of Minnesota Press, Minneapolis), pp. 37–98. Reprinted in C. G. Hempel, 1965, *Aspects of Scientific Explanation* (The Free Press, New York), pp. 173–226.
Hempel, C. G., 1970, *On the "standard conception" of scientific theories*, in: Minnesota Studies in the Philosophy of Science, vol. 4, eds. M. Radner and S. Winokur (University of Minnesota Press, Minneapolis), pp. 142–163.
Hesse, M., 1961, *Forces and Fields* (Thomas Nelson, London).
Hesse, M., 1963, *Models and Analogies in Science* (Sliced and Ward, London).
Kuhn, T. S., 1970, *The Structure of Scientific Revolutions*, 2nd edition (University of Chicago Press, Chicago).
Kyburg, H. E., 1968, *Philosophy of Science: A Formal Approach* (The Macmillan Company, New York).
Nagel, E., 1961, *The Structure of Science* (Harcourt, Brace and World, New York).
Quine, W. V., 1963, *Carnap and logical truth*, in: The Philosophy of Rudolf Carnap, ed. P. A. Schilpp (Open Court, La Salle, Illinois), pp. 385–406.
Schlick, M., 1925, *Allgemeine Erkenntnislehre*, 2nd edition (Springer, Berlin).
Stegmüller, W., 1970, *Theorie und Erfahrung* (Springer, Berlin).

11

On the "Standard Conception" of Scientific Theories

1. Theories: A Preliminary Characterization

Theories, it is generally agreed, are the keys to the scientific understanding of empirical phenomena: to claim that a given kind of phenomenon is scientifically understood is tantamount to saying that science can offer a satisfactory theoretical account of it.[1]

Theories are normally constructed only when prior research in a given field has yielded a body of knowledge that includes empirical generalizations or putative laws concerning the phenomena under study. A theory then aims at providing a deeper understanding by construing those phenomena as manifestations of certain underlying processes governed by laws which account for the uniformities previously studied, and which, as a rule, yield corrections and refinements of the putative laws by means of which those uniformities had been previously characterized.

Prima facie, therefore, the formulation of a theory may be thought of as calling for statements of two kinds; let us call them *internal principles* and *bridge principles* for short. The internal principles serve to characterize the theoretical setting or the "theoretical scenario": they specify the basic entities and processes posited by the theory, as well as the theoretical laws that are assumed to govern them. The bridge principles, on the other hand, indicate the ways in which the scenario is linked to the previously examined phenomena which the theory is intended to explain. This general conception applies equally, I think, to the two types of theory which Nagel, following Rankine, distinguishes in his thorough study of the subject,[2] namely, "abstractive" theories, such as the Newtonian theory of gravitation and motion, and "hypothetical" theories, such as the kinetic theory of heat or the undulatory and corpuscular theories of light.

If I and B are the sets of internal and bridge principles by which a theory T is characterized, then T may be represented as the ordered couple of those sets:

(1a) $\qquad T = (I, B).$

Or alternatively, and with greater intuitive appeal, T may be construed as the set of logical consequences of the sum of the two sets:

(1b) $$T = c(I \cup B)$$

'The formulation of the internal principles will typically make use of a *theoretical vocabulary* V_T, i.e., a set of terms not employed in the earlier descriptions of, and generalizations about, the empirical phenomena which T is to explain, but rather introduced specifically to characterize the theoretical scenario and its laws. The bridge principles will evidently contain both the terms of V_T and those of the vocabulary used in formulating the original descriptions of, and generalizations about, the phenomena for which the theory is to account. This vocabulary will thus be available and understood before the introduction of the theory, and its use will be governed by principles which, at least initially, are independent of the theory. Let us refer to it as the *pretheoretical*, or *antecedent, vocabulary*, V_A, relative to the theory in question.

The antecedently examined phenomena for which a theory is to account have often been conceived as being described, or at least describable, by means of an observational vocabulary, i.e., a set of terms standing for particular individuals or for general attributes which, under suitable conditions, are accessible to "direct observation" by human observers. But this conception has been found inadequate on several important counts.[3]

The distinction I have suggested between theoretical and antecedent vocabularies hinges on no such assumption. The terms of the antecedent vocabulary need not, and indeed should not, generally be conceived as observational in the narrow sense just adumbrated, for the antecedent vocabulary of a given theory will often contain terms which were originally introduced in the context of an earlier theory, and which are not observational in a narrow intuitive sense. Let us look at some examples.

In the classical kinetic theory of gases, the internal principles are assumptions about the gas molecules; they concern their size, their mass, their large number; and they include also various laws, partly taken over from classical mechanics, partly statistical in nature, pertaining to the motions and collisions of the molecules, and to the resulting changes in their momenta and energies. The bridge principles include statements such as that the temperature of a gas is proportional to the mean kinetic energy of its molecules, and that the rates at which different gases diffuse through the walls of a container are proportional to the numbers of molecules of the gases in question and to their average speeds. By means of such bridge principles, certain microcharacteristics of a gas, which belong to the scenario of the kinetic theory, are linked to macroscopic features such as temperature, pressure, and diffusion rate; these can be described, and generalizations concerning them can be formulated, in terms of an antecedently available vocabulary, namely, that of classical thermodynamics. And some of the features in question might well be regarded as rather directly observable or measurable.

Take, on the other hand, the theoretical account that Bohr's early theory of the hydrogen atom provided for certain previously established empirical laws, such as these: the light emitted by glowing hydrogen gas is limited to certain characteristic discrete wavelengths, which correspond to a set of distinct lines in the emission spectrum of hydrogen; these wavelengths conform to certain general mathematical formulas, the first and most famous of which was Balmer's

$$\lambda = b \frac{n^2}{n^2 - 4}$$

Here, b is a numerical constant; and when n is given the values 3, 4, 5, ..., the formula yields the wavelengths of the lines that form the so-called Balmer series in the spectrum of hydrogen.

Now let us look briefly at the *internal principles* and the *bridge principles* of the theory by which Bohr explained these and other empirical laws concerning the hydrogen spectrum.

The *internal principles* formulate Bohr's conception that a hydrogen atom consists of a nucleus about which an electron circles in one or another of a set of discrete orbits with radii $r_1, r_2, r_3, \ldots$, where r_1 is proportional to i^2; that when the electron is in the ith orbit, it has an energy E_1 characteristic of that orbit and proportional to $(-1/r_i)$; that the electron can jump from a narrower to a wider orbit, or vice versa, and that in this process it absorbs or emits an amount of energy that equals the absolute difference between the energies associated with those orbits.

The *bridge principles*, which connect these goings-on with the optical phenomena to be explained, include statements such as these: (a) the light given off by glowing hydrogen gas results from the emission of energy by those atoms whose electrons happen to be jumping from outer to inner orbits; (b) the energy released by an electron jump from the ith to the jth orbit ($i > j$) is given off in the form of monochromatic electromagnetic waves with the wavelength $\lambda = (h \cdot c)/(E_1 - E_j)$, where h is Planck's constant and c the velocity of light.

As is to be expected, these bridge principles contain, on the one hand, certain theoretical terms such as 'electronic orbit' and 'electron jump,' which were specifically introduced to describe the theoretical scenario; on the other hand, they contain also certain antecedently available terms, such as 'hydrogen gas,' 'spectrum,' 'wavelength of light,' 'velocity of light,' and 'energy.' And clearly at least some of these terms—for example, 'wavelength of light' and 'hydrogen gas'— are not observational terms in the intuitive sense mentioned earlier. Nonetheless, the terms are antecedently understood in the sense indicated above; for when Bohr proposed his theory of the hydrogen atom, principles for their use, including principles for the measurement of optical wavelengths, were already available; they were based on antecedent theories, including wave optics.

2. The Construal of Theories as Interpreted Calculi

In the analytic philosophy of science, theories have usually been characterized in a manner rather different from the one just outlined; and, at least until recently, this characterization was so widely accepted that it could count as the "standard," or the "received," philosophical construal of scientific theories.[4] On this construal, too, a theory is characterized by two constituents, which, moreover, have certain clear affinities to what were called above its internal principles and its bridge principles.

The first constituent is an axiomatized deductive system—sometimes referred to as a calculus—of uninterpreted formulas, the postulates of the system corresponding to the basic principles of the theory. Thus, roughly speaking, the postulates of the calculus may be thought of as formulas obtained by axiomatizing the internal principles of the theory and then replacing the primitive theoretical terms in the axioms by variables or by dummy constants.

The second component is a set of sentences that give empirical import or applicability to the calculus by interpreting some of its formulas in empirical terms—namely, in terms of the vocabulary that serves to describe the phenomena which the theory is to explain. These sentences, which evidently are akin to the bridge principles mentioned above, were characterized by Campbell and by Ramsey as forming a "dictionary" that relates the theoretical terms to pretheoretical ones;[5] other writers have referred to them as "operational definitions" or "coordinative definitions" for the theoretical terms, as "rules of correspondence," or as "interpretative principles."

The standard conception, then, may be schematized by representing a theory as an ordered couple of sets of sentences:

(2) $$T = (C, R)$$

where C is the set of formulas of the calculus and R the set of correspondence rules.

Whereas the bridge principles invoked in our initial characterization of a theory are conceived as a subset of the class of sentences asserted by the theory, the status of the correspondence rules in the standard construal is less clear. One plausible construal of them would be as terminological rules belonging to the metalanguage of the theory, which stipulate the truth, by definition or more general terminological convention, of certain sentences (in the language of the theory) that contain both theoretical and pretheoretical terms. For this reason, no immediate analogue to (1b) is available as an alternative schematization of the standard view. The status of the correspondence rules will be examined further in section 6 below.

It would be a task of interest both for the history and for the philosophy of science to locate the origins of the standard conception and to trace its development in some detail. Such a study would surely have to take account of Reichenbach's characterization of physical geometry (i.e., the theory of the geometrical structure of physical space) as an abstract, uninterpreted system of "pure" or mathematical geometry, supplemented by a set of coordinative definitions for the primitives,[6] and it would have to consider Poincaré's and Einstein's views on the geometrical structure of physical space.

Campbell and some other proponents of the standard conception make provision for a third constituent of a theory—Campbell calls it an analogy, others

(Nagel among them) call it a model—which is said to characterize the basic ideas of the theory by means of concepts with which we are antecedently acquainted, and which are governed by familiar empirical laws that have the same form as some of the basic principles of the theory. The role of models in this sense will be considered later; until then, the standard conception of theories will be understood in the sense of schema (2). I have myself relied on the standard construal in several earlier studies,[7] but I have now come to consider it misleading in certain philosophically significant respects, which I will try to indicate in the following sections.

3. The Role of an Axiomatized Calculus in the Formulation of a Theory

My misgivings do not concern the obvious fact that theories as actually stated and used by scientists are almost never formulated in accordance with the standard schema; nor do they stem from the thought that a standard formulation could at best represent a theory quick-frozen, as it were, at one momentary stage of what is in fact a continually developing system of ideas. These observations represent no telling criticisms, I think, for the standard construal was never claimed to provide a descriptive account of the actual formulation and use of theories by scientists in the ongoing process of scientific inquiry; it was intended, rather, as a schematic explication that would clearly exhibit certain logical and epistemological characteristics of scientific theories.

This defense of the standard conception, however, naturally suggests this question: What are the logical and epistemological characteristics of theories that schema (2) serves to exhibit and illuminate? Let us consider in turn the various features which the schema attributes to a theory, beginning with the axiomatized calculus.

What is to be said in support of assuming axiomatization? It might quite plausibly be argued that an axiomatic exposition is an indispensable device for an unambiguous statement of a theory. For a theory has to be conceived as asserting a set of sentences that is closed under the relation of logical consequence in the sense that it contains all logical consequences (expressible in the language of the theory) of any of its subsets. A theory will therefore amount to an infinite set of sentences. In order to specify unambiguously the infinite set of sentences that a proposed theory is intended to assert, it will be necessary to provide a general criterion determining, for any sentence S, whether S is asserted by the theory. Axiomnatization yields such a criterion: S is asserted by the theory just in case S is deducible from the specified axioms or postulates.

This criterion determines membership in the intended set of sentences unambiguously, but it does not provide us with a general method of actually finding out whether a given sentence belongs to the set; for in general there is no effective decision procedure which, for any given sentence S, determines in a finite number of steps whether S is deducible from the axioms. But in any event, the standard construal assumes axiomatization only for the formulas of the uninterpreted calculus C rather than for all the sentences asserted by T,[8] so that the proposed supporting argument does not actually apply here.

One of the attractions the standard construal has had for philosophers lies no doubt in its apparent ability to offer neat solutions to philosophical problems concerning the meaning and the reference of theoretical expressions. If the characteristic vocabulary of a theory represents "new" concepts, not previously employed, and designed specifically to describe the theoretical scenario, then it seems reasonable, and indeed philosophically important, to inquire how their meanings are specified. For if they should have no clearly determined meanings, then, it seems, neither do the theoretical principles in which they are invoked; and in that case, it would make no sense to ask whether those principles are true or false, whether events of the sort called for by the theoretical scenario do actually occur, and so forth.[9] The answer that the standard construal is often taken to offer is, broadly speaking, that the meanings of theoretical terms are determined in part by the postulates of the calculus, which serve as "implicit definitions" for them; and in part by the correspondence rules, which provide them with empirical content. But this conception is open to various questions, some of which will be raised as we proceed.

As for the merits of axiomatization, its enormous significance for logic and mathematics and their metatheories needs no acknowledgment here. In some instances, axiomatic studies have served also to shed light on philosophical problems concerning theories in empirical science. One interesting example is Reichenbach's axiomatically oriented, though not strictly formalized, analysis of the basis and structure of the theory of relativity.[10] This analysis, which was undertaken some forty years ago, is technically distinctly inferior to more recent rigorous axiomatic formalizations; but it was nonetheless philosophically stimulating and illuminating, for it sought to clarify—much in the spirit of Einstein, I think—the roles of experience and convention in physical theorizing about space, time, and motion and the physical basis of the relativistic theory of spatial and temporal distances, of simultaneity, and so forth. More fundamentally, Reichenbach's investigations were intended as a critique, based on a specific case study, of the Kantian notion of *a priori* knowledge. Again, an axiomatic approach played an important role in von Neumann's argument[11] that it is impossible to supplement the formalism of quantum mechanics by the introduction by hidden parameters in a way that yields a deterministic theory.

Some contemporary logicians and philosophers of science consider the axiomatization of scientific theories so important for the purposes of both science and philosophy that they have expended much effort and shown remarkable ingenuity in actually constructing such axiomatic formulations. Some of these, such as those developed by Kyburg,[12] are small and relatively simple fragments of scientific theories in first-order logic; others, especially those constructed by Suppes and his associates, deal with richer, quantitative theories and formalize these with the more powerful apparatus of set theory and mathematical analysis.[13]

But some of the claims that have been made in support of axiomatizing scientific theories are, I think, open to question. For example, Suppes has argued that formalizing and axiomatizing scientific concepts and theories is "a primary method of philosophical analysis," and thus helps to "clarify conceptual problems and to make explicit the foundational assumptions of each scientific discipline," and that to "formalize a connected family of concepts is one way of bringing out their meaning in an explicit fashion."[14]

In what sense can an uninterpreted axiomatization be said to "bring out the meanings" of the primitive terms? The postulates of a formalized theory are often said to constitute "implicit definitions" of the primitives, requiring the latter to stand for kinds of entities and relations which jointly satisfy the postulates. If axiomatization is to be viewed as somehow *defining* the primitives, then it is logically more satisfactory to construe axiomatization, with Suppes, as yielding an explicit definition of a higher order set-theoretical predicate. In either case, the formalized theory is then viewed in effect as dealing with just such kinds of entities and relations as make the postulates true.[15]

This construal may have some plausibility for axiomatized purely mathematical theories—Hilbert adopted it in regard to his axiomatization of euclidean geometry—but it is not plausible at all to hold that the primitive terms of an axiomatized theory in empirical science must be understood to stand for entities and attributes of which the postulates, and hence also the theorems, are true; for on this construal, the truth of the axiomatized theory would be guaranteed *a priori*, without any need for empirical study.[16]

There are indeed cases in which axiomatization may be said to have contributed very significantly to the analytic clarification of a system of concepts. Suppes rightly mentions Kolmogorov's axiomatization of probability theory as an outstanding example.[17] But it should be noted that Kolmogorov's formal system admits of such diverse interpretations as Carnap's logical or inductive probability, Savage's personal probability, and the empirical construal of probability in terms of long-run relative frequencies. The latter, of central importance in empirical science, has presented vexing difficulties to philosophical efforts at a satisfactory explication. Von Mises, Reichenbach, Popper, Braithwaite, and others all have sought to explicate the concept of statistical probability, or to specify the principles governing its scientific use. Some of these principles concern the pure calculus of probability, with which alone Kolmogorov's axiomatization is concerned; others—and indeed the philosophically most perplexing ones—concern its application. And Kolmogorov's analysis does not touch at all on this second part of the problem of "bringing out the meaning" of the term 'probability' "in an explicit fashion."[18]

Generally speaking, the formalization of the internal principles as a calculus sheds no light on what in the standard construal is viewed as its interpretation; it sheds light at best on part of the scientific theory in question. And as for the claim that formalization makes explicit the foundational assumptions of the scientific discipline concerned, it should be borne in mind that axiomatization is basically an expository device, determining a set of sentences and exhibiting their logical relationships, but not their epistemic grounds and connections. A scientific theory admits of many different axiomatizations, and the postulates chosen in a particular one need not, therefore, correspond to what in some more substantial sense might count as the basic assumptions of the theory; nor need the terms chosen as primitive in a given axiomatization represent what on epistemological or other grounds might qualify as the basic concepts of the theory; nor need the formal definitions of other theoretical terms by means of the chosen primitives correspond to statements which in science would be regarded as definitionally true and thus analytic. In an axiomatization of Newtonian mechanics, the second law of motion can be given the status of a definition, a postulate, or a theorem, as one pleases; but the role it is thus assigned within the axiomatized system does not indicate whether in its scientific use it functions as a definitional truth, as a basic theoretical law, or as a derivative one (if indeed it may be said to have just one of these functions).

Hence, whatever philosophical illumination may be obtainable by presenting a theory in axiomatized form will come only from axiomatization of some particular and appropriate kind rather than just any axiomatization or even a formally especially economic and elegant one.

4. The Role of Pretheoretical Concepts in Internal Principles

The assumption, in the standard construal, of an axiomatized uninterpreted calculus as a constituent of a theory seems to me, moreover, to obscure certain important characteristics shared by many scientific theories. For that assumption suggests that the basic principles of a theory—those corresponding to the calculus—are formulated exclusively by means of a "new" theoretical vocabulary, whose terms would be replaced by variables or by dummy constants in the axiomatized calculus C. In this case, the conjunction of the postulates of C would be an expression of the type $\phi(t_1, t_2, \ldots t_n)$, formed from the theoretical terms by means of logical symbols alone. Actually, however, the internal principles of most scientific theories employ not only "new" theoretical concepts but also "old," or pretheoretical, ones that are characterized in terms of the antecedent vocabulary. For the theoretical scenario is normally described in part by means of terms that have a use, and are understood, prior to, and independently of, the introduction of the theory. For example, the basic assumptions of the classical kinetic theory of gases attribute to atoms and molecules such characteristics as masses, volumes, velocities, momenta, and kinetic energies, which have been dealt with already in the antecedent study of macroscopic objects; the wave theory of light uses such antecedently available concepts as those of wavelength and wave frequency; and so forth. Thus, the internal principles of a theory—and hence also the corresponding calculus C—have to be viewed, in general, as containing pretheoretical terms in addition to those of time theoretical vocabulary. Accordingly, the conjoined postulates of C would form an expression of the type $\psi(t_1, t_2, \ldots t_k, p_1, p_2, \ldots, p_m)$, where the t's again correspond to "new" theoretical terms, while the p's are pretheoretical, previously understood ones. Consequently, the theoretical calculus that the standard conception associates with a theory is not, as a rule, a totally uninterpreted system containing, apart from logical and mathematical symbols, only new theoretical terms.

It might be objected, from the vantage point of a narrow operationism, that in this new context, the "old" terms $p_1, p_2, \ldots p_m$ represent new concepts, quite different from those they signify in their pretheoretical employment. For the use of such terms as 'mass,' 'velocity,' and 'energy' in reference to atoms or subatomic particles requires entirely new operational criteria of application, since at the atomic and subatomic levels the quantities in question cannot be measured by means of scales, electrometers, and the like, which afford operational criteria for their measurement at the pretheoretical level of macroscopic objects. On the strict operationist maxim that different criteria of application determine different concepts, we would thus have to conclude that, when used in internal principles, the terms $p_1, p_2, \ldots p_m$ stand for new concepts, and that it is therefore improper to use the old pretheoretical terms in theoretical contexts: that they should be replaced here by appropriate new terms, which, along with $t_1, t_2, \ldots, t_k$, would then belong to the theoretical vocabulary.

But differences in operational criteria of application, as is well known, cannot generally be regarded as indicative of differences in the concepts concerned; otherwise, it would have to be held impossible to measure "one and the same quantity" in a particular instance—such as the temperature or the density of a given body of gas—by different methods, or even with different instruments of like construction; as a consequence, the diversity of methods of measuring a quantity, already at the macroscopic level, would call for a self-defeating endless proliferation and distinction of concepts of temperature, of concepts of density, and so forth.

Moreover, as long as we allow ourselves to use the notoriously vague and elusive notion of meaning, we will have to regard the meanings of scientific terms as reflected not only in their operational criteria of application, but also in some of the laws or theoretical principles in which they function. And in this context, it seems significant to note that some of the most basic principles that govern the pretheoretical use (relative to the classical kinetic theory, let us say) of such terms as 'mass,' 'velocity,' and 'energy' are carried over into their theoretical use. Thus, in the classical kinetic theory, mass is taken to be additive in the sense that the mass of several particles taken jointly equals the sum of the masses of the constituents, exactly as for macroscopic bodies. Similarly, the conservation laws for mass, energy, and momentum and the laws of motion are—at least initially—carried over from the pretheoretical to the theoretical level.

In fact, the principle of additivity of mass is here used not only as a pretheoretical and as an internal theoretical principle, but also as a bridge principle. In the latter role, it implies, for example, that the mass of a body of gas equals the sum of the masses of its constituent molecules; it thus connects certain features of the theoretical scenario with corresponding features of macroscopic systems that can be described in pretheoretical terms. Those different roles of the additivity principle are clearly presupposed in the explanation of the laws of constant and of multiple proportions, and in certain methods of determining Avogadro's number. These considerations suggest that the term 'mass' and others can hardly be taken to stand for quite different concepts, depending on whether they are applied to macroscopic objects or to atoms and molecules.

In support of the same point, it might be argued also that classical mechanics imposes no lower bounds on the size or the mass of the bodies to which the concepts of mass, velocity, kinetic energy, etc., can be significantly applied, and the laws governing these concepts are subject to no such restrictions either.[19] This suggests a further response to the operationist objection considered a moment ago: the application of classical mechanical principles indicates that macroscopic methods using mechanical precision scales, etc., are not sufficiently sensitive for weighing atoms, but that certain indirect procedures will provide operational means for determining their masses. Accordingly, the need for different methods of measurement indicates, not a conceptual difference in the meanings of the word 'mass' as used in the two contexts, but a large substantive difference in mass between the objects concerned.

Analogous arguments, however, are not applicable in every case where pretheoretical terms are used in the formulation of theoretical principles. According to current theory, for example, the mass of an atomic nucleus is less than the sum of the masses of its constituent protons and neutrons; thus the principles of additivity—and of conservation—of mass are abandoned at the subatomic level. Are we to say that this "theoretical change" indicates a change in the meaning of the term 'mass,' or rather that there has been a change in certain previously well-entrenched general laws which, before the advent of the new theory, had been erroneously believed to hold true of that one quantity, mass, to which both the new theory and the earlier one refer?

This question has received much attention in the debate over the ideas of Feyerabend, Kuhn, and some others concerning theoretical change in science and the theory-dependence of the meanings of scientific terms.[20] As the debate has shown, however, a satisfactory resolution of the issue would require a more adequate theory of the notion of sameness of meaning than seems yet to be at hand.

5. The Role of a Model in the Specification of a Theory

As mentioned earlier, some adherents of the standard construal regard a theory as having a third component, in addition to the calculus and the rules of correspondence, namely, "a model for the abstract calculus, which supplies some flesh for the skeletal structure in terms of more or less familiar conceptual or visualizable materials."[21]

In Bohr's theory of the hydrogen atom, for example, the postulates of the calculus would be the basic mathematical equations of the theory, expressed in terms of uninterpreted symbols such as 'i', 'r_i,' 'E_i.' The model specifies the conception, referred to earlier, of a hydrogen atom as consisting of a nucleus circled by an electron in one or another of the orbits available to it, etc. In this model, 'r_i' is interpreted as the radius of the ith orbit, 'E_i' as the energy of the electron when in the ith orbit, etc. The correspondence rules, finally, link the theoretical notion of energy emission associated with an orbital jump to the experimental concept of corresponding wavelengths or spectral lines, and they establish other linkages of this kind.

In discussing these three components of Bohr's theory, Nagel remarks that as a rule the theory is embedded in a model rather than being formulated simply as an abstract calculus and a set of correspondence rules because, among other reasons, the theory can then be understood with greater ease than the inevitably more complex formal exposition.[22] It seems, however, that in some cases the significance of models in Nagel's sense goes further than this, as I will try to indicate briefly.

The term 'model' has been used in several different senses in the philosophy of science. One of these pertains to what might be called analogical models, such as the mechanical or hydrodynamic representations of electric currents or of the luminiferous ether that played a considerable role in the physics of the late nineteenth and early twentieth centuries. Models of this kind clearly are not intended to represent the actual microstructure of the modeled phenomena. They carry an implicit 'as if' clause with them; thus, electric currents behave in certain respects as if they consisted in the flow of a liquid through pipes of various widths and under various pressures; the analogy lies in the fact that phenomena of the two different kinds are governed by certain laws that have the same mathematical form. Analogical models may be of considerable heuristic value; they may make it easier to grasp a new theory, and they may suggest possible implications and even promising extensions of it; but they add nothing to the content of the theory and are, thus, logically dispensable.

But this verdict does not seem to me to apply to what Nagel would call the models implicit in such theories as the kinetic theory of gases, the classical wave and particle theories of light, Bohr's theory of the hydrogen atom, the molecular-lattice theory of crystal structure, or recent theories of the molecular structure of genes and the basis of the genetic code. All these claim to offer, not analogies, but tentative descriptions of the actual microstructure of the objects and processes under study. Gases are claimed actually to consist of molecules moving about and colliding at various high speeds, atoms are claimed to have certain subatomic constituents, and so forth. To be sure, these claims, like those of any other scientific hypothesis, may subsequently be modified or discarded; but they form an integral part of the theory. For example, as I suggested earlier, if a model in Nagel's sense characterizes certain theoretical variables as masses, velocities, energies, and the like, this may be taken to indicate that certain laws which are characteristic of masses, velocities, and energies apply to those variables, and that, if some of those laws are suspended in the theory, the requisite modifications will be made explicit. This happened, for example, in Bohr's model, where—in contrast to classical electromagnetic theory—an orbiting electron is assumed to radiate no energy. Hence, the specification of the model determines in part what consequences may be derived from the theory and, hence, what the theory can explain or predict.

More specifically, it seems that when a scientific theory is axiomatized, the process is limited to the mathematical connections that the theory assumes between quantitative features of the scenario; other theoretically relevant aspects of the scenario arc specified by means of a model. I therefore agree with Sellars who remarks in a very similar vein that "in actual practice . . . the conceptual texture of theoretical terms in scientific use is far richer and more finely grained than the texture generated by the explicitly listed postulates," and that, in particular, the "thingish or quasi-thingish character of theoretical objects, their conditions of identity . . . are some of the more familiar categorical features conveyed by the use of models and analogies."[23] Thus, a model in the sense here considered is not only of didactic and heuristic value: The statements specifying the model seem to me to form part of the internal principles of a theory and as such to play a systematic role in its formulation.

It must be acknowledged that this way of formulating part of the internal principles of a theory is not fully specific and precise, that it does not provide an unequivocal characterization of exactly what statements the theory is meant to assert. But axiomatization, in the form of a "calculus," of part of a theory does not satisfy this desideratum, either; for it does not cover the correspondence rules; and for these, too, it seems virtually impossible to provide a formulation that could be regarded as adequate and complete. Indeed, as Nagel remarks,

> theories in the sciences . . . are generally formulated with painstaking care and . . . the relations of theoretical notions to each other . . . are stated with great precision. Such care and precision are essential if the deductive consequences of theoretical assumptions arc to be rigorously explored. On the other hand, rules of correspondence for connecting theoretical with experimental ideas generally receive no explicit formulation; and in actual practice the coordinations are comparatively loose and imprecise.[24]

6. The Status of Correspondence Rules

In the standard construal, schematized by (2) above, R is conceived as a class of sentences that assign empirical content to the expressions of the calculus; and their designation as operational *definitions*, coordinative *definitions*, or *rules of correspondence* conveys the suggestion that they have the status of metalinguistic principles which render certain sentences true by terminological convention or legislation. The sentences thus declared true—let us call them interpretative sentences—would belong to an object language containing both the calculus and the pretheoretical terms employed in its interpretation. The theoretical terms in the calculus are then best thought of as "new" constants that are being introduced into the object language by means of the correspondence rules for the purpose of formulating the theory. The interpretative sentences might have the form of explicit definition sentences (biconditionals or identities) for theoretical terms, or they might be of a more general type, providing only a partial specification of meaning for theoretical sentences, perhaps in the manner of Carnap's reduction sentences or still more flexible devices.[25] But at any rate they would be sentences whose truth is guaranteed by the correspondence rules.

But such a conception of correspondence rules is untenable for several reasons, among them the following:

First, scientific statements that are initially introduced by "operational definitions" or more general rules of application for scientific terms—such as the statements characterizing length by reference to measurement with a standard rod, or temperature in terms of thermometer readings—usually change their status in response to new empirical findings and theoretical developments. 'They come to be regarded as statements which are simply false in their original generality, though perhaps very nearly true within a restricted range of application, and possibly only under additional precautionary conditions. Most sentences warranted by operational definitions or criteria of application are eventually qualified as, strictly speaking, false by the very theories in whose development they played a significant role. Much the same point is illustrated by the following example: To "define" in experimental terms equal intervals of time, some periodic process may be chosen to serve as a standard clock, such as the swinging of a pendulum or the axial rotation of the earth as reflected in the periodic, apparent daily motion of some fixed star. The time intervals marked off by the chosen process are then equal by convention or stipulation. Yet it may happen that certain laws or theoretical principles originally based on evidence that includes the readings of standard clocks give rise to the verdict that those clocks do not mark off strictly equal time intervals. One striking example is the use of ancient astronomical reports of an almost purely qualitative character—concerning the date and very roughly the time of day when a certain total solar eclipse was observed at a given place—to establish a very slow deceleration of the earth's axial rotation, with a consequent slow lengthening of the mean solar day (by no more than .003 seconds in a century).[26]

Thus, even though a sentence may originally be introduced as true by stipulation, it soon joins the club of all other member statements of the theory and becomes subject to revision in response to further empirical findings and theoretical developments. As Quine has said, "conventionality is a passing trait, significant at the moving front of science, but useless in classifying the sentences behind the lines."[27]

These considerations might invite the following reply: Of course a theory—including its correspondence rules—may well undergo changes in response to new empirical findings; the question at issue, however, does not concern the possible effects of scientific change on correspondence rules, but rather the epistemic status of the interpretative sentences of a given theory, "frozen," as it were, at a particular point of its development. If such a theory is systematically characterized by means of a calculus and a set of interpretative sentences, do not the latter have the character of terminological conventions?

Here it should be recalled, first of all, that a theory usually links a given theoretical concept to several distinct kinds of phenomena that are characterizable in terms of the antecedently available vocabulary. For example, contemporary physical theory provides for several different ways of determining Avogadro's number or the charge of an electron or the velocity of light. But not all the interpretative sentences thus provided for a given theoretical term can be true by convention; for they imply statements to the effect that if one of the specified methods yields a certain numerical value for the quantity in question, then the alternative methods will yield the same value, and whether this is in fact the case is surely an empirical matter and cannot be settled by terminological convention. This point has indeed been stressed by some proponents of the standard conception. Thus, Carnap pointed out in his theory of reduction sentences that when a term is introduced (or interpreted, as we might say) by means of several reduction sentences, the latter taken jointly normally have empirical implications.[28]

Moreover, it is not clear just what claim is being made, in the context of a systematic exposition of a theory, by qualifying certain of its sentences as "true by convention." As was noted above, such designation may serve to make a historical point about the way in which those sentences came first to be admitted into the theory—but that is of no significance for a systematic characterization of the theory.

Nor, despite initial plausibility, can it be said that to qualify a sentence as true by rule or convention is to mark it as immune to revision in the eventuality that the theory should encounter adverse evidence. For, with the possible exception of the truths of logic and mathematics, no statement enjoys this kind of absolute immunity, as is illustrated by our preceding considerations, and as has been made very clear especially by Quine's critique of the analytic-synthetic distinction.[29]

The concept of bridge principle as invoked in our initial characterization of theories does not presuppose the analytic-synthetic distinction and treats the bridge principles as part of the theory, on a par with its internal principles. In fact, it should be explicitly acknowledged now that no precise criterion has been provided for distinguishing internal principles from bridge principles. In particular, the dividing line cannot be characterized syntactically, by reference to the constituent terms; for, as has been noted, both internal principles and bridge principles contain theoretical as well as antecedently available terms. Nor is the difference one of epistemic status, such as truth by convention versus empirical truth. The distinction is, thus, admittedly vague. But no sharp dividing line was needed for the use here made of the intuitive construal (1), namely, as a vantage point for a critical scrutiny of the standard conception.

7. On "Specifying the Meanings" of Theoretical Terms

Our critical scrutiny, however, has suggested no solution to one central question which the standard construal sought to answer, namely, the question of how the meanings of the "new" terms in a theory are specified. We found difficulties both with the conception that the postulates of the uninterpreted calculus provide implicit definitions for the theoretical terms and with the idea of correspondence rules as principles of empirical interpretation; but no alternative answer to the question has been offered. I believe now that the presumptive problem "does not exist," as Putnam has said and argued,[30] or, as I would rather say, that it is misconceived. In conclusion, I will briefly suggest some considerations in support of this view.

What reasons are there for thinking that the "new" concepts introduced by a theory are—or at least should be—specifiable by means of the antecedently available vocabulary? One consideration that influenced my earlier concern with the problem is, briefly, to this effect: A theory purports to describe certain facts, to make assertions that are either true or false. But a sentence will qualify for the status of being either true or false only if the meanings of its constituent terms are fully determined; and if we want to understand a theory, or to examine the truth of its claims, or to apply it to particular situations, we must understand the relevant terms, we must know their meanings. Thus, an adequate statement of a theory will require a specification of the meanings of its terms—and what other means is there for such a specification than the antecedently available vocabulary?

But even if, for the sake of argument, we waive questions about the concept of meaning here invoked, these considerations are not compelling. On the contrary; when at some stage in the development of a scientific discipline a new theory is proposed, offering a changed perspective on the subject matter under study, it seems highly plausible that new concepts will be needed for the purpose, concepts not fully characterizable by means of those antecedently available. This view seems to me to derive support from those studies of the language of science—especially in the logical empiricist tradition—which have led to a steady retrenchment of the initial belief in, or demand for, full definability of all scientific terms by means of some antecedent vocabulary consisting of observational predicates or the like. The reasons that led to countenancing the introduction of new terms by means of reduction sentences, interpretative systems, or probabilistic criteria of application all support the idea that the concepts used in a new scientific theory cannot be expected always to be fully characterizable by antecedently available ones.

But the very relaxation of the requirements for the introduction of new scientific terms gave rise to such questions as whether we can claim to understand such partially interpreted terms; whether the sentences containing them can count as significant assertions or can be regarded at best as an effective, but inherently meaningless, machinery for inferring significant statements, couched in fully understood terms, from other such statements; and whether reliance on incompletely interpreted theoretical terms could be entirely avoided in science.

But this way of looking at the issue presupposes that we cannot come to understand new theoretical terms except by way of sentences specifying their meanings with the help of previously understood terms; and surely this notion is untenable. We come to understand new terms, we learn how to use them properly, in many ways besides definition: from instances of their use in particular contexts, from paraphrases that can make no claim to being definitions, and so forth. The internal principles and bridge principles of a theory, apart from systematically characterizing its content, no doubt offer the learner the most important access to an "understanding" of its expressions, including terms as well as sentences.

To be sure, all these devices still leave unanswered various questions concerning the proper use of the expressions in question; and this may seem to show that, after all, the meanings of those expressions have not been fully specified, and that the expressions therefore are not fully understood. But the notion of an expression that has a fully specified meaning or an expression that is fully understood is obscure; besides, even for terms that are generally regarded as quite well understood there are open questions concerning their proper use. For example, there are no sharp criteria that would determine, for any strange object an astronaut might encounter on another planet, or indeed for any object that might be produced in a test tube on earth, whether it counts as a living organism. Theoretical concepts, just like the concept of living organism, are "open-ended"; but that, evidently, is no bar to their being adequately understood for the purposes of science.

NOTES

1. This essay develops further, and modifies in certain respects, some ideas set forth in an earlier paper, "On the Structure of Scientific Theories," published in *The Isenberg Memorial Lecture* Series, 1965–66 (East Lansing: Michigan State University Press, 1969), pp. 11–38. I am indebted to the Michigan State University Press for permission to include some passages from that essay in the present one.
2. E. Nagel. *The Structure of Science* (New York: Harcourt, Brace and World, 1961), pp. 125–129.
3. See the following discussions of the subject, which also give further references to the literature: H. Putnam, "What Theories Are Not," in E. Nagel, P. Suppes, A. Tarski, eds., *Logic, Methodology and Philosophy of Science* (Stanford, Calif.: Stanford University Press, 1962), pp. 240–251; R. Jeffrey's comments on this essay in *Journal of Philosophy*, 61 (1964), 80–84; G. Maxwell, "The Ontological Status of Theoretical Entities," in H. Feigl and G. Maxwell, eds., *Minnesota Studies in the Philosophy of Science*, vol. III (Minneapolis: University of Minnesota Press, 1962), pp. 3–27; P. Achinstein, *Concepts of Science* (Baltimore: Johns Hopkins Press, 1968), chapter 5.
4. The appellation "the 'received view' on the role of theories" is Putnam's ("What Theories Are Not," p. 240). Some characteristic stages in the evolution of this construal of scientific theories are represented by the following works: N. R. Campbell, *Physics: The Elements* (Cambridge: Cambridge University Press, 1920; reprinted as *Foundations of Science*, New York: Dover, 1957), chapter 5; F. P. Ramsey,

"Theories" (1929), in Ramsey, *The Foundations of Mathematics* (London: Routledge and Kegan Paul, 1931); R. Carnap, *Foundations of Logic and Mathematics* (Chicago: University of Chicago Press, 1939), especially sections 21–25; R. B. Braithwaite, *Scientific Explanation* (Cambridge: Cambridge University Press, 1953), chapters I–III; R. Carnap, "The Methodological Character of Theoretical Concepts," in H. Feigl and M. Scriven, eds., *Minnesota Studies in the Philosophy of Science*, vol. I (Minneapolis: University of Minnesota Press, 1956), pp. 38–76; R. Carnap, *Philosophical Foundations of Physics*, ed. M. Gardner (New York, London: Basic Books, 1966), part V; Nagel, *The Structure of Science*, chapters 5 and 6.

5. Campbell, *Foundations of Science*, p. 122; Ramsey, "Theories," p. 215.
6. This idea is set forth very explicitly in chapter 8 of H. Reichenbach, *The Rise of Scientific Philosophy* (Berkeley and Los Angeles: University of California Press, 1951).
7. For example, in my essay "The Theoretician's Dilemma," in H. Feigl, M. Scriven, and G. Maxwell, eds., *Minnesota Studies in the Philosophy of Science*, vol. II (Minneapolis: University of Minnesota Press, 1958), pp. 37–98.
8. Note by contrast that in the investigations by Ramsey (in "Theories") and by Craig concerning the avoidability of theoretical terms in favor of pretheoretical ones, axiomatization of the entire theory is presupposed. For a careful exposition and appraisal of those investigations, see I. Scheffler, *The Anatomy of Inquiry* (New York: Knopf, 1963), pp. 193–222. On Ramsey's method, see also Carnap, *Philosophical Foundations of Physics*, chapter 26; I. Scheffler, "Reflections on the Ramsey Method," *Journal of Philosophy*, 65 (1968), 269–274; and H. Bohnert, "In Defense of Ramsey's Elimination Method," *Journal of Philosophy*, 65 (1968), 275–281.
9. Various facets of this problem are carefully presented and explored in Nagel, *The Structure of Science*, chapter 6, and in Scheffler, *The Anatomy of Inquiry*, part II.
10. See H. Reichenbach, *Axiomatik der relativistischen Raum-Zeit-Lehre* (Braunschweig: Friedrich Vieweg und Sohn, 1924); and also Reichenbach's article, "Ueber die physikalischen Konsequenzen der relativistischen Axiomatik," *Zeitschrift für Physik*, 34 (1925), 32–48. In this article, to which Professor A. Grünbaum kindly called my attention, Reichenbach sets forth the main objectives of his axiomatic efforts; on pp. 37–38, he rejects as irrelevant to his enterprise Hermann Weyl's objection that Reichenbach's axiomatization is too complicated and opaque from a purely mathematical point of view.
11. J. von Neumann, *Mathematical Foundations of Quantum Mechanics* (Princeton, N.J.: Princeton University Press, 1955), chapter 4.
12. H. E. Kyburg, *Philosophy of Science: A Formal Approach* (New York: Macmillan, 1968).
13. A lucid and copiously illustrated introduction to this method of axiomatization by definition of a set-theoretical predicate is given in chapter 12 of P. Suppes, *Introduction to Logic* (New York: Van Nostrand, 1957).
14. P. Suppes, "The Desirability of Formalization in Science," *Journal of Philosophy*, 65 (1968), 651–664; quotations from pp. 653 and 654.
15. For a careful and illuminating critical examination of the construal of postulates as implicit definitions for the primitives see chapter II of R. Grandy, "On Formalization and Formalistic Philosophies of Mathematics" (doctoral dissertation, Princeton University, 1967). Concerning the restrictions that the requirement of truth for the postulates imposes on the permissible interpretations of the primitives, Grandy notes that it "is not a restriction on the constants alone but on the set of constants

plus the universe of discourse. A paraphrase of this is: The postulates implicitly define, if anything, the constants plus the quantifiers" (p. 41).

16. Kyburg therefore divides the axioms of a theory into "material axioms" and meaning postulates (in the sense of Carnap and Kemeny) and stresses that "we cannot lump [these] together and regard them as an *implicit definition* of the terms that occur in them" (*Philosophy of Science*, p. 124). It is presumably the meaning postulates alone that provide implicit definitions; but the distinction of two kinds of axioms is beset by the same difficulties as the analytic-synthetic distinction.
17. Suppes, "The Desirability of Formalization in Science," p. 654.
18. Suppes himself acknowledges that the "difficulty with the purely set-theoretical characterization of Kolmogorov is that the concept of probability is not sufficiently categorical" (ibid.), and he stresses that the interpretation of a formalized theory is logically much more complex than the talk of correspondence rules in "the standard sketch of scientific theories" would suggest (P. Suppes, "What Is a Scientific Theory?" in S. Morgenbesser, ed, *Philosophy of Science Today*, New York and London: Basic Books, 1967, pp. 55–67).
19. This point is made also by Achinstein, *Concepts of Science*, p. 114; indeed his discussion, on pp. 106–119, of ways in which theoretical terms are introduced in science presents many illuminating observations and illustrations which accord well with the view expressed in this section, and which lend further support to it.
20. See, for example, T. S. Kuhn, *The* Structure of *Scientific Revolutions* (Chicago: University of Chicago Press, 1962); P. K. Feyerabend, "Explanation, Reduction, and Empiricism," in Feigl and Maxwell, eds., *Minnesota Studies in the Philosophy of Science*, vol. III, pp. 28–97; P. K. Feyerabend, "Reply to Criticism," in R. S. Cohen and M. W. Wartofsky, eds.. *Boston Studies in the Philosophy of Science*, vol. II (New York: Humanities, 1965), pp. 223–261; N. R. Hanson, *Patterns of Discovery* (Cambridge: Cambridge University Press, 1958). For illuminating critical and constructive discussions of these ideas, and for further bibliographic references, see P. Achinstein, "On the Meaning of Scientific Terms," *Journal of Philosophy*, 61 (1964), 497–510; Achinstein, *Concepts of Science*, pp. 91–105; H. Putnam, "How Not to Talk about Meaning," in Cohen and Wartofsky, eds., *Boston Studies in the Philosophy of Science*, II, 205–222; D. Shapere, "Meaning and Scientific Change," in R. G. Colodny. ed., *Mind and Cosmos* (Pittsburgh: University of Pittsburgh Press, 1966), pp. 41–85; I. Scheffler, *Science and Subjectivity* (Indianapolis: Bobbs-Merrill, 1967), especially chapters 1, 3, 4.
21. Nagel, *The Structure of Science*, p. 90.
22. Ibid., p. 95. Nagel's detailed discussion of the subject (chapter 5 and pp. 107–117) calls attention also to other functions of models in his sense. among them their heuristic role.
23. W. Sellars, "Scientific Realism or Irenic Instrimentalism," in Cohen and Wartofsky, eds.. *Boston Studies in the Philosophy of Science*, II, 171–204; quotations from pp. 178–179.
24. Nagel, *The Structure of Science*, p. 99.
25. Such, perhaps, as interpretative systems of the kind I suggested in section 8 of "The Theoretician's Dilemma."
26. See N. Feather, Mass, *Length and Time* (Baltimore: Penguin, 1961), pp. 54–55.
27. W. V. O. Quine, "Carnap and Logical Truth," reprinted in Quine, *The Ways of Paradox and Other Essays* (New York: Random House, 1966), pp. 100–125; quotation from p. 112.

28. Cf. R. Carnap, "Testability and Meaning," *Philosophy of Science*, 3 (1936), 419–471, and 4 (1937), pp. 1–40, especially pp. 444 and 451. An analogous comment applies to interpretative systems, of course; see my "The Theoretician's Dilemma," p. 74.
29. Perhaps Quine's earliest detailed attack on the distinction is mounted in his classical "Two Dogmas of Empiricism" (1951). reprinted in W. V. O. Quine, *From a Logical Point of View*, 2nd ed. (Cambridge. Mass.: Harvard University Press, 1961). Another early critique is given in M. G. White, "The Analytic and the Synthetic: An Untenable Dualism," in S. Hook, ed., *John Dewey: Philosopher of Science and of Freedom* (New York: Dial, 1950).
30. Putnam, "What Theories Are Not," p. 241.

12

Limits of a Deductive Construal of the Function of Scientific Theories

1. The Deductivist Construal

The goal and the proudest achievement of basic scientific inquiry is the construction of comprehensive theories which enable us to understand large sectors of the world, to predict, to retrodict, to explain what occurs in them. Any theory, however far-reaching and successful, eventually proves wanting in some respects and comes to be replaced by a superior alternative. The search of basic science is unending.

This essay will not be concerned, however, with the ways in which scientific theories are arrived at and eventually changed. Rather, I propose to consider with you what is meant by saying that a given theory *applies* to a certain range of phenomena, or that it *explains*, *predicts*, or *retrodicts* them.

Specifically, I intend to discuss a widely held view, the so-called *hypothetico-deductive model*, according to which, briefly, to explain or describe a phenomenon by means of a theory is to deduce a sentence describing the phenomenon from the theory taken in conjuction with sentences providing certain initial information. I intend to argue that this conception has serious limitations which raise some fundamental problems concerning the character of scientific knowledge.

The best-known elaboration of this general conception is provided by the so-called *standard empirieist construal* of theories and their application. It views a theory T as analyzable into two components: a set C containing the basic principles of the theory and a set I of interpretative statements.

The sentences, or formulas, of C serve to characterize the specific entities and processes posited by the theory (e.g., elementary particles and their interactions) and to state the basic laws to which they are assumed to conform. The sentences of C will be formulated with the help of a theoretical vocabulary, V_C, that is characteristic of C and that refers to the kinds and characteristics of the theoretical entities and processes posited by the theory.

The sentences of the interpretative set I serve to link the theoretical scenario represented by C to the empirical phenomena to which the theory is to be applied. These phenomena are taken to be formulated in a vocabulary V_A which is antecedently understood, i.e., which is available and understood independently of, or logically prior to, the theory. Thus, the sentences of I are said to provide partial interpretations, though not necessarily full definitions, of the theoretical terms in V_C by means of the well-understood terms of V_A. So-called operational definitions and reduction sentences in Carnap's sense might be viewed as special kinds of interpretative sentences.

Schematically:

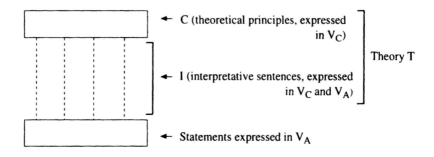

The application of T to phenomena described in terms of V_A, is then effected by means of a deductive argument of this type:

$$S_A^1 \xrightarrow{I^1} S_C^1 \xrightarrow{C^1} S_C^2 \xrightarrow{I^2} S_A^2$$

It leads from initial information S_A^1, which is expressed in V_A, via suitable interpretative sentences I^1. to a statement S_C^1 couched in terms of V_C; from there by means of basic theoretic laws C^1 to another statement S_C^2 containing theoretical terms, and finally, with the help of further interpretative sentences I^2, to a statement S_A^2 expressed in V_A that describes the phenomenon to be predicted or explained.

By way of a simple example, assume that T is an elementary theory of magnetism whose theoretical vocabulary V_C contains such terms as "magnet," "north pole," "south pole," and whose theoretical principles include the laws of magnetic attraction and repulsion and the law that the parts of a magnet are also magnets; the set I would include some operational criteria for the terms of the theory. This theory yields a prediction by virtue of the following argument:

$$\underbrace{\text{b is a metal bar to which iron filings cling}}_{S_A^1} \xrightarrow{I^1}$$

$$\underbrace{\text{b is a magnet}}_{S_C^1} \xrightarrow{C^1} \underbrace{\text{If b is cut in two, both parts are magnets}}_{S_C^2} \xrightarrow{C^2}$$

Opposite poles will attract each other $\underrightarrow{\quad I^2 \quad}$

$$S_C^3$$

When suspended, the parts will align themselves in a straight line

$$S_A^2$$

Deductivist construals in this general vein have been put forward, for example, by N. R. Campbell, R. Braithwaite, and—in an especially liberal version—by Carnap.

Carnap and other logical empiricists assumed that the vocabulary V_A, which serves to describe the phenomena to he explained by the theory, is *observational* at least in a broad sense, i.e., that its terms refer to features of the world whose presence or absence can be established by means of more or less direct observation. In recognition of the difficulties that face the notion of observability, I do not want to make any such assumption concerning V_A, here. Indeed, I want to provide specifically for cases in which, as often happens, the vocabulary V_A was originally introduced in the context of an earlier theory. All that the standard construal needs to assume is that the phenomena for which the theory is to account are described by means of a vocabulary V_A that is "antecedently available" in the sense that it is well understood and is used with high intersubjective agreement by the scientists in the field. The interpretative sentences in I may then be viewed as interpreting the new terms introduced by the theory, i.e., those in its theoretical vocabulary V_C, by means of the antecedently understood terms V_A.

At first blush, this kind of deductivist construal may appear eminently plausible, especially when one considers the rigorously deductive standards that govern the application of theories having precise quantitative form.

Yet this deductivist construal faces two basic difficulties. I will call them the problem of inductive ascent and the problem of provisos.

I will now spell out these two problems in turn and will consider some of their implications.

2. Theoretical or Inductive Ascent

Consider the first inferential step in our schematic argument about the bar magnet. It claims that with the help of interpretative sentences belonging to part I of the theory of magnetism, S_C^1 is deducible from S_A^1. Actually, however, the theory of magnetism surely contains no interpretative principle to the effect that when iron filings cling to a metal bar, then the bar is a magnet. The theory does not preclude the possibility, for example, that the bar is made of lead and is covered with an adhesive to which the filings stick, or that the filings are held in place by a magnet hidden under a wooden board supporting the lead bar. Thus, the theory does not warrant a deductive step from S_A^1 to S_C^1. It is more nearly correct to say that the theory contains an interpretative principle which is the *converse* of the one just considered, namely, that if a bar is a magnet, then iron filings will cling to it. But even this is not strictly correct, as I will argue in a moment.

Hence, the transition from S_A^1 to S_C^1 is *not deductive*, even if the entire theory of magnetism is used as an additional premise. Rather, the transition involves what I will call *inductive or theoretical ascent*, i.e., a transition from a data sentence expressed in V_A to a theoretical hypothesis S_C^1 which, by way of the theory of magnetism, would *explain* what the data sentence describes.

This illustrates *one* of the two previously mentioned problems that face a deductivist construal of the systematic connections which, on the standard construal, a theory establishes between V_A-sentences, i.e., between sentences describing empirical phenomena in terms of V_A. This problem of inductive or theoretical ascent has been widely discussed, and various efforts have been made to resolve it by constructing theories of inductive reasoning that would govern such theoretical ascent. I will not consider those efforts further but will rather turn to the problem of provisos, which has not, it seems to me, been formulated and investigated in nearly the same detail.

3. Provisos

Consider the third step in our example. the transition from S_C^1 to S_A^2. Again, the theory of magnetism does not provide hypotheses—I^2 in our example—which would turn this into a strictly deductive inference. The theory clearly allows for the possibility that two bar magnets, suspended by fine threads close to each other at the same level, will not arrange themselves in a straight line; for if a strong magnetic field of suitable direction should be present in addition, then the bars would orient themselves so as to be parallel to each other; similarly, a strong air current would foil the prediction, and so forth.

The theory of magnetism does not guarantee the absence of such disturbing factors. Hence, the inference from S_C^3 to S_A^2 presupposes as an additional premise the assumption that the suspended pieces of the bar are subject to no disturbing influence or, to put it positively, that their rotational motions are subject only to the magnetic forces they exert upon each other.

Incidentally, the explanatory inference I mentioned above, from S_C^1 to $S^1{}_A$, presupposes an analogous tacit premise; that is the reason why I hedged in describing it as deductive.

I will use the term *"provisos"* to refer to *assumptions* of the kind just illustrated, *which are essential, but generally unstated, presuppositions of theoretical inferences.*

Provisos are presupposed also in ostensibly deductive *theoretical* inferences leading from one V_1–sentence to another. This is the case, for example, in the inference from S_C^1 to S_C^2 in the case of the magnet: for example, if the breaking of the magnet takes place at a high temperature, the pieces may become demagnetized.

Or consider the application of the Newtonian theory of gravitation and of motion to a system of physical bodies like our solar system. In predicting, from a specification of the state of the system at a time t_0, subsequent changes of state, the basic idea is that the force acting on any one of the bodies is the vector sum of the gravitational forces exerted on it by the other bodies in accordance with the law of gravitation. That force then determines, via the second law of motion ($f=ma$), the resulting change of velocity and of position for the given body. But the quantity f in the second law is understood to be the *total* force acting on the given body; and the envisaged application of the theory therefore presupposes a proviso to the effect that the constituent bodies of the system are subject to no forces other than their mutual gravitational attraction. This proviso precludes not only gravitational forces that might be exerted by bodies outside the system, but also any electric, magnetic, frictional, and other forces to which the bodies in the system might be subject.

The absence of such forces is not, of course, vouchsafed by the principles of Newton's theory, and it is for this reason that the proviso is needed.

4. Escape by Interpretative Sentences of Probabilistic Form?

Our examination of the first and the last step in the inferential application of a theory indicates that a theory does not, in general, state any necessary conditions or sufficient conditions of applicability for theoretical expressions in terms of the antecedently available vocabulary. In particular, no definitional reduction of V_C to V_A is feasible.

This observation might suggest the idea of construing the interpretative sentences of a theory as expressing only probabilistic rather than strictly general connections between theoretically described states or events and certain associated manifestations, or indicator phenomena, described in antecedently available terms. Such a construal might seem to come closer to scientific usage and at the same time to obviate the need for provisos: for with probabilistic interpretation sentences, a theory would establish only probabilistic connections between V_A-sentences. And what would otherwise appear as occasional violations of provisos would be automatically anticipated by the merely probabilistic character of the theoretical inference.

Probabilistic interpretation sentences have indeed been envisaged by several writers. Thus, Carnap argues that interpretative sentences (he calls them rules of correspondence) might have the form of statistical laws involving the concept of statistical probability. He notes that the inference from observation sentences to a sentence containing a scientific disposition term (e.g., "electrically charged") is in general only a probabilistic inference, and that the same holds for *most* theoretical concepts.[1]

But this program faces several serious difficulties.

First, it obviously abandons the idea of a deductivist construal of the application of theories.

Second, the need for provisos also arises in the application of probabilistic laws. Thus, for example, the probabilistic laws assigning definite half-lives to the atoms of various radioactive elements are subject to the proviso that the atoms are not being bombarded by high-velocity subatomic particles.

Third, and most serious, scientific theories do not normally provide probabilistic laws of the kind envisaged to obviate the need for provisos.

Consider, for example, the character of the interpretative sentences that would be required for the term "magnet." They would have to take the form "in cases where iron filings stick to a metal bar, the probability of the bar being a magnet is P_1"; or, for inferences in the opposite direction: "Given that a metal bar is magnetic, the probability that iron filings will cling to it is P_2." But surely, the theory of magnetism contains no sentences of this kind; it is a matter quite beyond its scope to state how frequently air currents, disturbing additional magnetic fields, or other factors will interfere with the effect in question. It seems to me that no scientific theory can provide probabilistic statements of this sort.

The same basic consideration applies also, I think, where no well-developed and sharply formulated theories are available. For example, probabilification cannot avoid the need for provisos in the application of theoretical sentences linking psychological states or events to their behavioral manifestations.

5. Consequences for Elimination Programs and Instrumentalism

The conclusion that a scientific theory even of nonprobabilistic form does not, in general, establish deductive bridges between V_A-sentences has significant consequences for other topics in the philosophy of science.

I propose briefly to consider two of those topics, namely (1) the so-called elimination programs for theoretical terms, and (2) the instrumentalist construal of scientific theories.

The elimination programs are of special interest for what I will call a "conscientious analytic philosopher": he is concerned about the fact that the terms in the theoretical vocabulary V_C are not, in general, fully definable by means of the antecedently understood vocabulary V_A. This fact raises the question of how theoretical sentences can be understood at all, on what grounds, if any, they can be regarded as true or false, and whether the theoretical terms can claim to refer to definite "theoretical entities and characteristics," which the theory posits as features of the world.

The ingenious and logically impeccable methods proposed by Ramsey and by Craig[2] show—to put it very briefly—that if a theory T consisting of C and I does yield deductive connections between certain sentences expressed in terms of V_A, then it is possible to formulate another theory, T_A, such that

(a) T_A is logically implied by T;
(b) T_A is expressed exclusively in V_A;
(c) T_A establishes exactly the same deductive connections among V_A-sentences as does T.

From this result, the conscientious analytic philosopher might derive the comforting thought that the use of theoretical terms is, in principle, always avoidable in favor of a "functionally equivalent" theory T_A, which contains none of those troublesome theoretical terms. In that case, there would be no need to worry about the meanings of theoretical terms and the existence of theoretical entities: talk in terms of theoretical entities would be just a convenient *façon de parler* about matters expressible in the vocabulary V_A that is antecedently understood.

For analogous reasons it might further seem that all the problems about theoretical ascent and about provisos simply disappear when T is replaced by T_A.

All this philosophical comfort is seen to be a chimera, however, when we recall that the theory T_A constructed, à la Ramsey or Craig, from T, will yield deductive connections between V_A-sentences only if T yields such connections; and as I have tried to argue, that is not generally the case. The elimination procedures, therefore, are not strictly applicable to scientific theories.

The deductive sterility of a theory T with respect to V_A-sentences could be avoided if the provisos required for applying T to individual cases could themselves be precisely expressed, each time, as sentences containing extralogical terms only from V_A.

For if P_A is such a proviso, the transition from S^1_A to S^2_A would then take the form

$$P_A \cdot S^1_A \xrightarrow{T} S^2_A$$

and the theory T would then still establish deductive inference bridges between certain sentences in V_A. But in general the requisite provisos *cannot* he expressed in terms of V_A: as is illustrated by our earlier examples, the provisos may assert, for example, the absence of other *magnetic fields*, or of disturbing *forces*, etc., and will then require at least the use of terms from V_C for their formulation.

The considerations just outlined analogously militate against the instrumentalist conception, which construes theories as mere inferential devices that, from an input in the form of V_A-sentences, produce an output of other V_A-sentences. The need for provisos shows that theories do *not* render this service. In each particular case, the applicability of the theoretical instrument would be subject to the condition that the pertinent provisos are fulfilled; but the assertion that they are fulfilled could not just be added to the input into the theoretical calculating machine, for that assertion would not generally be expressible in V_A.

Thus, if a theory is to be thought of as a calculating instrument that generates new V_A-sentences from given ones, then it must be conceived as supplemented by an instruction booklet that says: "This instrument should be used only in cases in which certain provisos are satisfied, namely, the assumption that no disturbing factors of certain kinds are present"; and the formulation of these provisos will make use of V_C and perhaps even of terms not contained in V_C. Thus, one has to check whether certain empirical conditions not expressible in V_A are satisfied, and that surely provides a tug away from instrumentalism and in the direction of realism concerning theoretical entities.

6. Further Thoughts on the Character of Provisos

It might seem that provisos are nothing other than *ceteris paribus* clauses; but this association is not illuminating for the idea of a *ceteris paribus* clause is itself elusive. What does it mean to say "other things being equal, such-and-such is the case"? Equal to what? Just what is that clause supposed to express; how is it to function in theoretical reasoning?

Provisos might rather be viewed as *assumptions of completeness*. The provisos required for a theoretical inference from one sentence, S^1, to another, S^2, asserts, broadly speaking, that in the given case (e.g., in the case of the metal bar considered earlier) no factors other than those specified in S^1 are present which could affect the event described by S^2.

For example, in the application of Newtonian theory to a double star, it is presupposed that the components of the system are subject to no forces other than their mutual gravitational attraction and hence, that the specification given in S^1 of the initial and boundary conditions which determine that gravitational attraction is a *complete* or *exhaustive* specification of all the forces affecting the components of the system.

The completeness expressed by a proviso is of a special kind. It differs sharply, for example, from a familiar completeness requirement pertaining to inductive-probabilistic reasoning, namely, the requirement of complete or total evidence. The latter is an epistemological condition to the effect that in a probabilistic inference concerning, say, a future occurrence, the total both of evidence available at the time must be chosen as the evidential basis for the inference.

A proviso, on the other hand, calls *not for epistemic, but for ontic completeness*: the specifies expressed by S^1 must include not all the information available at the time (information which may well include false items), but rather all the factors present in the given case which in fact affect the outcome to be predicted by the theoretical inference. The factors in question might be said to be those which are *"nomically relevant"* to the outcome, i.e., those on which the outcome depends in virtue of nomic connections.

Provisos, then, as briefly noted before, might be regarded as additional premises in deductive, theory-mediated inferences.

Consider once again the use of Newtonian theory to deduce, from a specification S^1 of the state of a binary star system at time t_1, a specification S^2 of its state at t_2. Let us suppose, for simplicity, that S^1 and S^2 are couched in the language of the theory; this enables us to bypass for the moment the hurdle of the inductive ascent from astronomical observation data to a theoretical redescription in terms of masses, positions, and velocities of the two objects.

The theoretical inference might then be schematized thus:

$$P \cdot S^1 \cdot T \to S^2$$

where the proviso P expresses the assumption that between t_1 and t_2, the two bodies are subject to no influences from within or from outside the system that would affect their motion. The proviso must thus imply the absence of electric, magnetic, and frictional forces, of radiation pressure, and of any telekinetic, angelic, or diabolic influences.

One might well wonder whether this proviso can even be expressed in the language of celestial mechanics, or in the combined languages of mechanics and other physical theories. For neither singly nor jointly do those theories assert that forces of the kinds they deal with are the only kinds by which the motion of a physical body can be affected. A scientific theory propounds an account of certain kinds of empirical phenomena, but it does not pronounce on what other kinds there are. The theory of gravitation neither asserts nor denies the existence of nongravitational forces, and it offers no means of characterizing or distinguishing them.

It might seem, therefore, that the formulation of the proviso transcends the conceptual resources of the theory whose deductive applicability it is to secure. That, however, is not the case, as a graduate student at Princeton once pointed out to me. For in Newton's second law, $f = ma$, by which the motion of a body is determined, f stands for the *total* force impressed on the body. Our proviso can therefore be expressed by asserting that the total force acting on each of the two bodies equals the gravitational force exerted upon it by the other body; and the latter force is determined by the law of gravitation.

I do not think that a similarly direct and elegant formulation is available in all cases. But even when it is, the reliance on provisos raises one further question: if the theoretical inference

$$P \cdot S^1 \xrightarrow{T} S^2$$

is to be used, say, to predict S^2, then we will need a way of getting the inference off the ground by first ascertaining whether the premises S^1 and P are true.

In the example of the double star, we are able to do so for S^1—the sentence describing the initial state of the system; we can support it, at any rate inductively, by reference to astronomical measurements. But how can we check the proviso P?

We might just wait and check S^2 at t_2; if it turns out to be true, we might assume that P was satisfied; if S^2 proves false, we might take this to show that P was false.

It is well known, however, that as a general policy this procedure would be self-defeating. It would involve what Popper has called a conventionalist stratagem that would enable us to protect our theory against any predictive failure, but at the price of depriving it entirely of effective predictive import.

Nevertheless there are particular cases in which this procedure is applied in science. Take the controversy between Felix Ehrenhaft and Arthur Millikan over the magnitudes of the electrical charges on oil droplets, which Millikan had measured in his famous experiment, and which he offered in support of the hypothesis that all electric charges are integral multiples of a certain minimum value, the charge of the electron. Ehrenhaft protested that in similar experiments, he had found charges that were anything but integral multiples of Millikan's value. Millikan suggested that those deviations could be due to flaws in Ehrenhaft's experiment. Ehrenhaft repeated his experiment with special precautions aimed at excluding such disturbing factors and still found charges that did not conform to Millikan's hypothesis. Eventually, Ehrenhaft's claims received no further attention, although no clearly relevant flaws had been identified in his experiment.

Though humanly speaking this was a tragedy, it may have been quite reasonable from the point of view of scientific inquiry, as Millikan's idea was sustained by various other successful tests and it contributed to the explanation of diverse physical phenomena. For this reason, it would have been unwise to abandon it for conflicting with the results obtained by one type of experiment which, incidentally, had not even proved to be generally reproducible.

It is of some interest for our present concern to note a sequel to this story, which has recently been told by Gerald Holton, who had access to Millikan's laboratory diaries.[3] In these, Millikan, too, had recorded several sets of quite deviant charge measurements; concerning them, Millikan had noted that there must be some disturbing factor, whereas next to other sets of measurements he had noted "Excellent: Must publish"; and he omitted the deviant values from his published results.

There are various situations, however, in which violation of a proviso can be established on grounds other than the mere failure of a theoretical prediction, as when the gravitational pull of a previously unknown planet can be shown to account for certain perturbations in the orbit of a known planet, which could not be correctly accounted for by Newtonian theory in terms of the attraction exerted by the Sun and by (lie previously known planets.

Generally one can say, I think, that, at least in periods of what Kuhn calls *normal science*, a search for disturbing influences will be governed by the methodological maxim that only factors of such kinds need be considered as are specifically countenanced by one or another of the currently accepted scientific theories as being nomically relevant to the phenomenon under consideration.

Thus, if a prediction based on Newtonian mechanics fails, one might look for disturbing gravitational, electric, magnetic, and frictional forces, and even for some other kinds, but not for telekinetic or diabolic ones. Indeed, since there are no currently accepted theories for such forces, we would be unable to tell under what conditions and in what manner they act; and consequently, there is no way of checking for their presence or absence in a given case.

The methodological maxim just mentioned is clearly followed also in experiments that require the screening-out of disturbing outside influences—for example, in experimental studies of the frequency with which a certain kind of subatomic event occurs under specified conditions. What outside influences—such as cosmic rays—would affect the frequency in question, and what shielding devices can serve to block them and thus to ensure satisfaction of the relevant proviso, is usually determined in the light of available scientific knowledge, which again would indicate no way of screening out telekinetic influences.

If a theory persistently fails to yield correct predictions for a repeatable phenomenon by reference to factors it qualifies as relevant, then certain changes may be made within that theory, introducing a new kind nomically relevant factor. Roentgen's discovery of a photographic plate that had been blackened while lying in a closed desk drawer is, I think, a case in point; it led to the acknowledgment of a new kind of radiation.

Finally, in the case of serious and persistent failures, an entire theory may come to be replaced by a new one which accounts for the recalcitrant phenomena by putting them into a quite new perspective rather than by introducing an additional feature into the going theory. The explanation by general relativity theory of the precession of the perihelion of Mercury, which resisted an explanation in Newtonian terms, would be a case in point.

But this kind of revolutionary change no longer has a bearing on the preceding considerations, which were simply meant to show that there are ways, in the context of theoretical inference, of bringing evidence to bear on the question of whether the requisite provisos are satisfied—and this is essential if proviso-dependent theoretical inferences are to get off the ground.

However, in connection with this very issue, I must now add a quasi-metaphysical cautionary remark. I said earlier that a proviso pertaining to a particular application of a theory makes an *ontological* claim to the effect that, apart from the factors explicitly noted in the given theoretical inference, there *are* no further factors present that are nomically relevant to (or: that could affect) the outcome to be predicted. The proviso should not be construed as requiring only that, according to our current information, no other nomically relevant and thus potentially disturbing conditions are present.

But what can be established in accordance with the methodological maxim I mentioned is at best that other than those factors specifically mentioned in the theoretical inference, no factors arc present that current scientific theories acknowledge as nomically relevant to the outcome. This clearly is not an ontological claim, but an *epistemological* one; it refers to our current scientific knowledge or, rather, to the set of currently accepted theories and particular statements.

What is wanted, however, is a proviso that tells us that there are no disturbing factors present, whether or not they are of a kind envisaged by our theories.

Is this a reasonable demand? Consider a theoretical inference that might have been offered some 250 years ago on the basis of the caloric fluid theory of heat or the phlogiston theory of combustion. The relevant provisos would then have to assert, for example, that apart from the factors explicitly taken into account in the inference, no other factors are present that affect, say, the flow of caloric fluid between bodies or the degree of dephlogistication of a body. But from our present vantage point, we have to say that there are no such substances as caloric fluid or phlogiston, and that therefore there could be no true proviso claim of the requisite kind at all. Considerations of this kind seem to push one toward the view that, in general, a precise formulation of ontological provisos transcends the expressive power of any theory.

I think that in judging whether there are occurrences to which the principles of a given theory apply and further, whether in a particular inferential application of the theory the relevant provisos are satisfied, we have to base our answer on the best information available to us—and that, will normally be the system of currently accepted theories.

On what there is, or what the world is like, we can make no more reasonable judgment than that based on the best world view, or the best theoretical system, we have so far been able to devise.

7. Concluding Remarks

The construal of theory-based reasoning in science as a procedure that links given empirical statements to others by a set of strictly deductive steps is untenable for at least two reasons: (1) the need for inductive ascent from a description of the initial data to a theoretical recharacterization, and (2) the need for provisos. The former point has long been acknowledged in analytic methodology; the latter does not seem to have been explored in similar detail.

Nothing I have said is meant to imply that elaborate deductively organized theoretical arguments do not play a highly important role in science, or that the precision and rigor of the deduction of specific quantitative implications from mathematically formulated theories is somehow illusory: the contrary is usually the case.

The limitations of deductivist procedures present themselves, rather, in the context of applying the theoretical apparatus to empirical subject matter. Kuhn has repeatedly insisted that the attaching of symbolic labels to nature often is not subject to, or learned by means of, precise rules, but is acquired by the scientist in the course of his professional training and career, somewhat in the manner in which we learn to master large parts of ordinary language without the benefit of rules, by observation and imitation of the way others use the language.

There exists, I think, an affinity between this context and the context in which the question of provisos arises. One might argue that scientists acquire certain shared ways of judging not actually whether in a given case a specific proviso is satisfied, but rather whether to apply a theory-based inference, without any explicit mention of provisos, to a concrete situation, or what to do if the prediction fails.

What I have tried to do is to give some explication, in a broadly analytic spirit, of considerations that function in such judgments.

NOTES

1. See R. Carnap, "The Methodological Character of Theoretical Concepts," in *The Foundations of Science and the Concepts of Psychology and Psychoanalysis*, ed. H. Feigl and M. Scriven, Minneapolis: University of Minnesota Press. 1965. pp. 38–76. A probabilistic conception was propounded also by A. Pap; cf. his article "Reduction Sentences and Disposition Concepts," in *The Philosophy of Rudolf Carnap*, ed. P. H. Schilpp, La Salle, Ill.: Open Court; London: Cambridge University Press, 1963, pp. 559–597. Carnap comments on Pap on p. 950 of the same book.
2. For bibliographic references and a fuller discussion of these perspectives see, for example, section 9 of my essay "The Theoretician's Dilemma," in *Minnesota Studies in the Philosophy of Science*, Vol. II, ed. H. Feigl. M. Scriven, and G. Maxwell, Minneapolis: University of Minnesota Press, 1958, pp. 37–98.
3. G. Holton, *The Scientific Imagination: Case Case Studies*, "Subelectrons, Presuppositions, and the Millikan–Ehrenhaft Dispute," Cambridge: Cambridge University Press, 1978, pp. 25–83 (cf. especially pp. 58–63).

IV

EXPLANATIONS OF BEHAVIOR

13

Logical Positivism and the Social Sciences

1. Introduction

The principal task of philosophy, according to logical positivism, or logical empiricism, is the analysis of the concepts, theories, and methods of the various branches of scientific inquiry, ranging from logic and mathematics through physics, chemistry, and biology to psychology, the social sciences, and historiography. Curiously, however, most of the analytic studies undertaken by logical empiricists have been concerned either with logic and mathematics or with the physical sciences; biology, psychology, and the social and historical disciplines have received much less extensive and detailed attention. This difference is undoubtedly attributable, in large measure, to the professional backgrounds and interests of the most influential among the logical positivists. The majority of them were not "pure philosophers" by training, but had devoted a large part of their academic studies—often including their doctoral work—to logic and mathematics, to physics, or to a combination of these subjects. This is true, for example, of Rudolf Carnap, Herbert Feigl, Philipp Frank, Hans Hahn, Richard von Mises, Hans Reichenbach, Moritz Schlick, Friedrich Waismann, and others. It is hardly surprising, therefore, that these men devoted a great deal of their philosophical effort to the development of logic—partly as a theoretical discipline, partly as a tool for rigorous philosophical analysis—to the philosophy of logic and mathematics, and to the methodology and philosophy of the physical sciences.

Only one among the influential logical empiricists had a specialized knowledge of the social sciences: Otto Neurath (1882–1945). He was an economist and sociologist by training, and many of his writings were devoted specifically to questions concerning the subject matter, the methods, the theories, and the history of the social sciences. Next to him, Carnap has dealt rather extensively, but in less specific detail, with the logic of psychology and the social sciences. Philipp Frank's works include perceptive observations about social and political factors affecting scientific inquiry and about the ideological exploitation of some scientific theories; several of Gustav Bergmann's articles deal specifically with methodological and philosophical issues concerning psychology; and Herbert Feigl has contributed extensively to the analysis of concept formation and theory construction in psychology.[1]

Naturally, one is inclined to adduce here also Sir Karl Popper's work in the philosophy of the social sciences and of history; but, though Popper carried on an intensive and fruitful exchange of ideas with various logical positivists, and, although there were important affinities between his views and theirs,[2] he has consistently represented himself as an outside critic of the movement, and he cannot, therefore, be reckoned among its proponents.

I will now attempt to survey and appraise the principal conceptions that were developed by logical empiricist writers, especially in the 1930s and 1940s, concerning the nature and status of the social sciences and their relation to other branches of scientific inquiry. I shall begin with Neurath's ideas and then turn to contributions made by Carnap, Feigl, and others.

2. Neurath on the Social Sciences

Neurath's special fields of interest were economics, political theory, and history, and his bibliography includes many publications on these subjects. But he was also an activist and planner of tremendous energy; he contributed extensively to the organization of the various congresses at which representatives of logical empiricism presented their ideas and debated with their critics; and—with Carnap, Frank, Charles Morris, and others—he played an essential role in founding the Institute for the Unity of Science and in planning and organizing the publication of the *International Encyclopedia of Unified Science*. In contrast to his more introverted, cerebrotonic associates in Vienna, such as Schlick, Carnap, and Waismann, Neurath was extroverted, ebullient, and extremely vivacious. He was a heavy but very dynamic and most engaging man, who often signed letters to his friends with a drawing of a cheerful elephant with Neurath's initials branded on his hindquarter, holding a bunch of flowers for the addressee in his trunk.

Neurath had deep social and political interests. His ideas in these matters were of a basically Marxist persuasion; but, as Carnap notes in his recollections on Neurath's role in the Vienna Circle, "he was not a dogmatic Marxist; for him every theory must be further developed by constant criticism and re-examination."[3] Neurath was keenly aware and wary of the possibilities of social and political misuse of metaphysical and theological ideas; his strong concern with this danger is reflected in some of the more extreme features of his philosophical and methodological views.

Neurath's conception of psychology and the social sciences was basically a sharp, materialistically inspired antithesis to certain idealistic views, much in vogue during the earlier decades of this century, which conceived of scientific inquiry as comprising two fundamentally different major branches: the *Geisteswissenschaften* or *Kulturwissenschaften*, and the *Naturwissenschaften*. In this view, the mental or cultural disciplines, in contrast to the natural sciences, are concerned with phenomena in which the human mind or spirit plays an essential role; and, for this reason, the proper modes of ascertaining, describing, and understanding such phenomena differ essentially from the methods of the natural sciences. Thus, observation in the natural sciences is contrasted, by this school of thought, with procedures requiring empathy and insight in the cultural disciplines; and explanation by causes, presumed to be characteristic of the natural sciences, is contrasted with a supposedly quite different procedure in the cultural sciences, namely, the understanding of human actions and of social and historical changes in terms of immaterial reasons or similar "meaningful" connections.

Neurath emphatically rejected the idea of any fundamental differences between the natural sciences on one hand and psychology and the social and historical disciplines on the other; but, unlike Carnap and other logical empiricists, he did so in a rather summary and not very closely reasoned manner. Indeed, Neurath's writings on this subject[4] often seem more like political manifestoes, like programs, both for analysis and for action, than like carefully reasoned analytic studies. Neurath stresses, in particular, that there is no fundamental difference in subject matter between the natural sciences and the psychological and sociological disciplines, because human individuals and societies are basically nothing other than more or less complex physical systems. Thus he remarks that a living human being may be "more precisely defined" by the expression "heap of cells whose individual cells may exhibit certain large electric potential differences within a very small space, and whose temperature differences between brain and body may display certain oscillations."[5] A state is said to be describable as "a conglomeration of people, streets, houses, prisons, rifles, factories, and so on," and also of "objects of art, buildings, pictures, sculptures, religious books or religious speeches, scientific books and lectures, facial expressions, gestures, love behavior, and so on."[6] And, in Neurath's view, what psychology and sociology are properly concerned with are the behavior patterns of physical systems of these kinds, and the ways in which those patterns change. For the discipline thus envisaged, Neurath coined the name "behavioristics," so as to distinguish it explicitly from behaviorism as a doctrine within psychology; for, while he was in agreement with the general outlook of behaviorists such as J. B. Watson and I. P. Pavlov, he did not wish to endorse all the specific theses of this school of thought.[7]

From the vantage point of behavioristics, what significance could be attributed to the many statements of psychology and the social sciences that are couched in mentalistic terms—statements that speak of thoughts, beliefs, feelings, wants, hopes, fears, purposes, and decisions, for example? Neurath was extremely wary of them and counseled avoidance of mentalistic locutions in favor of a "physicalistic" terminology, which will be considered shortly. In fact, he drew up, for his own use and as a reminder for others, a list of terms that he thought should be eschewed because they were likely to lead to metaphysical pitfalls and confusions; and, in allusion to a more widely known and more influential list, he called it his *index verborum prohibitorum*.[8] His index of forbidden words included mentalistic terms such as "mind," "mental," "motive," and "meaning," but also others, such as "matter," "cause and effect," and "fact." He objected strongly to saying that statements express facts,[9] and he regarded as metaphysical the early Wittgenstein's ideas about the structure of propositional expressions mirroring the structure of the corresponding facts. I have been told—and the story sounds entirely credible to me—that, at the meetings of the Vienna Circle, in which Wittgenstein was often discussed, Neurath again and again protested that the participants were indulging in metaphysics; and when Schlick became impatient with these frequent interruptions Neurath asked permission simply to call out "M!" each time he thought the debate was becoming metaphysical. But after this had been going on for a while Neurath turned to Schlick once more and proposed: "to minimize my interruptions, let me rather call out 'non-M' whenever you are *not* talking metaphysics." And, while Neurath did not commit himself to all the details of Watson's behaviorism, he would, no doubt, have agreed heartily with Watson's pronouncement on the impact of behavioristic thinking on philosophy—a discipline that had so long been struggling in vain with such intractable issues as the mind-body problem: "With the behavioristic point of view now becoming dominant, it is hard to find a place for what has been called philosophy. Philosophy is passing—has all but passed, and unless new issues arise which will give a foundation for a new philosophy, the world has seen its last great philosopher."[10]

Indeed, Neurath put mentalistic terms such as "mind" and "motive" on his *index* on the grounds that they tended to be construed as standing for immaterial agencies and that this kind of reification gave rise to much stultifying perplexity concerning the relation of those mental agencies to the physical world. Some of his observations on this issue are strikingly suggestive and remind one of ideas that Gilbert Ryle was later to develop much more subtly and fully in his book *The Concept of Mind*.

For example, Neurath argued against conceiving of a man's mind, or of his will or personality, as a homunculus, a "second little man" who plays the role of a machinist controlling the man's behavior.[11] To illustrate the conceptual dangers of such a reification of mentalistic concepts, Neurath uses an analogy concerning the "Gang," that is, the running, of a watch.[12] His version does not lend itself to idiomatic translation, and I will therefore modify it slightly, speaking of the accuracy or precision of a watch instead. Consider a fine watch that keeps time very accurately, neither losing nor gaining more than a minute per month. How does it do that? Well, it's a certified precision chronometer; its close time-keeping is due to its high precision. But that precision surely is not a material component of the watch: however carefully we may examine its works, we won't find it there—in fact, it has no spatial location at all. But how can a nonmaterial, nonspatial agency affect and control a material physical process, the running of the mechanism and the movement of the hands? And what happens to the precision when the clock stops running and when, eventually, its material components turn into rust and dust? The analogy to the psychophysical problem is clear: how can the mind, a nonmaterial agency that has no spatial location, influence and control the behavior of the human organism, which is a material system with spatial location? And what happens to the mind after the body dies and decays?

It is only one step from here to Ryle's conclusion: The mind, and such mental features as desires and beliefs, must not be viewed as Cartesian ghosts in the body machine any more than the precision of a watch can be regarded as a ghostlike agency. To ascribe high precision or accuracy to a watch is not to say that its running is governed by an immaterial agency; it is rather to attribute to the watch a certain behavioral disposition, namely, the disposition, when properly wound and oiled, to keep time precisely or accurately. Similarly, to ascribe to a person a mental trait, say, intelligence, is to attribute to him a complex disposition to behave, under relevant conditions, in certain characteristic ways commonly described as behaving intelligently.

Neurath himself does not explicitly offer this kind of dispositional construal; indeed, he presents no general schema for a physicalistic interpretation of mentalistic terms. His ideas on this issue, as on many others, are typically stated in bold outlines, in a manner that is colorful and suggestive, but frustratingly vague on points of systematic detail. Broadly speaking, he holds that, insofar as psychological and sociological statements containing mentalistic terms possess objective scientific significance, it is possible to "replace" them by statements couched in "physicalistic terminology," or in "the physicalistic unitary language," which he considered to be adequate for the formulation of all scientific statements. The physicalistic vocabulary is conceived of as including, besides the terms of physics proper, everyday terms for material objects and processes, and, moreover, expressions like "cell aggregate," "stimulus," "amount of cattle," "behavior of chieftains." Thus, in Neurath's science of behavioristics, statements about phenomena of consciousness and about mental processes would be replaced by statements about spatiotemporally localizable occurrences such as macroscopic behavior (including gestures and speech acts) and about physiologically or physicochemically described processes in the brain and in the nervous system.[13]

Now, philosophical programs or theses suggesting the replacement of certain kinds of linguistic expressions by certain others are usually propounded with the claim that the replacement should or does preserve certain important characteristics of the expressions concerned, while eliminating certain undesirable features. The undesirable feature that Neurath proposes to avoid is the use of mentalistic or teleological or otherwise nonphysicalistic terminology; about what is to be preserved Neurath is not very explicit. I think he has in mind what might be called the observational content of the replaced statement, as characterized by the "protocol sentences" (observation sentences) to which the statement "leads" or can be "reduced," as he says.[14] This construal seems to accord well with Neurath's conception of science as being primarily concerned with the predicting of new observable occurrences on the basis of given observation statements.

Neurath, then, does not attempt to state general rules for the construction of physicalistic counterparts of sentences containing dangerous or "prohibited" terms; indeed, the very thought of general rules of this kind would probably have been suspect to him. He remarks, for example, that some sociological sentences containing expressions such as "spirit of a nation," "ethos of a religion," or "ethical forces" "can hardly be connected with observation-statements and have to be dropped as parts of metaphysical speculations," whereas we may regard other such sentences as metaphorical and may "transform them into physicalistic statements" because "dropping them would often imply an impoverishment of our argument."[15] It was in this spirit that Neurath acknowledged the great importance of Max Weber's work while objecting to Weber's talk of "rational economic ethos" as a force over and above actual behavior, and to his treatment of protestantism "as a reality that acts upon people." Neurath stressed that there could be no question of routinely transforming Weber's pronouncements into scientific (i.e., physicalistically worded) statements: a strictly scientific sociology, he remarks, would have to describe the behavior, the customs, the modes of life and production of the people under study, and then to inquire how new customs arise from the given ones in interaction with other prevailing conditions.[16]

Ultimately, Neurath's program for the elimination of misleading locutions envisaged a reformulation of all of empirical science in a unitary physicalistic language. Neurath held that, in principle, any child could learn to use such a language right from the beginning, without first acquiring the usual natural languages with their metaphysical pitfalls; and he regarded this kind of training as highly desirable.[17] In the same vein, he thought it an important desideratum that a "lexicon of special scientific terminologies" be prepared which would be a valuable tool for stating the findings of psychology, sociology, and other fields in physicalistic terms; he was astonishingly optimistic concerning the possibility of producing such a lexicon and called for a cooperative effort to this effect.[18]

Neurath noted, quite rightly, that in the field of psychology such important doctrines as Gestalt theory, psychoanalytic theory, and behavioristic theories do not permit comparison or useful combination with one another, because they use different and ostensibly unrelated terminologies; and he expected that a lexicon making it possible to reformulate the empirical substance of those theories in a unitary physical language would very happily serve to fuse and consolidate what was significant in those doctrines and would at the same time considerably enhance their predictive potential.[19] Some years after suggesting this program, however, Neurath offered a more realistic conception of "Unified Science as Encyclopedic Integration,"[20] which no longer envisaged the physicalistic lexicon; indeed, in order to get the project of the *International Encyclopedia of Unified Science* off the ground, Neurath did not even insist that all contributors should aim at restricting themselves to a physicalistic terminology.

Neurath's conception of the scientific enterprise had strong pragmatic and instrumentalist overtones; *prediction* of empirical phenomena is frequently cited in his writings as the prime objective of the scientific enterprise;[21] theoretical understanding of the world as an aim in itself does not seem to have figured significantly in his view. And he stressed that, also for the sake of effective prediction, it was essential that the various scientific disciplines be organized into one coherent system of unitary science. He was fond of pointing out that in order to predict particular events, such as the outcome of a forest fire or the place where a windblown ten-dollar bill will end up, we often have to combine knowledge drawn from several of the traditionally distinguished branches of empirical science: physics, chemistry, and biology, as well as psychology and sociology; for we will have to take into account the availability, organization, and effectiveness of fire fighters, or, in the other case, the economic status, alacrity, and inhibitions of the persons who happen to notice the ten-dollar bill being blown into the street.

Neurath noted that, in an effort to master conceptually the multitude of social changes, the sociologist must try to formulate general laws that govern those changes and by means of which particular events may then be predicted. His observations concerning such laws are much more sober and realistic than his idea of a physicalistic lexicon of special terminologies in science. Neurath noted that some of the laws established by sociological research may well apply only to rather special social formations and thus may be analogous to certain laws that might be said to apply only to lions or only to ants. And he added that "doubtless there are only few general sociological laws: in general, one has to go back to laws of individual behavior and other things."[22] As for the extent to which the behavior of social groups can be derived from that of its constituent individuals, Neurath declared that that was for the sociologists to find out; thus he refrained from making any general claims concerning the realizability of the program of methodological individualism.[23] His views on this issue were shared, I think, by his fellow empiricists.

Speaking of the limitations of prediction in the social sciences, Neurath noted the problems of self-fulfilling and self-stultifying predictions, and already in his *Empirische Soziologie* of 1931 he argued that innovations in science and technology cannot be predicted in very specific detail, on the ground that to make such predictions would be to anticipate the innovations. To be able to predict Einstein's computations, he says, one would have to predict Einstein's computations, he says, one would have to be Einstein oneself, and, he adds, "here lies an essential limit of all sociological prognoses."[24]

3. The Elaboration of Physicalism by Carnap

Let us now turn to Carnap's views on the character of psychology and the social sciences. Carnap was in substantial agreement with the basic ideas of Neurath's physicalism; but he formulated, defended, and gradually modified those ideas in accordance with his own characteristic style and standards of philosophizing, which are very different from Neurath's. Neurath conveyed his ideas in broad strokes, in vigorous and suggestive, but imprecise and often elusive, language. He cautioned against overestimating the ideal of linguistic precision and held that artificially constructed formal languages, on which Carnap relied extensively in his work, had only limited value for the clarification of philosophical issues. Accordingly, he did not conceive of the unitary physicalistic language of science as having a precise vocabulary and a rigorously specifiable formal structure; to lend emphasis to this point, he referred to the language of empirical science as a "Universal Jargon" which cannot be built up from scratch as a precise system, which "will always be in the making," and which will always contain certain clusters of "vaguely defined assumptions and assertions."[25] Also, Neurath frequently offered no systematic arguments in support of his ideas; his theses and programs often appear to be prompted by ideological considerations, especially by the desire to pull the terminological rug out from under the pronouncements of idealistic philosophers and theologians.

Carnap, on the other hand, was and has remained a highly rigorous and systematic thinker, insisting on precision and explicitness in the formulation of philosophical ideas, and on careful arguments in support of them. While he, and other members of the Vienna Circle, appreciated Neurath's social, political, and ideological concerns and often shared them, Carnap held that they were logically irrelevant to the problems of physicalism and should therefore be left out of consideration in the philosophical discussion of this subject.[26]

Carnap's own construal and defense of physicalism has gone through several stages, which can be traced here only in outline. A first, prephysicalistic stage is that of *Der logische Aufbau der Welt*,[27] where Carnap argued in considerable detail that all the concepts of empirical science, including those of psychology and the social sciences, can be defined—essentially by iterative use of extensive abstraction—in terms of one basic phenomenalistic concept. The ingeniously designed procedure was to lead from this basis, by definitional steps, first to concepts for sense qualities, then to concepts pertaining to physical objects, and finally to the concepts of psychology and sociology.[28] And Carnap not only set forth a program; he actually formulated detailed definitions for a considerable set of concepts pertaining to the sensory realm. Thus, as Quine has said, Carnap "was the first empiricist who, not content with asserting the reducibility of science to terms of immediate experience, took serious steps toward carrying out the reduction."[29] The construction of the concepts on higher levels, however, was presented only in outline; Carnap later abandoned the view that all those concepts can be introduced by explicit definitions on a phenomenalistic basis.

Already in the *Aufbau* Carnap remarked that a "logical construction of the world" did not require a phenomenalistic basis, but could be achieved on a physical basis as well, that is, that the concepts of empirical science could all be defined in terms of some suitably chosen set of physical concepts; and he briefly suggested several possible choices for such physical definition bases.[30] In the early 1930s, partly under Neurath's influence,[31] Carnap developed this idea in greater detail in the form of the thesis that the language of physics affords a unitary language for all of empirical science, a language in which all the concepts and statements of empirical science can be expressed. At that time he considered a "physicalistic" basis as preferable to a phenomenalistic one, especially because of its intersubjective, public character, that is, because "the events described in this language are in principle observable by all users of the language."[32]

By physical language, Carnap here understands roughly the same thing as Neurath's physicalistic language, namely, the language "in which we speak about physical things in everyday life or in physics."[33] Its vocabulary is taken to include not only the technical terms of physics but also words like "cold," "heat," "liquid," "blue," "taller than," "soluble in," which occur already at the level of what Carnap later called the thing language: "that language which we use in every-day life in speaking about the perceptible things surrounding us."[34]

The principal thesis of physicalism as Carnap construed it at this stage was that the physical language is a universal language, that is, that any sentence of any branch of empirical science can be translated into it "without change of content."[35]

By sentences of empirical science, Carnap here understands only sentences that have empirical meaning, or empirical content, in the sense of what I will call *the wider empiricist criterion of meaning*. This is, strictly speaking, a criterion of meaningfulness; it specifies conditions under which a sentence is empirically meaningful or has empirical content. This must be distinguished from *the narrower empiricist criterion of meaning*, which purports to characterize "the meaning" or "the content" of a sentence, and which will be considered shortly.

The wider criterion qualifies a sentence as having empirical meaning if, at least "in principle," it is testable by means of observational data, that is, if it "implies," in a suitable sense, observation sentences or "protocol sentences" describing potential observational findings. This criterion of testability was meant to bar from the class of scientific statements untestable metaphysical formulations that might speak of entelechies, *Zeitgeist*, or a historical figure's "manifest destiny": to such "pseudostatements" the thesis of translatability into physical terms was not, of course, meant to apply.[36]

Carnap's explication and justification of the physicalistic thesis also presupposed, however, a narrower, more stringent version of the empiricist meaning criterion: according to it, two sentences that imply the same observation sentences or protocol sentences "have the same content. They say the same thing, and may be translated into one another."[37] Thus, the empirical meaning or content of a sentence here is understood to be determined by the class of its observational implications. Hence, the second meaning criterion implies the first, but not conversely; this is why I referred to them as the narrower and the wider empiricist criterion of meaning.

Carnap's arguments in support of the physicalistic thesis were aimed at showing that for any sentence of empirical science there exists a sentence in the physical language which implies the same protocol sentences and hence has the same content. Thus, Carnap reasons that the psychological sentence P_1, "Mr. A is now excited," has the same content as a physical sentence P_2, to the effect that Mr. A's body is in a physical state characterized by rapid pulse and breathing and by a disposition to react, under specifiable physical stimulus conditions, with characteristic kinds of behavior which can be described in physical terms.[38] And he adds, in effect, that, if P_1 were held to contain a further component not shared with any physical sentence P_2, then that component would lack observational implications and hence empirical content; for any of its observational implications would always be describable in physical terms and thus would be shared with some physical sentence.[39]

Thus, as early as 1932 Carnap anticipated one of Ryle's ideas by arguing that sentences such as "A is excited" or "A has a headache," which attribute psychological characteristics to an individual, are analogous in their logical character to physical sentences like "This object is plastic" or "This wooden support is firm." The latter, Carnap says, are sentences "about a physical property, defined as a disposition to behave (or respond) in a specific manner under specific circumstances (or stimuli)."[40] The psychological terms in question are thus explicitly construed as pertaining to more or less complex behavioral dispositions.

Yet already at this stage Carnap noted that it is a further aim of science to define psychological concepts, not in terms of behavioral dispositions, but in terms of microstructures of the human body with which the dispositions are associated. This would "enable us to replace dispositional concepts by actual properties," just as the characterization of temperature as a disposition to produce certain thermometer readings has been replaced, in kinetic theory, by a microstructural definition in terms of the molecular kinetic energy.[41] Carnap stressed that a sentence S_1, containing psychological terms, and its counterpart S_2 in the physical language often will not be logically equivalent, but only "P-equipollent," that is, derivable from each other only by means of empirical laws. And, indeed, those behavioral symptoms which are generally associated with a given psychological feature will often be determined by empirical investigations leading to empirical laws rather than by an aprioristic reflection upon the meaning of the psychological terms in question. That every psychological sentence has *some* physical counterpart, however, is guaranteed by the fact that the constituent psychological terms must have some observational, and thus physically describable, criteria of application; otherwise, the psychological sentence would have no content.[42]

In arguing for this physicalistic construal of psychology Carnap considers several likely objections, among them the claim that a purely physicalistic analysis cannot give an adequate account of "meaningful" behavior: that it cannot, for example, distinguish between a mere reflex movement of an arm and a "meaningful" movement such as a beckoning gesture. Carnap replies that whatever meaning can scientifically—and thus testably—be attributed to an arm movement must be fully determined, and hence fully characterizable, by physical features of the moving arm, the rest of the body, and the environment. For, indeed, he reasons, if we were to watch a film of the arm movement, we would be able to ascertain its "meaning" just as well as if we were witnessing the event itself. But the film records only physical phenomena; hence, the physical aspects of the arm movement must be sufficient to determine its beckoning character or whatever other "meaning" can significantly be ascribed to it.[43]

Carnap acknowledges that it is not as yet possible to give explicit characterizations of "meaningful" kinds of behavior in terms of the concepts of physical theory; but he takes the view that expressions such as "assertive nod" or "beckoning gesture" may themselves be regarded as physical terms of a crude sort because, by the argument just outlined, their criteria of application pertain exclusively to physical aspects of behavior.

At the stage we have been considering, then, Carnap maintained that all scientific terms were fully definable in physical terms and all scientific sentences fully translatable into physical ones, with the understanding that the definitions and translations in question were not always based solely on logical or analytic truths, but rested, in some cases, on empirical laws.

In the mid-thirties Carnap began to weaken this claim. Briefly, he argued that "we know the meaning (designation) of a term if we know under what conditions we are permitted to apply it in a concrete case and under what conditions not,"[44] and he went on to show that, even within physics, the criteria of application for a term, which in this sense specify its meaning, do not in general provide a full definition for it, at least not when stated in a language using only extensional logical connectives. He argued this particularly for dispositional terms, such as "elastic": the apparently plausible definition

x is elastic $\equiv$ (x is stretched and released $\supset x$ contracts)

would qualify as elastic not only those objects which contract on being stretched and released, but also those which never are stretched or released at all, because, for these, the antecedent of the definiens is false and, hence, the definiens is true. In order to avoid this difficulty, Carnap suggested that sentences specifying criteria of application for dispositional terms be construed as having the form of what he called reduction sentences, which would play the role of partial rather than full definitions. Using obvious abbreviations, the reduction sentence for our example could be written as follows:

(R) $SR(x) \supset (E(x) \equiv C(x))$.

This sentence is logically equivalent to the conjunction of the following two:

(R_1) $E(x) \supset (SR(x) \supset C(x))$.

and

(R_2) $(SR(x) \cdot C(x)) \supset E(x)$.

The first of these states a necessary condition for "$E(x)$"; the second, a different sufficient one; whereas a full definition of a term specifies one condition which is both necessary and sufficient. The reduction sentence R has the character of a partial definition in this sense: it specifies a necessary and sufficient condition of elasticity—namely, contraction—for just those things which are subjected to the stretch-and-release test; in reference to all other things, no criteria of applicability for the word "elastic" are provided, no "meaning" is given to it. The area of indeterminancy thus left may be reduced by means of further reduction sentences specifying additional criteria of application; but all the criteria available for a scientific term will not, in general, amount to a full definition for it.

Reduction sentences seem to lend themselves well to the formulation of "operational" criteria of application for psychological and sociological terms, as based on specified procedures of testing, rating, or evaluating. And it is clear also that these procedures afford only partial criteria of application. For example, an intelligence test that presupposes literacy on the part of the subject is not applicable to just any subject and thus offers no general definition; an analogous remark applies to a rating scale for socioeconomic status whose criteria of application require a count of the electric appliances owned by the family to be rated.

In accordance with these considerations, Carnap's modified physicalistic thesis asserts that any term of empirical science can be linked by means of reduction sentences, but not generally by means of definitions, to the vocabulary of physics, and indeed to a narrow subset of it, namely, observational "thing predicates," that is, terms which stand for directly observable properties or relations of physical bodies. Carnap's argument in support of this claim was briefly to this effect: Any scientific term must refer to a kind of thing or event or state Q whose presence can, under suitable circumstances T, be ascertained on the basis of publicly observable symptoms R, such as instrument readings in physics and behavioral manifestations in the case of psychological or sociological terms. (Otherwise, the term would have no empirical meaning.) Being publicly observable, these symptoms must be describable by means of observational thing predicates; and the resulting criteria of application will be expressible by means of reduction sentences of this kind: "When observable test conditions T are realized, a thing (state, event) of kind Q is present if and only if an observable response of kind R occurs."

As for psychological terms, Carnap did not deny that we often apply them to ourselves "introspectively," without observing our own behavioral symptoms. His claim was merely that, for every such term, there are behavioral criteria which, under suitable conditions, make it possible to determine whether a person is in the kind of psychological condition to which the term refers. For example, the utterances of a person describing his ("introspected") pain or joy or anger are among the potential symptoms for the presence of the states in question; and so are other characteristic kinds of behavior which a person in such a state will display under suitable conditions. Thus, Carnap's theory of reduction sentences enabled him to give a logically subtler expression to the construal of psychological terms as complex bundles of dispositions. And it is of interest to note with Feigl that Carnap's ideas—and, I should add, Feigl's as well—have met with considerable interest among psychologists and have found applications in the work of such investigators as S. Koch; Tolman, Ritchie, and Kalish; and MacCorquodale and Meehl.[45]

Having abandoned the thesis of full physicalistic definability of all scientific terms, Carnap also withdrew the claim of the translatability of all scientific statements into the language of physics.[46] The result, as Feigl notes in his contribution to the present volume, was a considerably attenuated version of physicalism, which asserted only the reducibility of all scientific terms to physical terms, and indeed to a narrow subset of these, namely, the observational thing predicates.[47]

More recently, Carnap has taken the view, closely related to ideas developed by Feigl, that a strictly dispositional construal of psychological terms still is too confining and fails to do justice to their scientific use.[48] For, if a psychological characteristic Q is construed as an invariable disposition to show a certain kind of response R under test conditions or stimulus conditions of a specified kind S, then the nonoccurrence of R in a case where conditions S are met would establish conclusively the absence of Q in that case. Thus, Carnap illustrates, if "an IQ above 130" is understood strictly as a person's disposition to respond to a certain kind of test with answers scoring above 130, then a subject's failure to score appropriately must count as establishing conclusively that he does not have an IQ above 130; whereas, Carnap argues, psychologists would in fact allow for the possibility that the subject does have that high an IQ, but that depression, fatigue, or other factors interfered with his test performance.

In order to allow for this possibility, Carnap now proposes to assign to the psychological expressions in question the status of theoretical terms. Such terms—whether in physics or in psychology—he construes as being introduced, not simply by definitions or universal reduction sentences providing observational criteria of application, but rather by specifying (a) a set of theoretical principles in which they function and (b) a set of "correspondence rules" that afford partial observational criteria of application for some of the theoretical terms. The correspondence rules may take the form of reduction sentences, but Carnap does not insist on this particular form; indeed, he mentions even the possibility of correspondence rules of probabilistic-statistical form.[49] In any event he conceives of the meanings of theoretical terms as depending not only on the correspondence rules but also on the theoretical principles in question, with the result that psychological terms thus construed can no longer be held to stand simply for behavoral dispositions. This construal abandons the notion that every psychological term possesses (necessary or sufficient) conditions of application that can be stated in an observational vocabulary. Concomitantly, it gives up two ideas concerning the significance of scientific terms that are closely related to the empiricist criteria of empirical significance for statements; namely, first, the principle that a term has empirical significance only if it has necessary or sufficient observational conditions of application; and, second, the narrower operationist maxim that the meaning of a term is fully determined by those conditions of application. In order to make sure that his conception of theoretical terms would not sanction the admission of pseudoconcepts and pseudohypotheses, Carnap proposed a new, considerably liberalized "criterion of significance for theoretical terms" and a corresponding "criterion of significance for theoretical sentences," whose somewhat complicated details[50] we need not consider here, however. It suffices to note that the idea of the reducibility of all scientific terms to a "reduction basis" consisting of observational predicates was abandoned.

4. Linguistic and Ontological Aspects of Physicalism

This result certainly represents a considerable weakening of one physicalistic claim. But the question whether every scientific term has necessary or sufficient conditions of application expressible in purely observational predicates is hardly the root issue that lent philosophical interest to physicalism. That root issue, I think, lies in questions raised by the old distinctions between the organic and the inorganic, between the mental and the physical, between *Kulturwissenschaft* and *Naturwissenschaft*. Broadly speaking, physicalism was exciting because it promised to shed light on the question whether, as Neurath said, living and thinking organisms, as well as societies and nations, are nothing other than peculiar combinations of physical bodies that exhibit more or less complex kinds of physical features, or whether there is something more to them, something that falls outside the domain of physics. Thus stated, the issues are formulated in what Carnap called the material mode of speech, that is, in sentences that ostensibly concern empirical items such as physical, biological, psychological, and sociological systems and processes. But, quite early in the development of his ideas on physicalism, Carnap argued that these and other philosophic problems expressed in the material mode were not actually concerned with the subject matter of empirical science, but rather with its language, and that the first step toward philosophical clarification should consist in reformulating the issues accordingly—in giving them a "linguistic turn," to use Rorty's apt phrase.[51]

A first step toward a linguistic restatement of the theses just mentioned is suggested by this consideration: the claim that all biological and sociological systems and processes, and all psychological states and events, are just physical systems or occurrences would seem to amount to two theses: (1) any biological, psychological, or sociological characteristic of any system, state, or event can be described by means of purely physical concepts alone; and (2) any phenomenon that can be explained with the help of biological, psychological, or sociological laws or theoretical principles can be explained by physical laws alone.

But, inasmuch as biological, psychological, and social characteristics would be described by means of biological, psychological, and sociological terms, the first thesis might be given this linguistic turn: (1a) All biological, psychological, and sociological terms can be defined by means of physical terms. Similarly, the second thesis might be turned into: (2a) All biological, psychological, and sociological laws can be deduced from purely physical laws.

These two theses, and certain variants of them, have in fact played a central role in the efforts made by logical empiricists to express physicalism as a thesis about the language of science. As for (1a), we saw above that it was part of Carnap's physicalism in an early stage of its development. Let us note that the recent construal—by Carnap, Feigl, and others—of psychological terms as theoretical terms, some of which may lack observational criteria of application, is quite compatible with the thesis that, nonetheless, all psychological (and similarly sociological, etc.) terms are fully definable by means of physical terms—some of which would then have to belong, however, to the "nonobservational," "theoretical," vocabulary of physics. Let us note also that the "definitions" envisaged in (1a) were not conceived of as expressing synonymies; rather, as Carnap has pointed out at an early stage, they would normally be based on empirical laws. This is the case, for example, when a certain hormone, originally characterized in biological terms by reference to the gland secreting it, is later "defined" in physicochemical terms by reference to its molecular structure. In this case the biological term and the corresponding physicochemical expression might be said to be coextensive by law, or nomically coextensive.

A thesis to the effect that psychological terms are thus nonanalytically coextensive with physical expressions has been developed in detail by Feigl, whose conception takes explicit account also of the theoretical role of psychological and physical terms. Briefly, Feigl holds that for every psychological predicate there is a corresponding physical—more specifically, a neurophysiological—expression with which it is coextensive by virtue of fundamental theoretical principles, much as the term "temperature" as used "macroscopically" in classical thermodynamics is coextensive, by virtue of theoretical principles, with an expression in the kinetic theory of matter, namely, "(a disjunction of) microdescriptions in terms of molecular motions."[52] Feigl argues that, when a psychological term is thus theoretically coextensive with a physical expression, the two may be viewed as having the same referent, and he speaks of this identity of the referents as systemic or theoretical identity. In this sense Feigl's "physicalistic identity theory" holds that every kind of psychological state or event is theoretically identifiable with a kind of physical state or event which is describable in neurophysiological terms. Feigl argues that the assumption of the theoretical coextensiveness of all psychological terms with physical expressions, far from being a philosophical a priori truth, is made increasingly plausible by the advances of scientific knowledge in the relevant fields, and that the identification of associated referents recommends itself on methodological and philosophical grounds.

In one of his recent essays Feigl presents this conception in the context of a critical appraisal and restatement of physicalism. He summarizes the basic tenets of his version of physicalism in two theses, which Carnap, in his response to Feigl, endorses in a slightly modified form.[53] The first thesis is to the effect that whatever is subjectively or privately confirmable is also intersubjectively or publicly confirmable, so that anything I can know introspectively to be the case (e.g., that I have a headache) can in principle be ascertained indirectly by others on the basis of publicly observable symptoms or manifestations. This thesis attempts to make a claim similar to, but more cautious than, that expressed by (1a) or by its later variant to the effect that all scientific terms are reducible to physical terms. The second thesis is closely akin to (2a). As stated by Carnap it asserts that "all laws of nature, including those which hold for organisms, human beings, and human societies, are logical consequences of the physical laws, i.e., of those laws which are needed for the explanation of inorganic processes," so that whatever empirical occurrences can be explained at all by means of laws or theories can be explained by means of physical laws and theories alone. Carnap, like Feigl, stresses that these two physicalistic theses are not *a priori* truths: they reflect, he says, certain very general empirical facts of a kind sometimes referred to as all-pervasive features of the world and of the language in which the world can be described.[54]

Clearly, these recent formulations no longer purport to construe physicalism strictly as a set of theses concerning the language of empirical science. In fact, that construal was implicitly abandoned much earlier, by emphasizing that the "translatability" of biological, psychological, or sociological expressions into physical ones depended, in general, on appropriate empirical laws. Thus, the latest statement of physicalism presents again what might be called an ontological aspect.

The ontological reconstrual makes physicalism a rather elusive thesis—more elusive, perhaps, than its formulation might at first suggest. In the second thesis, for example, the notions of physical—or inorganic—process and of physical law gave rise to questions. Perhaps it may be assumed that, in regard to processes that can be described in terms of the vocabulary of contemporary science, there would be fairly good agreement as to which of them count as inorganic and which do not; and perhaps a similar consensus might be attainable as to which of the laws expressible in contemporary scientific terms are physical laws and which are not. But suppose that further scientific research were to lead to a well-substantiated, high-level, new theory T that would unite physics, biology, and psychology in the way in which Einstein hoped to unite the theories of gravitation and electromagnetism by a unified field theory. The basic principles or laws of T might then be formulated in terms having no purely and exclusively physical, or biological, or psychological interpretation, but they might permit the derivation of various more specific general statements, some of which we would readily classify as physical laws, others as biological laws, and yet others as psychological ones. The fundamental principles or laws of T would thus explain inorganic as well as other kinds of processes; but would they constitute physical or nonphysical laws? Would such a theoretical development strengthen or weaken the second thesis of physicalism?

Feigl takes cognizance of difficulties of this kind, pointing out that the notion of physical explanation has undergone, and will no doubt continue to undergo, radical changes. He therefore construes the claim of the second thesis—that all empirical facts and laws can be given a "physical" explanation—as asserting that those facts and laws can be explained by means of a theory whose terms are partially interpreted in observational terms and whose postulates are conceived "according to the paradigm of modern physical theory"; this qualification is taken to preclude, for example, teleological explanations of the vitalistic variety (even if testable) and interactionist theories of the relation of the mental to the physical.[55] Thus understood, the thesis of the universality of physical explanation is quite vague, as Feigl is the first to acknowledge. Moreover, the thesis would seem to qualify as physical even those explanations which our fictitious unifying theory T would provide for biological and psychological phenomena. Thus, the latest versions of physicalism look pretty pale and innocuous by comparison with the sturdy ontological formulations of materialism and mechanism, which appear to make strong and clear claims about the purely physical character of biological and social systems and events and of mental states and processes. But this throws into relief one of the insights afforded by the evolution of physicalism: those claims are strong and clear in appearance only, as is evidenced by the fact that the steadily refined empiricist efforts to explicate the tenable substance of those theses have yielded formulations that have been increasingly cautious, elusive, and weak.[56]

5. On the Methodological Unity of Science

The unity of science as envisaged by logical empiricism had three major aspects, of which we have so far considered two, sometimes referred to as the unity of language and the unity of laws. In conclusion, at least brief mention should be made of the third aspect, the unity of method. The idea of the methodological unity of science was developed in opposition to the view that the mental or cultural disciplines are distinguished from the natural sciences by fundamental differences in the methods that are required to ascertain and to explain the facts with which these disciplines are concerned.

The thesis of the methodological unity of science states, first of all, that, notwithstanding many differences in their techniques of investigation, all branches of empirical science test and support their statements in basically the same manner, namely, by deriving from them implications that can be checked intersubjectively and by performing for those implications the appropriate experimental or observational tests. This, the unity of method thesis holds, is true also of psychology and the social and historical disciplines. In response to the claim that the scholar in these fields, in contrast to the natural scientist, often must rely on empathy to establish his assertions, logical-empiricist writers stressed that imaginative identification with a given person often may prove a useful heuristic aid to the investigator who seeks to guess at a hypothesis about that person's beliefs, hopes, fears, and goals. But whether or not a hypothesis thus arrived at is factually sound must be determined by reference to objective evidence: the investigator's empathic experience is logically irrelevant to it. As Feigl puts it, "Quite generally, the significance of intuition, insight, empathetic understanding consists in the power of these processes to *suggest* hypotheses or assumptions, which, however, could *not* be established, i.e., confirmed as scientific statements except by intersubjective methods."[57]

Similarly, the logical-empiricist conception of the unity of science rejects the claim that certain psychological facts can be ascertained only by the method of introspection; it holds instead that for any privately ascertainable kind of state or event there are some publicly observable symptoms which make it possible intersubjectively to confirm the presence of the state or event in question; this, in fact, is the point of the first thesis of physicalism as recently stated by Feigl and Carnap.

As for basic differences in the methodology of explanation, one of the best-known claims holds that the explanation of human actions in terms of reasons differs fundamentally from the kind of explanation found in the natural sciences, and especially from causal explanation. Here the construal of psychological terms as dispositional or theoretical terms provided the basis for a reply to the effect that explanations of human actions by reference to reasons are strictly analogous to explanations of physical events by reference to dispositional properties or theoretical characteristics of the objects involved. Thus, to illustrate the central idea sketchily, the explanation that Henry passed the cookies because (a) he had been asked to and (b) he was polite, and was thus disposed to comply, is analogous to the explanation that the window pane broke because (a) it was struck by a stone and b) it was brittle and was thus disposed to shatter under a sharp impact.

While a dispositional construal of psychological terms was explicitly set forth by Carnap in the mid-thirties, I believe it was Ryle who first argued in careful detail that to explain an action by motives or other psychological factors is to subsume it under a "lawlike proposition," attributing to the agent a general behavioral disposition characteristic of the motive in question. Thus he says: "The imputation of a motive for a particular action is not a causal inference to an unwitnessed event but the subsumption of an episode proposition under a lawlike proposition. It is therefore analogous to the explanation of reactions and actions by reflexes and habits, or to the explanation of the fracture of the glass by reference to its brittleness."[58] To adduce the brittleness of the glass pane in the latter explanation is not to cite a cause for its shattering, but to indicate a lawlike proposition expressing the general disposition of the pane to break under impact. This proposition, however, explains the breaking of the pane only in combination with another proposition mentioning the impact of the stone, which plays the role of the cause. Analogously, in Ryle's analysis, to specify a motive for an action is not to cite its cause, but to indicate a lawlike proposition concerning a behavioral disposition of the agent; however, "the general fact that a person is disposed to act in such and such ways in such and such circumstances does not by itself account for his doing a particular thing at a particular moment. . . . As the impact of the stone at 10 P.M. caused the glass to break, so some antecedent of an action causes or accasions the agent to perform it when and where he does so."[59] For example, being asked to pass the cookies might be what caused Henry to do so, in accordance with his politeness, that is, his general disposition to act politely.

This argument seems to me quite effectively to dispose of the claim that explanation by motives or other psychological factors is fundamentally different from the explanation of physical events. But, as was noted earlier, the psychological factors in question cannot generally be construed as behavioral dispositions nor even, in a manner envisaged by Ryle,[60] as large or endless bundles of such dispositions. This becomes clear also when we consider the logic of explaining an action by the reasons that led the agent to perform it. Those reasons will normally have to include the agent's relevant desires and beliefs; and it is readily seen that neither a desire nor a belief by itself can be construed as a disposition to show a characteristic kind of publicly observable (macro- or micro-) behavior under physically characterizable stimulus conditions. For, the behavior in which a given desire (e.g., a desire for food) manifests itself under given stimulus conditions (e.g., when food is placed before the hungry person) will depend essentially on the person's beliefs—for example, on whether he believes that what is put before him is in fact edible, not prohibited by religious or medical considerations, and so forth. Conversely, the behavior in which a given belief manifests itself will depend on the given person's desires (as well as on other beliefs he may entertain, on his moral standards, etc.). Thus, only certain *combinations*, so to speak, of beliefs with desires and possibly with other factors will be associated with more or less specific behavioral dispositions.

Considerations such as these suggest that explaining an action by means of reasons (and similar kinds of psychological explanation) is analogous to explaining a physical event by means of theoretically characterized factors (and theoretical principles governing them): such factors also cannot usually be conceived of as general dispositions or as bundles of such.

An interest in exploring the extent of the methodological unity of empirical science is manifested also in further studies, growing out of the logical-empiricist tradition, that examine the form, the force, and the modes of validation of various other kinds of explanation which can be found in the social sciences and in historiography—among them, the functional and the typological analysis of social phenomena and genetic and other types of historical explanation. One of the basic features that all these types of explanation have been held to share with explanation in the natural sciences is the explicit or implicit reliance on general laws or theoretical principles of universal or of probabilistic-statistical form.[61]

6. Conclusion

A fair-minded appraisal of the accomplishments of logical positivism should not focus on the bold and naturally oversimplified devices its adherents wrote upon their banners, but on the quality of the detailed logical and methodological studies carried out under those banners: it should examine the standards of clarity and rigor those studies exemplify, the stimulation they provided for the work of others, and the light they shed on philosophical issues of importance. Thus judged, logical positivism will be found, I think, to have been a strong and fruitful influence in recent systematic philosophy.

NOTES

1. Frank's observations can be found at various places in his books *Modern Science and Its Philosophy* (Cambridge, Mass.: Harvard University Press, 1949) and *Philosophy of Science* (Englewood Cliffs, N.J.: Prentice-Hall, 1957). Bergmann's contributions include: "The Logic of Psychological Concepts," *Philosophy of Science*, 18 (1951): "Psychoanalysis and Experimental Psychology," *Mind*, 53 (1944); and an article written jointly with the psychologist K. W. Spence, "Operationism and

Theory in Psychology," *Psychological Review*, 48 (1941). Feigl's writings will be referred to later.

2. See Carnap's remarks on this point on pp. 31–32 of his autobiography in Paul A. Schilpp, ed., *The Philosophy of Rudolf Carnap* (La Salle, Ill.: Open Court, 1963). Feigl makes a similar point on pp. 641–42 of his engaging and informative essay "The Wiener Kreis in America" (in D. Fleming and B. Bailyn, eds., *Perspectives in American History*, vol. 2: *The Intellectual Migration: Europe and America, 1930– 1960* [Cambridge, Mass., 1968]). Feigl's essay offers a historical account, based to a large extent on his personal experiences, of the development of logical empiricism in Europe and its migration to the United States; it includes interesting information also about several other figures referred to in the present study, among them Carnap and Neurath.

3. In Schilpp, *The Philosophy of Rudolf Carnap*, p. 24. For fuller details see the article "Neurath, Otto," by R. S. Cohen, in *The Encyclopedia of Philosophy* (New York: Macmillan and The Free Press, 1967), 5: 477–79. This article also refers to a forthcoming volume, *selected Papers of Otto Neurath*, ed. Marie Neurath and R. S. Cohen, which is to include a biographical memoir and a complete list of Neurath's writings.

4. Among them: *Empirische Soziologie* (Vienna: Springer, 1931): *Einheitswissenschaft und Psychologie* (Vienna: Gerold & Co. 1933): *Foundations of the Social Sciences* (Chicago: University of Chicago Press, 1944); "Sociology and Physicalism" (translation of an article originally published in *Erkenntnis*, 2 [1931–32]), in A. J. Ayer, ed., *Logical Positivism* (New York: The Free Press, 1959). Passages quoted in this essay from the first two of these works are translated by the present writer.

5. *Einheitswissenschaft und Psychologie*, p. 17.

6. *Empirische Soziologie*, p. 44.

7. See, for example, Neurath, *Foundations of the Social Sciences*, p. 17; *Einheitswissenschaft und Psychologie*, p. 17; *Empirische Soziologie*, pp. 63, 83.

8. *Einheitswissenschaft und Psychologie*, p. 12; for further instances of "dangerous terms," see, for example, *Foundations of the Social Sciences*, pp. 4, 5, 18, and p. 55 ("Expressions avoided in this monograph").

9. He insisted, for example, that the test—or, as he called it, the assaying—of a scientific hypothesis should not be conceived of or described as involving a comparison of the hypothesis with relevant "facts," but rather as involving a comparison of the hypothesis statement with certain other statements, namely, "observation statements" or "protocol statements." (*Foundations of the Social Sciences*, pp. 4–5; cf. sec. I of "Sociology and Physiicalism" and *Einheitswissenschaft und Psychologie*, p. 29.)

10. J. B. Watson, *The Ways of Behaviorism* (New York: Harper and Brothers, 1928), p. 14.

11. *Einheitswissenschaft und Psychologie*, p. 16; see also *Empirische Soziologie*, p. 65.

12. *Einheitswissenschaft und Psychologie*, p. 11; see also "Sociology and Physicalism," p. 299.

13. Cf. *Einheitswissenschaft und Psychologie*, pp. 7, 18–19; *Foundations of the Social Sciences*, pp. 2–4; *Empirische Soziologie*, p. 60.

14. *Einheitswissenschaft und Psychologie*, pp. 6, 12–13.

15. *Foundations of the Social Sciences*, p. 4; see also p. 10.

16. *Empirische Soziologie*, p. 57; see also *Foundations of the Social Sciences*, pp. 16–17.

17. *Einheitswissenschaft und Psychologie*, p. 13.

18. Ibid., p. 27.

19. Ibid., pp. 26–28.
20. *International Encyclopedia of Unified Science*, vol. 1, no. 1 (Chicago: University of Chicago Press, 1938), pp. 1–27.
21. See, for example, *Empirische Soziologie*, pp. 11, 13; "Sociology and Physicalism," pp. 285, 293.
22. *Empirische Soziologie*, p. 77.
23. Ibid., pp. 65, 77.
24. Ibid., p. 130; see also *Foundations of the Social Sciences*, pp. 28–30.
25. *Foundations of the Social Sciences*, pp. 2, 3.
26. See Carnap in Schilpp, *The Philosophy of Rudolf Carnap*, pp. 22–24, 51.
27. 1st ed. (Berlin-Schlachtensee: Weltkreisverlag, 1928); 2d ed. (Hamburg: Felix Meiner, 1962). An English translation of the second edition, which also includes a smaller pamphlet originally published in 1928, has appeared under the title *The Logical Structure of the World and Pseudoproblems in Philosophy* (Berkeley and Los Angeles: University of California Press, 1967).
28. For a detailed account and critical appraisal of the procedure, see Chap. V of N. Goodman, *the Structure of Appearance*, 2d. ed. (Indianapolis: Bobbs-Merrill, 1966), as well as Goodman's essay "The Significance of *Der Logische Aufbau der Welt*," and Carnap's reply, in Schilpp, *The Philosophy of Rudolf Carnap*. Further reflections by Carnap on the *Aufbau* and on physicalism will be found on pp. 16–17 and 50–53 of the Schilpp volume and in the "Preface to the Second Edition" of *The Logical Structure of the World*.
29. W. V. O. Quine, *From a Logical Point of View*, 2d. ed. (Cambridge, Mass.: Harvard University Press, 1961), p. 39.
30. See secs. 59 and 62 of *The Logical Structure of the World*.
31. See, for example, pp. 74–75 of Carnap, *The Unity of Science* (translated with an introduction by M. Black) (London: Kegan Paul, 1934), and Carnap's observations on pp. 50–53 of Schilpp, *The Philosophy of Rudolf Carnap*.
32. Carnap in Schilpp, *the Philosophy of Rudolf Carnap*, p. 52. This stage in the development of Carnap's physicalism is reflected especially in two articles he published in *Erkenntnis* in 1932: "Die physikalische Sprache als Universalsprache der Wissenschaft" and "Psychologie in physikalischer Sprache." Translations of these, with slight revisions, appeared, respectively, in Carnap, *The Unity of Science*, and, under the title "Psychology in Physical Language," in Ayer, *Logical Positivism*, pp. 165–98.
33. Carnap, *Philosophy and Logical Syntax* (London: Kegan Paul, 1935), p. 89. Evidently this characterization is not very satisfactory. For plants, animals, and people might well be counted among physical things, especially from the general vantage point of physicalism; and, in this case, biological and psychological terms, which we use in speaking about such things, would qualify as belonging to the physicalistic language, and physicalism would be trivially true. I think that the problem of characterizing with sufficient clarity the notions of physical object, event, and characteristic, and those of physical term and physical law, has constituted a persistent difficulty for physicalism in all stages of its development, although it was not often explicitly recognized as such. One of the exceptions is I. Scheffler's article "The New Dualism: Psychological and Physical Terms," *The Journal of Philosophy*, 47 (1950): 737–52. Scheffler here examines, more specifically, a number of ways in which logical-empiricist writers had sought to differentiate physical from psychological terms; he finds them all inadequate and concludes that the search for a clear

general distinction between physical and psychological terms is futile, and that a reconstruction of science must be "based on a unitary approach to all scientific terms" (ibid., p. 752).
34. Carnap, "Testability and Meaning," *Philosophy of Science*, 3 (1936): 419–71, and 4 (1937): 1–40; quotation from 3: 466. See also pp. 52–53 of Carnap, "Logical Foundations of the Unity of Science," in *International Encyclopedia of Unified Science*, vol. 1, no. 1.
35. See, for example, *The Unity of Science*, p. 67; *Philosophy and Logical Syntax*, p. 89; "Psychology in Physical Language," p. 166.
36. See Carnap's explicit exclusion of "pseudostatements" and "pseudoconcepts" in *The Unity of Science*, pp. 70–75.
37. Carnap, "Psychology in Physical Language," p. 166.
38. Ibid., pp. 170–71, 186–87.
39. Ibid., p. 174.
40. Ibid., p. 170.
41. Ibid., pp. 186–87.
42. See *Philosophy and Logical Syntax*, pp. 88–92.
43. "Psychology in Physical Language," p. 182.
44. "Logical Foundations of the Unity of Science," p. 49.
45. See H. Feigl, "Principles and Problems of Theory Construction in Psychology," in W. Dennis, ed., *Current Trends in Psychological Theory* (Pittsburgh: University of Pittsburgh Press, 1951), pp. 179–213.
46. "Testability and Meaning," p. 467.
47. The most important statements of this stage of Carnap's physicalism are contained in "Testability and Meaning" (especially sec. 15) and in "Logical Foundations of the Unity of Science."
48. See Carnap, "The Methodological Character of Theoretical Concepts," in H. Feigl and M. Scriven, eds., *Minnesotsa Studies in the Philosophy of Science*, vol. 1 (Minneapolis: University of Minnesota Press, 1956), pp. 38–76. Among Feigl's studies dealing with theoretical concepts and theoretical principles and with their significance in psychology are: "Existential Hypotheses: Realistic *vs*. Phenomenalistic Interpretations," *Philosophy of Science*, 17 (1950): 35–62; "Principles and Problems of Theory Construction in Psychology"; and "Physicalism, Unity of Science, and the Foundations of Psychology," in Schilpp, *The Philosophy of Rudolf Carnap*, pp. 277–67.
49. "The Methodological Character of Theoretical Concepts," p. 49; the IQ example is discussed on pp. 71–72.
50. Ibid., secs. VI, VII, and VIII.
51. For detailed statements of this conception of philosophical problems as concerning linguistic issues, see pt. III of Carnap's *Philosophy and Logical Syntax* and pt. V of his *The Logical Syntax of Language* (New York: Harcourt, Brace and Co., 1937). While Carnap here thought of the philosophically relevant linguistic questions as being strictly syntactic, he later broadened his view; but the details need not be considered here. *The Linguistic Turn* is the title of a collection of articles edited by R. Rorty (Chicago: The University of Chicago Press, 1967) which set forth or criticize various modes of linguistic construal and clarification of philosophical problems.
52. Feigl, "Physicalism, Unity of Science, and the Foundations of Psychology," p. 256. A much fuller presentation of Feigl's ideas is given in his essay "The 'Mental' and

the 'Physical,'" in H. Feigl, M. Scriven, and G. Maxwell, eds., *Minnesota Studies in the Philosophy of Science*, vol. 2 (Minneapolis: University of Minnesota Press, 1958), pp. 370–497; this essay, supplemented by a chapter, "Postscript After Ten Years," has since been republished as a separate monograph (Minneapolis: University of Minnesota Press, 1967).
53. See Feigl, "Physicalism, Unity of Science, and the Foundations of Psychology," pp. 241–42, 265–66; Carnap, "Herbert Feigl on Physicalism," in Schilpp, *The Philosophy of Rudolf Carnap*, pp. 882–86.
54. Carnap, "Herbert Feigl on Physicalism," pp. 882, 883; see also Feigl, "Physicalism, Unity of Science, and the Foundations of Psychology," p. 241.
55. Feigl, "Physicalism, Unity of Science, and the Foundations of Psychology," pp. 242–43, 266, 253.
56. The problems here adumbrated are examined more fully in my essay "Reduction: Ontological and Linguistic Facets," in Sidney Morgenbesser, Patrick Suppes, and Morton White, eds., *Philosophy, Science and Method: Essays in Honor of Ernest Nagel* (New York: St. Martin's Press, 1969).
57. Feigl, "Physicalism, Unity of Science, and the Foundations of Psychology," p. 258 (italics quoted).
58. G. Ryle, *The Concept of Mind* (London, 1949), p. 90.
59. Ibid., p. 113.
60. Ibid., p. 44.
61. This is argued, for example, in the articles "The Function of General Laws in History" (1942), "Studies in the Logic of Explanation" (with Paul Oppenheim, 1948), "Typological Methods in the Natural and the Social Sciences" (1952), "The Logic of Functional Analysis" (1959), and "Aspects of Scientific Explanation," all of which are included in my book *Aspects of Scientific Explanation and Other Essays in the Philosophy of Science* (New York: Macmillan [The Free Press], 1965). This book also contains extensive references to further literature on the subject, including various criticisms of the basic idea of a unity of explanatory method.

14

Explanation in Science and in History

1. Introduction

Among the diverse factors that have encouraged and sustained scientific inquiry through its long history are two pervasive human concerns which provide, I think, the basic motivation for all scientific research. One of these if man's persistent desire to improve his strateigc position in the world by means of dependable methods for predicting and, whenever possible, controlling the events that occur in it. The extent to which science has been able to satisfy this urge is reflected impressively in the vast and steadily widening range of its technological applications. But besides this practical concern, there is a second basic motivation for the scientific quest, namely, man's insatiable intellectual curiosity, his deep concern to *know* the world he lives in, and to *explain*, and thus to *understand*, the unending flow of phenomena it presents to him.

In times past questions as to the *what* and the *why* of the empirical world were often answered by myths; and to some extent, this is so even in our time. But gradually, the myths are displaced by the concepts, hypotheses, and theories developed in the various branches of empirical science, including the natural sciences, psychology, and sociological as well as historical inquiry. What is the general character of the understanding attainable by these means, and what is its potential scope? In this essay I will try to shed some light on these questions by examining what seem to me the two basic types of explanation offered by the natural sciences, and then comparing them with some modes of explanation and understanding that are found in historical studies.

First, then, a look at explanation in the natural sciences.

2. Two Basic Types of Scientific Explanation

2.1. Deductive-Nomological Explanation

In his book, *How We Think*,[1] John Dewey describes an observation he made one day when, washing dishes, he took some glass tumblers out of the hot soap suds and put them upside down on a plate: he noticed that soap bubbles emerged from under the tumblers' rims, grew for a while, came to a standstill, and finally receded inside the tumblers. Why did this happen? The explanation Dewey outlines comes to this: In transferring a tumbler to the plate, cool air is caught in it; this air is gradually warmed by the glass, which initially has the temperature of the hot suds. The warming of the air is accompanied by an increase in its pressure, which in turn produces an expansion of the soap film between the plate and the rim. Gradually, the glass cools off, and so does the air inside, with the result that the soap bubbles recede.

This explanatory account may be regarded as an argument to the effect that the event to be explained (let me call it the explanandum-event) was to be expected by reason of certain explanatory facts. These may be divided into two groups: (i) particular facts and (ii) unifromities expressed by general laws. The first group includes facts such as these: the tumblers had been immersed, for some time, in soap suds of a temperature considerably higher than that of the surrounding air; they were put, upside down, on a plate on which a puddle of soapy water had formed, providing a connecting soap film, etc. The second group of items presupposed in the argument includes the gas laws and various other laws that have not been explicitly suggested concerning the exchange of heat between bodies of different temperature, the elastic behavior of soap bubbles, etc. If we imagine these various presuppositions explicitly spelled out, the idea suggests itself of construing the explanation as a deductive argument of this form:

(D) $$\frac{C_1, C_2, \ldots, C_k}{L_1, L_2, \ldots, L_r}{E}$$

Here, $C_1, C_2, \ldots, C_k$ are statements describing the particular facts invoked; $L_1, L_2, \ldots, L_r$ are general laws: jointly, these statements will be said to form the explanans. The conclusion E is a statement describing the explanandum-event; let me call if the explanandum-statement, and let me use the word "explanandum" to refer to either E or to the event described by it.

The kind of explanation thus characterized I will call *deductive-nomological explanation*; for it amounts to a deductive subsumption of the explanandum under principles which have the character of general laws: it answers the question "Why did the explanandum event occur?" by showing that the event resulted from the particular circumstances specified in $C_1, C_2, \ldots, C_k$ in accordance with the laws $L_1, L_2, \ldots, L_r$. This conception of explanation, as exhibited in schema (D), has therefore been referred to as the covering-law model, or as the deductive model, of explanation.[2]

A good many scientific explanations can be regarded as deductive-nomological in character. Consider, for example, the explanation of mirror-images, of rainbows, or of the appearance that a spoon handle is bent at the point where it emerges from a glass of water: in all these cases, the explanandum is deductively subsumed under the laws of reflection and refraction. Similarly, certain aspects of free fall and of planetary motion can be accounted for by deductive subsumption under Galileo's or Kepler's laws.

In the illustrations given so far the explanatory laws had, by and large, the character of empirical generalizations connecting different observable aspects of the phenomena under scrutiny: angle of incidence with angle of reflection or refraction, distance covered with falling time, etc. But science raises the question "why?" also with respect to the uniformities expressed by such laws, and often answers it in basically the same manner, namely, by subsuming the uniformities under more inclusive laws, and eventually under comprehensive theories. For example, the question, "Why do Galileo's and Kepler's laws hold?" is answered by showing that these laws are but special consequences of the Newtonian laws of motion and of gravitation; and these, in turn, may be explained by subsumption under the more comprehensive general theory of relativity. Such subsumption under broader laws or theories usually increases both the breadth and the depth of our scientific understanding. There is an increase in breadth, or scope, because the new explanatory principles cover a broader range of phenomena; for example, Newton's principles govern free fall on the earth and on other celestial bodies, as well as the motions of planets, comets, and artificial satellites, the movements of pendulums, tidal changes, and various other phenomena. And the increase thus effected in the depth of our understanding is strikingly reflected in the fact that, in the light of more advanced explanatory principles, the original empirical laws are usually seen to hold only approximately, or within certain limits. For example, Newton's theory implies that the factor g in Galileo's law, $s = \frac{1}{2} gt^2$, is not strictly a constant for free fall near the surface of the earth; and that, since every planet undergoes gravitational attraction not only from the sun, but also from the other planets, the planetary orbits are not strictly ellipses, as stated in Kepler's laws.

One further point deserves brief mention here. An explanation of a particular event is often conceived as specifying its *cause*, or causes. Thus, the account outlined in our first illustration might be held to explain the growth and the recession of the soap bubbles by showing that the phenomenon was *caused* by a rise and a subsequent drop of the temperature of the air trapped in the tumblers. Clearly, however, these temperature changes provide the requisite explanation only in conjunction with certain other conditions, such as the presence of a soap film, practically constant pressure of the air surrounding the glasses, etc. Accordingly, in the context of explanation, a cause must be allowed to consist in a more or less complex set of particular circumstances; these might be described by a set of sentences: $C_1, C_2, \ldots, C_k$. And, as suggested by the principle "Same cause, same effect," the assertion that those circumstances jointly caused a given event—described, let us say, by a sentence E—implies that whenever and wherever circumstances of the kind in question occur, an event of the kind to be explained comes about. Hence, the given causal explanation implicitly claims that there are general laws—such as $L_1, L_2, \ldots, L_r$ in schema (D)—by virtue of which the occurrence of the causal antecedents mentioned in $C_1, C_2, \ldots, C_k$ is a sufficient condition for the occurrence of the event to be explained. Thus, the relation between causal factors and effect is reflected in schema (D): causal explanation is deductive-nomological in character. (However, the customary formulations of causal and other explanations often do not explicitly specify all the relevant laws and particular facts: to this point, we will return later.)

The converse does not hold: there are deductive-nomological explanations which would not normally be counted as causal. For one thing, the subsumption of laws, such as Galileo's or Kepler's laws, under more comprehensive principles is clearly not causal in character: we speak of causes only in reference to *particular facts* or events, and not in reference to *universal facts* as expressed by general laws. But not even all deductive-nomological explanations of particular facts or events will qualify as causal; for in a causal explanation some of the explanatory circumstances will temporally precede the effect to be explained: and there are explanations of type (D) which lack this characteristic. For example, the pressure which a gas of specified mass possesses at a given time might be explained by reference to its temperature and its volume at the same time, in conjunction with the gas law which connects simultaneous values of the three parameters.[3]

In conclusion, let me stress once more the important role of laws in deductive-nomological explanation: the laws connect the explanandum event with the particular conditions cited in the explanans, and this is what confers upon the latter the status of explanatory (and, in some cases, causal) factors in regard to the phenomenon to be explained.

2.2. Probabilistic Explanation

In deductive-nomological explanation as schematized in (D), the laws and theoretical principles involved are of *strictly universal form*: they assert that in *all* cases in which certain specified conditions are realized an occurrence of such and such a kind will result; the law that any metal, when heated under constant pressure, will increase in volume, is a typical example; Galileo's, Kepler's, Newton's, Boyle's, and Snell's laws, and many others, are of the same character.

Now let me turn next to a second basic type of scientific explanation. This kind of explanation, too, is nomological, i.e., it accounts for a given phenomenon by reference to general laws or theoretical principles; but some or all of these are of *probabilistic-statistical form*, i.e., they are, generally speaking, assertions to the effect that if certain specified conditions are realized, then an occurrence of such and such a kind will come about with such and such a statistical probability.

For example, the subsiding of a violent attack of hay fever in a given case might well be attributed to, and thus explained by reference to, the administration of 8 milligrams of chlortrimeton. But if we wish to connect this antecedent event with the explanandum, and thus to establish its explanatory significance for the latter, we cannot invoke a universal law to the effect that the administration of 8 milligrams of that antihistamine will invariably terminate a hay fever attack: this simply is not so. What can be asserted is only a generalization to the effect that administration of the drug will be followed by relief with high statistical probability, i.e., roughly speaking, with a high relative frequency in the long run. The resulting explanans will thus be of the following type:

> John Doe had a hay fever attack and took 8 milligrams of chlortrimeton. The probability for subsidence of a hay fever attack upon administration of 8 milligrams of chlortrimeton is high.

Clearly, this explanans does not deductively imply the explanandum, "John Doe's hay fever attack subsided"; the truth of the explanans makes the truth of the explanandum not certain (as it does in a deductive-nomological explanation) but only more or less likely or, perhaps "practically" certain.

Reduced to its simplest essentials, a probabilistic explanation thus takes the following form:

$$(P) \quad \left. \begin{array}{c} F_i \\ \hline p(O, F) \text{ is very high} \\ \hline O_i \end{array} \right\} \text{makes very likely}$$

The explanandum, expressed by the statement "O_i," consists in the fact that in the particular instance under consideration, here called i (e.g., John Doe's allergic attack), an outcome of kind O (subsidence) occurred. This is explained by means of two explanans-statements. The first of these, "F_i," corresponds to $C_1, C_2, \ldots, C_k$ in (D); it states that in case i, the factors F (which may be more or less complex) were realized. The second expresses a law of probabilistic form, to the effect that the statistical probability for outcome O to occur in cases where F is realized is very high (close to 1). The double line separating explanandum from explanans is to indicate that, in contrast to the case of deductive-nomological explanation, the explanans does not logically imply the explanandum, but only confers a high likelihood upon it. The concept of likelihood here referred to must be clearly distinguished from that of statistical probability, symbolized by "p" in our schema. A statistical probability is, roughly speaking, the long-run relative frequency with which an occurrence of a given kind (say, F) is accompanied by an "outcome" of a specified kind (say, O). Our likelihood, on the other hand, is a relation (capable of gradations) not between kinds of occurrences, but between statements. The likelihood referred to in (P) may be characterized as the strength of the inductive support, or the degree of rational credibility, which the explanans confers upon the explanandum; or, in Carnap's terminology, as the *logical*, or *inductive* (in contrast to statistical), *probability* which the explanandum possesses relative to the explanans.

Thus, probabilistic explanation, just like explanation in the manner of schema (D), is nomological in that it presupposes general laws; but because these laws are of statistical rather than of strictly universal form, the resulting explanatory arguments are inductive rather than deductive in character. An inductive argument of this kind *explains* a given phenomenon by showing that, in view of certain particular events and certain statistical laws, its occurrence was to be expected with high logical, or inductive, probability.

By reason of its inductive character, probabilistic explanation differs from its deductive-nomological counterpart in several other important respects; for example, its explanans may confer upon the explanandum a more or less high degree of inductive support; in this sense, probabilistic explanation admits of degrees, whereas deductive-nomological explanation appears as an either-or affair: a given set of universal laws and particular statements either does or does not imply a given explanandum statement. A fuller examination of these differences, however, would lead us far afield and is not required for the purposes of this essay.[4]

One final point: the distinction here suggested between deductive-nomological and probabilistic explanation might be questioned on the ground that, after all, the universal laws invoked in a deductive explanation can have been established only on the basis of a finite body of evidence, which surely affords no exhaustive verification, but only more or less strong probability for it; and that, therefore, all scientific laws have to be regarded as probabilistic. This argument, however, confounds a logical issue with an epistemological one: it fails to distinguish properly between the *claim* made by a given law-statement and the *degree of confirmation*, or *probability*, which it possesses on the available evidence. It is quite true that statements expressing laws of either kind can be only incompletely confirmed by any given finite set—however large—of data about particular facts; but law-statements of the two different types make claims of different kind, which are reflected in their logical forms: roughly, a universal law-statement of the simplest kind asserts that *all* elements of an indefinitely large reference class (e.g., copper objects) have a certain characteristic (e.g., that of being good conductors of electricity); while statistical law-statements assert that in the long run, a specified proportion of the members of the reference class have some specified property. And our distinction of two types of law and, concomitantly, of two types of scientific explanation, is based on this difference in claim as reflected in the difference of form.

The great scientific importance of probabilistic explanation is eloquently attested to by the extensive and highly successful explanatory use that has been made of fundamental laws of statistical form in genetics, statistical mechanics, and quantum theory.

3. Elliptic and Partial Explanations: Explanation Sketches

As I mentioned earlier, the conception of deductive-nomological explanation reflected in our schema (D) is often referred to as the covering-law model, or the deductive model, of explanation: similarly, the conception underlying schema (P) might be called the probabilistic, or the inductive-statistical, model of explanation. The term "model" can serve as a useful reminder that the two types of explanation as characterized above constitute ideal types or theoretical idealizations and are not intended to reflect the manner in which working scientists actually formulate their explanatory accounts. Rather, they are meant to provide explications, or rational reconstructions, or theoretical models, of certain modes of scientific explanation.

In this respect our models might be compared to the concept of mathematical proof (within a given theory) as construed in metamathematics. This concept, too, may be regarded as a theoretical model: it is not intended to provide a descriptive account of how proofs are formulated in the writings of mathematicians: most of these actual formulations fall short of rigorous and, as it were, ideal, metamathematical standards. But the theoretical model has certain other functions: it exhibits the rationale of mathematical proofs by revealing the logical connections underlying the successive steps; it provides standards for a critical appraisal of any proposed proof constructed within the mathematical system to which the model refers; and it affords a basis for a precise and far-reaching theory of proof, provability, decidability, and related concepts. I think the two models of explanation can fulfill the same functions, if only on a much more modest scale. For example, the arguments presented in constructing the models give an indication of the sense in which the models exhibit the rationale and the logical structure of the explanations they are intended to represent.

I now want to add a few words concerning the second of the functions just mentioned; but I will have to forgo a discussion of the third.

When a mathematician proves a theorem, he will often omit mention of certain propositions which he presupposes in his argument and which he is in fact entitled to presuppose because, for example, they follow readily from the postulates of his system or from previously established theorems or perhaps from the hypothesis of his own theorem, if the latter is in hypothetical form; he then simply assumes that his readers or listeners will be able to supply the missing items if they so desire. If judged by ideal standards, the given formulation of the proof is elliptic or incomplete; but the departure from the ideal is harmless: the gaps can readily be filled in. Similarly, explanations put forward in everyday discourse and also in scientific contexts are often *elliptically formulated*. When we explain, for example, that a lump of butter melted because it was put into a hot frying pan, or that a small rainbow appeared in the spray of the lawn sprinkler because the sunlight was reflected and refracted by the water droplets, we may be said to offer elliptic formulations of deductive-nomological explanations; an account of this kind omits mention of certain laws or particular facts which it tacitly takes for granted, and whose explicit citation would yield a complete deductive-nomological argument.

In addition to elliptic formulation, there is another, quite important, respect in which many explanatory arguments deviate from the theoretical model. It often happens that the statement actually included in the explanans, together with those which may reasonably be assumed to have been taken for granted in the context at hand, explain the given explanandum only *partially*, in a sense which I will try to indicate by an example. In his *Psychopathology of Everyday Life*, Freud offers the following explanation of a slip of the pen that occurred to him: "On a sheet of paper containing principally short daily notes of business interest, I found, to my surprise, the incorrect date, 'Thursday, October 20th,' bracketed under the correct date of the month of September. It was not difficult to explain this anticipation as the expression of a wish. A few days before I had returned fresh from my vacation and felt ready for any amount of professional work, but as yet there were few patients. On my arrival I had found a letter from a patient announcing her arrival on the 20th of October. As I wrote the same date in September I may certainly have thought 'X. ought to be here already; what a pity about the whole month!,' and with this thought I pushed the current date a month ahead."[5]

Clearly, the formulation of the intended explanation is *at least incomplete* in the sense considered a moment ago. In particular, it fails to mention any laws or theoretical principles in virtue of which the subconscious wish, and the other antecedent circumstances referred to, could be held to explain Freud's slip of the pen. However, the general theoretical considerations Freud presents here and elsewhere in his writings suggests strongly that his explanatory account relies on a hypothesis to the effect that when a person has a strong, though perhaps unconscious, desire, then if he commits a slip of pen, tongue, memory, or the like, the slip will take a form in which it expresses, and perhaps symbolically fulfills, the given desire.

Even this rather vague hypothesis is probably more definite than what Freud would have been willing to assert. But for the sake of the argument let us accept it and include it in the explanans, together with the particular statements that Freud did have the subconscious wish he mentions, and that he was going to commit a slip of the pen. Even then, the resulting explanans permits us to deduce only that the slip made by Freud would, *in some way or other*, express and perhaps symbolically fulfill Freud's subconscious wish. But clearly, such expression and fulfillment might have been achieved by many other kinds of slip of the pen than the one actually committed.

In other words, the explanans does not imply, and thus fully explain, that the particular slip, say s, which Freud committed on this occasion, would fall within the narrow class, say W, of acts which consist in writing the words "Thursday, October 20th"; rather, the explanans implies only that s would fall into a wider class, say F, which includes W as a proper subclass, and which consists of all acts which would express and symbolically fulfill Freud's subconscious wish *in some way or other*.

The argument under consideration might be called a *partial explanation*: it provides complete, or conclusive, grounds for expecting s to be a member of F, and since W is a subclass of F, it thus shows that the explanandum, i.e., s falling within W, accords with, or bears out, what is to be expected in consideration of the explanans. By contrast, a deductive-nomological explanation of the form (D) might then be called *complete* since the explanans here does imply the explanandum.

Clearly, the question whether a given explanatory argument is complete or partial can be significantly raised only if the explanandum sentence is fully specified; only then can we ask whether the explanandum does or does not follow from the explanans. Completeness of explanation, in this sense, is relative to our explanandum sentence. Now, it might seem much more important and interesting to consider instead the notion of a complete explanation of some *concrete event*, such as the destruction of Pompeii, or the death of Adolf Hitler, or the launching of the first artificial satellite: we might want to regard a particular event as completely explained only if an explanatory account of deductive or of inductive form had been provided for all of its aspects. This notion, however, is self-defeating; for any particular event may be regarded as having infinitely many different aspects or characteristics, which cannot all be accounted for by a finite set, however large, of explanatory statements.

In some cases, what is intended as an explanatory account will depart even further from the standards reflected in the model schemata (D) and (P) above. An explanatory account, for example, which is not explicit and specific enough to be reasonably qualified as an elliptically formulated explanation or as a partial one, can often be viewed as an *explanation sketch*: it may suggest, perhaps quite vividly and persuasively, the general outlines of what, it is hoped, can eventually be supplemented so as to yield a more closely reasoned argument based on explanatory hypotheses which are indicated more fully, and which more readily permit of critical appraisal by reference to empirical evidence.

The decision whether a proposed explanatory account is to be qualified as an elliptically formulated deductive or probabilistic explanation, as a partial explanation, as an explanation sketch, or perhaps as none of these is a matter of judicious interpretation; it calls for an appraisal of the intent of the given argument and of the background assumptions that may be assumed to have been tacitly taken for granted, or at least to be available, in the given context. Unequivocal decision rules cannot be set down for this purpose any more than for determining whether a given informally stated inference which is not deductively valid by reasonably strict standards is to count nevertheless as valid but enthymematically formulated, or as fallacious, or as an instance of sound inductive reasoning, or perhaps, for lack of clarity, as none of these.

4. Nomological Explanation in History

So far, we have examined nomological explanation, both deductive and inductive, as found in the natural sciences; and we have considered certain characteristic ways in which actual explanatory accounts often depart from the ideal standards of our two basic models. Now it is time to ask what light the preceding inquiries can shed on the explanatory procedures used in historical research.

In examining this question, we will consider a number of specific explanatory arguments offered by a variety of writers. It should be understood from the beginning that we are here concerned, not to appraise the factual adequacy of these explanations, but only to attempt an explication of the claims they make and of the assumptions they presuppose.

Let us note first, then, that some historical explanations are surely nomological in character: they aim to show that the explanandum phenomenon resulted from certain antecedent, and perhaps, concomitant, conditions; and in arguing these, they rely more or less explicitly on relevant generalizations. These may concern, for example, psychological or sociological tendencies and may best be conceived as broadly probabilistic in character. This point is illustrated by the following argument, which might be called an attempt to explain Parkinson's Law by subsumption under broader psychological principles:

> As the activities of the government are enlarged, more people develop a vested interest in the continuation and expansion of governmental functions. People who have jobs do not like to lose them; those who are habituated to certain skills do not welcome change; those who have become accustomed to the exercise of a certain kind of power do not like to relinquish their control—if anything, they want to develop greater power and correspondingly greater prestige. . . . Thus, government offices and bureaus, once created, in turn institute drives, not only to fortify themselves against assault, but to enlarge the scope of their operations.[6]

The psychological generalizations here explicitly adduced will reasonably have to be understood as expressing, not strict uniformities, but strong *tendencies*, which might be formulated by means of rough probability statements; so that the explanation here suggested is probabilistic in character.

As a rule, however, the generalizations underlying a proposed historical explanation are largely left unspecified; and most concrete explanatory accounts have to be qualified as partial explanations or as explanation sketches. Consider, for example, F. J. Turner's essay "The Significance of the Frontier in American History,"[7] which amplifies and defends the view that

> up to our own day American history has been in a large degree the history of the colonization of the Great West. The existence of an area of free land, its continuous recession, and the advance of American settlement westward explain American development. . . . The peculiarity of American institutions is the fact that they have been compelled to adapt themselves . . . to the changes involved in crossing a continent, in winning a wilderness, and in developing at each area of this progress, out of the primitive economic and political conditions of the frontier, the complexity of city life.[8]

One of the phenomena Turner considers in developing his thesis is the rapid westward advance of what he calls the Indian trader's frontier. "Why was it," Turner asks, "that the Indian trader passed so rapidly across the continent?"; and he answers, "The explanation of the rapidity of this advance is bound up with the effects of the trader on the Indian. The trading post left the unarmed tribes at the mercy of those that had purchased firearms—a truth which the Iroquois Indians wrote in blood, and so the remote and unvisited tribes gave eager welcome to the trader. . . . This accounts for the trader's power and the rapidity of his advance."[9] There is no explicit mention here of any laws, but it is clear that this sketch of an explanation presupposes, first of all, various particular facts, such as that the remote and unvisited tribes had heard of the efficacy and availability of firearms, and that there were no culture patterns or institutions precluding their use by those tribes; but in addition, the account clearly rests also on certain assumptions as to how human beings will tend to behave in situations presenting the kinds of danger and of opportunity that Turner refers to.

Similar comments apply to Turner's account of the westward advance of what he calls the farmer's frontier:

> Omitting those of the pioneer farmers who move from the love of adventure, the advance of the more steady farmer is easy to understand. Obviously the immigrant was attracted by the cheap lands of the frontier, and even the native farmer felt their influence strongly. Year by year the farmers who lived on soil, whose returns were diminished by unrotated crops, were offered the virgin soil of the frontier at nominal prices. Their growing families demanded more lands, and these were dear. The competition of the unexhausted, cheap, and easily tilled prairie lands compelled the farmer either to go West . . . or to adopt intensive culture.[10]

This passage is clearly intended to do more than describe a sequence of particular events: it is meant to afford an understanding of the farmers' westward advance by pointing to their interests and needs and by calling attention to the facts and the opportunities facing them. Again, this explanation takes it for granted that under such conditions normal human beings will tend to seize new opportunities in the manner in which the pioneer farmers did.

Examining the various consequences of this moving-frontier history, Turner states that "the most important effect of the frontier has been in the promotion of democracy here and in Europe,"[11] and he begins his elaboration of this theme with the remark that "the frontier is productive of individualism. . . . The tendency is anti-social. It produces antipathy to control, and particularly to any direct control":[12] and this is, of course, a sociological generalization in a nutshell.

Similarly, any explanation that accounts for a historical phenomenon by reference to economic factors or by means of general principles of social or cultural change are nomological in import, even if not in explicit formulation.

But if this be granted there still remains another question, to which we must now turn, namely, whether, in addition to explanations of a broadly nomological character, the historian also employs certain other distinctly historical ways of explaining and understanding whose import cannot be adequately characterized by means of our two models. The question has often been answered in the affirmative, and several kinds of historical explanation have been adduced in support of this affirmation. I will now consider what seem to me two especially interesting candidates for the role of specifically historical explanation; namely first, genetic explanation, and secondly, explanation of an action in terms of its underlying rationale.

5. Genetic Explanation in History

In order to make the occurrence of a historical phenomenon intelligible, a historian will frequently offer a "genetic explanation" aimed at exhibiting the principal stages in a sequence of events which led up to the given phenomenon.

Consider, for example, the practice of selling indulgences as it existed in Luther's time. H. Boehmer, in his work, *Luther and the Reformation*, points out that until about the end of the nineteenth century, "the indulgence was in fact still a great unknown quantity, at sight of which the scholar would ask himself with a sigh: 'Where did it come from?'"[13] An answer was provided by Adolf Gottlob,[14] who tackled the problem by asking himself what led the popes and bishops to offer indulgences. As a result, ". . . origin and development of the unknown quantity appeared clearly in the light, and doubts as to its original meaning came to an end. It revealed itself as a true descendant of the time of the great struggle between Christianity and Islam, and at the same time a highly characteristic product of Germanic Christianity."[15]

In brief outline,[16] the origins of the indulgence appear to go back to the ninth century, when the popes were strongly concerned with the fight against Islam. The Mohammedan fighter was assured by the teachings of his religion that if he were to be killed in battle his sould would immediately go to heaven; but the defender of the Christian faith had to fear that he might still be lost if he had not done the regular penance for his sins. To allay these doubts, John VII, in 877, promised absolution for their sins to crusaders who should be killed in battle. "Once the crusade was so highly thought of, it was an easy transition to regard participation in a crusade as equivalent to the performance of atonement . . . and to promise remission of these penances in return for expeditions against the Church's enemies."[17] Thus, there was introduced the indulgence of the Cross, which granted complete remission of the penitential punishment to all those who participated in a religious war. "If it is remembered what inconveniences, what ecclesiastical and civil disadvantages the ecclesiastical penances entailed, it is easy to understand that the penitents flocked to obtain this indulgence."[18] A further strong incentive came from the belief that whoever obtained an indulgence secured liberation not only from the ecclesiastical penances, but also from the corresponding suffering in purgatory after death. The benefits of these indulgences were next extended to those who, being physically unfit to participate in a religious war, contributed the funds required to send a soldier on a crusade: in 1199, Pope Innocent III recognized the payment of money as adequate qualification for the benefits of a crusading indulgence.

When the crusades were on the decline, new ways were explored of raising funds through indulgences. Thus, there was instituted a "jubilee indulgence," to be celebrated every hundred years, for the benefit of pilgrims coming to Rome on that occasion. The first of these indulgences, in 1300, brought in huge sums of money; and the time interval between successive jubilee indulgences was therefore reduced to 50, 33, and even 25 years. And from 1393 on the jubilee indulgence was made available, not only in Rome, for the benefit of pilgrims, but everywhere in Europe, through special agents who were empowered to absolve the penitent of their sins upon payment of an appropriate amount. The development went even further: in 1477, a dogmatic declaration by Sixtus IV attributed to the indulgence the power of delivering even the dead from purgatory.

Undeniably, a genetic account of this kind can enhance our understanding of a historical phenomenon. But its explanatory role, far from being *sui generis*, seems to me basically nomological in character. For the successive stages singled out for consideration surely must be qualified for their function by more than the fact that they form a temporal sequence and that they all precede the final stage, which is to be explained: the mere enumeration in a yearbook of "the year's important events" in the order of their occurrence clearly is not a genetic explanation of the final event or of anything else. In a genetic explanation each stage must be shown to "lead to" the next, and thus to be linked to its successor by virtue of some general principle which makes the occurrence of the latter at least reasonably probable, given the former. But in this sense, even successive stages in a physical phenomenon such as the free fall of a stone may be regarded as forming a genetic sequence whose different stages—characterized, let us say, by the position and the velocity of the stone at different times—are interconnected by strictly universal laws; and the successive stages in the movement of a steel ball bouncing its zigzaggy way down a Galton pegboard may be regarded as forming a genetc sequence with probabilistic connections.

The genetic accounts given by historians are not, of course, of the purely nomological kind suggested by these examples from physics. Rather, they combine a certain measure of nomological interconnecting with more or less large amounts of straight description. For consider an intermediate stage mentioned in a genetic account: some aspects of it will be presented as having evolved from the preceding stages (in virtue of connecting laws, which often will be no more than hinted at); while other aspects, which are not accounted for by information about the preceding development, will be descriptively added because they are relevant to an understanding of subsequent stages in the genetic sequence. Thus, schematically speaking, a genetic explanation will begin with a pure description of an initial stage; thence, it will proceed to an account of a second stage, part of which is nomologically linked to, and explained by, the characteristic features of the initial stage; while the balance is simply described as relevant for a nomological account of some aspects of the third stage; and so forth.[19]

In our illustration the connecting laws are hinted at in the mention made of motivating factors: the explanatory claims made for the interest of the popes in securing a fighting force and in amassing ever larger funds clearly presuppose suitable psychological generalizations as to the manner in which an intelligent individual will act, in the light of his factual beliefs, when he seeks to attain a certain objective. Similarly, general assumptions underly the reference to the fear of purgatory in explaining the eagerness with which indulgences were bought. And when, referring to the huge financial returns of the first jubilee indulgence, Schwiebert says "This success only whetted the insatiable appetite of the popes.

The intervening period of time was variously reduced from 100 to 50, to 33, to 25 years ...,"[20] the explanatory force here implied might be said to rest on some principle of reinforcement by rewards. As need hardly be added, even if such a principle were explicitly introduced, the resulting account would provide at most a partial explanation; it could not be expected to show, for example, why the intervening intervals should have the particular lengths here mentioned.

In the genetic account of the indulgences, those factors which are simply described (or tacitly presupposed) rather than explained include, for example, the doctrines, the organization, and the power of the Church; the occurrence of the crusades and their eventual decline; and innumerable other factors which are not even explicitly mentioned, but which have to be understood as background conditions if the genetic survey is to serve its explanatory purpose.

The general conception here outlined of the logic of genetic explanation could also be illustrated by reference to Turner's studies of the American frontier; this will be clear even from the brief remarks made earlier on Turner's ideas.

Some analysts of historical development put special emphasis on the importance of the laws underlying a historical explanation; thus, e.g., A. Gerschenkron maintains, "Historical research consists essentially in application to empirical material of various sets of empirically derived hypothetical generalizations and in testing the closeness of the resulting fit, in the hope that in this way certain uniformities, certain typical situations, and certain typical relationships among individual factors in these situations can be ascertained,"[21] and his subsequent substantive observations include a brief genetic survey of patterns of industrial development in nineteenth-century Europe, in which some of the presumably relevant uniformities are made reasonably explicit.

6. Explanation by Motivating Reasons

Let us now turn to another kind of historical explanation that is often considered as *sui generis*, namely, the explanation of an action in terms of the underlying *rationale*, which will include, in particular, the ends the agent sought to attain, and the alternative courses of action he believed to be open to him. The following passage explaining the transition from the indulgence of the Cross to the institution of the jubilee indulgence illustrates this procedure:

> ... in the course of the thirteenth century the idea of a crusade more and more lost its power over men's spirits. If the Popes would keep open the important source of income which the indulgence represented, they must invent new motives to attract people to the purchase of indulgences. It is the merit of Boniface VIII to have recognized this clearly. By creating the jubilee indulgence in 1300 he assured the species a further long development most welcome to the Papal finances.[22]

This passage clearly seeks to explain the establishment of the first jubilee indulgence by suggesting the reasons for which Boniface VIII took this step. If properly spelled out, these reasons would include not only Boniface's objective of ensuring a continuation of the income so far derived from the indulgence of the Cross, but also his estimate of the relevant empirical circumstances, including the different courses of action open to him, and their probable efficacy as well as potential difficulties in pursuing them and adverse consequences to which they might lead.

The kind of explanation achieved by specifying the rationale underlying a given action is widely held to be fundamentally different from nomological explanation as found in the natural sciences. Various reasons have been adduced in support of this view; but I will limit my discussion largely to the stimulating ideas on the subjects that have been set forth by Dray.[23] According to Dray, there is an important type of historical explanation whose features "make the covering law model peculiarly inept"; he calls it "rational explanation," i.e., "explanation which displays the *rationale* of what was done," or, more fully, "a reconstruction of the agent's *calculation* of means to be adopted toward his chosen end in the light of the circumstances in which he found himself." The object of rational explanation is not to subsume the explanandum under general laws, but "to show that what was done was the thing to have done for the reasons given, rather than merely the thing that is done on such occasions, perhaps in accordance with certain laws." Hence, a rational explanation has "an element of *appraisal*" in it: it "must exhibit what was done as appropriate or justified." Accordingly, Dray conceives a rational explanation as being based on a standard of appropriateness or of rationality of a special kind which he calls a "*principle of action*," i.e., "a judgment of the form 'When in a situation of type $C_1, C_2, \ldots C_n$ the thing to do is X.'"

Dray does not give a full account of the kind of "situation" here referred to; but to do justice to his intentions, these situations must evidently be taken to include, at least, items of the following three types: (i) the end the agent was seeking to attain; (ii) the empirical circumstances, as seen by the agent, in which he had to act; (iii) the moral standards or principles of conduct to which the agent was committed. For while this brief list requires considerable further scrutiny and elaboration, it seems clear that only if at least these items are specified does it make sense to raise the question of the appropriateness of what the agent did in the given "situation."

It seems fair to say, then, that according to Dray's conception a rational explanation answers a question of the form "Why did agent A do X?" by offering an explanans of the following type (our formulation replaces the notation "$C_1, C_2, \ldots C_n$" by the simpler "C", without, of course, precluding that the kind of situation thus referred to may be extremely complex.):

(R)   A was in a situation of type C
         In a situation of type C, the appropriate thing to do is X

But can an explanans of this type possibly serve to explain A's having in fact done X? It seems to me beyond dispute that in any adequate explanation of an empirical phenomenon the explanans must provide good grounds for believing or asserting that the explanandum phenomenon did in fact occur. Yet this requirement, which is necessary though not sufficient[24] for an adequate explanation, is not met by a rational explanation as conceived by Dray. For the two statements included in the contemplated explanans (R) provide good reasons for believing that the appropriate thing for A to do was X, but not for believing that A did in fact do X. Thus, a rational explanation in the sense in which Dray appears to understand it does not explain what it is meant to explain. Indeed, the expression "the thing to do" in the standard formulation of a principle of action, "functions as a value term," as Dray himself points out: but then, it is unclear, on purely logical grounds, how the valuational principle expressed by the second sentence in (R), in conjunction with the plainly empirical, nonvaluational first sentence, should permit any inferences concerning empirical matters such as A's action, which could not be drawn from the first sentence alone.

To explain, in the general vein here under discussion, why A did in fact do X, we have to refer to the underlying rationale not by means of a normative principle of action, but by descriptive statements to the effect that, at the time in question A was a rational agent, or had the disposition to act rationally; and that a rational agent, when in circumstances of kind C, will always (or: with high probability) do X. Thus construed, the explanans takes on the following form:

(R') (a) A was in a situation of type C
      (b) A was disposed to act rationally
      (c) Any person who is disposed to act rationally will, when in a situation of type C, invariably (with high probability) do X

But by this explanans A's having done X is accounted for in the manner of a deductive or of a probabilistic nomological explanation. Thus, in so far as reference to the rationale of an agent does explain his action, the explanation conforms to one of our nomological models.

An analogous diagnosis applies, incidentally, also to explanations which attribute an agent's behavior in a given situation not to rationality and more or less explicit deliberation on his part, but to other dispositional features, such as his character and emotional make-up. The following comment on Luther illustrates this point:

> Even stranger to him than the sense of anxiety was the allied sense of fear. In 1527 and 1535, when the plague broke out in Wittenberg, he was the only professor besides Bugenhagen who remained calmly at his post to comfort and care for the sick and dying. . . . He had, indeed, so little sense as to take victims of the plague into his house and touch them with his own hand. Death, martyrdom, dishonor, contempt . . . he feared just as little infectious disease.[25]

It may well be said that these observations give more than a description: that they shed some explanatory light on the particular occurrences mentioned. But in so far as they explain, they do so by presenting Luther's actions as manifestations of certain personality traits, such as fearlessness; thus, the particular acts are again subsumed under generalizations as to how a fearless person is likely to behave under certain circumstances.

It might seem that both in this case and in rational explanation as construed in (R'), the statements which we took to express general laws—namely, (c) in (R'), and the statement about the probable behavior of a fearless person in our last illustration—do not have the character of empirical laws at all, but rather that of analytic statements which simply express part of what is *meant* by a rational agent, a fearless person, or the like. Thus, in contrast to nomological explanations, these accounts in terms of certain dispositional characteristics of the agent appear to presuppose no general laws at all. Now, the idea of analyticity gives rise to considerable philosophical difficulties; but let us disregard these here and take the division of statements into analytic and synthetic to be reasonably clear. Even then, the objection just outlined cannot be upheld. For dispositional concepts of the kind invoked in our explanations have to be regarded as governed by entire clusters of general statements—we might call them symptom statements—which connect the given disposition with various specific manifestations, or symptoms, of its presence (each symptom will be a particular mode of "responding," or acting, under specified "stimulus" conditions); and the whole cluster of these symptom statements for a given disposition will have implications which are plainly not analytic (in the intuitive sense here assumed). Under these circumstances it would be arbitrary to attribute to some of the symptom statements the analytic character of partial definitions.

The logic of this situation has a precise representation in Carnap's theory of reduction sentences.[26] Here, the connections between a given disposition and its various manifest symptoms are assumed to be expressed by a set of so-called reduction sentences (these are characterized by their logical form). Some of these state, in terms of manifest characteristics, sufficient conditions for the presence of the given disposition; others similarly state necessary conditions. The reduction sentences for a given dispositional concept cannot, as a rule, all be qualified as analytic; for jointly they imply certain nonanalytic consequences which have the status of general laws connecting exclusively the manifest characteristics; the strongest of the laws so implied is the so-called *representative sentence*, which "represents, so to speak, the factual content of the set" of all the reduction sentences for the given disposition concept. This representative sentence asserts, in effect, that whenever at least one of the sufficient conditions specified by the given reduction sentences is satisfied, then so are all the necessary conditions laid down by the reduction sentences. And when A is one of the manifest criteria sufficient for the presence of the a given disposition, and B is a necessary one, then the statement that whenever A is present so is B will normally turn out to be synthetic.

So far then, I have argued that Dray's construal of explanation by motivating reasons is untenable; that the normative principles of action envisaged by him have to be replaced by statements of a dispositional kind; and that, when this is done, explanations in terms of a motivating rationale, as well as those referring to other psychological factors, are seen to be basically nomological.

Let me add a few further remarks on the idea of rational explanation. First: in many cases of so-called purposive action, there is no conscious deliberation, no rational calculation that leads the agent to his decision. Dray is quite aware of this; but he holds that a rational explanation in his sense is still possible; for "in so far as we say an action is purposive at all, no matter at what level of conscious deliberation, there is a calculation which could be constructed for it: the one the agent would have gone through if he had had time, if he had not seen what to do in a flash, if he had been called upon to account for what he did after the event, etc. And it is by eliciting some such calculation that we explain the action."[27] But the explanatory significance of reasons or "calculations" which are "reconstructed" in this manner is certainly puzzling. If, to take Dray's example, an agent arrives at his decision "in a flash" rather than by deliberation, then it would seem to be simply false to say that the decision can be accounted for by some argument which the agent might have gone through under more propitious circumstances, or which he might produce later if called upon to account for his action; for, by hypothesis, no such argument was in fact gone through by the agent at the crucial time; considerations of appropriateness or rationality played no part in shaping his decision; the rationale that Dray assumes to be adduced and appraised in the corresponding rational explanation is simply fictitious.

But, in fairness to Dray, these remarks call for a qualifying observation: in at least some of the cases Dray has in mind it might not be fictitious to ascribe the action under study to a disposition which the agent acquired through a learning process whose initial stages did involve conscious ratiocination. Consider, for example, the various complex maneuvers of accelerating, braking, signaling, dodging jaywalkers and animals, swerving into and out of traffic lanes, estimating the changes of traffic lights, etc., which are involved in driving a car through city traffic. A beginning driver will often perform these only upon some sort of conscious deliberation or even calculation; but gradually, he learns to do the appropriate thing automatically, "in a flash," without giving them any conscious thought. The habit pattern he has thus acquired may be viewed as consisting in a set of dispositions to react in certain appropriate ways in various situations; and a particular performance of such an appropriate action would then be explained, not by a "constructed" calculation which actually the agent did not perform but by reference to the disposition just mentioned and thus, again, in a nomological fashion.

The method of explaining a given action by "constructing," in Dray's sense, the agent's calculation of means faces yet another, though less fundamental, difficulty: it will frequently yield a rationalization rather than an explanation, especially when the reconstruction relies on the reasons the agent might produce when called upon to account for his action. As G. Watson remarks, "Motivation, as presented in the perspective of history, is often too simple and straightforward, reflecting the psychology of the Age of Reason.... Psychology has come ... to recognize the enormous weight of irrational and intimately personal impulses in conduct. In history, biography, and in autobiography, especially of public characters, the tendency is strong to present 'good' reasons instead of 'real' reasons."[28] Accordingly, as Watson goes on to point out, it is important, in examining the motivation of historical figures, to take into account the significance of such psychological mechanisms as reaction formation, "the dialectic dynamic by which stinginess cloaks itself in generosity, or rabid pacifism arises from the attempt to repress strong aggressive impulses."[29]

These remarks have a bearing also on an idea set forth by P. Gardiner in his illuminating book on historical explanation.[30] Commenting on the notion of the "real reason" for a man's action, Gardiner says: "In general, it appears safe to say that by a man's 'real reasons' we mean those reasons he would be prepared to give under circumstances where his confession would not entail adverse consequences to himself." And he adds, "An exception to this is the psychoanalyst's usage of the expression where different criteria are adopted."[31] This observation might be taken to imply that the explanation of human actions in terms of underlying motives is properly aimed at exhibiting the agent's "real reasons" in the ordinary sense of the phrase, as just described; and that, by implication, reasons in the psychoanalyst's sense require less or no consideration. But such a construal of explanation would give undue importance to considerations of ordinary language. Gardiner is entirely right when he reminds us that the "language in which history is written is for the most part the language of ordinary speech";[32] but the historian in search of reasons that will correctly explain human actions will obviously have to give up his reliance on the everyday conception of "real reasons" if psychological or other investigations show that real reasons, thus understood, do not yield as adequate an account of human actions as an analysis in terms of less familiar conceptions such as, perhaps, the idea of motivating factors which are kept out of the agent's normal awareness by processes of repression and reaction formation.

I would say, then, first of all, that historical explanation cannot be bound by conceptions that might be implicit in the way in which ordinary language deals with motivating reasons. But secondly, I would doubt that Gardiner's expressly tentative characterization does justice even to what we ordinarily mean when we speak of a man's "real reasons." For considerations of the kind that support the idea of subconscious motives are quite familiar in our time, and we are therefore prepared to say in ordinary, nontechnical discourse that the reasons given by an agent may not be the "real reasons" behind his action, even if his statement was subjectively honest, and he had no grounds to expect that it would lead to any adverse consequences for him. For no matter whether an explanation of human actions is attempted in the language of ordinary speech or in the technical terms of some theory, the overriding criterion for what-if-anything should count as a "real," and thus explanatory, reason for a given action is surely not to be found by examining the way in which the term "real reason" has thus far been used, but by investigating what conception of real reason would yield the most satisfactory explanation of human conduct; and ordinary usage gradually changes accordingly.

7. Concluding Remarks

We have surveyed some of the most prominent candidates for the role of characteristically historical mode of explanation; and we have found that they conform essentially to one or the other of our two basic types of scientific explanation.

This result and the arguments that led to it do not in any way imply a mechanistic view of man, of society, and of historical processes; nor, of course, do they deny the importance of ideas and ideals for human decision and action. What the preceding considerations do suggest is, rather, that the nature of understanding, in the sense in which explanation is meant to give us an understanding of empirical phenomena, is basically the same in all areas of scientific inquiry; and that the deductive and the probabilistic model of nomological explanation accommodate vastly more than just the explanatory arguments of, say, classical mechanics: in particular, they accord well also with the character of explanations that deal with the influence of rational deliberation, of conscious and subconscious motives, and of ideas and ideals on the shaping of historical events. In so doing, our schemata exhibit, I think, one important aspect of the methodological unity of all empirical science.

NOTES

1. See Dewey, John. *How We Think*. Boston, New York, Chicago, 1910; Chapter VI.
2. For a fuller presentation of the model and for further references, see, for example, Hempel, C. G. and P. Oppenheim, "Studies in the Logic of Explanation," *Philosophy of Science* 15: 135–175 (1948). (Secs. 1–7 of this article, which contain all the fundamentals of the presentation, are reprinted in Feigl, H. and M. Brodbeck (eds.), *Readings in the Philosophy of Science*. New York, 1953.) The suggestive term "covering law model" is W. Dray's; cf. his *Laws and Explanation in History*. Oxford, 1957; Chapter I. Dray characterizes this type of explanation as "subsuming what is to be explained under a general law" (p. 1), and then rightly urges, in the name of methodological realism, that "the requirement of a *single* law be dropped" (p. 24; italics, the author's): it should be noted, however, that, like the schema (D) above, several earlier publications on the subject (among them the article mentioned at the beginning of this note) make explicit provision for the inclusion of more laws than one in the explanans.
3. The relevance of the covering-law model to causal explanation is examined more fully in sec. 4 of Hempel, C. G., "Deductive-Nomological vs. Statistical Explanation." In Feigl, H., et al. (eds.), *Minnesota Studies in the Philosophy of Science*, vol. III. Minneapolis, 1962.
4. The concept of probabilistic explanation, and some of the peculiar logical and methodological problems engendered by it, are examined in some detail in Part II of the essay cited in note 3.
5. Freud, S. *Psychopathology of Everyday Life*. Translated by A. A. Brill. New York (Mentor Books) 1951; p. 64.
6. McConnell, D. W., et al., *Economic Behavior*. New York, 1939; pp. 894–95.
7. First published in 1893, and reprinted in several publications, among them: Edwards, Everett E. (ed.), *The Early Writings of Frederick Jackson Turner*. Madison, Wisconsin, 1938. Page references given in the present article pertain to this book.

8. Ibid., pp. 185–86.
9. Ibid., pp. 200–201.
10. Ibid., p. 210.
11. Ibid., p. 219.
12. Ibid., p. 220.
13. Boehmer, H. *Luther and the Reformation*. Translated by E. S. G. Potter. London, 1930; p. 91.
14. Gottlob's study, *Kreuzablass und Almosenablass*, was published in 1906; cf. the references to the work of Gottlob and other investigators in Schwiebert, E. G., *Luther and His Times*. St. Louis, Missouri, 1950, notes to Chapter 10.
15. Boehmer, *Luther and the Reformation*, p. 91.
16. This outline follows the accounts given by Boehmer, ibid., Chapter III and by Schwiebert, *Luther and His Times*, Chapter 10.
17. Boehmer, *Luther and the Reformation*, p. 92.
18. Ibid., p. 93.
19. The logic of genetic explanations in history is examined in some detail in E. Nagel's recent book, *The Structure of Science*. New York, 1961; pp. 564–568. The conception outlined in the present essay, though arrived at without any knowledge of Nagel's work on this subject, accords well with the latter's results.
20. Schwiebert, *Luther and His Times*, p. 304.
21. Gerschenkron, A. "Economic Backwardness in Historical Perspective," in Hoselitz, B. F. (ed.), *The Progress of Underdeveloped Areas*. Chicago, 1952; pp. 3–29.
22. Boehmer, *Luther and the Reformation*, pp. 93–94.
23. Dray, W. *Laws and Explanation in History*, Chapter V. All quotations are from this chapter; italics in the quoted passages are Dray's.
24. Empirical evidence supporting a given hypothesis may afford strong grounds for believing the latter without providing an explanation for it.
25. Boehmer, *Luther and the Reformation*, p. 234.
26. See especially Carnap's classical essay, "Testability and Meaning," *Philosophy of Science* 3: 419–71 (1936) and 4: 1–40 (1937); reprinted, with some omissions, in Feigl and Brodbeck, *Readings in the Philosophy of Science*. On the point here under discussion, see sec. 9 and particularly sec. 10 of the original essay or sec. 7 of the reprinted version.
27. Dray, *Laws and Explanation in History*, p. 123.
28. Watson, G. "Clio and Psyche: Some Interrelations of Psychology and History." In Ware, C. F. (ed.), *The Cultural Approach to History*. New York, 1940, pp. 34–47; quotation from p. 36.
29. Watson, ibid.
30. Gardiner, P. *The Nature of Historical Explanation*. Oxford, 1952.
31. Ibid., p. 136.
32. Ibid., p. 63.

15

Reasons and Covering Laws in Historical Explanation

1. Deductive and Probabilistic Explanation by Covering Laws

As a background for the following discussion of historical explanation, which is prompted mainly by Professor Dray's substantial and stimulating essay,[1] I propose to present first a brief sketch of, and some amplificatory comments on, the covering-law analysis of explanation.

The suggestive term "covering-law model of explanation" was introduced by Professor Dray in his monograph, *Laws and Explanation in History*, in which, after a very fair-minded presentation of this conception of explanation, he develops a number of intresting arguments against its general adequacy, particularly in the field of historical inquiry.

In his book, Mr. Dray used the term "covering-law model" to refer to the construal of an explanation as a deductive subsumption under covering laws. In an explanation of this kind, a given empirical phenomenon—in this essay, I will normally take it to be a particular event—is accounted for by deducing the *explanandum* statement, which describes the event in question, from a set of other statements, called the *explanans*. This set consists of some general laws and of statements describing certain particular facts or conditions, which usually are antecedent to or simultaneous with the event to be explained. In a causal explanation, for example—to mention one important variety of deductive explanation by covering laws—an individual event (e.g., an increase in the volume of a particular body of gas at a particular place and time) is presented as the "effect" of certain other particular events and conditions (e.g., heating of that body of gas under conditions of constant pressure), from which it resulted (from whose realization its occurrence can be inferred) in accordance with certain general laws (e.g., gas laws).

In explanations of the deductive, or "deductive-nomological," kind the covering laws are all of strictly universal form; i.e., schematically speaking, they are statements to the effect that in *all* cases where a certain complex F of conditions is satisfied, an event or state of kind G will come about; in symbolic notation: $(x)\,(Fx \supset Gx)$.

But there is a second, logically quite different, kind of explanation, which plays an important role in various branches of empirical science, and which I will call "covering-law explanation" as well. The distinctive feature of this second type, to which Mr. Dray briefly alludes in his essay, is that some of the covering laws are of probabilistic-statistical form. In the simplest case, a law of this form is a statement to the effect that under conditions of a more or less complex kind F, an event or "result" of kind G will occur with statistical probability—i.e., roughly: with long-run relative frequency—q; in symbolic notation: $ps\,(G,F) = q$. If the probability q is close to 1, a law of this type may be invoked to explain the occurrence of G in a given particular case in which conditions F are realized. By way of a simple illustration, suppose that after one particular rolling of a given set of four dice, the total number of dots facing up is greater than 4. This might be explained by the following information (whose factual correctness is, of course, an empirical matter and subject to empirical test; it would not be true, for example, if one of the dice were loaded): (i) For every one of the dice, the statistical probability for any particular face to show up as a result of a rolling is the same as for any other face, and (ii) the results yielded by the individual dice, when rolled jointly, are statistically independent of each other; so that the statistical probability for a joint rolling (R) of all four dice to yield a total of more than four dots (M) is: $ps(M,R) = 1295/1296 = 0.9992\ldots$. This general probability statement, combined with the information that the particular occurrence under consideration, say i, was a case of joint rolling of the four dice (or briefly that Ri), does not logically imply that in the particular case i, the total number of eyes facing up will be more than four (or that Mi, for short): but the two statements provide strong inductive grounds, or strong inductive support, or, as it is sometimes put, high inductive probability, for the assumption or expectation that Mi. The logical character of this explanatory argument may be represented by the following schema:

(Explanans) $\left\{ \begin{array}{l} ps(M,R) = 1295/1296 \\ Ri \end{array} \right\}$ confers high inductive probability on

(Explanandum) Mi

The probability which the *explanans* is here said to confer upon the *explanandum* is clearly not of the statistical kind; it does not represent an empirically determined quantitative relation between two kinds of event, such as R and M; rather, it is a logical relation between two statements—in our case, between the conjunction of the *explanans* statement on one hand and the *explanandum* statement on the other. This relation of inductive-logical support or probability is the central concept of the logical theories of probability developed by Keynes and by Carnap, to mention two outstanding examples. Carnap's theory, which is applicable to formalized languages of certain kinds, in fact provides ways of giving an explicit definition of logical probability in quantitative terms. To what extent these systems of inductive logic are applicable to actual scientific contexts is still a subject of study and debate; but that does not affect the basic thesis that in an explanation by means of probabilistic-statistical laws, the "subsumption" of the *explanandum* statement under the "covering laws" rests, not on a deductive implication, but on a relation of inductive support between the *explanans* and the *explanandum* statement. I will therefore refer to this kind of explanation as *probabilistic or inductive explanation*. Explanations of this kind play an important role in several areas of scientific inquiry; for example, the irreversibility of certain macrophenomena, such as the mixing of coffee and cream, is probabilistically explained by the assumption of certain purely statistical uniformities at the level of the underlying microevents.

2. A Necessary Condition of Adequacy for Explanations

The two kinds of explanation by covering laws have this feature in common: they explain an event by showing that, in view of certain particular circumstances and general laws, its occurrence was to be expected (in a purely logical sense), either with deductive certainty or with inductive probability.[2] In virtue of this feature, the two modes of explanation clearly satisfy what is, I submit, a general *condition of adequacy* for any account that is to qualify as a rationally acceptable explanation of a given event. The condition is that any such explanation, i.e., any rationally acceptable answer to a question of the type "Why did X occur?", must provide information which constitutes good grounds for the belief that X did in fact occur.[3] To state the point a little more fully: if the question "Why did X occur?" is answered by "Because Z is, or was, the case," then the answer does not afford a rationally adequate explanation of X's occurrence unless the information that Z is, or was, the case constitutes good grounds for expecting or believing the X did occur; otherwise, the explanatory information would provide no adequate grounds for saying, as it were: "That explains it—that does show why X occurred!"

Two amplificatory remarks may be indicated. First: the condition of adequacy just stated is to be understood as a necessary condition for an adequate explanation, not as a sufficient one; certain kinds of information—such as the results of a scientific test—may provide excellent grounds for believing that X occurred without in the least explaining why.

Secondly, the covering-law concepts of explanation, as schematically represented by the models, refer to the logic, not to the psychology, of explanation, just as metamathematical concepts of proof refer to the logic, not to the psychology, of proving mathematical theorems. Proofs and explanations that are adequate in the psychologic-pragmatic sense (which is of interest and importance in its own right) of making someone "understand" whatever is being proved or explained may well be achieved—and are in fact often achieved—by procedures that do not meet the formal standards for the concepts of proof or explanation construed in a nonpragmatic, metatheoretical sense. For example, it may be sufficient to call to a person's attention just one particular fact or just some general principle he had overlooked or forgotten or not known at all: taken in combination with other items in his background knowledge, this may make the puzzling item, X, fall into place for him: he will "understand why" X is the case. And since the proofs and explanations offered by mathematicians and empirical scientists in their writings, lectures, and informal conversations are normally formulated with some particular kind of audience in mind, they are accordingly elliptic in varying degrees. But this surely does not show that attempts to construct nonpragmatic metatheoretical concepts of proof and explanation are either mistaken in principle or at any rate bound to be unilluminating and theoretically unprofitable. In the case of proof-theory, the contrary is well known to be the case. And while the logical theory of explanation cannot claim achievements comparable in depth and importance to those of recent proof-theory, it has led to some significant insights. For example, certain results by Ramsey and by Craig illuminate the role of "theoretical entities" in scientific theories and shed light on the possibility of avoiding reference to such entities in scientific theories without loss of explanatory import; and problems such as these clearly concern the logic, not the psychology, of explanation.

As I mentioned a moment ago, the explanatory accounts actually offered by investigators in various fields of empirical inquiry, ranging from physics to historical research, will often fail to meet the condition of adequacy set forth above, and yet those accounts might intuitively be quite satisfactory. Clearly, in appraising the logical adequacy of a proposed explanation we must in fairness take into account not only what it explicitly tells us in the *explanans*, but also what it omits as not requiring mention, as tacitly taken for granted and presumed to be understood. Of course, it is not the task of a logical theory of explanation to tell us how to carry out an appraisal of this kind—any more than it is the task of a logical theory of inference to tell us how to judge whether a proposed argument that falls short of the formal standards of deductive validity is to be qualified as invalid or as deductively valid but elliptically formulated. The parallel to the case of mathematical proof is clear.

The condition of adequacy here proposed conflicts with a claim that has been made particularly, but not exclusively, with respect to historical explanation, namely, that sometimes an event can be quite adequately explained by pointing out that such-and-such antecedent conditions which are necessary but not sufficient for its occurrence, were realized. As Mr. Dray mentions in his survey of various modifications that have been suggested for the covering-law construal of explanation, this idea has been put forward by Frankel and by Gallie; it has also been strongly endorsed by Scriven, who offers this illustration in support of his view:[4] paresis occurs only in persons who have previously suffered from syphilis; and the occurrence of paresis in a given patient can therefore be properly explained by antecedent syphilitic infection—and thus by reference to an antecedent which constitutes a necessary but far from sufficient condition; for in fact, only quite a small percentage of syphilitics develop paresis. This "explanation" clearly violates the condition of adequacy proposed above. Indeed, as Scriven is the first to point out, on the information that a person has had syphilis, "we must . . . predict that [paresis] will *not* occur."[5] But precisely because the statistical probability for syphilis to lead to paresis is so small, and because therefore on the given information we must rationally expect the given person *not* to have developed paresis, the information that the patient has had syphilis (and that only syphilitics can develop paresis) clearly no more explains the actual occurrence of paresis in this case than a man's winning the first prize in the Irish Sweepstakes is explained by the information that he had bought a ticket (and that only a person who has bought a ticket can win the first prize).

3. Individual Events and "Complete" Explanation

In his essay, Mr. Dray touches briefly upon a question that has received a good deal of attention in the literature, namely, whether any individual event admits of a *complete* explanation, and in particular, whether such an explanation could possibly be achieved by means of covering laws. I would like to comment briefly on this issue.

In any covering-law explanation of an individual event, the event in question is always characterized by a *statement*, the *explanandum* statement. Thus, when we ask why a given body of gas, g, increased in volume between 5:00 and 5:01 P.M. or why the particular rolling i of our four dice yielded a total of more than four dots facing up, the *explanandum* events are described by the statements "the body of gas, g, increased in volume between 5:00 and 5:01 P.M." and "the particular rolling i of the four dice yielded a total of more than four dots facing up." Clearly then, only individual events in this sense, as described by statements, can possibly be explained by means of covering laws. (This is not to say, of course, that every such event can actually be so explained: the covering-law analysis of explanation presents a thesis about the logical structure of scientific explanation, but not about the extent to which individual occurrences in the world can be explained; that depends on what laws hold in the world and clearly cannot be determined just by logical analysis. In particular, therefore, the covering-law analysis of explanation does not presuppose or imply universal determinism.)

Quite frequently, however, the notion of an individual event is understood in a very different way. An event in this second sense is typically characterized, not by a statement describing it, but by an individual name or by a definite description, such as "the Children's Crusade," "the October Revolution," "the eruption of Mt. Vesuvius in A.D. 79," "the assassination of Julius Caesar," "the first solar eclipse of the 1960s," and the like. Individual occurrences thus understood cannot be explained by covering laws or in any other way; indeed, it is unclear what could be meant by explaining such an event. For any event thus understood has infinitely many aspects and thus cannot be even fully described, let alone explained. For example, the various aspects of Julius Caesar's assassination include the fact that it was plotted by Brutus and Cassius: that Brutus and his fellow conspirators were in such-and-such political positions and had such-and-such expectations and aspirations; that Caesar received such-and-such wounds; and—if I may trust an estimate theory—that with every breath we draw today, we inhale some of the molecules of oxygen and nitrogen that Caesar exhaled in his dying breath. Evidently, a complete characterization, let alone explanation, of an individual event in this sense is impossible.

For lack of a better expression, I will use the phrase 'concrete event' to refer to individual events understood in this latter sense. Individual events of the only kind admitting in principle of explanation by covering laws, i.e., events describable by statement, might then be said to constitute particular *aspects of, or facts about, concrete events.*[6]

I need hardly add that concrete events are not limited to the domain of the historian. An event such as the first total solar eclipse of the 1960s also exhibits infinitely many physical, chemical, biological, sociological, and yet other aspects and thus resists complete and, a fortiori, complete explanation. But certain particular aspects of it—e.g., that it is visible from such-and-such a region on the earth, that the duration of its totality is so many seconds, etc.—may well permit of explanation by covering laws.[7]

But it would be incorrect to say that an explanation by covering laws can explain only some *kind* of event rather than an individual event. For, first of all, a *kind* of event would be represented by a predicate-expression, such as "total solar eclipse visible from Alaska"; and since such expressions, not being statements, cannot be the conclusions of any deductive or inductive argument, a kind of event cannot be explained in accordance with the covering-law models. Secondly, what might be also explained is the *occurrence of an event of a certain kind in a particular space-time region;* for example, a lengthening of the mercury column in a particular thermometer at a particular place during a specified perod of time, or a particular individual developing yellow fever after being exposed to mosquitoes of a certain type. But what is thus explained is very definitely an individual event, of the sort that can be described by a statement. On this point, therefore, I agree with Mandelbaum, who rejects Hayek's thesis that explanation and prediction never refer to an individual event but always to phenomena of a certain kind, with the comment: "One would think that the prediction of a specific solar eclipse, or the explanation of that eclipse, would count as referring to a particular event even if it does not refer to all aspects of the event, such as the temperature of the sun, or the effect of the eclipse on the temperature of the earth, and the like."[8]

I said earlier that a concrete event, having infinitely many aspects, cannot be completely described, let alone explained. But there is at least one other sense in which the possibility of a complete explanation has recently been discussed and questioned, even in regard to individual events described by statements. Mr. Dray raises the issue in his essay when he asks whether an event can be *completely* explained by subsuming it under statistical rather than strictly universal laws, and thus without showing that it "had to happen." And indeed, as was noted earlier, the *explanans* of a statistical explanation confers upon the *explanandum* only a more or less high inductive probability, but does not imply it with deductive necessity, as is the case in deductive-nomological explanations. The latter might be said, in this sense, to be complete; probabilistic explanations, incomplete.[9] If the terms are thus understood, however, it is important to bear in mind that a more complete explanation of an event is not one that explains more aspects of it; in fact, the idea of completeness here under consideration applies only to the explanation of events described by statements, whereas the notion of aspects of an event was introduced in specific reference to concrete events.

Finally, it is now possible to specify a sense in which one might speak of partial explanations of concrete events and in which some of those explanations might be called more complete—in a third sense of the term—than others. First, any set of deductive-nomological explanations, each of which explains some aspect of a concrete event, might be called a partial deductive-nomological explanation of that event; and if the aspects explained in one of the sets form a proper subset of those in the other, the former set might be said to provide a less complete explanation of the concrete event than the latter. These notions can be generalized so as to apply also to sets containing probabilistic explanations, but this is not the place to enter into further details.

4. Explaining Actions by Reasons

4.1 Dray's Construal

I now turn to some comments on the central topic of Mr. Dray's essay, the concept of rational explanation. Dray holds that the method, widely used by historians among others, of explaining human actions in terms of underlying reasons cannot be construed as conforming to the covering-law pattern: to do so, he says, would be to give the wrong kind of reconstruction, it would get the form of such explanations wrong. In my opinion, Dray's arguments in support of this verdict, and his own alternative construal of such explanations, form a substantial contribution toward the formulation and clarification of the perplexing issues here at stake.

According to Dray, the object of explaining an action by reference to the reasons for which it was done is "to show that what was done was the thing to have done for the reasons given, rather than merely the thing that is done on such occasions, perhaps in accordance with certain laws."[10] The explanatory reasons will include the objectives the agent sought to attain and his beliefs concerning relevant empirical matters, such as the alternative courses of action open to him and their likely consequences. The explanation, according to Dray, then provides "a reconstruction of the agent's *calculation* of means to be adopted towards his chosen end in the light of the circumstances in which he found himself,"[11] and it shows that the agent's choice was appropriate, that it was the thing to do under the circumstances. The appraisal thus made of the appropriateness of what was done presupposes, not general laws, but instead what Dray calls a "principle of action," i.e., a normative or evaluative principle of the form "When in a situation of type C, the thing to do is X."[12]

4.2 The Problem of Criteria of Rationality

Before considering the central question whether, or in what sense, principles of this kind can explain an action, I want to call attention to what seems to me a problematic assumption underlying Dray's concept of a principle of action. As is suggested by the phrase "the thing to do," Dray seems to assume (i) that, given a specification of the circumstances in which an agent finds himself (including, I take it, in particular his objectives and beliefs), there is a clear and unequivocal sense in which an action can be said to be appropriate, or reasonable, or rational under the circumstances; and (ii) that, at least in many cases, there is exactly one course of action that is appropriate in this sense. Indeed, Dray argues that on this score rational explanation is superior to statistical explanation because the question why an action had to be done often permits an answer that involves the rational ruling out of all possible alternatives—a result that cannot be achieved in a probabilistic explanation.

But the two assumptions just listed seem to be unwarranted or at least highly questionable. First of all, it is by no means clear by what criteria of rationality "the thing to do" in a given situation is to be characterized. While several recent writers assume that there is one clear notion of rationality in the sense here required,[13] they have proposed no explicit definitions; and doubts about the possibility of formulating adequate general criteria of rationality are enhanced by the mathematical theory of decisions, which shows that even for some rather simple types of decision-situation several different criteria of rational choice can be formulated, each of which is quite plausible and yet incompatible with its alternatives.[14] And if this is so in simple cases, then the notion of *the* thing to do under given circumstances must be regarded as even more problematic when applied to the kinds of decision and action the historian seeks to explain. I think, therefore, that the presuppositions underlying the idea of a principle of action require further elaboration and scrutiny.

However, in order not to complicate the remainder of my discussion, I will disregard this difficulty from here on and will assume, for the sake of the argument, that the intended meaning of the expression "X is the appropriate, or rational, thing to do under circumstances of kind C" has been agreed upon and adequately specified by objective criteria.

4.3 The Explanatory Import of Citing Reasons for an Action

The question we have to consider then is this: How can a principle of action serve in an explanatory capacity? Dray's account, both in his essay and in his book, would seem to suggest that a rational explanation of why agent A did X would take the following form:

Agent A was in a situation of kind C.
When in a situation of kind C, the thing to do is X.

Therefore, agent A did X.

The first statement in the *explanans* specifies certain antecedent conditions; the second is a principle of action taking the place which, in a covering-law explanation, is held by a set of general laws.

Thus conceived, the logic of rational explanation does indeed differ decisively from that of covering-law explanation. But precisely because of the feature that makes the difference it cannot, I submit, explain why A did X. For by the general condition of adequacy considered earlier, an adequate *explanans* for A's having done X must afford good reasons for the belief or the assertion that A did in fact do X. But while the *explanans* just formulated affords good grounds for asserting that the appropriate thing for A to do under the circumstances was X, it does not provide good reasons for asserting or believing that A did in fact do X. To justify this latter assertion, the *explanans* would have to include a further assumption, to the effect that at the time in question A was a rational agent, and was thus disposed to do what was appropriate in the given situation. When modified accordingly, our *explanans* takes on a form which may be schematized as follows:

Agent A was in a situation of kind C.
A was a rational agent at the time.
Any rational agent, when in a situation of kind C, will invariably (or: with high probability) do X,

and it will then logically imply (or confer a high inductive probability on) the *explanandum*:

A did X.

Thus modified, the account will indeed provide an explanation of why A did in fact do X. But its adequacy for this purpose has been achieved by replacing Dray's evaluative principle of action by a descriptive principle stating what rational agents will do in situations of kind C. The result is a covering-law explanation, which will be deductive or inductive according as the general statement about the behavior of rational agents is of strictly universal or of probabilistic-statistical form. This construal of an explanation by reasons is evidently akin to Ryle's conception of an explanation by reference to dispositions;[15] for it presents A's action, as it were, as a manifestation of his general disposition to act in characteristic ways—in ways that qualify as appropriate or rational—when in certain situations.

It might be objected[16] to the broadly dispositional analysis here proposed that the "covering law" allegedly expressed by the third statement in the *explanans* is not really an empirical law about how rational agents do in fact behave, but an analytic statement of a definitional character, which expresses part of what is *meant* by a rational agent—so that the given action is not actually explained by subsumption under a general law. However, this objection does not, I think, do justice to the logical character of concepts such as that of a rataional agent. The reason, stated very briefly, is that such concepts are governed by large clusters of general statements—they might be called symptom statements—which assign to the dispositional characteristic in question various typical manifestations or symptoms; each symptom being a particular manner in wich a person who has the dispositional characteristic will "respond to" or "act under" certain specific ("stimulus-") conditions. The third statement in our *explanans* is then just one of many symptom statements for the concept of rational agent. But the totality of the symptom statements for a given broadly dispositional concept will normally have empirical implications, so that they cannot all be qualified as definitional or analytic; and it would be arbitrary to attribute to some of them—e.g., the one invoked in our *explanans*—the analytic character of partial definitions and to construe only the remaining ones as having empirical import.[17]

In sum, then, I think that Dray's very suggestively presented construal of explanations by reasons has a basic logical defect, which springs from the view that such explanations must be based on principles of action rather than on general laws. Dray explicitly makes a distinction between the two on the ground that the phrase "the thing to do," which characteristically occurs in a principle of action, "functions as a value-term," and that therefore there is a certain "element of *appraisal*" in a rational explanation, for it must tell us in what way an action "was *appropriate*."[18] But—and this seems to me the crux of the matter—to show that an action was the appropriate or rational thing to have done under the circumstances is not to explain why in fact it was done. Indeed, no normative or evaluative principle specifying what kind of action is appropriate in given circumstances can possibly serve to explain why a person acted in a particular way; and this is so no matter whether the action does or does not conform to the normative principle in question.

The basic point of the objection here raised has also been made by J. Passmore, who states it succinctly as follows: ". . . explanation by reference to a 'principle of action' or 'a good reason' is not, by itself, explanation at all. . . . For a reason may be a 'good reason'—in the sense of being a principle to which one *could* appeal in justification of one's action—without having in fact the slightest influence on us."[19]

It might perhaps be suspected that in arguing for a broadly dispositional analysis which presents explanations by reasons as having basically the logical structure of one or other of the covering-law models, we are violating a maxim of which Mr. Dray rightly reminds us in his essay, namely, that a sound logical analysis must refrain from forcing historical explanation on to the Procrustean bed of some preconceived general schema, and that instead it must take careful account of the practice generally agreed to be acceptable within the discipline concerned; that it must show sensitivity to the concept of explanation historians normally employ. No doubt a historian who adduces an agent's presumptive reasons in order to explain his actions, may well conceive it to be his main task to show that in the light of those reasons, the action was the appropriate thing to have done. But in giving his account, the historian undoubtedly also intends to show why in fact the agent acted as he did—e.g., to take Dray's example, why Louis XIV in fact withdrew military pressure from Holland. And this question cannot be answered by showing that the action was a (or even "the") reasonable thing to do, given Louis's objectives and beliefs; for after all, many agents on many occasions do not actually do the rational thing. This observation seems akin to an objection raised by Strawson, to which Dray refers in his essay. Dray agrees there that human action can fall short of the ideal of rationality and he stresses that his claim is only that the criterion of rational appropriateness does function for actions that are not judged to be defective in various ways. But this seems to me the crucial point: if an explanation by reasons invokes standards of rationality then, to have the desired explanatory force, it must in addition make the empirical assumption that the action was not defective in relevant ways, i.e., that the agent was at the time disposed to act in accordance with the standards invoked, and that the external circumstances did not prevent him from doing so.

And it seems clear to me that a historian would simply see no point in displaying the appropriateness or rationality of an action if he did not assume that the agent, at the time in question, was disposed to act rationally (as he might not be under conditions of extreme fatigue, under severe emotional strain, under the influence of drugs, and the like). And since, in an explanation by reasons, this essential presupposition will normally be taken for granted, it will not, as a rule, be explicitly mentioned; it is rather when departures from rationality are considered that the need is felt explicitly to specify disturbing circumstances. But while an elliptic formulation that forgoes explicit mention of the assumption of rationality may be quite satisfactory for practical purposes, i.e., in the pragmatic-psychological context of explanation, it obscures the logic of the explanatory argument; and surely, an analysis that makes explicit this essential assumption underlying the historian's account does not thereby force the method of explanation by reasons upon a Procrustean bed.

I think the broadly dispositional analysis I have outlined applies also to the intriguing case, invoked by Mr. Dray, of explaining one's own actions by reference to the reasons for which they were done. To be sure, in an account of the form "I did X for reasons R," explanation and justification are almost inextricably fused, and yet, we do distinguish between a genuine explanation and a mere rationalization in such contexts; and an account of the form "I did X for reasons R" would be suspected of being a rationalization if there were grounds to believe that I had not actually done X for the reasons given: e.g., that I had not in fact had the reasons and beliefs adduced in my account, or that I had been in a state in which I might well have tended not to take an action appropriate to my objectives and relevant empirical beliefs. Thus again, a statement given by me of the reasons for my action can have explanatory force only on the assumption of a disposition to act rationally in the given situation.

4.4 The "Rationality" of Nondeliberate Actions

A dispositional construal of rational explanation can also resolve a difficulty inherent in a view that Mr. Dray expresses in his book, and to which he briefly adverts again in his essay. According to Dray, there are certain actions that qualify as rational although they are decided upon without the benefit of actual deliberation or calculation. Indeed, in his book he argues that in so far as an action is purposive at all—no matter at what level of conscious deliberation—it is capable of rational explanation because "there is a calculation which could be constructed for it," a calculation the agent might have performed had he had the time, and which he might produce if questioned later.[20] But since, by hypothesis, no such deliberation or calculation did take place, since considerations of rationality actually played no role in the agent's action, an explanation of the latter by reference to such calculations seems to me to be simply fictitious.

Responding to an objection by Nowell-Smith which appears to be aimed at this point, Mr. Dray states again in his essay that the reasons adduced in a rational explanation need not actually have been considered by the agent in adopting his course of action, and he adds that our understanding of that action may arise out of our perception of a rational connection between the action and the motives and beliefs the rational explanation ascribes to the agent. But again, our awareness of such a logical connection surely cannot show why the action was taken by the agent, who, by hypothesis, took no account of that connection at all.

But I think Mr. Dray has a point in regarding some of those actions that are decided upon "in a flash," without reflection, as being akin to those which are prompted by careful deliberation. And it is possible to do justice to this idea by giving it a different—and again broadly dispositional—construal. Under this construal, a "rational explanation" of such an action is affected by ascribing to the agent certain behavioral dispositions acquired through a learning process whose initial phases did involve conscious reflection and deliberation. Consider, for example, the various intricate maneuvers required in using a typewriter, in driving a car through heavy traffic, in drilling and filling a patient's teeth: all these are learned in training processes that involve more or less elaborate deliberation in the beginning; but eventually they become "second nature" and are performed routinely, with little or no conscious reflection.

A particular act of this kind might then be explained, not by a reconstructed calculation or deliberation which the agent in fact did not perform, nor by pointing out that his action was appropriate to his putative objectives, but by presenting it as a manifestation of a general behavior pattern that the agent had learned in the manner just alluded to.[21] And clearly, this derivative kind of rational explanation would again be broadly dispositional, and hence of the covering-law variety.

To adopt the general conception I have presented here of explanation by reasons is by no means to deny that, as Mr. Dray rightly stresses, the historian adducing motivating reasons in explanation of an action normally does seek to show that the action "makes sense" when considered in the light of the purposes and the beliefs that presumably prompted it; nor is it to deny that perceiving an action as thus making sense can be a source of great intellectual satisfaction. What I have tried to argue is rather that—apart from the problematic status of the requiisite concept of appropriateness—the presentation of an action as being appropriate to the given situation, as making sense, cannot, for purely logical reasons, serve to explain why in fact the action was taken.

NOTES

1. William Dray, 'The Historical Explanation of Acting Reconsidered," in Sidney Hook, ed., *Philosophy of History: A Symposium* (New York: New York University Press, 1963), pp. 105–35.
2. For a fuller account of the deductive-nomological model see, for example, C. G. Hempel, "The Function of General Laws in History," *The Journal of Philosophy*, 39 (1942), 35–48. Reprinted in *Theories of History*, ed. P. Gardiner (Glencoe, Ill., 1959), pp. 344–56. See also C. G. Hempel and P. Oppenheim, "Studies in the Logic of Explanation," *Philosophy of Science*, 15 (1948), 135–75. Secs. 1–7 of this article are reprinted in *Readings in the Philosophy of Science*, eds. H. Feigl and M. Brodbeck (New York, 1953), pp. 319–52. The former of these articles also deals with the relevance of covering-law explanation to historical inquiry. A more detailed logical analysis of inductive-probabilistic explanation has been attempted in C. G. Hempel, "Deductive-Nomological vs. Statistical Explanation," in *Minnesota Studies in the Philosophy of Science*, ed. H. Feigl and G. Maxwell, III (Minneapolis, 1962), 98–169.
3. The condition can readily be formulated so as to cover also explanations that are intended to account, not for an individual event or state of affairs, but for some general uniformity, such as that expressed by Kepler's second law, for example. But explanations of this latter kind—which are discussed, for example, in the second and third of the articles mentioned in n. 2—need not be considered in this essay.
4. M. Scriven, "Explanation and Prediction in Evolutionary Theory," *Science*, 130 (1959), 480.
5. Ibid. (Italics the author's.)
6. At the end of the present section, a derivative sense will be suggested in which one might speak of more or less complete covering-law explanations of concrete events.
7. The gist of what I have so far said here about individual events and their explanation was briefly, but quite explicitly, stated already in Hempel, "The Function of General Laws in History," section 2.2.
8. M. Mandelbaum, "Historical Explanation: The Problem of 'Covering Laws,'" *History and Theory*, 1 (1961), 223, n. 6.
9. Completeness and incompleteness of explanation are obviously understood in this sense by J. Pitt, "Generalizations in Historical Explanation," *The Journal of Philosophy*, 56 (1959), 580–1.
10. W. Dray, *Laws and Explanation in History* (Oxford, 1957), p. 124.
11. Ibid., p. 122 (Italics the author's.)
12. Ibid., p. 132.
13. For example, Q. Gibson, in his stimulating study, *The Logic of Social Enquiry* (London and New York, 1960), asserts: "there may be various alternative ways of achieving an end. To act rationally . . . is to select what on the evidence is *the best*

way of achieving it" (p. 160; italics the author's); and he refers to "an elementary logical point—namely, that given certain evidence, there can only be one correct solution to the problem as to the best way of achieving a given end" (p. 162).
14. For a clear account and comparative analysis of such criteria, see, for example, R. D. Luce and H. Raiffa, *Games and Decisions* (New York, 1957), Ch. 13.
15. G. Ryle, *The Concept of Mind* (London, 1949). The construal here intended, which has been outlined only sketchily, differs, however, in certain respects from what I take to be Ryle's conception. To indicate this, I refer to the analysis here envisaged as "broadly dispositional." For a fuller account, see Hempel, "Rational Action," in *Proceedings and Addresses of the American Philosophical Association*, vol. 35 (Yellow Springs, Ohio, 1962); section 3.2 of that article, in particular, states and discusses the differences in question.
16. An objection to this effect was in fact raised in the discussion by Professor R. Brandt.
17. This idea is presented somewhat more fully in Hempel, "Explanation in Science and in History," in *Frontiers of Science and Philosophy*, ed. R. G. Colodny (Pittsburgh, Pa., 1962), section 6; also section 3.2 of Hempel, "Rational Action," has a direct bearing on this issue.
18. Dray, *Laws and Explanation in History*, p. 124. (Italics the author's.)
19. J. Passmore, "Review Article: Law and Explanation in History," *Australian Journal of Politics and History*, 4 (1958), 275. (Italics the author's.) Passmore then goes on to argue very briefly also that an explanation by reasons amounts to an explanation "by reference to a general statement," for to "take a 'reason' to be the actual explanation of anyone's conduct . . . is to assert, at least . . . the general statement: 'People of type X, in situation Y, act in such a way as to conserve the principle Z.'" (Ibid.)
20. Dray, *Laws and Explanation in History*, p. 123.
21. In a similar vein, I. Scheffler, in "Thoughts on Teleology," *British Journal for the Philosophy of Science*, 9 (1959), 269–75, has suggested that an interpretation in terms of learning may shed light on some types of nonpurposive teleological behavior.

16

Rational Action

1. Two Aspects of the Concept of Rational Action

To say of an action that it is rational is to put forward an empirical hypothesis and a critical appraisal. The hypothesis is to the effect that the action was done for certain reasons, that it can be *explained* as having been motivated by them; these reasons will include certain ends the agent sought to attain, and his beliefs about available means of attaining them. And the *critical appraisal* implied by the attribution of rationality is to the effect that, judged in the light of the agent's beliefs, his action constituted a reasonable or appropriate choice of means for the attainment of his ends.

Both the critical and the explanatory aspects of the concept of rational action give rise to various philosophical questions. The considerations that follow are an attempt to delineate and explore some of the most important among these.

2. Rationality of Action as a Critical Concept

2.1. General Characterization

Let us consider first the basic problem of explicating the critical, or normative, idea of rational action. This calls for the elaboration of precise criteria of rationality which might provide us with standards for appraising the rationality of particular actions, and which might thus also afford guidance in making rational decisions.

Rationality in the sense here intended is obviously a relative concept. Whether a given action—or the decision to perform it—is rational will depend on the objectives that the action is meant to achieve and on the relevant empirical information available at the time of the decision. Broadly speaking, an action will qualify as rational if, on the basis of the given information, it offers optimal prospects of achieving its objectives. I will now discuss more closely the key concepts invoked in this characterization: the concepts of the information basis and of the objectives of an action, and finally that of rationality relative to a given basis and given objectives.

2.2. The Information Basis of Rational Decision and Action

If we are to choose a rational course of action in pursuit of given ends, we will have to take into account all available information concerning such matters as the particular circumstances in which the action is to be taken; the different means by which, in these circumstances, the given ends might be attained; and the side-effects that may be expected to result from the use of different available means.

The total empirical information that is available for a given decision may be thought of as represented by a set of sentences, which I will call the *information basis* of the decision or of the corresponding action. This construal of the empirical basis for a decision takes account of an obvious but important point: to judge the rationality of a decision, we have to consider, not what empirical facts—particular facts as well as general laws—are actually relevant to the success or failure of the action decided upon, but what information concerning such facts is available to the decision-maker. Indeed, a decision may clearly qualify as rational even though it is based on incomplete or on false empirical assumptions. For example, the historian, precisely in order to present an action by a historical figure as rational, will often have to assume—and may well be able to show on independent grounds—that the agent was incompletely informed, or even entertained false beliefs, concerning relevant empirical matters.

But while the information basis of a rational action thus need not be true, should there not at least be good reasons for believing it true? Should not the basis satisfy a requirement of adequate evidential support? Some writers do consider this a necessary condition of rational action; and this view is indeed quite plausible; for example, as one of its recent advocates, Quentin Gibson, points out, if "someone were, carefully and deliberately, to walk round a ladder because he believed, without evidence, that walking under it would bring him bad luck, we would not hesitate to say that he acted irrationally."[1]

No doubt we often understand rationality in this restricted sense. But if we wish to construct a concept of rational action that might later prove useful in explaining certain types of human behavior, then it seems preferable not to impose on it a requirement of evidential support; for in order to explain an action in terms of the agent's reasons, we need to know what the agent believed, but not necessarily on what grounds. For example, an explanation of the behavior of Gibson's ladder-shunner in terms of motivating reasons would have to invoke the man's superstitious beliefs, but not necessarily the grounds on which he holds them; and the man may well be said to be acting quite reasonably, given his beliefs.

2.3. The Objective of Rational Decision or Action

From the information basis of a decision let me now turn to its objectives. In very simple cases, an action might be construed simply as intended to bring about a particular state of affairs, which I will call the end-state. But even in such simple cases, some of the courses of action which, according to the information basis, are available and are likely to achieve the end-state, may nevertheless be ruled out because they violate certain general constraining principles, such as moral or legal norms, contractual commitments, social conventions, the rules of the game being played, or the like. Accordingly, the contemplated action will be aimed at achieving the end-state without such violation; and what I will call its *total objective* may then be characterized by a set E of sentences describing the intended end-state, in conjunction with another set, N, of constraining norms.

Again, as in the case of the empirical basis, I will not impose the requirement that there must be "good reasons" for adopting the given ends and norms: rationality of an action will here be understood in a strictly relative sense, as its suitability, judged in the light of the given information, for achieving the specified objective.

2.4. Basic Criteria of Rationality of Action

How can such suitability be defined? For decision situations of the simple kind just contemplated, a characterization can readily be given: If the information basis contains general laws by virtue of which certain of the available courses of action would be bound to achieve the total objective, then, clearly, any one of those actions will count as rational in the given context. If the information basis does not single out any available course of action as a sufficient means for attaining the objective, it may yet assign a numerical probability of success to each of the different available actions, and in this case, any one of those actions will count as rational whose probability of success is not exceeded by that of any available alternative.

2.5. Broadened Construal of Objective and of Rationality

However, for many problems of rational decision, the available information, the objectives, and the criteria of rationality cannot be construed in this simple manner. Our construal becomes inapplicable, in particular, when the objective of a proposed action does not consist in attaining a specified end-state; and this is quite frequently the case, as we will see.

To begin with, even when a particular end-state is to be attained, the available information will often indicate that there are several alternative ways of definitely or probably attaining it, and that these alternatives would be attended by different incidental consequences, such side-effects, after-effects, and the like. Some of these might be regarded as more or less desirable, others as more or less undesirable. In a theoretical model of such situations the total goal must accordingly be indicated, not simply by describing the desired end-state, but by specifying the relative desirability of the different total outcomes that may result from the available courses of action.

In the mathematical theory of decision-making, various models of rational choice have been constructed in which those desirabilities are assumed to be specifiable in numerical terms, as the so-called utilities of the different total outcomes.

If the given information basis specifies the probabilities of the different outcomes, we have a case of what is called *decision-making under risk*. For this case, one criterion of rationality has gained wide acceptance, namely that of *maximizing expected utility*. The expected utility which, on the given information, is associated with a contemplated course of action is determined by multiplying, for each possible outcome of the action, its probability with its utility, and adding the products. An action then qualifies as rational if its expected utility is maximal in the sense of not being exceeded by the expected utility of any alternative action.

One more type of decision-situation deserves brief mention here because of its interesting philosophical implications. This is the case of *decision under uncertainty*. Here the formulation of the problem is assumed to specify the available courses of action, and for each of them its different possible outcomes with their utilities, but not their probabilities. By way of illustration, suppose that you are offered as a present a metal ball that you will obtain by one single drawing made, at your option, from one of two urns. You are given the information that the metal balls are of the same size, and that the first urn contains platinum balls and lead balls in an unspecified proportion; the second urn, gold and silver balls in an unspecified proportion. Suppose that the ulties you assign to platinum, gold, silver, and lead are in the ratio of 1000: 100: 10: 1; from which urn is it rational to draw? Interestingly, several quite different criteria of rational choice under uncertainty have been set forth in recent decision theory. Perhaps the best-known of them is *the maximin rule*; it directs us to maximize the minimum utility, that is to choose an action whose worst possible outcome is at least as good as the worst possible outcome of any alternative. In our example, this calls for a drawing from the second urn; for at worst, it will give you a silver ball, whereas the worst outcome of a drawing from the first urn would give you a lead ball. This rule clearly represents a policy of extreme caution, reflecting the pessimistic maxim: act on the assumption that the worst possible outcome will result from your action.

By contrast, the so-called *maximax rule* reflects an attitude of optimism; it directs us to act on the assumption that the best possible thing is going to happen, and hence to choose an action whose best possible outcome is at least as good as the best possible outcome of any alternative. In our example, the proper decision under this rule would be to draw from the first urn; for at best this will give us a platinum ball, whereas a drawing from the second urn can at best yield a gold ball.

Apart from the two rules just considered, several other rules of rational choice have been suggested for decision under uncertainty. The standards of rationality they reflect all have a certain plausibility, yet they conflict with one another: for one and the same decision situation, they will normally single out different choices as optimal.[2]

The mathematical models here briefly characterized do not offer us much help for a rational solution of the grave and complex decision problems that confront us in our daily affairs. For in these cases, we are usually far from having the data required by our models: we often have no clear idea of the available courses of action, nor can we specify the possible outcomes, let alone their probabilities and utilities. In contexts, however, where such information is available, mathematical decision theory has been applied quite successfully even to rather complex problems, for example in industrial quality control and some phases of strategic planning.

But whatever their practical promise, these models contribute, I think, to the analytic clarification of the concept of rational action. In particular, they throw into relief the complex, multiply relative, character of this concept; and they show that some of the characterizations of rational action which have been put forward in the philosophical literature are of a deceptive neatness and simplicity. For example, Gibson, in his careful and illuminating study, remarks: "there may be various alternative ways of achieving an end. To act rationally ... is to select what on the evidence is *the best* way of achieving it;[3] and he refers to "an elementary logical point—namely, that, given certain evidence, there can only be one correct solution to the problem as to the best way of achieving a given end."[4] Gibson offers no criterion for what constitutes the best solution. But surely, what he asserts here is not an elementary logical point, and indeed it is not true. For, first, even when the decision situation is of a kind for which one definite criterion of rational choice may be assumed to be available and agreed upon—for example, the principle of maximizing expected utility—then that criterion may qualify several different courses of action as equally rational. Secondly, there are various kinds of decision—for example, decisions under uncertainty—for which there is not even agreement on a criterion of rationality, where maximin opposes maximax, and both are opposed by various alternative rules.

It is important to bear in mind that the different competing criteria of rationality of decision or action do not reflect differences in the evaluation of the various ends which, on the given information, are attainable; all the competing rules here referred to presuppose that the utilities of those ends have been antecedently fixed. Rather, the different decision rules or criteria of rationality reflect different inductive attitudes; different degrees of optimism or pessimism concerning what to expect of the world; and accordingly different degrees of venturesomeness or caution in deciding upon a course of action.

The considerations here outlined concerning the critical or normative notion of rationality have important implications for the explanatory use of the idea of rational action. I now turn to this second topic of my essay.

3. Rational Action as an Explanatory Concept

Purposive human actions are often explained in terms of motivating reasons. The preceding discussion suggests that, if fully stated, a specification of such reasons will have to indicate the agent's objectives as well as his beliefs about such matters as the available means and their likely consequences. This conception is clearly reflected, for example, in R. S. Peters' remark that in such motivational explanations we "assume that *men are rational* in that they will take means which lead to ends if they have the information and want the ends."[5] Here, then, we have the idea of an *explanatory* use of the concept of rationality.

Let us now examine the logic of explanations by motivating reasons, and especially the role which the attribution of rationality to the agent plays in this context.

3.1. Dray's Concept of Rational Explanation

As our point of departure, let us choose Professor William Dray's stimulating and lucid analysis of this kind of explanation: Dray calls it *rational explanation* because, as he says, it "displays the *rationale* of what was done" by offering "a reconstruction of the agent's *calculation* of means to be adopted toward his chosen end in the light of the circumstances in which he found himself. To explain the action we need to know what considerations convinced him that he should act as he did."[6] But Dray attributes to rational explanation a further characteristic, which clearly assigns an essential role to the evaluative or critical concept of rationality. According to him, the "goal of such explanation is to show that what was done was the thing to have done for the reasons given, rather than merely the thing that is done on such occasions, perhaps in accordance with certain laws."[7] Hence, "Reported reasons, if they are to be explanatory in the rational way, must be *good* reasons at least in the sense that *if* the situation had been as the agent envisaged it ..., then what was done would have been the thing to have done."[8] To show that the agent had good reasons for his action, a rational explanation must therefore invoke, not a general empirical law, but a "*principle of action*," which expresses "a judgment of the form: 'When in a situation of type $C_1 \ldots C_n$ the thing to do is x'."[9] Thus, there is "an element of *appraisal* of what was done in such explanations."[10] And it is precisely in this reliance on a principle of action expressing an appraisal that Dray sees the essential difference between rational explanations and those explanatory accounts, familiar especially from the natural sciences, which explain a phenomenon by subsuming it under covering general laws that describe but do not appraise.

It appears then that according to Dray's conception a rational explanation answers the question 'Why did agent A do x?' by a statement of the form: 'A was in a situation of type $C_1 \ldots C_n$; and in a situation of that type, the thing to do is x'; or briefly: 'A was in a situation of type C (whose description would presumably include a specification of A's objectives and relevant beliefs), and in such a situation, the rational thing to do is x'.

Now, this construal of rational explanation clearly presupposes that there is a criterion of rationality which, for the given kind of decision situation, uniquely singles out one particular course of action as "*the* thing to do." However, this assumption seems to be untenable, for reasons indicated earlier.

But, more important, even if such a criterion were granted, an account of the form Dray attributes to a rational explanation cannot, it seems to me, do the job of explaining why A did x. For any adequate answer to the question why a certain event occurred will surely have to provide us with information which, if accepted as true, would afford good grounds for believing that that event did indeed occur—even if there were no other evidence for its occurrence. This seems clearly a necessary condition for an adequate explanation—though of course by no means a sufficient one: producing evidence for the occurrence of an event is not the same thing as explaining it. Now, information to the effect that agent A was in a situation of kind C, and that in such a situation the rational thing to do is x, affords grounds for believing that it would have been *rational for A to do x*; but not for believing that A did *in fact* do x. To justify this latter belief, we clearly need a further explanatory assumption, namely that—at least at the time in question—A was a *rational agent* and thus was *disposed* to do whatever was rational under the circumstances.

But when this assumption is added, the answer to the question 'Why did A do x?' takes on the following form:

(Schema R)
A was in a situation of type C
A was a rational agent
In a situation of type C any rational agent will do x
―――――――――――――――――――――――――
Therefore A did x.

This construal of rational explanation differs from Dray's in two respects: First, the assumption that A was a rational agent is explicitly added; and secondly, the evaluative or appraising *principle of action*, which tells us what is the thing to do in situation C, is replaced by a *descriptive generalization* telling us how a rational agent will act in situations of that kind: but this restores the covering-law form to the explanation.

In thus disagreeing with Dray's analysis of rational explanation, I do not wish to deny that an explanatory account in terms of motivating reasons may well have evaluative overtones: what I maintain is only that whether a critical appraisal is included in, or suggested by, a given account, is irrelevant to its explanatory force: and that an appraisal alone, by means of what Dray calls a principle of action, does not explain at all why A did in fact do x.

3.2. Explanation by Reasons as Broadly Dispositional

The alternative construal which I have so far sketched only in outline now requires a somewhat fuller statement.

The notion of rational agent invoked in Schema R above must of course be conceived as a descriptive-psychological concept governed by objective criteria of application; any normative or evaluative connotations it may carry with it are inessential for the explanatory force of the argument. To be sure, normative preconceptions as to how a truly rational person ought to behave may well influence the choice of descriptive criteria for a rational agent—just as the construction of tests providing objective criteria of intelligence, verbal aptitude, mathematical aptitude, and the like will be influenced by presystematic conceptions and norms. But the descriptive-psychological use of the term 'rational agent' (just like that of the terms 'IQ,' 'verbal aptitude,' 'mathematical aptitude,' et cetera) must then be governed by the objective empirical rules of application that have been adopted, irrespective of whether this or that person (for example, the proponent of a rational explanation or the person to whom it is addressed) happens to find those objective rules in accord with his own normative standards of rationality.

By whatever specific empirical criteria it may be characterized, rationality in the descriptive-psychological sense is what I will call a *broadly dispositional* trait: to say of someone that he is a rational agent is to attribute to him, by implication, a complex bundle of dispositions, each of them a tendency to behave in characteristic ways in certain kinds of situation (whose full specification would have to include information about the agent's objectives and beliefs, about other aspects of his psychological and biological state, about his environment, et cetera). To explain a given action by reference to the agent's reasons and his rationality is thus to present it as conforming to, as being an instance of, one of those general tendencies. Roughly speaking, therefore, explanations by motivating reasons have the character of dispositional explanations in the sense examined by Ryle in *The Concept of Mind*.[11] However, this rough characterization now must be elaborated a little and must also be qualified in certain respects.

To begin with, the dispositions implied by the psychological concept of rational agent are not simply dispositions to respond by certain characteristic overt behavior to specific external stimuli. They differ in this respect from at least some of the dispositions implied when we say of a person that he is allergic to ragweed pollen; for to say this is to imply, among other things, that he will exhibit the symptoms of a head cold when exposed to the pollen. When we call someone a rational agent, we assert by implication that he will behave in characteristic ways if he finds himself in certain kinds of situation; but—and this is a first point to note—those situations cannot be described simply in terms of certain environmental conditions and external stimuli; for characteristically they include the agent's having certain objectives and entertaining certain revelant beliefs. To mark this difference, we might say that the dispositions implied by attributing rationality to a person are *higher-order dispositions*; for the beliefs and ends-in-view in response to which, as it were, a rational agent acts in a characteristic way are not manifest external stimuli but rather, in turn, broadly dispositional features of the agent. Indeed, to attribute to someone a particular belief or end-in-view is to imply that in certain circumstances he will tend to behave in certain ways which are indicative or symptomatic of his belief or his end-in-view. When I say that belief-attributions or end-attributions "imply" certain dispositional statements, the implying in question will usually have to be conceived as being probabilistic in character; but in order not to complicate the discussion of our central problems, I will make no further explicit reference to this qualification.

There is yet another point I wish to indicate by saying that the ascription of a belief, of a goal, or of rationality to a person is only *broadly* dispositional in character; namely, that a statement expressing such an ascription may *imply, but is not tantamount to,* a set of other statements which attribute to the person certain clusters of dispositions. These dispositions constitute symptoms or indices of the person's beliefs, objectives, or rationality; but they do not suffice fully to specify the latter.

Let me try to support this view first by means of a parallel. To say of a physical body that it is electrically charged, or that it is magnetic, is to attribute to it, *by implication,* bundles of dispositions to respond in characteristic, or symptomatic, ways to various testing procedures. But this does not exhaust what is being asserted; for the concepts of electric charge, magnetization, and so on are governed by a network of theoretical principles interconnecting a large number of physical concepts. Conjointly, these theoretical principles determine an infinite set of empirical consequences, among them various dispositional statements which provide operational criteria for ascertaining whether a given body is electrically charged or magnetic or the like. Thus, the underlying theoretical assumptions contribute essentially to what is being asserted by the attribution of those physical properties. Indeed, it is only in conjunction with such theoretical background assumptions that a statement attributing an electric charge to a given body implies a set of dispositional statements; whereas the whole set of dispositional statements does not imply the statement about the charge, let alone the theoretical background principles.

Now, to be sure, the psychological concepts that serve to indicate a person's beliefs, objectives, moral standards, rationality, et cetera, do not function in a theoretical network comparable in scope or explicitness to that of electromagnetic theory. Nevertheless, we use those psychological concepts in a manner that clearly presupposes certain similar connections—we might call them *quasi-theoretical connections.* For example, we assume that the overt behavior shown by a person pursuing a certain *objective* will depend on his beliefs; and conversely. Thus, the attribution, to Henry, of the belief that the streets are slushy will be taken to imply that he will put on galoshes only on suitable further assumptions about his objectives and indeed about his further beliefs; such as that he wants to go out, wants to keep his feet dry, believes that his galoshes will serve the purpose, does not feel in too much of a hurry to put them on, et cetera: and this plainly reflects the assumption of many complex interdependencies between the psychological concepts in question. It is these assumptions which determine our expectations as to what behavioral manifestations, including overt action, a psychological trait will have in a particular case.

To reject the construal of those traits as simply bundles of dispositions is not to conjure up again the ghost in the machine, so deftly and subtly exorcised by Ryle and earlier—more summarily, but on basically similar grounds—by the logical behaviorism of Carnap. The point is rather than to characterize the psychological features in question, we have to consider not only their dispositional implications, which provide operational criteria for attributing certain beliefs and objectives to a person: we must also take account of the quasi-theoretical assumptions connecting them; for these, too, govern the use of those concepts, and they cannot be regarded as logical consequences of the sets of dispositional statements associated with them.

3.3. Epistemic Interdependence of Belief Attributions and Goal Attributions

The quasi-theoretical connections just referred to give rise to a problem that requires at least brief consideration. For our purposes it will suffice to examine one form of it which is of fundamental importance to the idea of rational explanation.

What sorts of dispositions do we attribute to a person by implication when we assert that he has such and such objectives or beliefs? To begin with objectives or ends-in-view: The statement that Henry wants a drink of water implies, among other things, that Henry is disposed to drink a liquid offered him—provided that he *believes* it to be potable water (and provided he has no overriding reasons for refusing to accept it, et cetera). Thus, ascription of an objective here has implications concerning characteristic overt behavior only when taken in conjunction with ascriptions of appropriate beliefs. Similarly, in our earlier example, the hypothesis that Henry *believes* the streets to be slushy implies the occurrence of characteristics of overt behavior only when taken in conjunction with suitable hypotheses about Henry's objectives.

And indeed it seems that generally a hypothesis about an agent's objectives can be taken to imply the occurrence of specific overt action only when conjoined with appropriate hypotheses about his beliefs; and vice versa. Hence, strictly speaking, an examination of an agent's behavior cannot serve to test assumptions about his beliefs or about his objectives separately, but only in suitable pairs, as it were; or briefly, belief attributions and goal attributions are *epistemically interdependent*.

This fact does not create insuperable difficulties in ascertaining a person's beliefs or his objectives. For often we have good antecedent information about one of the interdependent items and then a hypothesis about the other may be tested by ascertaining how the person acts in certain situations. For example, we may have good grounds for the assumption that our man is subjectively honest; then his answers to our questions may afford a reliable indication of his beliefs. Conversely, we are often able to test a hypothesis about a person's objectives by examining his behavior in certain critical situations because we have good reasons to assume that he has certain relevant beliefs.

But the epistemic interdependence of belief attributions and goal attributions does raise the question whether an explanation by motivating reasons ever requires the explanatory assumption that the acting person was, at least, at the time in question, a rational agent. How this question arises can be seen by a closer look at the criteria for belief and goal attributions.

Suppose we know an agent's beliefs and wish to test the hypothesis that he wants to attain goal G. Just what sort of action is implied by this hypothesis?

It seems clear that the criterion used in such cases is roughly this: If A actually wants to attain G, then he will follow a course of action which, in the light of his beliefs, offers him the best chance of success. In the parlance of our earlier discussion, therefore, the test and the justification of our goal attribution appears to presuppose the assumption that A will choose an action that is rational relative to his objectives and beliefs. This would mean that the way in which we use a person's actions as evidence in ascertaining his goals has the presupposition of rationality built into it. An analogous comment applies to the way in which the actions of a person whose objectives we know are normally used as evidence in ascertaining his beliefs. But this seems to discredit the construal of rational explanation as involving, in the manner suggested in Schema R, an explanatory hypothesis to the effect that the person in question was a rational agent. For the considerations just outlined suggest that this hypothesis is always made true by a tacit convention governing our attribution of motivating reasons—that is, objectives and beliefs—to the agent. If this is generally the case, then the assumption of rationality could not possibly be violated; any apparent violation would be taken to show only that our conjectures about the agent's beliefs, or those about his objectives, or both, were in error. And undeniably, such will in fact often be our verdict.

But will it always be so? I think there are various kinds of circumstances in which we might well leave our belief and goal attributions unchanged and abandon instead the assumption of rationality. First of all, in deciding upon his action, a person may well overlook certain relevant terms of information which he clearly knows or at least believes to be true and which, if properly taken into account, would have called for a different course of action. Secondly, the agent may overlook certain items in the total goal he is clearly seeking to attain, and may thus decide upon an action that is not rational as judged by his objectives and beliefs. Thirdly, even if the agent were to take into account all aspects of his total goal as well as all the relevant information at his disposal, and even if he should go through a deliberate "calculation of means to be adopted toward his chosen end" (to repeat an earlier quotation from Dray), the result may still fail to be a rational decision because of some logical flaw in his calculation. It is quite clear that there could be strong evidence, in certain cases, that an agent had actually fallen short of rationality in one of the ways here suggested; and indeed, if his decision had been made under pressure of time or under emotional strain, fatigue, or other disturbing influences, such deviations from rationality would be regarded as quite likely. (This reflects another one of the quasi-theoretical connections among the various psychological concepts that play a role in explanations by reasons or by motives.)

In sum then, rationality of human actions is not universally guaranteed by conventions governing the attribution of goals and beliefs in human agents; there may be very good grounds for ascribing to an agent certain goals and beliefs and yet acknowledging that his action was not rational as judged by those goals and beliefs.

3.4. Rational Action as an Explanatory Model Concept

So far I have argued three main points concerning the explanatory use of the concept of rational action, namely (1) that explanations by motivating reasons are broadly dispositional in character; (2) that therefore they conform to the general conception of an explanation as subsuming its explanadum under covering laws (the laws may be of strictly universal or of statistical form, and the subsumption will accordingly be deductive or inductive-probabilistic in character[12]); and (3) that in explanations by motivating beliefs and ends-in-view, the assumption that the acting individual was a rational agent is not, as it may appear to be, always made true by a tacit convention governing the attribution of beliefs and ends-in-view.

For further clarification of the role that the assumption of rationality plays in explanations by motivating reasons, it may be illuminating to ask whether the concept of rational agent might not be viewed as an idealized explanatory model comparable to the explanatory concept of an ideal gas, that is, a gas conforming exactly to Boyle's and Charles's laws. No actual gas strictly satisfies those laws; but there is a wide range of conditions within which many gases conform at least very closely to the account the model gives of the interrelations between temperature, pressure, and volume. Moreover, there are more general, but less simple laws, such as van der Waals's, Clausius's, and others, which explain to a large extent the deviations from the ideal model that are exhibited by actual gases under certain conditions.

Perhaps the concept of a rational agent can similarly be regarded as an explanatory model characterized by an "ideal law" to the effect that the agent's actions are strictly rational (in the sense of some specific criterion) relative to his objectives and beliefs. How could this programmatic conception be implemented? How could an explanatory model of rational action be precisely characterized, and how could it be applied and tested?

As noted earlier, the concept of rationality is by no means as clear and unequivocal as is sometimes implied in the literature on rational explanation. But let us assume that the proposed explanatory use of the concept of rational action is limited, to begin with, to cases of a relatively simple type for which some precise criterion of rationality can be formulated and incorporated into our model.

Then there is still the question of how to apply the model to particular instances, how to test whether a given action does in fact conform to the criterion of rationality the model incorporates. And this raises a perplexing problem. The problem is not just the practical one of how to *ascertain* an agent's beliefs and actions in a given case, but the conceptual one of what is to be *understood* by the beliefs and objectives of an agent at a given time, and what kind of logical device might serve to characterize them. Let me amplify this briefly.

First, a person must surely be taken to hold many beliefs which he is not consciously entertaining at the time, but which could be elicited by various means. Indeed, a person may be held to believe many things he has never thought of at all and perhaps never will think of as long as he lives. For example, if he believes that five and seven are twelve, we would surely take him to believe also that five speckled hens and seven more speckled hens make twelve speckled hens—although he might never consciously entertain this particular belief. Generally, a man will be taken to believe certain things that are consequences of other things he believes: but surely not all those consequences, since—to mention but one reason—his logical perspicacity is limited.

Hence, while in a theoretical model of the normative or critical concept of rational decision the information basis may be construed as a set of statements that is closed under an appropriate relation of logical derivability, this assumption definitely cannot be transferred to an explanatory model of rational decision. In particular, a person may well give his believing assent to one of a pair of logically equivalent statements but withhold it from the other—although, according to a familiar parlance, both express the same proposition. It seems clear, therefore, that the objects of a person's beliefs cannot be construed to be propositions each of which may be represented by any one of an infinite set of equivalent statements; in specifying an agent's beliefs, the mode of its formulation is essential. (This pecularity seems closely akin to what Quine has called the referential opacity of belief sentences.)[13]

Presumably, then, in an explanatory model concept of rational action, the agent's beliefs would have to be represented by some set of sentences that is not closed under logical derivability. But what set? For example: Should the belief-set for an agent at a given time be taken to include all sentences assent to which could be elicited from him by pertinent questions and arguments, no matter how numerous or complex? Clearly, such construal is unwarranted if we are interested in specifying a set of beliefs which can be regarded as motivating factors in explaining an action done by the agent. Where the boundary line of the belief-set is to be drawn—conceptually, not just practically—is a puzzling and obscure question.

Quite similar observations apply to the problem of how to characterize an agent's total objectives in a given decision situation. Consequently, though in a normative-critical model of rational decision rationality is always judged by reference to the total information basis and the total objective specified, it would be self-defeating to incorporate into an explanatory model of rational action the principle that a rational agent acts optimally, as judged by specified criteria, on the basis of his total set of objectives and beliefs: this latter notion is simply too obscure.

3.5. *The Model of a Consciously Rational Agent*

A way out seems to be suggested by the observation that many rational explanations present an action as rationally determined by considerations which presumably the agent took consciously into account in making his decision. Let us say that a person is a *consciously rational agent* (at a certain time) if (at that time) his actions are rational relative to those of his objectives and beliefs which he consciously takes into account in arriving at his decision.

This "ideal model" of a consciously rational agent seems to yield approximate explanatory and predictive accounts of at least some types of decision or action.

Consider, for example, a competent engineer who seeks an optimal solution to a problem of design or of industrial quality control for which the range of permissible solutions is clearly delimited, the relevant probabilities and utilities are precisely specified, and even the criterion of rationality to be employed (for example, maximization of expected utilities) has been explicitly stated. In this case, the objectives and beliefs which determine the engineer's decision may be taken to be fully indicated by the specification of the problem; and by applying to the engineer the explanatory model of a consciously rational agent (whose standard of rationality is that specified in the given problem), we can explain—or predict—that he will come up with that solution, or set of solutions, which is the theoretically correct one.

The idea of a consciously rational agent, with its very limited scope of application, does not, however, represent the only way in which a model concept of rational decision might be put to explanatory and predictive use. One interesting alternative has been suggested in a study by Davidson, Suppes, and Siegel.[14] These investigators present an empirical theory of human choice which is modeled on the mathematical model of decision under risk and incorporates the hypothesis that the choices made by human subjects will be rational in the precise sense of maximizing expected utilities.

As might be expected, the rigorously quantitative character of the theory has to be purchased at the price of limiting its applicability to decisions of a rather simple type which permit of strict experimental control. In the experiment designed by the authors to test the theory, the subjects had to make a series of decisions each of which called for a choice between two options. Each option offered the prospect of either gaining a specified small amount of money or losing some other specified small amount, depending on the outcome of a certain random experiment, such as rolling a regular die with peculiar markings on its faces. The random experiments, their possible outcomes, and the corresponding gains or losses were carefully described to the subjects, who then made their choices.

The results of the experiment conformed quite well to the hypothesis that subjects would choose the option with the greater expected utility, where the expected utility of an option is computed, in the standard manner, on the basis of theoretically postulated subjective probabilities and utilities which the different outcomes have for the choosing subject. The theory proposed by the authors provides an objective, if indirect, method for the simultaneous and independent measurement of such subjective probabilities and utilities for a given agent. Experimental study shows that the subjective probability which a specified outcome of a given random experiment possesses for a given subject is not, in general, equal to its objective probability, even though the subject may know the latter; nor are the subjective utilities proportional to the corresponding monetary gains or losses. Indeed, a person will normally be entirely unaware of the subjective probabilities and utilities which, on the theory under consideration, the possible outcomes possess for him.

Thus, insofar as the theory is correct, it gives a quite peculiar twist to the idea of rational action. Though the subjects make their choices in clearly structured decision situations, with full opportunity for antecedent deliberation and even calculation, they act rationally (in a precisely refined quantitative sense) relative to subjective probabilities and utilities which they do not know, and which, therefore, they cannot take into account in their deliberations; they act rationally in the sense of acting *as if* they were trying to maximize expected utilities. We seem to have here a type of conscious choice which is nonconsciously rational with quantitative precision. What might Freud have thought of this?

3.6. Concluding Remarks

Obviously, the more familiar instances of explanation by motivating reasons do not conform to this special theoretical model. By intent, at least, they come closer to invoking the model of a consciously rational agent. In particular, many of the rational explanations offered in historical writings seem to imply that the given action was the outcome of rational deliberation based on specific beliefs and objectives which the historian, often on very good evidence, attributes to the agent. But since it is impossible for the historian, even under the best of conditions, to ascertain all the considerations that may have entered into the agent's deliberation and may thus have influenced his decision, the most favorable construal that can be given to the explanatory import of such arguments appears to be this effect: The explanans includes the information that the agent had such and such goals and beliefs; and since he acted in a manner to be expected of a rational agent in these circumstances, it is plausible to suppose that whatever other considerations may have figured in his deliberation had no decisive influence on its outcome: in this sense, the agent's decision is accounted for by the specified beliefs and goals.

In explanations of this kind, the relevant sense of rationality is not explicitly defined; rather, it is left to our judgment to put an appropriate construal on the explanatory hypothesis of rationality and to recognize that what was done was rational relative to the adduced reasons.

Practically, this is no doubt often the best we can do by way of explaining an action. But I would not agree with the view that explanations of this kind are perfectly adequate for the purposes of history and that nothing further need be attempted. For since, in their explanations, historians make objective claims, they will have to take into account whatever relevant insights may be provided by the scientific study of motivation and action. And I think it likely that as a result the vague general procedure of explanation by reasons will gradually be replaced, at least in some areas, by the use of more specific explanatory hypotheses, in which our standard notions of rationality may play a less important role. The influence which some psychological theories, including the ideas of psychoanalysis, have had on the explanation of human action seems to me indicative of this trend.

If such theoretical developments should show that the explanatory power of the concept of rational action is in fact rather limited, we will have to accept this philosophically: after all, in the methodology of explanation we can ill afford to give a general advance endorsement to the saying: "Man is a rational being indeed: he can give reasons for *anything* he does."

NOTES

1. Quentin Gibson, *The Logic of Social Enquiry* (London and New York, 1960), p. 43.
2. For a lucid statement and comparative analysis of the criteria in question, see R. D. Luce and H. Raiffa, *Games and Decisions* (New York, 1957), Chap. 13.
3. Gibson, *The Logic of Social Enquiry*, p. 160 (italics the author's).
4. Ibid., p. 162.
5. R. S. Peters, *The Concept of Motivation* (London and New York, 1958), p. 4 (Italics supplied).
6. William Dray, *Laws and Explanation in History* (Oxford, 1957), pp. 124 and 122 (italics the author's).
7. Ibid., p. 124.
8. Ibid., p. 126 (italics the author's).
9. Ibid., p. 132 (italics the author's).
10. Ibid., p. 124 (italics the author's).
11. Gilbert Ryle, *The Concept of Mind* (London, 1949).
12. These two types of explanation by covering laws are discussed more fully in my essay, "Deductive-Nomological vs. Statistical Explanation," in H. Feigl and G. Maxwell (eds.), *Minnesota Studies in the Philosophy of Science*, Vol. III (Minneapolis, 1962).
13. See, for example, W. V. Quine, *Word and Object* (New York, 1960), Sec. 30. This section and the subsequent ones through Sec. 45 contain incisive analyses of the basic logical problems raised by belief attributions and goal attributions and an illuminating discussion of recent philosophical literature on this subject.
14. Donald Davidson, Patrick Suppes, and Sidney Siegel, *Decision Making: An Experimental Approach* (Stanford, 1957).

V

SCIENTIFIC RATIONALITY

17

Science Unlimited?

1. Introduction: The Problem

Our age is often called an age of science, and with good reason: the advances made during the past few centuries by the natural sciences and more recently by psychology and the social sciences have vastly broadened our knowledge and deepened our understanding of the world we live in and of our fellow men; and the basic soundness of the insights achieved by science is eloquently attested to by the striking success of their applications, both constructive and destructive, which have radically changed the quality of life on our planet and have left their characteristic imprint on every aspect of contemporary civilization.

Contemplating the increasingly rapid expansion of scientific knowledge, one may well wonder how far this process can continue; whether all questions about the world can ultimately be answered by scientific inquiry, or whether there are absolute limits beyond which scientific knowledge and understanding can never go. This is the problem to be considered in my essay.

My discussion is meant to apply to all branches of scientific and scholarly inquiry, from the physical sciences through biology and psychology to the social and historical disciplines; thus, when the terms 'science' or 'scientific' are used, they are meant to refer comprehensively to all of empirical science. I shall be concerned, however, only with pure, or basic, research and not with the technological applications of scientific knowledge or procedure; accordingly, I shall not discuss the limits of the technologically attainable control of empirical phenomena.

Pure scientific research has two principal objectives, which I will briefly call knowledge and understanding: science seeks to ascertain *what* goes on in this world of ours, and to understand *why*. I propose to consider the question of limits first with respect to the pursuit of scientific knowledge and then with respect to the quest for scientific understanding.

It will help further to circumscribe our central topic if we acknowledge at the outset one limitation that is shared by scientific knowledge and by scientific understanding. Obviously, there are many questions of empirical fact and many problems of explanation which science is unable to answer at present. Moreover, since the life span of humanity is presumably finite, scientific inquiry as carried out by human beings will never raise, let alone answer, more than finitely many questions of fact and of explanation, whereas the set of all logically possible such questions is infinite. Hence, infinitely many questions aimed at knowledge or understanding will, in fact, never be answered by science.

But this limitation is, if I may say so, only of a practical kind; it does not bear on the philosophically interesting problem whether scientific knowledge and understanding are limited "in principle," i.e., in ways that would remain unaffected even if scientific research could go on indefinitely. Such limits-in-principle would have to stem from certain general logical and methodological characteristics of scientific knowledge and understanding. It is with the question of such limits that we will now be concerned.

2. The Inductive Uncertainty of Scientific Knowledge

There is one pervasive characteristic of scientific knowledge that may well appear to constitute an insuperable limitation, namely, its inductive uncertainty. Consider any of the statements that purport to express scientific knowledge, such as Galileo's law of free fall or Kepler's laws of planetary motion or contemporary theories in the natural and the social sciences, or statements about particular facts, such as that the Moon has no atmosphere, or that King George III of England suffered from porphyria: on what grounds can they be claimed to express knowledge, and thus, to be true? How does empirical science certify its knowledge claims?

As is well known, the answer—or at least an important part of the answer—is that the statements asserted by science are ultimately certified by reference to relevant empirical "data," i.e., evidence obtained by experimentation, by systematic observation, by sociological field work, by the careful scrutiny of historical records or relics, and so forth. What can such evidence show concerning the truth of the statement, or "hypothesis," under scrutiny?

Let us assume, for the sake of the argument, that there are no errors or uncertainties about the recording of the evidence, so that the "data sentences" or "evidence sentences" describing the evidence for a given scientific assertion are all true. Even then, the evidence can show at best that in the particular cases thus far examined, the experimental or observational data were in accord with the hypothesis under investigation. But that does not suffice to guarantee the truth of the hypothesis. Galileo's law, for example, concerns not only those cases of free fall which have been scientifically examined so far; it also makes a general assertion about the relation between elapsed time and distance covered in any case of free fall, including the fall of prehistoric meteorites that flashed across the sky unseen and unexamined by any human observer, as well as all cases of free fall that are as yet in the future. And clearly, however many instances of free fall may have been observed and found to conform to Galileo's law, they afford no guarantee that all the unexamined cases did or do or will equally conform to that law.

This consideration can be generalized. No matter how extensively an empirical hypothesis has been tested, and no matter how well it has been borne out by the test findings, it may yet fail in cases that have not been examined. The supporting evidence does not suffice to verify the hypothesis, i.e., to establish its truth with deductive certainty; it can only lend the hypothesis more or less strong "inductive support."

Can this inductive uncertainty properly be regarded as a shortcoming of scientific knowledge? Such criticism would reflect the view that the quest for knowledge is a quest for certainty, and that scientific ways of establishing empirical claims are essentially defective because they fall short of this ideal of certainty. But where does this ideal stem from, and can it ever be attained? If there is any field in which such certainty is possible, it surely is the nonempirical field of pure mathematics.

Let us consider briefly the proverbial certainty of mathematical propositions such as the theorems of Euclidean geometry. These propositions are established by mathematical proof, which shows that they follow logically from the axioms of Euclidean geometry; so that if the postulates are true, the theorems are certain to be true as well. (More explicitly: for any interpretation of the primitive geometrical terms that makes the postulates true, the theorems are true as well.) The certainty of the theorems is thus relative to the postulates. And indeed, in a non-Euclidean geometry, some of the Euclidean theorems do not hold: they are not certain relative to the postulates in question; i.e., they do not follow deductively from the latter.

But if a sentence T follows logically from certain other sentences $P_1, P_2, \ldots, P_n$—let us call their conjunction P for short—then T simply restates part of what is asserted by P. For if P logically implies T, then, as is readily verified, P is logically equivalent to the conjunction of T and the conditional sentence $T \supset P$; or briefly:

If $P \to T$, then $P \leftrightarrow T \cdot (T \supset P)$

The certainty of T relative to P is thus purchased at the price that T asserts nothing that is not implicitly asserted already by P. If a statement is to be certain relative to the grounds by which it is supported, then its content cannot go beyond that of the statements expressing those grounds.

But the content of a hypothesis in empirical science does go far beyond the content of the data sentences adduced in its support. Scientific laws, for example, while supported by statements about instances examined in the past, make assertions about countless other, unexamined, instances. Indeed, it is an important objective of science to arrive at general laws and theories that yield predictions concerning future events and retrodictions concerning unexamined past ones: and such laws and theories evidently assert more than the data sentences by which they are supported. Precisely because science seeks knowledge that reaches far beyond the supporting evidence, its empirical claims cannot consistently be required or expected to be certain relative to that evidence; the ideal of empirical knowledge with certainty is logically self-contradictory. And surely, science cannot be held to be limited because of its failure to meet a standard of perfection which is logically inconsistent.

Let us briefly consider the issue from a somewhat different vantage point. Empirical science, we noted, supports its knowledge claims by means of evidence obtained by careful empirical tests. Could the acknowledged fallibility of this procedure be adjudged a limitation on the ground that there are, or might be, superior alternative methods that would yield knowledge-with-certainty? Perhaps, such knowledge could be attained by a gifted clairvoyant or by the use of procedures not tied to the requirements of empirical evidence?

Such sources may well yield some true statements, as may indeed blind guessing. Suppose that two persons consider the sentences

(1) There will be a manned landing on Mars before the year 2000
(2) There will be no manned landing on Mars before the year 2000,

and suppose that one of them, by simply guessing or flipping a coin, selects the first sentence and asserts it, while, by the same method, the second person is led to select and assert the second sentence. Then it is certain that one of them has asserted a true statement.

But the crucial question in our context is not, of course, whether some alternative to the method of empirical testing may yield true statements, but rather whether it can be known with certainty that all statements obtained by that method are true. And the answer to that question is no; for the certainty in question is not, of course, to consist in a subjective sense of confidence or conviction, but will have to be objectively guaranteed by supporting grounds. But whatever data we may have concerning the effectiveness of the method in question, they will concern past instances of its application; and, for reasons previously considered, these do not suffice to ensure the truth of the statements the method may yield in future applications. The claim concerning the superior effectiveness of alternative methods is itself subject to inductive uncertainty; no alternative to the procedures of empirical science can give us factual knowledge with certainty.

3. On Limits of the Potential Scope of Scientific Knowledge

From the issue of limits to the *reliability* of scientific knowledge claims let us now turn to the subject of limits to the *scope* of scientific knowledge. Are there any questions of fact that are in principle beyond the reach of a scientific answer?

In a book published some years ago that was intended to deflate various claims made for the power and efficacy of science, there occurred an argument that science cannot disprove the existence of ghosts since to point out that no scientist had ever observed a ghost does not suffice to show there are none; for ghosts might be allergic to scientists and refrain from appearing in their presence. Such reasoning evidently rests on a too simple-minded view of scentific testing. The hypothesis that there are ghosts who do appear, but only in the absence of scientists, might well be indirectly testable by means of suitable automatic recording instruments. Suppose we press the argument further by offering the hypothesis that there are ghosts who do appear, but only in the absence of scientists and scientific instruments. Even this might still be testable, for example by observers or recording equipment stationed at some distance, or by later examination of the scene itself, which might show traces of ghostly appearances.

To make it impossible in principle for science to find out about the existence of our ghosts, we would have to propound an even more fanciful proposition, namely: There are ghosts whose existence cannot be ascertained by any observer or any test equipment; they produce no effects which even in the most indirect manner manifest themselves in anything accessible to human experience.

But this is tantamount to saying that there are ghosts, but everything in the world of our actual and potential experience proceeds exactly as if there were none. And this sentence makes no empirical assertion at all because for a significant empirical claim, it must be possible to indicate what it asserts, however indirectly, about some ascertainable aspect of the world of our experience; it must be possible to specify what kinds of experiential data would lend it support, what others would tend to disconfirm it. Since, then, the given sentence makes no empirical assertion, the fact that scientific inquiry can neither establish nor disconfirm it does not reflect any limit of the potential scope of scientific knowledge.

Let me put the point a bit differently: Granting that scientific inquiry can give us no knowledge concerning those ghosts, is there any other source of insight that can? On what grounds could such claims to knowledge be based? Not on empirical evidence, since our fanciful hypothesis specifically precludes all empirical manifestations. Nor could the claim be based on premonitions or other intimations experienced, perhaps, by individuals endowed with a special sensitivity: for reports by "psychic" individuals about such intimations would constitute indirect evidence concerning those ghosts. Nor, of course, can the ghost-hypothesis be known to be true on purely logical or conceptual grounds. Hence, it is quite unclear what could even be *meant* by "knowing" whether the hypothesis is true; hence, the case surely reveals no limits of potential scientific knowledge.

4. Scientific Explanation: General Characteristics

I now turn to the second of my two principal topics, namely, the limits of scientific understanding.

The understanding of empirical phenomena that science can give us is conveyed by scientific explanations. Broadly speaking, science explains why a given event came about by showing that it occurred in certain particular circumstances (in the natural sciences often called initial and boundary conditions) in accordance with certain general laws of nature or well-established theoretical principles. For example, an explanation of a rainbow occurring on a particular occasion may note, first certain particular conditions obtaining at the time, especially the presence of spherical water drops suspended in the atmosphere and of sunlight impinging on them; secondly, it shows that, as a consequence of the laws of optics, rays of sunlight are refracted and reflected in the drops in such a way as to produce the characteristic rainbow pattern. Thus, the phenomenon is explained by showing that, under the given particular conditions, it "had to" occur according to the specified laws. The explanatory account can accordingly be conceived as a deductive argument whose premises—jointly referred to as the *explanans*—consist of the relevant laws and of descriptions of the particular circumstances, while the conclusion, the so-called *explanandum* sentence, describes the phenomenon to be explained. The argument enables us to understand the phenomenon by showing that, given the laws and the particular circumstances, its occurrence "was to be expected" in the sense that it could have been inferred from the explanatory information.

Scientific explanation is not restricted to individual events, however, such as the appearance of a rainbow or of a lunar eclipse on a particular occasion; science also seeks to explain the general uniformities expressed by the laws of nature. This is usually done by showing that they hold in virtue of a well-established comprehensive theory from which they can be derived, as when the uniformities expressed by the laws of refraction and reflection are explained by means of the electromagnetic wave theory of light.

In the cases considered so far, the explanatory laws or theoretical principles are of strictly universal form; they make a claim of the type: Whenever and wherever conditions of kind F are realized, an occurrence of kind G takes place. There is another mode of explanation, which relies on laws of a probabilistic-statistical form: Under circumstances of kind F, the probability that an occurrence of kind G will take place is r. Explanations by such laws have come to play an increasingly important role in science, for example, in statistical mechanics and in the theory of radioactive decay. An explanation of this kind cannot show, of course, that the phenomenon to be explained "had to" occur in the sense that the explanandum sentence describing it is deducible from the explanans and thus must be true if the explanans is true. Rather, a probabilistic-statistical explanation, such as Boltzmann's statistical explanation of the irreversibility of certain physical processes, can show only that, given the information provided by the explanans, the occurrence of the explanandum phenomenon has a high probability.[1]

In dealing with the question of limitations of scientific understanding, let us distinguish, much as in our discussion of scientific knowledge, two different issues: first, the question of irremediable shortcomings of the kind of understanding that scientific explanation can afford; second, the question whether the scope of scientific explanation is limited in the sense that some aspects of our world are inaccessible in principle to scientific understanding.

5. How versus Why

It is sometimes said, by scientists as well as by philosophers, that science can at best tell us *how* empirical phenomena come about, but can never explain *why*. Unfortunately, proponents of this view do not say clearly what they take to be the difference between showing how and explaining why; they do not indicate what kind of an account would afford a genuine explanation-why, and by what means such explanations might be achieved if science cannot provide them.

In some cases, the view is associated with the idea that a proper explanation-why requires resort to trans-scientific conceptions. Thus, it has been claimed[2] that a genuine understanding of gravitational attraction is afforded by the conception that the attraction is a manifestation of certain "appetites or natural tendencies" closely related to love, which are inherent in physical bodies and make their "natural movements intelligible and possible." But the kind of affinity called love differs from gravitational attraction in many respects: It is not the case that everybody loves everybody else, nor does one's love for another regularly increase with the loved one's mass, nor does it decrease rapidly with distance. A defender of the idea might reply that these objections take the kinship between love and gravitational affinity too literally; that gravitational affinity should be conceived as a tendency of physical bodies to attract each other in the manner specified by the law of gravitation. But that all physical bodies have *that* tendency is asserted by the law of gravitation itself; and to call it a natural appetite akin to love is not to explain gravitational attraction, but to redescribe it in terms of an anthropomorphic metaphor. Such redescription may give us an enhanced sense of familiarity with gravitation, but the impression that we thus gain an understanding of the phenomenon is entirely deceptive.

6. Explanatory Incompleteness

Another charge of deficiency sometimes brought against scientific explanation is considerably clearer and more substantial. I shall call it the charge of essential incompleteness. In response to the question "Why does this phenomenon occur?", an explanation typically cites certain general laws and particular circumstances and shows that in view of these, the phenomenon "had to" occur. The explanatory facts adduced are simply stated but not explained in turn. For example, the explanation of a particular occurrence of a rainbow tells us *that* there is sunlight impinging on water droplets and *that* light rays conform to the laws of refraction and reflection, but it does not tell us *why* this is so. A scientific explanation is thus always incomplete in the sense that the explanatory facts it adduces are left unexplained and thus ununderstood. It may even seem that, as a consequence, an explanation in science never does more than reduce the problem of explaining one fact to the problem of explaining several others.

One facet of the view that the incompleteness here considered sets insuperable limits to scientific understanding was presented by the German physiologist Emil Du Bois-Reymond in his famous address "Ueber die Grenzen der Naturerkennens."[3] Du Bois-Reymond's argument in this case, and in the case of another problem soon to be considered, was aimed at showing that the problem in question could not be solved even in the most perfect state of scientific knowledge.

In Du Bois-Reymond's view, all scientific understanding of nature amounts to reducing the changes occurring in the physical world to the motions of atoms as performed under the influence of central forces, such as gravitational attraction, which they exert upon each other. The most perfect state of scientific understanding would then be represented by what he calls "Laplace's Spirit," i.e., that superhuman intelligence imagined by Laplace in his characterization of mechanistic determinism. That a superscientist would know all the force laws and would have combined them into one single universal formula; he would also be a superobserver, capable of ascertaining, at any moment, the state of the universe as represented by the masses, positions, and velocities, of all atoms at that moment; and, from the momentary state thus ascertained, he would be able to infer, by means of the universal formula, the state of the universe at any other time. To this Spirit, as Laplace puts it, nothing would be uncertain; both past and future would be present before his eyes. Du Bois-Reymond notes that such "astronomical knowledge" of material systems, as he calls it, is an ideal that human science can never fully attain; but he adds that there are certain problems which remain unsolvable even for the Laplacean Spirit, and which are thus unsolvable in principle by science. The Spirit, for example, would be able to infer "the original state" of the universe: but if he should find matter in that state to be at rest and unevenly distributed, he would not know whence the uneven distribution; if he should find matter in the original state to be in motion, he would not be able to account for the latter.[4]

Max Weber makes a similar point in commenting on the limited understanding afforded by the "astronomical knowledge" that Du Bois-Reymond and many other scientists of his time regarded as the ideal of scientific attainment. Speaking of the explanation, by general laws, of the state of a system at a given time by reference to an earlier state, Weber remarks: "Every individual constellation which it 'explains' ... is causally explicable only as the consequence of another equally individual constellation which has preceded it. As far back as we may go into the grey mist of the far-off past, the reality to which the laws apply always remains equally *individual*, equally *undeducible* from laws."[5]

Both Du Bois-Reymond and Weber limit themselves to pointing out that the scientific explanation of particular states or events requires reference to other particular states or events, which are left unexplained. As we have noted, explanatory incompleteness can be argued also on the ground that the general laws invoked are left unexplained.

Does the incompleteness here acknowledged mark a limitation of the understanding attainable through scientific explanation?

When the appearance of a rainbow on a particular occasion is explained in the way mentioned earlier, the explanation answers perfectly well the question to which it is addressed, namely, "Why is there a rainbow over there?" The question plainly does not ask why there are water droplets suspended in the air or why sunlight conforms to the laws of refraction and reflection. To be sure, the explanation may evoke such further questions in the mind of the person to whom it is addressed; but this is a psychological matter, and the problems thus raised are new explanatory problems.

For much the same reasons, an explanation does not in any logical or systematic sense "reduce" the problem of accounting for the original explanandum to that of accounting for the explanatory facts adduced. If, for example, we wish to explain a particular event as having resulted by law from certain other particular events, we have to show *that* those other events did take place and *that* the laws invoked do hold, but we need not show *why*.

Perhaps these remarks still leave the impression that scientific explanation is essentially defective because it falls short of the ideal of conveying a complete understanding of its subject—an understanding that would be afforded only by an explanatory account that explained every fact it asserted, thus answering every explanatory problem that can be raised within its context. But is such completeness a reasonable standard for appraising the adequacy of a mode of explanation?

If an explanation E explains a given phenomenon, as well as the facts it cites for the purpose, as well as the facts cited to explain the latter, and so forth, then it clearly involves an infinite regress, and the total set A_E of statements it contains will be infinite. This in itself raises questions about the appropriateness of the ideal of completeness, for we will normally require of an explanation that it be expressed in a finite number of statements.

A more serious objection against the ideal of completeness is this: Each of the nonempty subsets of the set A_E of statements contained in an explanation E will represent a "fact asserted by E." Completeness would require that for each of those subsets, B, there be another subset, say C, of A_E which provides an explanation for the fact represented by B. But one of the subsets B of A_E is A_E itself; and for it, A_E cannot contain an explanation because what A_E asserts cannot be explained simply by reasserting A_E No empirical fact as described by some set of sentences can be explained by means of the same sentences; no account of the type 'p because p' is an explanation.

The argument just outlined clearly does not depend on the assumption that the explanations in question are all of the "covering-law" type. It applies equally to any presumptive mode of explanation in which the fact to be explained as well as the explanatory answer is expressed in the form of *sentences*—no matter how the relationship between explanans sentences and explanandum sentence is conceived, and no matter whether the explanation is couched scientific, metaphysical, or religious terms. No conceptualized account can explain anything without asserting something that it does not explain.

It is a fundamental misconception, therefore, to appeal to the ideal of completeness on the ground that "ultimate explanation must be possible in theory;

otherwise, even partial explanation must fail"[6]; hence, failure to meet this ideal cannot count as a shortcoming or a limitation of scientific understanding.

7. Riddles of Explanation

Now let us turn to the question whether the *scope* of scientific understanding is limited, whether there are any aspects of the world that are in principle beyond the reach of scientific explanation.

An emphatically affirmative answer was put forward by Du Bois-Reymond in the lecture mentioned earlier. Du Bois-Reymond here distinguishes two kinds of unsolved scientific problems. Those of the first kind, which he calls *riddles of the material world*, are questions which, though not answered at the time, are answerable in principle and may well yield to further scientific inquiry. Those of the second kind, on the other hand, which he calls *transcendent riddles*, are incapable in principle of a solution by scientific means. One of the problems which he assigns to this class is the relation between mind and matter, or, more specifically, the question why it is that certain configurations in the human brain give rise to consciousness, thought, feeling, and other mental phenomena. As in the case considered before, Du Bois-Reymond refers to the Laplacean Spirit and argues that even this superscientist with his astronomical knowledge of the brain would be able to establish no more than *that* certain mental states or events are associated with certain atomic configurations or events in the brain; the question why this is so would evidently be left entirely unanswered; hence, it constitutes a transcendent riddle.

Du Bois-Reymond stresses the difference between this absolutely unsolvable riddle and such problems as the determination of the brain-mechanical processes associated with the performance of an arithmetical computation, which are far beyond the reach of science at present, but which might conceivably be solved by future research. In regard to such riddles of the material world, he remarks, the scientist has long been used, in manly resignation, to acknowledge: *Ignoramus*—we do not know, at least as yet; but in regard to transcendent riddles, like that of the of mind/matter, the scientist has to decide upon a verdict much more difficult to render: *Ignorabimus*—we shall never know.[7]

A strikingly similar argument had been presented a few years earlier by John Tyndall in his address "Scientific Materialism," delivered in 1868 before the British Association for the Advancement of Science.[8] Tyndall remarks that if the powers of our senses and our minds were vastly strengthened, we might be able to discover that "the consciousness of love . . . [is] associated with a right-hand spiral motion of the molecules of the brain, and the consciousness of hate with a left-hand spiral motion": we would then know *that* this is so, "but the 'WHY?' would remain as unanswerable as before."[9] He concludes: "The problem of the connection between body and soul is as insoluble, in its modern form, as it was in the prescientific ages."[10] "Science is mute in reply to these questions," he acknowledges; and, faithful to his scientific materialism, he extends this verdict to the competition: "But ... who else is prepared with a solution? ... Let us lower our heads, and acknowledge our ignorance, priest and philosopher, one and all."[11]

Although Du Bois-Reymond's narrowly mechanistic view of scientific explanation has long since been abandoned, the psychophysical problem has continued to be widely regarded as inaccessible to a scientific solution. In a presidential address delivered before the British Association for the Advancement of Science almost 100 years after Tyndall's lecture,[12] the chemist Sir Cyril Hinshelwood remarks that despite the great advances made by molecular biology, it "remains utterly incomprehensible ... how and why the brain becomes the vehicle of consciousness."[13] Hinshelwood sees here an "inscrutable mystery" that is probably indicative of "the inherent limitations of human understanding."[14]

Is this view sound? Are the aupporting arguments compelling? Let us note first that psychological phenomena are often explained by reference to physical occurrences, on the basis of suitable connecting laws; as when certain hallucinatory states are explained as brought on by drugs or by sensory deprivation, loss of consciousness by an extreme lowering of the blood sugar level, the occurrence of afterimages by certain kinds of optical stimulation, and so forth. The covering laws invoked in these cases are basically of this type:

At any time t when an organism x is exposed to physical conditions of kind P, then psychological, or mental, phenomena of kind M occur:

$(x)(t)(Pxt \supset Mxt)$

Adherents of the view expressed by Tyndall, Du Bois-Reymond, and Hinshelwood would no doubt insist that the unsolvable riddle remains totally unaffected by these remarks; that there is no possibility of explaining *why* it is that physical occurrence of kind P are generally associated with mental phenomena of kind M. but even this claim is debatable. For some specific psychophysical connections can certainly be explained at least in a modest fashion, namely, by subsumption under other, more comprehensive, psychophysical connections. Consider, for example, the law that ingestion of phenobarbital induces drowsiness. The psychophysical connection it asserts can be explained by pointing out that phenobarbital is a barbiturate, that all barbiturates are depressants of the central nervous system, and that depression of the central nervous system is accompanied by drowsiness.

Without doubt, the discovery of such more comprehensive psychophysical laws is generally considered as advancing our understanding of the hypnotic effect of phenobarbital. But the objectors would argue that such discoveries do not bring us one whit closer to the solution of the basic riddle; for at every stage of the process, certain general psychophysical connections are invoked, and we never come to understand why certain purely physical conditions can give rise to mental states or events.

This problem is analogous, it seems to me, to that of explaining why one sort of physical occurrence, such as movement with friction, gives rise to a qualitatively quite different one, such as heat. In physics such connections are often explained by suitable theories. For example, the kinetic theory explains the generation of heat by friction by pointing out that heating a body is tantamount to increasing the kinetic energy of the random motion of its molecules, and that friction increases that energy for the molecules at the interface of the bodies concerned. Similarly, the theory of electromagnetic fields may be said to explain the connection between the flow of an electric current and the qualitatively quite different movement of a compass needle in the vicinity.

To understand why certain kinds of physical occurrence are associated with certain mental phenomena, we would then have to look for a theory exhibiting the connecting linkage, as it were.

It has often been argued in the philosophical literature that mental states and events should simply be regarded as indentical with the corresponding brain states or events, much as in the kinetic theory, the temperature of a body is identified with the mean kinetic energy of the random motions of its molecules. But a theory effecting such a linkage by identification would hardly satisfy the proponents of our riddle. For the identification of a mental state M with a physical state P_M involves the establishment of a general law to the effect that M occurs when and only when P_M does; and in regard to this law, the mind-matter riddle can be raised again.

A theory suited to answer the riddle would presumably have to employ a new set of basic concepts, which could not be qualified as strictly physical nor as strictly psychological; and the fundamental laws of the theory would have to link those new basic concepts to each other, as well as to psychological concepts on the one hand and physical concepts on the other, in a manner that would permit the derivation of specific psychophysical connections. Such a theory would then explain those connections by subsuming them under theoretical laws which themselves are not of a psychophysical kind.

I do not know whether a unifying psychophysical theory of this kind will ever be developed, but the arguments offered by Du Bois-Reymond and others certainly do not preclude that possibility.

As one further potential candidate for the status of a transcendent riddle, let us briefly consider a question that has been raised by various philosophers, and that seems both profound and utterly unanswerable by scientific means: Why is there anything at all and not rather nothing?

A literal-minded defender of science might argue that science can explain for example, why there is ice on the pond after a cold night, and that to show why there is ice on the pond is ipso facto to show why there is something. But this answer seems to miss the point of the riddle; for the explanation mentioned accounts for the presence of ice at a certain place by reference to the earlier presence there of water. The explanation thus already assumes that there is something rather than nothing, and no light seems to have been shed on the riddle.

But what kind of an answer would be appropriate? What seems to be wanted is an explanatory account which does not assume the existence of something or other. But such an account, I would submit, is a logical impossibility. For generally, when the question "Why is it the case that A?" is answered by "Because

B is the case," then surely that answer can be adequate only if the assertion that *B* is the case affords good grounds for the assertion that *A* is the case, so that it is rational to say: If *B* is the case, then indeed it is to be expected that *A* is the case. This clearly is a necessary, though not a sufficient, condition for the adequacy of any explanation.[15]

But an answer to our riddle which made no assumptions about the existence of anything cannot possibly provide grounds for saying: It is to be expected that there is something rather than nothing. The riddle has been constructed in a manner that makes an answer logically impossible: and scientific explanation can hardly be held to be limited because it cannot satisfy a logically inconsistent requirement. Nor, indeed, can any other discipline or source of insight provide an answer to the riddle; an answer in terms of a *prima causa*, for example, presupposes the existence of that agency and thus the existence of something. No theory, no conceptual scheme, can explain the existence of anything without assuming the existence of something.

8. Conclusion

We have considered a variety of considerations which appeared to prove that scientific knowledge and understanding are flawed by insuperable defects and limitations. I have tried to show that those arguments fall short of their mark. While my critical comments have been directed explicitly only against the specific limitation claims here considered, I think they can be extended to any other arguments purporting to show that certain problems of knowledge or understanding concerning empirical subject matter are incapable in principle of a scientific solution.

As for the limits of scientific understanding, however, I have not argued that all empirical phenomena can in fact be scientifically explained. Whether that is so is an empirical question which cannot be settled by philosophical arguments *a priori*. For the answer depends on the nomic structure of the world. Suppose, for example, that, as assumed in contemporary physics, radioactive decay is subject to no deterministic laws, but only to probabilistic-statistical ones. Then there can be no explanation of why, in a given sample of a radioactive element with a half-life of, say, 100 years, this particular atom decays now, while that one survives for another 10,000 years. Each of the occurrences has an extremely low probability, given the relevant half-life law; neither can be explained by showing that according to that statistical law it is to be expected with high probability.[16] But then, if science could determine that the given half-life law was an ultimate law of nature for events of the kind considered, it would have said all that can be said by way of accounting for individual cases.

This kind of example, however, does not prove that the decay events referred to are "in principle" beyond the reach of scientific explanation; for to establish that claim, it would have to be proved once and for all that there are no other laws governing the events in question: and such a proof, of course, is impossible.

18

Turns in the Evolution of the Problem of Induction

1. The Standard Conception: Inductive "Inference"

Since the days of Hume's skeptical doubt, philosophical conceptions of the problem of induction and of ways in which it might be properly solved or dissolved have undergone a series of striking metamorphoses.

In my essay, I propose to examine some of those turnings, which seem to me to raise particularly important questions about the nature of empirical knowledge and especially scientific knowledge.

Many, but by no means all, of the statements asserted by empirical science at a given time are accepted on the basis of previously established evidence sentences. Hume's skeptical doubt reflects the realization that most of those indirectly, or inferentially, accepted assertions rest on evidence that gives them no complete, no logically conclusive, support. This is, of course, the point of Hume's observation that even if we have examined many occurrences of A and have found them all to be accompanied by B, it is quite conceivable, or logically possible, that some future occurrence of A might not be accompanied by B. Nor, we might add, does our evidence guarantee that past or present occurrences of A that we have not observed were—or are—accompanied by B, let alone that all occurrences ever of A are, without exception, accompanied by B.

Yet, in our everyday pursuits as well as in scientific research we constantly rely on what I will call the method of inductive acceptance, or MIA for short: we adopt beliefs, or expectations, about empirical matters on logically incomplete evidence, and we even base our actions on such beliefs—to the point of staking our lives on some of them.

The *problem of induction* is usually understood as the question of what can be said in justification of this procedure.

Any attempt to answer that question requires, first of all, a clear characterization of the *method* of inductive acceptance, presumably in terms of rules that specify under what conditions a given hypothesis may be inductively inferred from, or inductively accepted on the basis of, a given body of evidence. Only if

MIA has been characterized by reference to such rules can the question of justification be significantly raised.

Such rules have indeed been proposed in the literature. Consider first a familiar and simple type, which seems to me to reflect induction as seen by Hume and, of course, by many later thinkers:

To argue from 'All examined instances of A have been B' to 'All A are B'.

This formulation seems quite plausible; but it is fundamentally defective: it does not make clear at all just what claim is being made for this rule, or what it means to say that this is a valid rule of inductive reasoning. For a rule of *deductive* reasoning, such as *modus tollens*, the claim is clear: if the premises to which the rule is applied are true, then invariably so will be the conclusion.

But this claim of deductive validity is too strong, of course, for a rule of induction. Suppose, then, we were to read the rule as saying that under the specified conditions, it is rational to *accept* the conclusion, or perhaps: to act *as if* the conclusion were known to be true.

Now, the notion of accepting a hypothesis is surely in need of clarification: this is an issue to which I will address myself later. But even without entering into the details, it can be shown that on this construal our rule is untenable because it would oblige us to accept logically incompatible hypotheses on one and the same body of evidence.

Suppose, for example, that we have measured the length of a given silver bar at different temperatures and now plot the associated values as data points in a rectangular coordinate system. We then can draw different curves through the data points, representing l as different functions, $f_1, f_2, \ldots,$ of T.

Now, our rule entitles us to argue from:

All examined associated values of T and l satisfy f_1
All examined associated values of T and l satisfy f_2

to:

It is reasonable to accept the hypothesis that *all* associated values of T, l satisfy f_1

and

It is reasonable to accept the hypothesis that *all* associated values of T, l satisfy f_2

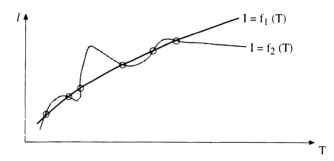

But the two general hypotheses thus inductively inferred are logically incompatible.

As this example illustrates, one and the same body of evidence can be described in different ways which, via our rule of induction, yield logically incompatible generalizations and also logically incompatible hypotheses concerning particular past, present, or future cases.

This point is also illustrated by Goodman's "New riddle of induction," exemplified by his "grue-bleen" paradox. That paradox raises yet another issue, however, which I will have to pass by here.

2. Probabilistic Construal of Inductive Reasoning

A way out of the difficulty illustrated by my example is suggested by a different construal of the basic form of inductive reasoning—a construal that has, indeed, come into wide acceptance. According to it, rules of inductive inference should not be conceived of as assigning certain specific "inductive conclusions" to a given body of evidence, but rather as principles which require that both the evidence and a specific contemplated hypothesis be *given*, and which then assign a certain probability or "rational credibility" to the given hypothesis relative to the given evidence.

A rule of inductive inference, thus construed, does not entitle us, as did our earlier rule, to argue from a given evidence sentence e to a hypothesis h specified by the rule. Rather, the rule now contemplated presupposes that both e and h are given and then specifies the probability of h on e:

$$p(h, e) = r$$

On this understanding, one body of evidence—e.g., a set of meteorological data available today—can assign certain probabilities to each of two incompatible hypotheses, e.g., 0.75 to 'Rain tomorrow', 0.25 to 'No rain tomorrow'. But no contradiction arises since our rule does not entitle us to assert either of the hypotheses, let alone both.

Indeed, in contrast to the case of deductive reasoning, there is no rule of detachment for inductive arguments which would entitle us, given that e is true, to detach, and assert without reference to e, an inductive conclusion such as 'probably h', or 'h holds with probability r'.

As these observations show, the new, probabilistic, construal of inductive inference avoids the earlier contradictions at a price: the new construal provides us with no principles of inductive acceptance or belief concerning empirical hypotheses.

But in all our practical and theoretical pursuits, we must eventually decide which, if any, of a set of alternative hypotheses under consideration we should accept as a basis for our expectations and actions.

3. A New Turn: Two Types of Rules of Induction

At this point, the problem of formulating rules of induction seems to split into two distinct issues:

(i) the problem of formulating *rules determining* the *probabilities* to be assigned to hypotheses on the basis of given evidence;
(ii) the problem of formulating *rules of acceptance* which would determine what hypotheses to accept, on given evidence, as a basis for our expectations and actions.

The first of these represents what Carnap considers as the central task of inductive logic. And indeed, Carnap's *Logical Foundations of Probability* offers an impressive formal theory of the probability of hypotheses on given evidence, for sentences expressible in a formalized language with the structure of first-order logic. Carnap and others subsequently changed and extended the approach taken in that work, and certain other conceptions of inductive probability have been developed as well; but it is certainly an open problem whether an adequate general explication of inductive probability can be given at all.

But let us set this question aside and assume for the moment that a suitable solution of the first problem is available. Then we are still left with the second task, that of formulating *rules of acceptance* for hypotheses on given evidence.

At first glance, it may seem plausible to think, and it has in fact often been suggested, that the acceptability of a hypothesis might be defined in terms of its probability, by a rule of this type: a hypothesis is to be accepted, on the relevant evidence available, just in case its probability on that evidence is greater than ½, or greater than some other fixed value, say 0.99.

But this rule runs afoul of the so-called lottery paradox, which shows that on certain types of evidence, several logically incompatible hypotheses would have to be adopted—just as in the case of the simple induction rule which we considered at the beginning.

This difficulty can be overcome in light of a very fruitful idea which was first developed precisely in the context of mathematical decision theory.

4. Induction and Valuation

Briefly, the idea is that adequate criteria for the rational acceptability of a hypothesis must take account not only of the relevant evidence available and the support it lends to the hypothesis, but also of the values attached to avoiding the mistakes of accepting the hypothesis when it is, in fact, false; or of rejecting it when it is true. This idea gives a new turn to the conception of induction by attributing to it both a cognitive and a valuational component.

This point was emphasized forcefully in a provocative article by Richard Rudner, published in 1953 in *Philosophy of Science* under the title "The Scientist *qua* Scientist Makes Value Judgments," which attracted a great deal of attention and generated a fruitful controversy. Briefly, Rudner argues as follows: Scientists, in the course of their research, do again and again accept or reject hypotheses. But the evidence their decisions are based on is generally incomplete: therefore,

> in accepting a hypothesis the scientist must make the decision that the evidence is *sufficiently* strong or that the probability is *sufficiently* high to warrant the acceptance of the hypothesis. Obviously our decision . . . respecting how strong is "strong enough," is going to be a function of the *importance*, in the typically ethical sense, of making a mistake in accepting or rejecting the hypothesis.

Rudner illustrates the point by an example, which I will modify slightly for convenience in later reference. Suppose that a pharmaceutical firm has produced a large quantity of tablets, intended for the treatment of a certain disease. In too large amounts, the active ingredient is toxic; in too small amounts, ineffective. The hypothesis to be considered before releasing the whole batch for sale is to the effect that the amount of the active ingredient in each tablet falls within an interval that makes the drug both safe and effective. The evidence consists of findings obtained by chemical analysis of a random sample drawn from the entire production. For acceptance of the hypothesis that the pills are safe and effective, we would then require a very high degree of evidential support because the consequences of a mistake could be extremely grave by our moral standards.

If, by contrast, the hypothesis were to the effect that a given large quantity of machine-stamped belt buckles are nondefective—i.e., fall within a specified range in regard to certain required characteristics—a mistake would not be nearly so serious, and a considerably lower level of evidential support would suffice for the acceptability of the hypothesis.

In sum, then, Rudner argues that the inductive acceptability of a hypothesis depends on two factors: the relevant evidence available and value judgments about the importance of avoiding mistakes.

Rudner's argument suffers from a serious ambiguity, however. This was pointed out by Richard Jeffrey in a reply, "Valuation and Acceptance of Scientific Hypotheses" (*Philosophy of Science*, 1956). Jeffrey notes that, as Rudner's examples show, Rudner is concerned, not with the acceptance of a given hypothesis *tout court*, but rather with the adoption of a practical course of action based on the hypothesis, such as releasing the tablets for the treatment of human patients, or releasing the belt buckles for sale to belt makers. But, Jeffrey notes, the scientist *qua* scientist is not concerned with giving advice, or making decisions, on contemplated courses of practical action.

Furthermore, Jeffrey argues, on Rudner's view of the matter there can be no *one* level of evidential support that is high enough for the acceptance of a given hypothesis, since one and the same hypothesis might be put to use in different courses of action, for which the moral seriousness of a mistake may differ considerably. For example, the tablets might be used for treating a sick child or a sick household pet; and it seems clear that accepting, and acting on, the hypothesis of safety and effectiveness would, by our moral standards, require a higher level of evidential support in the former case than in the latter. There is no *one* degree of moral importance that attaches to the avoidance of mistakes in all possible practical applications of a given hypothesis.

In the spirit of these considerations, Jeffrey holds—along with a number of other leading thinkers in the field—that the notion of acceptance is not properly applicable to empirical hypotheses at all, but only to courses of action—specific practical applications of a given hypothesis, as we might say.

Jeffrey concludes: "it is not the business of the scientist as such . . . to accept or reject hypotheses"; rather, "the scientist's proper role is to provide the rational agents in the society which he represents with probabilities for the hypotheses which, on [Rudner's] account he simply accepts or rejects."

This is a startling conception indeed. In the context of the scientific pursuit of pure, unapplied, knowledge of the world, it entirely eliminates the classical problem of induction: if the pure scientist never accepts or rejects any hypotheses, then the problem of formulating rules of inductive acceptance, and the further problem of justifying those rules, simply does not arise.

There arises, however, a proxy problem, namely, the question of rules for the assignment of probabilities to proposed hypotheses, and the question of how to justify those rules. But for the moment, let us leave this issue on one side.

Jeffrey himself is quick to point out that his conception faces serious difficulties of its own; among them, the following:

(i) it presupposes that an adequate theory can be provided for the assignment of probabilities to hypotheses on given evidence; this is part of the proxy problem just mentioned, and Jeffrey seems to regard it as dubious.

(ii) the role here assigned to the scientist "bears no resemblance to our ordinary conception of science. Books on electrodynamics, for example, simply list Maxwell's equations as laws; they do not add a degree of confirmation."

(iii) Rudner had antipatively stressed that on a view of the type later proposed by Jeffrey, the pure scientist *still* has to accept certain hypotheses, namely those of the type 'the probability, or the evidential support, of hypothesis h on the available evidence in p'. Jeffrey suggests that it is not the task of the scientist to accept even hypotheses of this type, but he acknowledges that Rudner's objection is one of the weightiest difficulties for a probabilistic view of science.

(iv) Finally, I would add one further difficulty: Even if the scientist limits himself to determining probabilities for hypotheses, he must perform tests to obtain the evidence on the basis of which to calculate those probabilities. He must, therefore, it seems, accept certain empirical statements after all, namely, the evidence sentences by which he judges the probability of contemplated hypotheses.

5. Induction and Epistemic Utility

In sum, then, neither Rudner's nor Jeffrey's account of the matter can be quite right. But it seems to me that there is indeed something right about Rudner's claim concerning the relevance of valuation to scientific inquiry; therefore, I would like to consider the possibility of modifying Rudner's idea a bit in an effort to overcome some of the difficulties it faces.

Let me outline briefly one decision-theoretical approach to problems of the kind considered by Rudner concerning the acceptance or the rejection of a hypothesis. Take the pharmaceutical case. Suppose a random sample of the total production of tablets has been tested and that, on the evidence E thus obtained, our theory of evidential support assigns to the hypothesis H that all tablets in the whole batch are safe and effective the probability $p(H, E) = 0.75$.

A rational decision as to whether to accept or reject H (or rather, the corresponding course of action) will then depend on the importance, or the value, attached to achieving or to avoiding certain possible outcomes of our action.

The desirability or undesirability attached to different possible outcomes are often assumed to be expressible numerically as positive or negative "utilities." In our example, the relevant utilities might be as follows:

	H is true	H is false
Accept H	+200	−1000
Reject H	−60	+300

Then, given the probabilities $p(H, E) = 0.75$, $p(-H, E) = 0.25$, each of the possible outcomes can be assigned an expectable utility:

$$U(\text{Acc } H, E) = (0.75 \times 200) - (0.25 \times 1000) = 150 - 250 = -150$$

$$U(\text{Rej } H, E) = (0.75 \times -60) + (0.25 \times 300) = -45 + 75 = +30$$

And one often invoked rule for rational decision directs us to reject H, as the course of action which offers the greater expectable utility.

In case no probabilities are available, and in particular if the entire idea of assigning probabilities to hypotheses should be ruled out, decision theory still offers rules for what is called "decision-making under uncertainty." One such rule directs us to choose the course of action for which the minimum payoff (utility) is maximal; in our example, this maximin rule would again call for rejection of H, since $-60 > -1000$.

Now, it seems to me that despite Jeffrey's strictures, there is a tenable, and indeed important, point in Rudner's view of the matter. I think that the acceptance of a hypothesis in pure or basic science may be construed as an action, too, though not as an action of the practical kind considered in our previous example.

The action consists in including the given hypothesis into the corpus, K, of previously accepted hypotheses; and its purpose and potential value lies, not in solving any practical or technological problems, but in the increase of scientific knowledge. Since science aims at establishing true hypotheses, the addition of a hypothesis to the corpus of accepted statements might plausibly be assigned a positive utility in case the hypothesis is true; otherwise a negative utility of the same numerical magnitude. That magnitude itself could plausibly be taken to be given by the proportion of informational content that the hypothesis would add to the content of K.

I sketched this idea as a somewhat incidental point in an article published in 1962 ("Deductive-Nomological vs. Statistical Explanation"; *Minnesota Studies in the Philosophy of Science*, vol. III; section 12), in which I proposed the term 'epistemic utility' to refer to the utility which the acceptance or rejection of a new hypothesis possesses for the purposes of pure, basic research.

By means of this concept and a suitable theory of probability for hypotheses, one could then formulate *a rule of inductive acceptance for pure or basic research*: in deciding whether to accept or to reject a given hypothesis or to leave it in suspense, choose the course of action which has the highest expectable epistemic utility attached to it. The idea is strictly analogous to that for practical actions, but with epistemic utilities taking the place of the utilities of practical consequences. Thus, on this construal, *the scientist qua scientist does indeed make value judgments*, but they are not of a *moral* kind; rather, they reflect the value to basic research of constructing sound and information-rich accounts of the world; and these are what I would call *epistemic* values.

Since it is often said that science *presupposes* value judgment, let me stress that epistemic judgments of value do not enter into the *content* of scientific hypotheses or theories; Kepler's laws, for example, do not presuppose or imply any value judgments at all—either epistemic or of other kinds. But epistemic valuation does enter into the *acceptance* of hypotheses or theories in this sense: the assertion that a given hypothesis H is *acceptable* in a given knowledge situation implies that the acceptance of H possesses a greater expectable epistemic value for science than does the acceptance of any rival hypothesis that may be under consideration.

That valuational considerations should play an essential role here is hardly surprising: how could a procedure like the adoption of a hypothesis be qualified as appropriate or rational, except in consideration of the objectives of scientific inquiry, i.e., in consideration of the contribution that the adoption of the hypothesis is likely to make to furthering the objectives of inquiry? And that contribution is expressed in the epistemic utilities or values assigned to the possible scientific consequences of the adoption.

On the simple construal I sketched a moment ago, the epistemic value of accepting a hypothesis would depend only on its truth value and on the proportion of new information it adds to the set K of previously accepted hypotheses. But, as I showed in my article, that interpretation, combined with the rule of maximizing expectable utility, yields an inductive acceptance rule that is intuitively unreasonable; indeed, I should add, it leads again to the lottery paradox.

Yet, I believe that there is something fundamentally right about the idea of epistemic value, and that the failure of the utility measure just considered may be attributable to a too narrow construal of the objectives of basic research. Science is interested not only in questions of truth and information content, but also in the simplicity of the total system of accepted hypotheses, in its explanatory and predictive powers, and other factors, all of which a theory of inductive acceptance would have to take into account.

6. The Pragmatist Construal of Theory Choice in Science

Just those factors have in fact been given considerable prominence in what I think may be viewed as the latest major turn in the evolution of the problem of induction. This turn is one important aspect of the historic-pragmatist construal of scientific inquiry of which Thomas Kuhn is the leading proponent.

The problem of induction arises here in an especially comprehensive and momentous context, namely, in regard to scientific revolutions, in which an established theory is eventually abandoned in favor of a new rival. In his account of the kind of choice between rival theories that occurs in this context, Kuhn considers the question whether there are general precise criteria of preferability for theories that determine the outcome of the choice.

To this end, he surveys a variety of characteristics that are widely acknowledged as desirable characteristics of scientific hypotheses and theories; let me call them *desiderata* for short. Kuhn holds that theory choice in science is definitely influenced by a comparison of rival theories in regard to the extent to which they fulfill such desiderata; and he further discusses the question whether those desiderata admit of precise objective definitions which might make it possible to construct explicit criteria of preferability (we might say: epistemic utility) that are objective in the sense that different scientists, provided with the same relevant information, would arrive at the same decision as to which of two competing theories, if any, is to be accepted.

One familiar desideratum is wide scope of application: it is highly desirable that a theory should account for many quite different kinds of phenomena. Newton's theory is a good example: it covers phenomena as diverse as free fall, the swinging of pendulums, the motions of planets, comets, double stars, artificial satellites, and so on.

But what exactly should be taken as the principle of individuation for *kinds* of phenomena? Why, for example, should all instances of free fall on the earth count as one kind in this context: why should not the fall of metal bodies count as a kind of event different from the fall of bodies of wood or of glass? Perhaps it might be said that irrespective of their composition, physical bodies fall in accordance with one and the same law, which can be (approximatively) derived from Newton's laws. But then, all the diverse kinds of events that Newton's theory as a whole accounts for behave in accordance with the basic laws of that theory: so, should they not count as belonging to one and the same kind of event;

and where would this leave the putative diversity of the phenomena falling within the range of Newton's theory? For these and other reasons, it is highly questionable whether a satisfactory precise explication of the notion of the range or the scope of a theory can be constructed.

A related desideratum recently proposed in Laurence Laudan's book, *Progress and Its Problems*, is that a theory should solve a large number of important problems. This idea is certainly plausible and reflects scientific preference, but there are no clear, unambiguous ways of differentiating and counting the problems solved by a theory, and of assessing their importance; the reasons are closely related to those just considered.

Compatibility, or more loosely, "fitting together," with well-established theories in neighboring fields also counts as a desideratum. For example, a hypothesis sometimes suggested for telepathic phenomena construes these as brought about by a special kind of radiation emitted by the sender of a telepathic message; but this idea has been objected to on the ground that it does not fit together with a certain characteristic of radiation theories in physics, namely, that the energy transmitted by the energy source to a given receiver decreases with the square of the distance of the receiver, whereas according to evidence claimed to describe telepathic communication, no such decrease in strength or clarity of the message is generally observed.

But again, no clear criterion is in sight for this notion of "fitting together" with well-established neighboring theories—a notion which plainly is stronger than just logical compatibility.

Similar difficulties face the attempt to formulate precise and general explications of concepts invoked in other desiderata affecting theory choice, among them the simplicity of a theory, the closeness of fit between experimental data and the corresponding implications of a theory, and the power of a theory to predict novel phenomena.

Thus, Kuhn holds, there are no precise criteria for the comparison of the merits of competing theories in regard to any of these desiderata. And even if such sharp criteria were available, there would remain the problem of combining them all into one precise overall criterion which would determine which, if either, of two competing theories to accept. But while scientists do agree about the importance of the various desiderata, there is no unanimity at all as to the relative weights that are to be assigned to them in the overall comparison of theories.

The prospects, then, of formulating precise explications of those desiderata that might provide an objective characterization of epistemic utility seem very dim indeed.

In fact, Kuhn's view implies a radical rejection of that conception of the problem of induction which calls, first of all, for the formulation of precise rules of acceptance and, secondly, for a justification of those rules.

Kuhn offers instead a fundamentally different, historic-pragmatist conception of scientific theory choice. He holds, to put it briefly, that the choice between competing theories lies in the hands of the specialists in the field. These specialists are all committed to what Kuhn calls the values of simplicity, accuracy, large scope, etc. in theory choice. And even though they do not share precise criteria of application for those concepts, and even though their preference is determined in part also by idiosyncratic factors, there does in fact, in the course of a controversy over the merits of two competing theories, eventually emerge a consensus leading to the acceptance of one of the competitors.

And though no explicit rules of acceptance are countenanced by Kuhn, he nevertheless offers what amounts to a justification for theory choice thus construed as resulting from the efforts of a group of professional specialists. Kuhn argues here that scientific research behavior as exemplified in theory choice serves an essential function in improving scientific knowledge, and that in the absence of an alternate mode of behavior that would serve the same function, "scientists should behave essentially as they do if their concern is to improve scientific knowledge" ("Reflections on My Critics," in I. Lakatos and A. Musgrave (eds.), *Criticism and the Growth of Knowledge*, 1970, p. 237). Kuhn regards this consideration also as an argument in favor of the claim that scientific behavior is rational, and is indeed the best example we have of rationality.

I think that this conception faces several difficulties:

First, it is clear that Kuhn's pragmatist account does not specify what counts as the essential aspects of scientific behavior, those which the aspiring scientist should emulate if he wants to improve scientific knowledge.

Second, the account offers no specific indication of what counts as "improving scientific knowledge."

Third, on Kuhn's view, scientific theory choice (and other facets of scientific behavior) are not affected by means of procedures that are deliberately adopted by the scientific community as a presumably optimal means for advancing scientific knowledge; and I would think that any kind of action, including scientific theory choice, can be called *rational* only if it can be causally linked to deliberation or reasoning aimed at achieving specific ends. Scientific theory choice as characterized by Kuhn would not be rational in this sense, but would rather be akin to what in anthropology are called *latently functional* behavior patterns, which serve a function they were never *chosen* to fulfill.

For these reasons I think that Kuhn's pragmatist account of theory choice is not entirely satisfactory. However, I consider it as an important and illuminating corrective for an approach, perhaps most elaborately developed by analytic empiricism, which seeks to "explicate" scientific procedures, including induction, as governed by explicit and precise rules. For it does seem highly unlikely that the more comprehensive modes of induction, especially those involved in theory change, can be characterized by general rules that would be acknowledged as binding by scientists and would be observed by them.

On the other hand, for certain narrow and specific inductive problems, such as measuring a quantity, or testing a statistical hypothesis, fairly precise rules can be stated that are acknowledged and observed in scientific practice.

7. Some Remarks on Justification

And how might such *explicit* rules of inductive acceptance be *justified*? Let me refer here to the ideas of a thinker who would generally be thought an analytic philosopher, but whose views concerning justification show a suggestive and illuminating kinship to the pragmatist approach—namely, Nelson Goodman. To quote a passage from his *Fact, Fiction, and Forecast* (1955; p. 67):

> ... deductive inferences are justified by their conformity to valid general rules, and ... general rules are justified by their conformity to valid inferences. But this circle is a virtuous one. The point is that rules and particular inferences alike are justified by being brought into agreement with each other. *A rule is amended if it yields an inference we are unwilling to accept; an inference is rejected if it violates a rule we are unwilling to amend.* The process of justification is the delicate one of making mutual adjustments between rules and accepted inferences; and in the agreement achieved lies the only justification needed for either.
>
> All this applies equally well to induction.

It seems to me that, somewhat surprisingly, this view of the justification of inductive procedures has a significant kinship both with the analytically oriented ideas of Carnap and with the pragmatist account of Kuhn and kindred thinkers.

Carnap holds that the reasons to be given in support, or justification, of the basic principles of inductive reasoning "are based upon our intuitive judgments concerning inductive validity" (P. A. Schilpp (ed.), *The Philosophy of Rudolf Carnap*, p. 978) and surely, there is a close relationship between this idea and that of considering what particular inductive arguments we are willing to give up, and what general rules we are willing to amend.

Kuhn leaves the ultimate decisions not to "our" intuitions, to "our" dispositions to accept or amend, but to those of the specialists professionally concerned and equipped to make decisions about acceptability.

Goodman's construal of justification, however, as the quoted passage shows, pertains only to modes of reasoning that are governed by explicitly stated rules; it cannot be straightforwardly extended to scientific theory choice as characterized by Kuhn.

As for that procedure, we briefly considered a different justification, offered by Kuhn. It argues that theory choice as practiced in science serves to improve scientific knowledge and that in the absence of a better alternative scientists should go about it essentially in the manner they do, if their aim is the improvement of scientific knowledge.

We noted some difficulties facing this idea, among them the point that no independent characterization had been provided of the notion of "improving scientific knowledge," so that it remained unclear just how to assess the effectiveness of scientific or alternate modes of theory choice in the pursuit of improved knowledge.

But this particular difficulty can be avoided by a change of perspective which does make it possible to attribute rationality to theory choice affected by reference to desiderata. The imposition of desiderata may be regarded, at least schematically, as the use of a set of means aimed at the improvement of scientific knowledge. But instead of viewing such improvement as a research goal that must be characterizable independently of the desiderata, we might plausibly conceive the goal of scientific inquiry to *be* the development of theories that ever better satisfy the desiderata. On this construal, the desiderata are different constituents of the goal of science rather than conceptually independent means for its attainment, and it becomes a truism that replacing a theory by a competing one that better satisfies the desiderata will constitute an improvement of scientific knowledge and will thus be a rational procedure.

This schematic construal in no way disposes of the vagueness, stressed earlier, of the desiderata and their relative weights, and it offers no prospect of a precise explication, in the sense of analytic empiricism, of scientific theory choice. But vagueness need not prevent a set of concepts from being epistemologically illuminating in certain contexts. The idea here outlined shows, I think, that if scientific inquiry aims at theories which ever better satisfy certain desiderata, then no matter how the latter may be construed in detail, the choice among theories by reference to the chosen desiderata is, in its basic structure, a rational enterprise.

19

Scientific Rationality

Normative versus Descriptive Construals

1. Two Conceptions of the Methodology of Science

In the course of the past few centuries, scientific inquiry has vastly broadened man's knowledge and deepened his understanding of the world he lives in; and the striking successes of the technologies based on the insights thus gained are eloquent testimony to the basic soundness of scientific modes of research. In view of these achievements, scientific inquiry has come to be widely acknowledged as the exemplar of rationality in the pursuit of reliable knowledge. But there is no unanimity among students of the methodology and the history of science when it comes to the question whether, or to what extent, it is possible to specify precisely a set of procedural rules or standards which are characteristic of scientific inquiry, and adherence to which qualifies science as the exemplar of rationality in the pursuit and acquisition of knowledge.

This problem has in recent years been the focus of an intense and fruitful controversy between two schools of thought which I will refer to as the analytic-empiricist and the historic-sociological, or pragmatist, school. By the former, I understand here a body of ideas which, broadly speaking, developed out of logical positivism and the work of kindred thinkers; among the protagonists of the more recent historic-sociological approach, I have in mind particularly Thomas Kuhn and Paul Feyerabend. Let me briefly and roughly sketch the background of the controversy.

In the view of analytic empiricism, it is indeed possible to formulate characteristic rules and standards of scientific procedure, and it is specifically the task of the methodology or of the philosophy of science to exhibit, by means of "logical analysis" or "rational reconstruction," the logical structure and the rationale of scientific inquiry. The methodology of science, thus understood, is concerned solely with certain logical and systematic aspects of science which form the basis of its soundness and rationality—in abstraction from, and indeed to the exclusion of, the psychological and historical facets of science as a social enterprise.

This construal of methodology is clearly analogous to the conception of formal logic as a discipline concerned solely with questions pertaining to the validity of arguments, the logical truth and the logical consistency of sentences, and the like, in deliberate abstraction from the genetic and psychological aspects of human reasoning. While formal logic thus does not afford an empirical, descriptive theory of "how we think," or "how logicians, mathematicians, and scientists think," the rules and criteria provided by logical theory can be employed prescriptively or normatively, i.e., as standards for a critical appraisal of particular inferences, claims to logical truth, and the like. Thus used, the principles of formal logic constitute, not categorical norms, but instrumental ones: conditions for the rational pursuit of certain objectives, such as guaranteeing the transfer of truth from premises to an inferred conclusion.

Analogously, one may say that, as understood by analytic empiricism, the principles established by the methodology of science could serve as conditions for the rational pursuit of empirical inquiry, as criteria of rationality for the formulation, test, and change of scientific knowledge claims. One well-known example of such normative-critical use of methodological maxims is the logical empiricists' rejection of neovitalism, not as false, but as being no empirical theory at all, on the ground that it violates the methodological requirement of testability-in-principle. Another example is Popper's refusal to grant the status of scientific theories to the doctrines of psychoanalysis and of Marxism on the ground that they violate certain requirements of his methodology, in particular those of falsifiability-in-principle and of avoidance of conventionalist stratagems.

The historic-sociological school, on the other hand, for reasons soon to be considered, rejects the idea of methodological principles arrived at by purely philosophical analysis, as it were: it insists that an adequate theory of scientific method must be based on a close study of the practice of scientific inquiry and should be able to explain at least some aspects of actual scientific theorizing, past as well as present.

The debate between the two schools of thought has been focused to a large extent on one fundamental and comprehensive issue, the problem of theory choice. This is the question of whether there are general principles governing the choice between competing theories in a field of inquiry, and if so, whether or to what extent such principles can be presented as conditions of rationality for scientific inquiry in the sense envisaged by analytic empiricism.

The issue arises in its most dramatic, and most widely discussed, form in reference to scientific revolutions in the sense of T. S. Kuhn, which eventually call for a choice between two comprehensive theoretical systems or paradigms, such as those represented by Newtonian and by relativistic mechanics.

Analytic empiricism might envision criteria of choice in the form of general rules determining which of two competing hypotheses or theories has the higher probability or "rational credibility," as judged by the results of experimental tests and by other relevant information available at the time. Carnap's theory of inductive logic, for example, is an impressive effort at formulating precise general criteria of this kind for the rational appraisal and comparison of scientific hypotheses, though not of complex theories.

Kuhn, on the other hand, considers the search for general and precise criteria of theory choice as basically misguided and doomed to failure. He acknowledges that there are certain general considerations, repeatedly noted also by earlier writers, which influence the decisions scientists make in the context of theory choice; he characterizes them as shared preferences or values of the scientific community; among them are a preference for theories of quantitative form whose predictions show a close fit with experimental findings; for theories covering a wide variety of phenomena; for theories that correctly predict novel phenomena; for fruitful theories, for simple theories rather than complex ones.[1]

But he notes that, for reasons we will consider later, those desiderata do not suffice unambiguously to single out one of two competing theories as superior to the other; he further insists that there just are no generally binding principles that compel a unique choice on the basis of "logic and experiment alone,"[2] and he presents the adoption of a new theory by the scientific community in the field as the result of a process which involves deliberation on the part of individual scientists, together with efforts at mutual persuasion, but whose final outcome depends also on a variety of other factors and thus is not uniquely determined by rules of rational procedure of the kind analytic empiricism might envision.[3]

Yet despite his naturalistic, socio-psychological account of theory choice, Kuhn calls science a *rational* enterprise. Thus he declares: "scientific behavior, taken as a whole, is the best example we have of rationality," and "if history or any other empirical discipline leads us to believe that the development of science depends essentially on behavior that we have previously thought to be irrational, then we should conclude not that science is irrational, but that our notion of rationality needs adjustment here and there."[4]

Thus, Kuhn views his construal of scientific theorizing as affording a *descriptive-explanatory* account of certain important characteristics of the actual development of science and as equally affording a *normative* or *prescriptive* account by exhibiting certain characteristics in virtue of which that development is to be qualified as rational. Indeed, in response to Feyerabend's question whether Kuhn's account is to be read as descriptive or as prescriptive, Kuhn declares unequivocally that it "should be read in both ways at once."[5]

2. Explanation versus Justification: The Janus Head of Methodology

Taken literally, Kuhn's pronouncement is surely untenable. Descriptive sentences are not prescriptive: the former purport to tell us what *is* the case; the latter what *ought* to be done, or what would be a right or appropriate or rational course of procedure in a given situation. For example, as we noted, the principles of logical theory may be said to furnish prescriptions, in the sense of criteria, for deductively valid reasoning; but they certainly are not descriptions of how people do in fact reason—or there could be no talk of people making logical mistakes. Analogously, a prescription or a criterion of rationality for theory choice cannot also be a description of how theory choices are in fact made by practicing scientists.

It is quite possible, however, to give a perfectly plausible interpretation to the idea, of which Kuhn's dictum is an example, of ascribing to methodological principles a Janus head with one descriptive and one prescriptive face. Consider, for example, the analytic-empiricist principle or condition, T, of testability-in-principle for scientific hypotheses or theories. On the interpretation I have in mind, the assertion that T is both descriptive and prescriptive would be a misleading conflation of two distinct claims, which I will call T_D and T_P, respectively.

T_D would be an empirical, descriptive claim to the effect that scientists do in fact share a commitment to the condition of testability and thus have a shared disposition to conform to the principle in their research. This empirical claim can evidently be invoked to *explain* why scientists in certain research situations proceed in such and such a manner, namely, why they bar from further consideration a proposed theory that they have come to consider as untestable-in-principle.

But—and here lies the confusion in the view of methodological principles as both prescriptive and descriptive—it is not the methodological norm requiring testability that explains the scientists' procedure, but the associated socio-psychological hypothesis T_D that the scientists are committed to that norm.

The second of the two claims conflated by the Janus head conception of methodological principles is an (instrumentally) prescriptive or normative one, T_P, to the effect that adherence to the testability condition is a condition of rationality for scientific inquiry.

Let us note that in so far as this second claim can be made good, it can serve to *justify* particular scientific research procedures or decisions by showing that they conform to the specified conditions of scientific rationality.

Methodological principles for which both the associated empirical claim and the associated normative claim are sound can therefore serve to "*account for*" particular instances of actual scientific behavior in the double sense of the ambiguous term "account." We can give an *explanatory account* of a particular case of scientific procedure or decision by pointing out that the scientists involved were committed to acting in accordance with the methodological norms; and those principles can also provide a *justificatory account* of the scientists' procedure by showing that it conforms to certain conditions of scientific rationality.

There are certain kinds of human decision and action which do, to some degree of approximation, admit of such a two-faced account by reference to pertinent methodological considerations.

Take, for example, the case of an engineer in charge of quality control who has to decide whether a given large quantity of hormone tablets or of ball bearings manufactured by his firm is to be released for sale or is rather to be reprocessed or discarded because of excessive deviations from specified requirements. His decision is based on the results obtained by performing quantitative tests on a random sample drawn from the whole batch. Given the test results, the decision made by the engineer may well be *explainable*, and indeed *predictable*, by the empirical assumption that the engineer generally employs such and such specific decision-theoretical criteria in situations of this kind. On the other hand, the criteria here invoked—or some more general principles of mathematical decision theory, from which the criteria can be derived—can provide a *justificatory* account of his decisions by exhibiting them as rational.

Similarly, the use of the double-blind method in testing a new drug for safety and effectiveness might be *justified* by arguing that the method is *rational* in the sense of offering better chances of avoiding certain kinds of error than do simpler tests; and its application by medical investigators in a particular study might be *explained* by pointing out that, in the course of their professional training, the investigators have acquired a disposition, a habit, to use the procedure in tackling research problems of the given kind.

The interpretation I have suggested for a "Janus-headed" conception of methodological principles seems to me the only plausible one. Yet, it does not seem to me to be in full accord with Kuhn's general characterization of scientific theory choice. Before turning to this issue, however, I will have to consider more closely the conception of methodological principles as prescriptive, i.e., as expressing conditions of rationality for scientific inquiry.

3. On the Notion of Scientific Rationality

It is interesting, but also somewhat perplexing that Feyerabend, Kuhn, Popper, Lakatos, and other protagonists in the recent methodological controversy have offered quite diverse pronouncements on the rationality or irrationality of various modes of inquiry without always giving a clear indication of the intended sense of "rationality." For example, we find Lakatos charging Kuhn's account of scientific theorizing with irrationalism and with appeal to mob psychology,[6] whereas Kuhn, as noted earlier, holds that his account presents scientific research behavior as a whole as the best example we have of rationality.

What is to be understood here by "rationality," and what kinds of consideration could be properly adduced in support of, or in opposition to attributions of rationality to science as a whole or to certain methodological rules and the corresponding modes of inquiry? I have no satisfactory general answers to these questions, but I would like to offer some tentative reflections on the subject.

To begin with, a given action or a mode of procedure cannot be qualified as rational or as irrational just by itself, but only in consideration of the goal that it is aimed at. For a man to jump fully clothed from a bridge into the river below may be rational if he intends to save a drowning swimmer and believes himself capable of doing so; it is irrational if he intends to get to the other side as fast as possible. In addition to the goal, an appraisal of the rationality of an action will have to take into account also the information available to the agent—or, more specifically, the beliefs entertained by the agent—concerning different courses of action available to him for the pursuit of his goal, and concerning the likelihood of their leading to the desired result. For example, a would-be rescuer is not acting rationally in jumping into the river if he believes that he cannot swim, so that his effort would be virtually certain to fail.

To put the point somewhat loosely: a mode of procedure is rational, relative to a certain goal and a given body of means-ends information, if, judged by that information, the procedure offers an optimal chance of attaining the goal.

In so far, then, as methodological principles express rules for scientific procedure, they do not constitute absolute or categorical norms, but relative or instrumental ones: they do not categorically tell us what to do but rather what way of proceeding is rational in the sense of offering the best chance of attaining a certain scientific objective.

In regard to the procedural rules laid down in laboratory manuals for the pursuit of certain limited and highly specific objectives, such as the measurement of particular quantities or the experimental testing of particular kinds of hypotheses, it can quite plausibly be argued that—given the current scientific knowledge in the field—they qualify as optimal, and thus as rational in the sense indicated.

4. Rationality in the Scientific Pursuit of Knowledge

It is a much more elusive task to formulate instrumental criteria of rationality for scientific inquiry in general. The first problem to consider here is that of specifying the goals by reference to which such rationality is to be characterized.

Scientific inquiry is often said to be the search for truth. We might imagine that all that is known to be true, or rather, believed to be true in science at a given time is expressed by means of a large class of statements. This class will continually change, sometimes quite radically, as a result of ongoing research. One might be tempted, accordingly, to see the ultimate goal of scientific research as knowledge of the truth, the whole truth, and nothing but the truth about the world, this complete knowledge being represented by a set of sentences which would describe "everything that is the case"—including particular occurrences in past, present, and future, as well as the ultimate network of the laws of nature that bind the particular facts together.

There is no need to belabor the point that this goal represents at best an idealization, that it is unattainable to frail, finite, fallible man. But it may be of interest to note that the ideal of total knowledge as just characterized is unattainable for purely logical reasons, and no being can achieve omniscience in this sense. For the sentences expressing such total knowledge would have to be formulated in some suitable language: but no matter how rich a language may be, there are always facts that cannot be expressed in it.[7] The contemplated conception of the goal of science is therefore untenable.

But this does not preclude the possibility of conceiving successive stages in the evolution of scientific knowledge as characterized by sets K_t of statements which are accepted at different times t, and which science has the goal of choosing in such a way that they represent a sequence of systems of empirical beliefs which increasingly satisfy such desiderata as accuracy, comprehensiveness, simplicity, and the like.

Let us note now that independently of how those desiderata may be construed in detail, this conception of the goals of science does impose certain necessary conditions of rationality on any set of sentences that can qualify as "acceptable."

First, in view of the goals of science as just adumbrated, a set of sentences would be rationally acceptable only if it is capable of test and has in fact been tested with success.[8]

Next, an acceptable set must not be known to be logically inconsistent since otherwise its sentences could not possibly all be true.

Also, every acceptable set must be deductively closed; i.e., if K' is a subset of an acceptable set K and S' is logically deducible from K', then S' must be included in K. The reason is that the deductive consequences of sentences that have been accepted as presumably true must be presumed true as well, and thus included in K.

These, then, are some modest necessary conditions of scientific rationality. They do not pertain to the issue of a rational choice between competing theories, but to the more basic question of what sets of sentences could possibly qualify as representing scientific knowledge at some time.

Note that these conditions of rationality are predicated upon the objectives of science as vaguely characterized by the desiderata mentioned above. If, instead of aiming at those objectives, we were rather seeking to formulate sets of sentences about the world that would afford us ever greater emotional security or esthetic satisfaction, then quite different standards of rationality would apply. For example, we might then do well not to accept all the deductive consequences of sentences we are accepting: for some might be disturbing or distasteful to us; similarly, rationality with respect to the alternative goal would not require us to judge the acceptability of sentences by the outcome of empirical tests: the question of factual accuracy is irrelevant to the objectives under consideration.

This point has a bearing on an idea put forward by Feyerabend. In his plea for methodological anarchy, Feyerabend maintains that "science as we know it today" may "create a monster," that "a reform of the sciences that makes it (sic) more anarchistic and more subjective . . . is therefore urgently needed,"[9] and that "we can change science and make it agree with our wishes. We can turn science from a stern and demanding mistress into an attractive and yielding courtesan who tries to anticipate every wish of her lover."[10]

Feyerabend here urges the replacement of the goals of science by another set of goals. But however one might feel about the latter, the modes of procedure appropriate to the pursuit of these alternative objectives are not appropriate, or rational, means of pursuing the goals of "science as we know it."

If it is our goal to obtain reliable knowledge about the world, knowledge that, among other things, enables us correctly to predict future occurrences; knowledge that may enable us to escape harm or to prevent it; knowledge that indicates means for achieving desired ends, then we will have to check our hypotheses and theories against carefully established data concerning the relevant features of the world—rather than follow, as Feyerabend puts it, "esthetic judgments, judgments of taste, and our own subjective wishes."[11]

Feyerabend suggests that a world in which "science as we know it . . . plays no role whatever . . . would be more pleasant to behold than the world we live in today, both materially and intellectually."[12] But surely, one who is seriously concerned to enhance the welfare and the happiness of mankind would still have to proceed by the standards of *scientific* rationality in the search for knowledge about suitable means to achieve those ends.

5. Rationality in Theory Choice

Let us now turn to the question of criteria for the rational comparison of competing theories. Such criteria would have to determine which of two competing theories—such as the caloric and the kinetic theories of heat, or Newtonian and relativistic mechanics—is rationally to be preferred to the other in consideration of the objectives of scientific theorizing.

I have repeatedly referred to certain familiar characteristics, noted by Kuhn and others, which scientists widely regard as desirable features of scientific theories: precise, preferably quantitative, formulation; accuracy, i.e., close agreement between theoretical predictions and empirical data; wide scope; simplicity; prediction of novel phenomena, and the like.

Given that these desiderata serve to characterize the goals of scientific theorizing, it is clear that they provide us with conditions of rationality for the comparison, adoption, and rejection of theories.

But these desiderata do not nearly suffice to provide an unequivocal and general criterion which will determine which of two competing theories is rationally preferable to the other. There are at least two reasons for this, as has been noted by Kuhn and to some extent by earlier writers, such as Ernest Nagel.[13]

First, not one of the desiderata has been characterized with sufficient precision to permit an unequivocal decision as to which of two competing theories satisfies the desideratum more fully. For example, none of the various efforts made by logicians and philosophers of science to explicate the notion of simplicity for theories has yielded a satisfactory generally applicable criterion for the comparison of theories in point of simplicity. Similar remarks apply to the idea of the scope of a theory; and there are considerable problems also for the comparison of theories in regard to the closeness of the fit between their implications and the available experimental data.

Second, even if precise criteria for each of the individual desiderata were available, there would remain the task of combining them all into one overall criterion of rational preferability for competing theories. But it may, and does, happen, that of two rival theories, one satisfies some of the desiderata to a higher degree, but others to a lower degree, than its rival: which of the two theories is then to be given preference? To secure one general standard of comparison, the various desiderata would have to be rank-ordered in point of relative importance: and there is no plausible way in sight to achieve such an ordering.[14]

Thus, the prospects seem bleak for a precise rational reconstruction or explication, in the sense intended by analytic empiricism, of a set of general principles of rational theory choice.

These considerations are certainly powerful. But they afford no proof, of course, of the impossibility of such rational reconstruction; and I think in fact that partial advances will be made, in the spirit of mathematical decision theory, in formulating precise principles of theory choice for more limited purposes and for theories of a less comprehensive kind than the paradigmatic ones Kuhn has in mind.

It should also be noted here that the analytic empiricist school was not much concerned with the analysis of theoretical *change*; Popper was a notable exception. The main concern of other members of the group was with such topics as induction, confirmation, probability, explanation, concept formation, and the structure and function of theories. There was no general doctrine as to how far the method of analytic explication might eventually reach—especially whether it would or could cover theory choice.

Let me return now to the pragmatist view of theory choice, as developed especially by Kuhn.

Since there are no precise general criteria of preference that are observed by all scientists, it is clear that actual theory choice in science cannot be *explained* by reference to a commitment of all scientists to such precise norms. Indeed, Kuhn stresses repeatedly that while scientists share a commitment to the desiderata mentioned, they will often understand them and their relative importance in somewhat different ways.

As for the adoption of one of the competing theories, which eventually resolves the conflict in the practice of science, Kuhn emphasizes that it is determined by the group of experts in the field. Indeed, he holds that "the very existence of science depends upon vesting the power to choose between paradigms in the members of a special kind of community," namely, the group of specialists in the field.[15] What is special about the members of this group is the high agreement in their shared standards and values—the values being of the kind of the desiderata we considered earlier. Those values are emphatically seen as not expressible in explicit precise rules determining unique preferabilities among paradigms; and Kuhn must be said, I think, to view theory choice, to a large extent, not as the conclusion of a reasoned application of explicit methodological principles, but as a nonreasoned effect, as it were, of shared attitudes, preferences, and values which the scientists have acquired, in a considerable measure by nonverbal clues, in the course of their specialized professional training.

It may be of interest to recall here that already in 1906, Pierre Duhem expressed a basically similar idea in connection with his famous argument that the outcome of a scientific experiment cannot refuse a theoretical assumption in isolation, but only a comprehensive set of assumptions. If the experimental findings conflict with predictions deducible from the set, then some change has to be made in the total set of assumptions; but no objective logical criteria determine uniquely what change should be made. That decision, says Duhem, must be left to the "good sense" of the scientists; and he adds that the "reasons of good sense do not impose themselves with the same implacable rigor that the prescriptions of logic do. There is something vague and uncertain about them. . . . Hence, the possibility of lengthy quarrels between the adherents of an old system and the partisans of a new doctrine, each camp claiming to have good sense on its side. . . ."[16]

In view of the considerations presented so far, I think there is no justification for charging Kuhn's account of theory choice, as has been done, with irrationalism and an "appeal to mob psychology" (referring to the role of the scientific community in theory choice). The charge of irrationalism would have to be supported by showing that Kuhn's account flaunts certain well-established and recognized standards of rationality; and I am not aware of any rule or standard that could be seriously held to be a binding requirement of scientific rationality that has been neglected or rejected by Kuhn.

6. Kuhn and Dewey on Scientific Rationality: Some Affinities

As noted earlier, Kuhn maintains that his descriptive account of scientific theorizing is also to be read as prescriptive, and that it exhibits science as the best example we have of rationality. Indeed, he gives a concise explicit characterization of the prescriptive import he attributes to his account: "The structure of my argument is simple and, I think, unexceptionable: scientists behave in the following ways; those modes of behavior have (here theory enters) the following essential functions; in the absence of an alternate mode *that would serve similar functions,* scientists should behave essentially as they do if their concern is to improve scientific knowledge."[17]

There seems to me to exist a clear basic affinity between Kuhn's view of scientific rationality as expressed in the quoted passage and the pragmatist views that John Dewey held on the subject.

Dewey characterizes knowledge as "the product of competent inquiries,"[18] and he comments on the characteristics of such inquiries as follows: "it may seem as if the criteria that emerge from the processes of continuous inquiry were only descriptive, and in that sense empirical. That they are empirical in one sense of that ambiguous word is undeniable. They have grown out of the experiences of actual inquiry. But they are not empirical in the sense in which 'empirical' means devoid of rational standing. Through examination of the *relations* which exist between means (methods) employed and conclusions attained as their consequence, reasons are discovered why some methods succeed and other methods fail . . . rationality is an affair of the relation of *means and consequences,* not of fixed first principles as ultimate premises or as contents of what the Neo-scholastics call *criteriology.*"[19] And: "Hence, from this point of view, the descriptive statement of methods that achieve progressively stable beliefs, or warranted assertibility, is also a *rational* statement...."[20]

In his editorial introduction to a volume of articles by and about Dewey, Sidney Morgenbesser offers the following illuminating observations on Dewey's position in this matter:

> Dewey took it to be evident ... that science is good at getting knowledge and also good at presenting us with reasons for changing our beliefs.... That being the case, it is reasonable for philosophers interested in knowledge to study the institution best suited, as far as we know, for getting it.
>
> Dewey seems to be saying that, were we to be asked to study firefighting, we would consider it reasonable to begin with the study of a well-run fire department justly renowned for its efficiency; we would study its history and the ways in which it had solved specific problems in the past.... We would not consider it reasonable to postpone inquiry because we had no clear criteria for fire, or reasonable to begin with a theory of ideal firefighters by reference to which we would judge and assess the work of the department in question.[21]

This passage throws into clear relief the strong similarities between Kuhn's views and those of John Dewey on this issue: both conceive scientific rationality in instrumental terms, as appropriateness for the acquisition of "warranted belief" (Dewey) or "improved scientific knowledge," as Kuhn puts it. Both seek to arrive at a clearer conception of rationality of means of a close empirical study of "competent inquiry" (Dewey), or of scientific research behavior, to use Kuhn's language; both hold that such empirical-descriptive study can yield insight into the way one *ought* to proceed in the rational pursuit of knowledge; and both voice strong skepticism (to put it mildly) concerning the characterization of rationality by means of "fixed first principles" (Dewey) established by more or less *a priori* philosophical analysis.

7. An Aside on Analytic-Empiricist "Explication"

As for the last of these points, I would like to note here, at least in passing, that the efforts of analytic empiricists to "explicate" norms for scientific inquiry, conditions of empirical significance, criteria of demarcation for scientific hypotheses, rules for the introduction of theoretical terms, and the like, were never undertaken in a purely *a priori* manner. Explications were always constructed with an eye on the practices and the needs of empirical science. Thus, for example, the early insistence that "empirically meaningful" sentences must be either verifiable or falsifiable by observational findings was abandoned for reasons that reflected close attention to the nature of scientific claims and procedures. The verifiability criterion, for example, was abandoned because it would deny the status of empirical hypotheses to any law of nature and to all sentences involving mixed quantification, such as "For every substance there is a solvent." The condition of falsifiability was given up for analogous logical reasons, and also for a methodological one that had already been noted by Duhem in his emphasis on the need for auxiliary hypotheses as additional premises in deriving testable consequences from theoretical hypotheses.[22]

Thus, explication in the sense of analytic empiricism has been guided to a considerable extent by close attention to salient features of actual scientific procedures and the logical means required to do justice to them. This process of rational reconstruction, as conceived especially by Carnap and some like-minded thinkers, does, it is true, lead to idealized and schematic models; but these are formulated in consideration of the kinds of scientific systems and procedures whose rationale they are intended to exhibit. In this sense, logical reconstruction, too, has a Janus head with one prescriptive and one descriptive face.

But logical reconstruction has limited itself to aspects of science that could be reflected in the syntactic and semantic features of a formalized model; whereas the pragmatist school introduces further considerations—such as shared values, scientific group processes, and the like—which lie outside the purview of explication as envisaged by analytic empiricism.

8. Rationality versus Adaptiveness in Kuhn's Account of Scientific Theorizing

Near the end of section 1, we mentioned Kuhn's insistence that his account of scientific theorizing should be read both as descriptive and as prescriptive. As I think is shown by the intervening discussion, this claim does not admit of the kind of interpretation suggested in section 2; for Kuhn denies the existence of a set of explicitly stateable general rules or norms which would fully determine rational scientific procedure, and in particular, rational theory choice.

And I have doubts about Kuhn's own construal of the claim, which was cited early in section 6. To indicate my reasons briefly: rationality seems to me intelligibly attributable only to behavior that is causally traceable to reasoning or deliberation about suitable means for attaining specified ends. But as we noted, Kuhn quite plausibly views theory choice as not fully determined by the reasoned application of instrumental methodological principles, but in part at least as a nonreasoned effect, as it were, of shared attitudes, preferences, expectations, and procedural dispositions which the specialists in a field acquire to a large extent in subtle *nonverbal* ways through their professional training and experience. That conditioning has equipped them with a shared flair for making procedural and theoretical judgments in similar but nonidentical ways without benefit of a full corresponding corpus of explicitly verbalized professional goals and methodological norms.

Thus, theory choice is clearly not presented as resulting from the reasoned application of procedural rules which, in light of the available information, are judged to specify optimal means of advancing scientific knowledge. How, then, can the process be called rational?

Kuhn does note that the peculiarities of such group choice may have certain advantages for the advancement of scientific knowledge. For example, there are then no unequivocal rules which would determine at what point in the conflict between two theories every rational scientist should shift his allegiance from one to the other. Thus, there is opportunity for different scientists to change their allegiance at different times. A few scientists will switch to the new rival of an old theory very early, and this is desirable if the potential of the new theory is to be properly explored. Yet if all scientists regularly jumped on the new bandwagon early on, the scientific enterprise would become too unstable and would eventually cease.[23]

But the benefits which thus accrue to the scientific enterprise are not, of course, objectives pursued by the reasoned adoption of a group procedure that has been deliberately chosen as an optimal means to the end of achieving those beneficial effects.

It seems to me therefore that, on Kuhn's account, group processes such as theory choice would have to be viewed as akin to certain other social institutions or behavior patterns which in anthropology and sociology are said to be "latently functional" on the ground that they fulfill certain requirements for the survival or the "success" of the group concerned, without, however, having been adopted by deliberate social choice as a means to that end. Now, such modes of behavior might be called *adaptive*, but surely not *rational*: they are not adopted as a result of goal-directed *reasoning*.

Similarly, certain traits or behavior patterns acquired by a biological group in the course of its evolution may be *adaptive*; but the acquisition of such features surely cannot be qualified as a rational process; and the familiar description of such biological traits as "purposive" is not, of course, meant to imply deliberate planning.

Interestingly, Kuhn does draw an analogy between the evolution of organisms and the evolution of scientific ideas. He describes the resolution of scientific revolutions—including the adoption of a new paradigmatic theory—as "the selection by conflict within the scientific community of *the fittest way* to practice future science.... Successive stages in that developmental process are marked by an increase in articulation and specialization. And the entire process may have occurred, as we now suppose biological evolution did, without benefit of a set goal...."[24] Elsewhere, he remarks further: "For me, therefore, scientific development is, like biological evolution, unidirectional and irreversible. One scientific theory is not as good as another for doing what scientists normally do."[25]

These remarks are suggestive, but they need elaboration and further support. I would not know, for example, how to construe the claim that theory choice as carried out by the scientific community selects the *fittest* way to practice further science, especially in the absence of definite objectives that might yield some explicit criteria of appraisal: the claim seems to me as elusive as the assertion that a certain kind of mimicry is the fittest mode of adaption for a given species.

To repeat: an adaptive process, even if very "successful," cannot, I think, be qualified as rational unless it is causally traceable to motivating reasoning that can be formulated discursively in the form of deliberations aimed at the attainment of specifiable ends.[26]

Otherwise, the process may be adaptive, it may be latently functional, but it cannot be viewed as based on reasoning for which the question of rationality can be significantly raised.

To the extent, then, that scientific research behavior cannot be accounted for as prompted by goal-directed reasoning—and this may be quite a large extent—scientific inquiry would have to be viewed neither as rational nor as irrational, but as arational.

NOTES

1. See Kuhn (1970a), pp. 155ff., 199; (1970b), pp. 261f.; (1977), pp. 321f.
2. Kuhn (1971), p. 144.
3. Kuhn (1970a), pp. 198–200; (1977), pp. 321–325.
4. Kuhn (1971), p. 144; cf. also (1970b), p. 264.
5. Kuhn (1970b), p. 237.
6. Lakatos (1970), p. 178.
7. One reason in support of this assertion is provided by the semantical paradoxes. Another is suggested by considerations of the following kind: Suppose that the center of gravity, C, of some physical body moves through a line segment of length 1 cm during a time interval of 1 sec. This process may then be said to comprise a superdenumerable set of facts, each consisting in the coincidence of C, at one of the superdenumerably many time points in the one-second interval, with a corresponding one among the superdenumerably many points of the one-centimeter line interval. If we consider a language L which, like all scientific languages, contains at most denumerably many primitive symbols and no sentences of infinite length, then, as follows by Cantor's diagonal argument, the set of all sentences expressible in L is only denumerably infinite and thus cannot contain a description of each of the superdenumerably many facts just mentioned. The argument can be extended to other languages.
8. This sketchy formulation glosses over some complicated questions of detail; but for the purposes of the present argument, those questions need not, I think, be entered into.
9. Feyerabend (1970), p. 76.
10. Feyerabend (1970), p. 92.
11. Feyerabend (1970), p. 90.
12. Feyerabend (1970), p. 90.
13. Nagel (1939), chapter 11, sec. 8.
14. See Kuhn's observations on these issues in (1970b), p. 262 and (1977), pp. 321–326.
15. Kuhn (1970a), p. 167; cf. also (1970b), p. 263, par. 1.
16. Duhem (1962), p. 217.
17. Kuhn (1970b), p. 237 (Italics in original).
18. Dewey (1938), p. 8.
19. Dewey (1938), p. 9 (Italics in original).

20. Dewey (1938), p. 10 (Italics in original).
21. Morgenbesser (1977), p. xxv.
22. Duhem (1962), pp. 183–190.
23. Kuhn (1970b), p. 262; (1977), p. 332.
24. Kuhn (1970a), pp. 172–173.
25. Kuhn (1970b), p. 264.
26. Such causal traceability has to be construed in a sufficiently liberal sense to accord rationality to actions that are performed in accordance with certain rules which were originally established by reasoning aimed at selecting suitable means for achieving specific ends, but which the agent has learned to conform to without being fully aware of the underlying rationale.

   This formulation is still sketchy and in need of fuller elaboration; but it is sufficient, I think, to make the point here intended.

REFERENCES

Dewey, John (1938): *Logic: The Theory of Inquiry*. New York: Holt.
Duhem, P. (1962): *The Aim and Structure of Physical Theory*. New York: Atheneum, 1962 (French original first published in 1906).
Feyerabend, P. (1970): "Against Method: Outline of an Anarchistic Theory of Knowledge." In Radner, M. and Winokur, S. (eds.), *Minnesota Studies in the Philosophy of Science*, vol. IV (1970), pp. 17–130.
Kuhn, T. S. (1970a): *The Structure of Scientific Revolutions*. Second edition. Chicago: The University of Chicago Press, 1970.
Kuhn, T. S. (1970b): "Reflections on my Critics." In Lakatos, I. and Musgrave, A. (eds.), *Criticism and the Growth of Knowledge*. Cambridge University Press, 1970, pp. 231–278.
Kuhn, T. S. (1971): "Notes on Lakatos." In Cohen, R. S. and Buck, R. C. (eds.) *PSA 1970, Boston Studies in the Philosophy of Science*, vol. VIII (1971), pp. 137–146.
Kuhn, T. S. (1977): "Objectivity, Value Judgment, and Theory Choice." In Kuhn, T. S., *The Essential Tension. Selected Studies in Scientific Tradition and Change*. Chicago and London: The University of Chicago Press, 1977; pp. 320–339.
Lakatos, I. (1970), "Falsification and the Methodology of Scientific Research Programs," in Lakatos, I. and Mussgrove, A. (eds.), *Criticism and the Growth of Scientific Knowledge* (Cambridge, UK: Cambridge University Press, 1970) pp. 91–195.
Morgenbesser, S. (1977): (ed.) *Dewey and His Critics. Essays from The Journal of Philosophy*. New York: The Journal of Philosophy, Inc.
Nagel, E. (1939): *Principles of the Theory of Probability*. The University of Chicago Press.

20

Valuation and Objectivity in Science

1. Introduction

The role of valuation in scientific research has been widely discussed in the methodological and philosophical literature. The interest in the problem stems to a large extent from the concern that value-dependence would jeopardize the objectivity of science. This concern is clearly reflected, for example, in Max Weber's influential writings on the subject.[1]

In my essay, I propose to consider principally some aspects of the problem which have come into prominence more recently. A discussion of these issues can contribute, I think, to clarifying the cognitive status of the methodology of science and, more generally, of epistemology.

2. Valuation as a Motivating Factor in Scientific Inquiry

The question of value-independence can be and has been raised concerning two quite different aspects of science, namely (1) the actual research behavior of scientists, and (2) the methodological standards for the critical appraisal and possibly the justification of scientific assertions and procedures.

There is no dispute about the important role that valuations of various kinds play in the first of these contexts. Moral norms, prudential considerations, and personal idiosyncrasies clearly can influence a scientist's choice of a field and of problems to explore; they can also affect what methods of investigation are used, what others eschewed. Social and political values can lead to the deployment of strong research efforts in particular problem areas; they can also encourage the advocacy of ill-founded theories. And, of course, the decision of scientific investigators to adopt or to reject a given hypothesis or theory will, as a rule, be strongly influenced by their commitment to what might be called epistemic values or norms, as reflected in their adherence to certain methodological standards of procedure.

In these contexts, valuations are 'involved' in scientific research in the sense of constituting important motivational factors that affect the conduct of inquiry. Such factors must therefore be taken into account in efforts, such as those made in the psychology, the sociology, and the history of science, to *explain* scientific research behavior.

Explanations of this kind are scientific explanations. While they refer to certain values espoused by the scientists in question, they do not themselves posit any value judgments. Rather, they descriptively attribute to the scientific investigators a commitment to certain values and thus the disposition to act in accordance with them. The given research behavior is then explained as a particular manifestation of general preferential dispositions.

To explain why scientists took a certain step, such as adopting or rejecting a given theory, is neither to justify it as sound nor to exhibit it as unsound scientific procedure: speaking broadly and programmatically, the latter task calls for a critical appraisal of the theory in light of the available evidential and other systematic grounds that have a bearing on its acceptability.

Grünbaum expresses basically the same idea when he says that both warranted and unwarranted beliefs have psychological causes, and that the difference between them must be sought in the peculiar character of the causal factors underlying their adoption: *"a warrantedly held belief . . . is one to which a person gave assent in response to awareness of supporting evidence. Assent in the face of awareness of a lack of supporting evidence is irrational, although there are indeed psychological causes in such cases for giving assent."*[2]

Applying this general idea to a topical example, Grünbaum argues in lucid detail that criticisms of various features of psychoanalytic theory cannot be invalidated by contending, as has not infrequently been done, that the critics have a subconsciously motivated resistance to the ideas in question. For, first of all, this explanatory contention presupposes psychoanalytic theory and may therefore be question-begging; and, more importantly, "the invocation of purely psychological, extraevidential explanations for *either* the rejection *or* the acceptance of the theory runs the risk of begging its validity, if only because *either attitude may well be prompted by relevant evidence!*"[3]

3. Normative versus Descriptive-Naturalistic Construals of Methodological Principles

The familiar idea here invoked of critically appraising the warrant or the rationality of scientific claims assumes that there are clear objective criteria governing such appraisals. These criteria are usually thought of as expressible in terms of logical relations of confirmation or of disconfirmation between the claim in question and the available evidence, and possibly also in terms of certain other objective factors, to be mentioned soon.

It is this conception, I think, which has given rise to the question of objectivity and value-neutrality of science in its recent, philosophically intriguing form: to what extent, and, for what reasons, can scientific inquiry and scientific knowledge claims be characterized as subject to such objective methodological standards?

To the extent that such characterization is possible, proper scientific inquiry and its results may be said to be objective in the sense of being independent of idiosyncratic beliefs and attitudes on the part of the scientific investigators. It then is possible to qualify certain procedures—perhaps the deliberate falsification or the free invention of empirical evidence—as 'violations' of scientific canons, and to seek motivational explanations for them in terms of an overriding commitment to extrascientific values, such as personal advancement, which conflict with the objective norms of proper scientific conduct.

In considering the question of objective standards for scientific inquiry, I will for convenience distinguish two extreme positions, to be called *methodological rationalism* and *methodological pragmatism*, or *naturalism*. These are ideal types, as it were. The views held by different thinkers in the field differ from those extremes in various ways, as will be seen later.

According to methodological rationalism, there are certain general norms to which all sound scientific claims have to conform. These are established largely on *a priori* grounds, by logical analysis and reconstruction of the rationale of the scientific search for knowledge. And they are expressible in precise terms, for example, as purely logical characterizations of the relations between scientific hypotheses and evidence sentences that confirm or disconfirm them.

Methodological naturalism, on the other hand, holds that characterizations of proper scientific procedure must be formulated so as to reflect actual scientific practice rather than aprioristic preconceptions we may have about rational ways of establishing knowledge claims. Thomas Kuhn voices this view when he says that "existing theories of rationality are not quite right and . . . we must readjust or change them to explain why science works as it does. To suppose, instead, that we possess criteria of rationality which are independent of our understanding of the essentials of the scientific process is to open the door to cloud-cuckooland."[4]

Earlier, John Dewey had in a similar spirit rejected an aprioristic conception of methodology as "an affair of . . . fixed first principles . . . of what the Neoscholastics call criteriology."[5]

Proponents of a pragmatist approach to methodology usually reject the conception that scientific inquiry is subject to standards that can be expressed in precise and fully objective terms.

I will now consider the two opposing views more closely and will try to show that there are stronger affinities between them than the controversies between their proponents might suggest.

4. Nonnaturalistic Construals of Methodological Norms

The strongest and most influential efforts made in the past 50 years to establish methodological principles for empirical science in a rationalist vein were those of the analytic empiricists and kindred thinkers.

Their analytic endeavors no doubt drew encouragement from a tempting analogy between methodology and metamathematics. The latter discipline, too, does not aim at giving a descriptive account of the mathematical enterprise, but rather at formulating in precise terms certain objective standards for the soundness of mathematical claims and procedures.

Carnap's conception of the philosophy of science as the logical analysis of the language of science, and his and Popper's exclusion of psychological and sociological issues from the domain of epistemology reflect a broadly similar view of the methodology of science.

Metamathematics does not provide precise procedural rules for the solution of all mathematical problems. There is no general algorithm which will automatically lead to the discovery of significant new theorems or which, for any given formula of an axiomatized mathematical theory, will decide whether the formula is a theorem of that theory. But there is an algorithmic procedure which, for any given formula and any proposed proof of it, will decide whether the proof is valid and thus, whether the formula is a theorem of the system.

Similarly, a precise normative methodology of science cannot provide general procedural rules for the discovery of a new theory: such discovery, as has often been emphasized, requires creative scientific imagination; and so does even the discovery of feasible ways of testing a proposed theory. But it might well seem possible to formulate precise objective criteria which, for any proposed hypothesis H and evidence sentence E, determine whether or to what degree E confirms H; or perhaps to state purely comparative criteria determining which of two alternative theories is rationally preferable to the other in consideration of the available evidence E and possibly certain other objective factors.

This was indeed the basic conception underlying analytic-empiricist efforts to develop formal theories of confirmation or of logical probability.[6] Popper's concept of degree of corroboration of a hypothesis[7] reflects a similar formal bent.

The effort to explicate methodological concepts in precise logical terms is evident also in the attempts made by analytically oriented empiricists to characterize genuine empirical hypotheses in terms of verifiability or testability or confirmability, and in Popper's falsifiability criterion for scientific hypotheses. The same objective is illustrated by analytic models of scientific explanation, which impose certain logical conditions on the explanatory sentences and the sentences expressing what is to be explained.

The conditions thus set forth for empirical hypotheses and theories and for scientific explanations were in fact often put to normative-critical use; for example, in declaring the doctrine of neovitalism to be devoid of empirical content and to lack the status of an explanatory theory, or in rejecting the idea of explaining human actions on the basis of empathy or with the help of certain norms of rationality.[8]

The same kind of analytic approach has been used also to formulate methodological norms for scientific concept formation. This is hardly surprising since theory formation and concept formation in science are two faces of the same coin. Theories are formulated in terms of concepts. Concepts are characterized by the theories in which they function. This point is clearly reflected in the stepwise liberalization of the methodology of concept formation developed in the analytic tradition. It led from explicit definition to the introduction of concepts by reduction sentences and on to a holistic method by which an entire system of concepts to be employed in a theory is specified by formulating an axiomatized version of the theory and its intended interpretation.[9] In this process, theory formation and concept formation become inextricably fused.

Again, the methodological principles of concept formation were put to normative use, for example in the rejection as nonempirical or nonscientific of the idea of entelechy or of vital force which plays a central role in neovitalist doctrines.

The ideal, referred to earlier, of the objectivity of science would call for methodological norms which are objective in the sense that they determine unambiguous answers to problems of critical appraisal, so that different scientists applying them will agree in their verdicts. The criteria we have briefly considered are couched largely in the terms of logical theory; this bodes well for their objectivity.

But it must be noted that the criteria also make use of certain nonlogical—more specifically: pragmatic—concepts, namely, those of observation sentence and of observational term. For those criteria characterize the testability, the rational credibility, and cognate characteristics of a hypothesis by certain logical relations between the hypothesis and a body of evidence consisting of so-called observation sentences or basic sentences. These are taken to describe phenomena whose occurrence or nonoccurrence can be established, with good intersubjective agreement, by means of direct observation. Similarly, the specification of scientific concepts by definition or reduction or by interpreted theoretical systems was taken to be ultimately effected with the help of so-called observational terms standing for directly and publicly observable characteristics of things or places or events.

It was just this intersubjective agreement here assumed in the use of observational terms and sentences that was seen as securing the objectivity of science at the evidential level. And methodological norms for the appraisal of scientific claims would then be objective and value-neutral since they called for precisely characterized logical relations between a hypothesis and a body of evidence that could be established with high intersubjective agreement by means of direct observation.

The criticism to which the notion of direct observabiity has been subjected in recent decades has necessitated considerable modifications in the analytic-empiricist construal of the evidential side of a critical appraisal, but this does not necessarily jeopardize the idea of the objectivity of science as characterized in Popper's remark that "the *objectivity* of scientific statements lies in the fact that they can be *inter-subjectively tested*."[10] I will revert to this issue later.

What is the cognitive status of methodological principles of the kind just considered? On what grounds are they propounded, and by what means can their adequacy be appraised? Let us consider first the views of some thinkers close to the rationalist position, especially Popper and Carnap.

5. Naturalistic and Valuational Facets of Popper's Methodology

Karl Popper rejects the 'naturalistic' conception which views methodology as a study of the actual research behavior of scientists, arguing among other things that "what is to be called a 'science' and who is to be called a 'scientist' must always remain a matter of convention or decision."[11] As for his own methodology, which characterizes scientific hypotheses by their falsifiability, and which sees scientific progress in the transition to ever more highly corroborated and ever better testable theories, Popper holds that its principles are *"conventions,"* which "might be described as the rules of the game of empirical science."[12]

In support of his methodology, Popper argues "that it is fruitful: that a great many points can be clarified and explained with its help," and that from the consequences of Popper's characterization of science "the scientist will be able to see how far it conforms to his intuitive idea of the goal of his endeavors."[13] He adds that the consequences of his definition enable us to detect inadequacies in older theories of knowledge. "It is by this method, if by any," Popper says, "that methodological conventions might be justified."[14]

Clearly, then, Popper's methodological conventions are not arbitrary: they are meant to meet certain justificatory requirements. Those I have just mentioned are rather unspecific; they could be applied also to a methodology of mathematics, for example. But Popper has more specific objectives in mind. As he tells us, his interest in methodology was stimulated by the thought that doctrines like astrology, Marxist theory of history, and Freud's and Adler's versions of psychoanalysis were unsatisfactory attempts at theorizing: they were protected against any possible empirical refutation by vagueness of formulation, by the use of face-saving conventionalist stratagems, or by being constructed so as to be totally untestable to begin with. In contrast, the general theory of relativity made far-reaching precise and specific predictions and thus laid itself open to severe testing and possible falsification.[15]

Popper considered these latter features as characteristic of genuine scientific theories and thus sought to construct a methodology that would systematically elaborate this conception, qualifying Einstein's and similar theories as scientific and excluding from the realm of science the unsatisfactory instances mentioned before. Popper's methodology therefore has a target: it is to exhibit the rationale of certain kinds of theories and theorizing which he judges to be scientific. Indeed, Popper notes that if we stipulate, as his methodology does, that science should "aim at *better and better testable* theories, then we arrive at a methodological principle . . . whose [unconscious] adoption in the past would rationally explain a great number of events in the history of science." "At the same time," he adds, the principle "gives us a statement of the task of science, telling us what should in science be regarded as *progress*."[16]

Thus, while Popper attributes to his methodological principles a prescriptive or normative character, he in effect assigns to them an empirical-explanatory facet as well. *This facet appears, not in the content of the norms, but in the justificatory claims adduced for them*; among them the claims that theories preanalytically acknowledged as scientific are qualified as such by the methodology; that others, preanalytically perceived as nonscientific are ruled out; and that important events in the actual history of science could be explained by the assumption that scientists in their professional research are disposed to conform to Popper's methodological norms.

The methodology of science as construed by Popper does therefore have a naturalistic facet in the sense that the justificatory claims made for it include empirical assertions. Indeed, Popper's methodology and similar ones have repeatedly been challenged on the ground that in important contexts scientists have not conformed to the stipulated methodological canons. I will not enter into those criticisms, however; my concern here is simply to note the naturalistic facet of Popper's methodology.

That methodology also has a valuational facet, as Popper, I think, would agree. His choice of methodological principles is prompted by his view that precise testability, large content, and the like, are characteristic features that properly scientific theories should possess. The valuation here involved is not moral or esthetic or prudential; it might rather be called an *epistemological* valuation, which, in the search for improved knowledge, assigns high value to susceptibility to severe tests and possible falsification.

A different epistemological valuation would be reflected in a set of methodological conventions that gives preference to a broad and suggestive but not very precise theory over a narrower, precise, and severely testable one.

Objectivity in the sense of intersubjective agreement among scientists on methodological decisions might well be preserved in spite of the valuational component—namely, to the extent that scientists *share* their epistemic values and the corresponding methodological commitments.

Both empirical information and epistemic valuations, then, are required for a justification or a critical appraisal of a methodology of the kind aimed at by Popper.[17] If his theory were formulated as a set of principles laid down strictly by convention to serve as the rules of a 'game of science' designed by Popper, it would have no methodological or epistemological interest. What lends it such interest is the fact that the rules are meant to afford an—assuredly idealized—account of a specific and very important set of procedures, namely, scientific research. It is by virtue of this claim that it has both the naturalistic and the valuational facets just indicated.

6. Carnap on the Explication of Methodological Concepts

Similar considerations are applicable to Carnap's views on the analytic elaboration of methodological concepts and principles. Carnap has applied the procedure to diverse philosophical issues, among them those concerning the standards for a rational appraisal of the credibility of empirical hypotheses. This is the object of his theory of inductive probability, which Carnap presents as offering a precise characterization of the vague preanalytic concept of the probability of a hypothesis.

Let us briefly consider the character of such precise characterizations and the grounds adduced in their support.

Carnap refers to conceptual clarification and refinement of the kind under discussion as *explication*.[18] He describes it as the replacement of a given, more or less inexact concept, the *explicandum*, by an exact one, the *explicatum*. He notes that the procedure is used also in science; for example, when the vague everyday concepts of hot and cold are replaced by a precise concept of temperature.

Explication plays an important role in analytic philosophy, where it has often been referred to as logical analysis or as rational reconstruction. All the accounts proposed by analytic empiricists for such notions as verification, falsification, confirmation, inductive reasoning, types of explanation, theoretical reduction, and the like are instances of explication, i.e., they propose explicit and precise reconstructions of vague concepts that play an important role in philosophical theories of knowledge.

Carnap lists four requirements which an adequate explication should satisfy:[19]

1. "The explicatum must be *similar to the explicandum*" in the sense that in most cases in which the explicandum has so far been used, the explicatum applies as well; but some considerable deviations are permitted.
2. The explicatum is to be characterized by rules of use which have "an *exact* form."
3. "The explicatum is to be a *fruitful* concept" in the sense of permitting the formulation of an extensive system of laws or theoretical principles.
4. "The explicatum should be as *simple* as possible."

The first of these requirements throws into relief what I called the descriptive facet of philosophical explication. A concern with descriptive fit is evident in the explicatory accounts that analytic empiricists offered of scientific testing, concept formation, explanation, theoretical reduction, and the like: these were formulated and often subsequently modified in consideration of actual scientific procedures. The complex system of explicatory definitions for empirical concepts that Carnap constructed in *The Logical Structure of the World*[20] incorporates a large amount of empirical knowledge, for example, about the structure of the color space. And scientific laws and theories are presupposed also in explications of scientific concepts along the lines of physicalism and logical behaviorism. Again, Carnap adduces certain modes of reasoning used in psychological research as showing that psychological concepts cannot generally be specified by means of reduction chains linking them to an observational vocabulary; he invokes this consideration to motivate a methodological conception which is quite close to holism, namely, that of specifying a system of scientific concepts by means of partially interpreted postulates containing the concepts in question.[21]

In the case of Carnap's explicatory theory of inductive probability, the descriptive facet is more elusive.

By way of a rough characterization of the explicandum concept, $P(H, E)$, of the logical probability of hypothesis H relative to evidence E, Carnap states, among other things, that $P(H, E)$ is to represent the degree to which a person is rationally entitled to believe in H on the basis of E: and that it is also to be a fair betting quotient for a bet on H for someone whose entire evidence concerning H is E.[22]

Carnap's explicatum is formulated in terms of an axiomatized theory in which $P(H, E)$ can then be defined as a quantitative, purely logical relation between the sentences H and E; the axioms ensure that $P(H, E)$ has all the characteristics of a probability function.

The justification Carnap offers for his explication is, briefly, to the effect that the axioms of his theory of rational credibility reflect our intuitive judgments about rational belief, about types of bets it would be irrational to engage in, and the like.[23] As Carnap puts it, "the reasons to be given for accepting any axiom of inductive logic . . . are based upon our intuitive judgments concerning inductive validity, i.e., concerning inductive rationality of practical decisions (e.g., about bets). . . . The reasons are *a priori* [i.e.] independent both of universal synthetic principles about the world . . . and of specific past experiences."[24]

This argument by reference to our intuitive judgments concerning rational decisions seems to me to have a clear affinity to Goodman's view concerning the justification of rules of deductive or of inductive reasoning. Goodman holds that particular inferences are justified by their conformity with general rules of inference and that general rules are justified by their conformity with valid particular inferences. "The point is," he says, "that rules and particular inferences alike are justified by being brought into agreement with each other. *A rule is amended if it yields an inference we are unwilling to accept; an inference is rejected if it violates a rule we are unwilling to amend.*"[25]

Now, when Carnap speaks of 'our' intuitive judgments concerning rationality, and when Goodman refers to particular inferences or to general rules which 'we' are unwilling to accept or to amend—who are 'we'? Surely, those intuitions and unwillingnesses are not meant to be just idiosyncratic; the idea is not: to everyone his own standards of rationality. The assumption must surely be that there is a body of widely *shared* intuitions and unwillingnesses, and that approximate conformity with them provides a justification for acknowledging as sound certain rules of deductive or inductive reasoning. Indeed, without such a body of shared ideas on sound reasoning, there would be no explicandum, and the question of an explicatory theory could not arise.

I think therefore that the grounds Carnap offers in support of his theory of rational inductive inference are not just *a priori*. To be sure, Carnap's formal theory of logical probability may be said to make no descriptive claims, but solely to provide, through its axioms, an "implicit definition," couched in purely logical terms, of its basic concept of logical probability. But the justificatory considerations adduced for accepting the theory are not simply *a priori*; for they make descriptive socio-psychological claims about shared intuitive judgments concerning the explicandum concept. Just as that concept is vague, so, admittedly, are those supporting claims. They do not specify, for one thing, exactly whose intuitions do, or are to, agree with the explication. But if, for example, some intuitive judgments adduced by Carnap were deemed counterintuitive by a large proportion of scientists, mathematical statisticians, and decision-theorists, then surely they could not be invoked for justificatory purposes. Indeed, in a remark on just this point, Carnap acknowledges that scientists do not as a rule explicitly assign numerical degrees of credibility to hypotheses; but, he adds,

> it seems to me that they show, in their behavior, implicit use of these numerical values. For example, a physicist may sometimes bet on the result of a planned experiment; and, more important, his practical decisions with respect to his investment of money and effort in a research project show implicitly certain features . . . of his credibility function . . . . If sufficient data about decisions of this kind made by scientists were known, then it would be possible to determine whether a proposed system of inductive logic is in agreement with these decisions.[26]

This passage comes very close to claiming that a theory of rational credibility should have the potential for providing at least an approximate descriptive and explanatory account of some aspects of the behavior of scientists on the basis of the degrees of rational credibility the theory assigns to scientific hypotheses.

The extent to which an explicatum meets Carnap's first requirement, demanding similarity to the explicans, will be constrained, however, by the three remaining requirements, which demand that the explicatum should function in a precise, comprehensive, and simple theory.[27]

It is particularly in these systematic requirements that the facet of epistemological valuation shows itself. Carnap presents some of them as general conditions of rationality. For example, he stipulates that the degree of rational credibility assigned to a hypothesis should be "dependent, not upon irrational factors like wishful or fearful thinking, but only on the totality of [the believer's] observational knowledge at the time. . . ."[28] This is a fundamental ideal of Carnap's theory, which implies that P must be a function of H and E alone. Another epistemological ideal is expressed in the area that the rational credibility of H on E should be definable exclusively in terms of purely logical attributes of H and E.

By reason of this restrictive requirement alone, Carnap's theory may be said to be adequate only to the critical appraisal of very simple kinds of hypotheses, but not to the complex considerations underlying the experimental testing of hypotheses or theories. The reason lies, briefly, in the Duhem–Quine argument, which has led to a holistic conception of scientific method and knowledge. The point of relevance to Carnap's view is that predictions of experimental findings cannot be deduced from the hypothesis under test alone, but only from the hypothesis taken in combination with an extensive system of other, previously accepted, hypotheses; broadly speaking, what evidence sentences are relevant to the hypothesis is determined by the entire theoretical system accepted at the time. If experimental findings conflict with theoretical predictions, some suitable adjustment has to be made, but not necessarily by rejecting the hypothesis ostensibly under test.

One consideration in choosing a suitable adjustment, it is often noted, is the desire to make a conservative change, one which changes the fundamental assumptions of the entire system as little as possible; another will be the concern to maintain or improve the simplicity and the systematic integration of the entire system. Thus, a rational decision as to whether the given hypothesis or another part of the system, or even the adverse experimental evidence itself, should count as discredited by a conflict between theory and new evidence will depend on considerations concerning the entire theoretical system and not only, as Carnap's requirement stipulates, on the hypothesis and the experimental evidence in question. The same difficulty faces, of course, the much narrower notions of verifiability and falsifiability.

It can be plausibly argued, however, that in various limited contexts of hypothesis-testing in science, such holistic considerations recede into the background, and the hypothesis is judged principally by the pertinent experimental evidence.[29]

Carnap's theory of rational credibility remains important as a carefully articulated explicatory model of the notion of rational credibility or of rational betting quotient for sufficiently simple testing or betting situations.

Whether adequate, more general and fully precise models can be constructed remains an open question. Thinkers favoring a pragmatist approach to the methodology of science have strong doubts on this score.

7. Pragmatist Approaches to Methodology

Let us now take a glance at the pragmatist perspective on methodology, especially in the form it has been given by Thomas Kuhn.

Here, the emphasis on the descriptive facet of methodological claims is dominant and massive. An adequate methodological theory must be informed, on this view, by a close study of the history, sociology, and psychology of actual scientific research behavior. A proper descriptive and explanatory account of this kind can also, it is argued, provide methodological norms or standards of rationality for empirical inquiry.

Kuhn makes this basic assumption concerning the rational pursuit of knowledge: "Scientific behavior, taken as a whole, is the best example we have of rationality. Our view of what it is to be rational depends in significant ways, though of course not exclusively, on what we take to be the essential aspects of scientific behavior." Hence, "if history or any other empirical discipline leads us to believe that the development of science depends essentially on behavior that we have previously thought to be irrational, then we should conclude not that science is irrational, but that our notion of rationality needs adjustment here and there."[30]

As for the normative side of his methodology, Kuhn argues, briefly, as follows: "If I have a theory of how and why science works, it must necessarily have implications for the way in which scientists should behave if their enterprise is to flourish." Now, in the pursuit of their research, scientists behave in ways explored by descriptive and explanatory methodological studies. Those modes of scientific behavior have certain "essential functions," in particular what Kuhn calls the improvement of scientific knowledge. Hence, "in the absence of an alternate mode *that would serve similar functions*, scientists should behave essentially as they do if their concern is to improve scientific knowledge."[31]

It is for this reason that, in response to the question whether his methodological principles are to be read as descriptions or as prescriptions, Kuhn states that "they should be read in both ways at once."[32]

But the assignment of a prescriptive reading to a descriptive account of scientific research is not quite as straightforward as that. It presupposes epistemological idealization and valuation no less than does the formulation of a methodological theory by way of analytic explication. There are at least two reasons for this.

First, Kuhn's basic assumption that science is the best example we have of rationality expresses one broad epistemological valuation, a judgment as to what is to count as exemplary of the rational pursuit of knowledge. This valuation of scientific research as ranking highest on the rationality scale is posited and not further argued. Kuhn seems to suggest just this by his remark that he takes that judgment "not as a matter of fact, but rather of principle."[33]

Secondly, the behavior of scientists in the context of their professional work often shows facets that one would surely not regard as contributing to the improvement of scientific knowledge, but rather as interfering with it. Take, for example, the widespread intensive competition among specialists working in the same problem area and the familiar tendency it engenders to conceal their methods of approach and their unpublished results from one another. I very much doubt that Kuhn would want to see this kind of behavior included in a descriptive methodological account that may properly be given a prescriptive turn. And there are other features of actual scientific research behavior whose suitability for a prescriptive reading would require careful appraisal in light of prior epistemological values or conceptions as to what is "essential," as Kuhn puts it, to scientific progress. Some examples are the widespread practice of 'fudging' the evidence for a hypothesis and the less frequent outright faking of purported experiments and of experimental findings.

Kuhn seems to acknowledge this point when, in a passage quoted earlier in this section, he says that our notion of rationality depends on what we take to be the essential aspects of scientific behavior. The term 'essential' here surely refers to antecedently assumed—perhaps intuitively held—epistemological standards or values.

Thus, the justificatory grounds for methodological theories both of the naturalist and of the analytic-explicatory varieties have a descriptive facet and a facet reflecting epistemological valuation. Neither of the two construals of the methodology of science is purely *a priori*, and neither is purely descriptive.

8. "Desiderata" as Imprecise Constraints on Scientific Theory Choice

Despite the basic affinities we have considered, Kuhn's prescriptive methodology differs significantly from Carnap's or Popper's characterizations.

Analytic explicators aim at formulating precise general criteria for such contexts as the critical testing of hypotheses or the comparative appraisal of competing hypotheses or theories. Carnap based his rules of appraisal on his theory of rational credibility; Popper propounds rules for the game of science that are expressed in terms of precise concepts of falsifiability, corroboration, and the like.

Kuhn's pragmatist account of scientific research behavior, on the other hand, does not admit of a prescriptive reading in the form of a system of precise methodological rules.

This is clearly shown by Kuhn's characterization of the ways in which scientists appraise competing theories and eventually make a choice between them. Kuhn discusses this subject particularly for choices required in the context of a scientific revolution, when a paradigmatic theory that has long dominated research in its field is encountering mounting difficulties and is opposed by a new rival theory that has overcome some of those difficulties. Kuhn's ideas on the resolution of such conflicts are well known, and I will mention here only a few points that have an immediate bearing on the character of the prescriptive principles that might be gleaned from Kuhn's descriptive account.

Kuhn argues that the choice between competing theories is a matter left to the specialists in the field, whose appraisals of the merits of those theories are strongly influenced by certain shared preferences or values which have been molded in the course of their scientific training and their professional experiences. In particular, scientists widely agree in giving preference to theories exhibiting certain characteristics which have often been referred to in the methodological literature as "marks of a good hypothesis"; I will call them *desiderata* for short. Among them are the following: a theory should yield precise, preferably quantitative, predictions; it should be accurate in the sense that testable consequences derivable from it should be in good agreement with the results of experimental tests; it should be consistent both internally and with currently accepted theories in neighboring fields; it should have broad scope; it should predict phenomena that are novel in the sense of not having been known or taken into account when the theory was formulated; it should be simple; it should be fruitful.[34]

Kuhn reasons that while scientists are in general agreement concerning the importance of these features and attach great weight to them in deliberating about theory choice, the desiderata cannot be expressed in the form of precise rules of comparative evaluation which unambiguously single out one of two competing theories as the rationally preferable one. He arrives at this conclusion by arguing (A) that the individual desiderata are too vague to permit of explication in terms of precise criteria of accuracy, simplicity, scope, fruitfulness, etc., and (B) that even if precise criteria could be formulated for a comparison of two theories in regard to each of the desiderata, one of two competing theories might be superior to the other in regard to some of the desiderata, but inferior in regard to others: to permit overall comparison of the theories with regard to the totality of the desiderata, a further rule would therefore have to be constructed which, in effect, would assign different weights or priorities to the different desiderata. And again, Kuhn argues that it is not possible to formulate a precise and unambiguous rule of that kind which does sufficient justice to theory choice as actually practiced in science.

A few words of amplification especially concerning (A). That the characterizations of the desiderata are vague is obvious. Some of them are also ambiguous. Kuhn notes, for example, that the requirement of simplicity favored the Copernican over the Ptolemaic theory if simplicity were judged by gross qualitative features, such as the number of circles required; but the two theories were substantially on a par if simplicity were judged by the ease of the computations required to predict the position of a planet.[35]

Or consider the desideratum of accuracy, which requires that "consequences deducible from a theory should be in demonstrated agreement with the results of existing experiments and observations."[36] Clearly, empirical data, which a theory does not fit in this sense, should not count against the theory if the experiments yielding them were affected by factors not taken into account in the deduction, such as faults in the equipment or interference by disturbing outside factors. But a judgment as to whether such interference may have been present will depend on current theories as to what kinds of factors can affect the outcome of the given experiment. Thus, the desideratum of accuracy will have to be understood as requiring agreement between theoretically predicted findings and experimental data which, as judged by currently available information, are not vitiated by disturbing factors. This consideration leads to an extended form of holism: when a hypothesis conificts with experimental evidence, then, as Duhem has pointed out, the conflict may be eliminated either by abandoning the hypothesis ostensibly under test, or by making changes elsewhere in the system of hypotheses accepted at the time; but, as just noted, there is also the possibility of rejecting the recalcitrant new evidence. It is not always the case that scientific theories are made to fit the observational or experimental data: often it is a well-established theory which determines whether given test findings can count as acceptable data.

This point is illustrated by the practice of eminent scientists. For example, the famous oil drop experiments by which Robert A. Millikan measured the charge of the electron yielded a number of instances in which the observed motion of the drops did not agree with Millikan's claim. He attributed the deviations to various possible disturbing factors and was in fact able, by screening some of these out, to obtain more uniform results. Yet, as Holton has shown by reference to Millikan's laboratory notes, there were observed cases of considerable deviation from the theoretically expected results, which Millikan did not include in his published data, assuming that something had gone wrong.[37]

The possibly disturbing factors mentioned by Millikan included fading of the battery that charged the condenser plates between which the oil drops were moving, partial evaporation and mass loss of an oil drop under observation, the observer's mistaking of dust particles for tiny oil drops, and several other possible occurrences. Broadly speaking, such sources of error could be checked and controlled by relevant knowledge that is quite independent of the hypothesis under test, which concerned the charge of the electron.

It might therefore seem reasonable to require that the attribution of adverse experimental findings to "disturbing factors" should never be based on the hypothesis under test, since otherwise any adverse evidence could be rejected simply on the ground that it conflicted with that hypothesis. Yet, this maxim is not generally adhered to in science. Thus, in certain deviant cases Millikan suggests, and offers some supporting reasons for, specific assumptions concerning the disturbing factors;[38] but there are other such cases recorded in his notebooks in which he simply comments "something wrong," "something the matter," "agreement poor," or the like.[39] And one might well argue that there was a good reason: in a large proportion of cases, the agreement between theoretically expectable and experimentally determined values was impressively close and gave grounds for the assumption that the hypothesis did exhibit a basic trait of nature.

These brief remarks were simply meant to illustrate that while in many cases, a judgment concerning the 'accuracy' of a theory may not pose great problems, it would be quite difficult to formulate a precise general criterion of accuracy of fit, which would take due account of the holistic character of scientific knowledge claims.[40]

The preceding considerations also have a bearing on Grünbaum's idea, mentioned earlier, that the warrant for a scientific claim lies in the evidence supporting it. A more detailed elaboration of this remark would call for consideration also of factors that are not evidence in the usual narrower sense; among them, I would think, features like the desiderata. Such a broad construal of the warrant of a hypothesis, however, leaves Grünbaum's point quite unaffected: a critic's hypothesized psychological resistance to some psychoanalytic doctrine is not a factor that has systematic relevance for the question whether the critic's objections are pertinent and well substantiated: indeed, as Grünbaum specifically notes in the passage quoted earlier, the critic's resistance might spring from an awareness of shortcomings in the systematic support that has been offered for the claims he is questioning.

As for the comments (A) and (B) outlined above concerning imprecision of the desiderata for good scientific theories, I think it of interest to note that quite similar views were expressed earlier by J. von Kries and by Ernest Nagel, and that Carnap agreed with them to some extent.[41]

Carnap (1950) expresses this view: "Inductive logic alone . . . cannot determine the best hypothesis on a given evidence, if the best hypothesis means that which good scientists would prefer. This preference is determined by factors of many different kinds, among them logical, methodological, and purely subjective factors" (p. 221). And he adds: "However, the task of inductive logic is not to represent all these factors, but only the logical ones; the methodological (practical, technological) and other nonlogical factors lie outside its scope" (p. 219). He then examines two among the factors mentioned by von Kries, which Carnap regards as purely logical, namely, the extension and the variety of the confirming evidence for a hypothesis, and he sketches ways in which they might be given exact quantitative definitions; but he acknowledges that great difficulties remain for an attempt to include them into one precise quantitative concept of degree of confirmation (pp. 226ff.).

Against the idea that "in the logical analysis of science we should not make abstractions but deal with the actual procedures, observations, statements, etc., made by scientists," Carnap acknowledges that a pragmatic study of methodology is highly desirable (p. 217), but he warns that for the achievement of powerful results concerning sound decision-making, "the method which uses abstract schemata is the most efficient one" (p. 218).

Then he adds with characteristic candor that those who prefer to use powerful abstract methods are subject to "the ever present temptation to overschematize and oversimplify . . . ; the result may be a theory which is wonderful to look at in its exactness, symmetry, and formal elegance, yet woefully inadequate for the tasks of application for which it is intended" (p. 218). (This is a warning directed at the author of this book by his critical superego.)

As we briefly noted, Carnap held that the question as to which of several hypotheses would be preferred by scientists on given evidence depended not only on logical characteristics of hypotheses and evidence, but also on methodological and on "purely subjective" factors. I do not know whether he thought it possible to offer precise explications of the relevant methodological considerations (which would require extralogical concepts as well as logical ones), nor whether he would have considered a search for a general normative-explicatory account of comparative preferability of hypotheses as a promising project.

9. Valuation, Vagueness, and the Objectivity of Science

Desiderata of the kind we have considered have the character of epistemological norms or values. They do not enter into the content of scientific theories, but they serve as standards of acceptability or preferability for such theories; thus they function in the critical appraisal or in the justification of scientific claims. It is not to be wondered at that standards of evaluation are needed in this context: the problem of justifying theoretical claims can be intelligibly raised only to the extent that it is clear what objectives are to be achieved by accepting, or by according preference to, a theory.

Science is widely conceived as seeking to formulate an increasingly comprehensive, systematically organized, worldview that is explanatory and predictive. It seems to me that the desiderata may best be viewed as attempts to articulate this conception somewhat more fully and explicitly. And if the goals of pure scientific research are indicated by the desiderata, then it is obviously rational, in choosing between two competing theories, to opt for the one which satisfies the desiderata better than its competitor.

The problem of formulating norms for the critical appraisal of theories may be regarded as a modern outgrowth of the classical problem of induction: it concerns the choice between competing comprehension theories rather than the adoption of simple generalizations and the grounds on which such adoption might be justified. And—disregarding for a moment the vagueness of the desiderata—the considerations sketched in the preceding paragraph might be viewed as *justifying* in a near-trivial way the choosing of theories in conformity with whatever constraints are imposed by the desiderata.

Note, however, that this kind of justification does not address at all what would be the central concern of the classical problem of induction, namely, the question whether there are any reasons to expect that a theory which, as judged by the desiderata, is preferable to its competitor, at a given time will continue to prove superior when faced with further, hitherto unexamined, occurrences in its domain.[42]

Since, at least so far, the desiderata can be formulated only vaguely, they do not unequivocally determine a choice between two theories. In particular, they do not yield an algorithm which, in analogy to mathematical algorithms, effectively determines a unique solution for every problem within its domain. Indeed, the theory choices made by individual scientists committed to the desiderata are influenced also by factors that may differ from person to person, among them the scientists' individual construals of the desiderata as well as certain other factors which lie outside science and which may be more or less idiosyncratic and subjective.[43]

Does this jeopardize the objectivity of science? To be sure, idiosyncratic factors of the kinds just mentioned, as well as a variety of physical and sociocultural conditions, can affect individual choice behavior; they may all be relevantly invoked in explanatory accounts of decisions arrived at by particular investigators.

But that is equally true in cases where scientists seek to solve problems for which correct solutions can be characterized by precise, and perhaps even effective, criteria. A scientist working on a computational problem for which alternative methods of algorithmic solutions are available, may have a preference for, and may therefore employ, a particular one of these; a mathematical purist may after much effort produce an ingenious proof of a theorem in number theory which, unlike all previously available proofs, avoids any recourse to real or complex analysis. Consideration of such idiosyncratic factors is essential in explaining the mathematician's procedure; but it is irrelevant for a critical appraisal of the correctness of the computation or the validity of the proof. The criteria appropriate for the latter purpose are objective in the sense of making no reference whatever to individual preferences or values or to external circumstances.

In examining the objectivity of scientific inquiry, we will similarly have to ask whether, even in consideration of the new perspectives provided by pragmatist studies, scientific procedures including theory choice can still be characterized by standards that do not depend essentially on purely idiosyncratic individual factors.

I now think that plausible reasons can be offered in support of an objectivist but 'relaxed' rational reconstruction according to which proper scientific procedures are governed by methodological norms some of which are explicit and precise, while others—including very important ones—are vague. The requirements of deductive closure and of logical consistency for acceptable theories would be of the former kind; many other desiderata governing theory choice, of the latter.

A construal of this kind cannot, of course, claim to be a descriptive account of the practices actually observed by practitioners of an important sociocultural pursuit broadly referred to as scientific research; the construal presupposes, as we saw other reconstructions must do, certain prior determinations, having the character of epistemic valuations, as to what peculiar features of that social enterprise are to count as characteristic of 'proper' science, as traits that make science scientific.

Let me briefly suggest some considerations that seem to me to favor a relaxed but objectivist construal of methodological principles.

To begin with—and this has to do with indicating important features of the explicandum—science is generally conceived as an objectivist enterprise where claims are subject to a critical appraisal in terms of standards that are not simply subjective and idiosyncratic; it is surely not regarded as a field in which 'everybody is entitled to his own opinion.'

Severe constraints are imposed by certain quite generally acknowledged norms. Among these are the demand for conformity with the standards of deductive logic, and the prohibition of logical inconsistencies: even though there is no general algorithmic test procedure for consistency, there is insistence on avoiding or in some way quarantining inconsistencies that may have been discovered. There are clear norms also for various methods of measurement and of testing statistical hypotheses.

And while norms like those represented by the desiderata are very much less explicit and precise, they surely do not license considerations that are idiosyncratic to some individual scientists as justificatory for theory choice. The various desiderata can be said, it seems to me, to have an objectivist intent and to be amenable to discussion and possible further clarification.[44]

Empirical science, too, sometimes employs concepts which are characterized only vaguely, but whose application is not for that reason entirely arbitrary or a matter of purely subjective choice. Take, for example, the social status scale proposed by Chapin for rating American homes in regard to their socioeconomic status.[45]

The total rating is based on appraisals of several component factors, each taken as a partial indicator of socioeconomic status. Some of the factors are characterized quite precisely by criteria referring to the presence or absence in the living room of specified items such as hardwood floors, radios, etc., and their state of repair. On these, trained investigators will readily come to a good agreement in their ratings for a given home. Other component factors, however, are characterized much more vaguely; among them one that requires the investigator to express on a numerical rating scale his "general impression of good taste" shown in the living room. Here, the appraisal calls for ratings such as "Bizarre, clashing, inharmonious, or offensive (–4)" or "Attractive in a positive way, harmonious, quiet, and restful (+2)". On points of this kind, the 'reliability' of appraisals, as measured by the correlation between the ratings of different investigators, was expectably found to be lower than for items of the former kind. But it is reasonable to expect that by further training of the appraisers the reliability of their judgments could be enhanced. And the relevant training might well be effected with the help of paradigmatic examples rather than by means of fully explicit and precise criteria.

In a similar vein, Kuhn has stressed that, to a large extent, it is not by being taught rigorous definitions and rules that scientists learn how to apply their technical concepts and the formal apparatus of their theories to concrete problems in their field: they acquire that ability in considerable measure by being exposed to characteristic examples and by picking up other relevant clues in the course of their academic training and their professional experience.[46]

A relaxed objectivist construal of theory choice of the kind here adumbrated evidently falls far short of Carnap's rigorous conception of an explication (cf. section 6 above). In particular, because of the vagueness of the desiderata, a relaxed account violates Carnap's requirement that the explicatum "be characterized by rules which have an *exact* form," and it does not satisfy well Carnap's condition that the explicatum be fruitful in the sense of permitting the construction of a precise and comprehensive theory. In these respects, a relaxed explication of rational theory choice bears no comparison to, say, Carnap's inductive logic as a precise explication of rational belief or of rational betting behavior. But it should be borne in mind that the virtues of that explication were achieved at the cost of strong idealization and simplification.

In some earlier essays,[47] I expressed the view that a methodological characterization of scientific theory choice as being essentially dependent on factors having the character of the desiderata does not warrant Kuhn's conception of scientific inquiry as being a rational procedure. I argued that a rational procedure consists in the pursuit of a specified goal in accordance with definite rules deliberately adopted on the ground that, in light of the available information, they offer the best prospects of attaining the goal. I concluded that insofar as scientific procedures are constrained only by considerations of the kind of the desiderata, they should be viewed as arational (though not as irrational), and I argued further that perhaps they might be qualified as latently functional practices in the sense of functionalist theories in the social sciences.[48]

That verdict would not apply, of course, to the many scientific procedures for which reasonably explicit and precise methodological standards can be formulated; some of these were mentioned earlier.

But in view of the considerations here outlined, it seems to me now that the characterization as arational or as latently functional does not do justice even to the broad process of theory choice; for it does not take sufficient account of the considerable role that precise and rule-governed reasoning does play in the critical appraisal of competing theories, which requires among other things a rigorous derivation of experimental implications and the performance of experimental tests that have to meet appropriate standards. But even the considerations adduced in appraising the satisfaction of the vaguer desiderata are typically perceived, it seems to me, as expressing not just individual taste or preference, but objective, if only roughly specified, traits of the competing theories. Thus, a less rigid construal of rationality may be indicated; the relaxed explication here adumbrated might be an attempt in this direction.

This conception leaves open the possibility that the methodology of science may gradually be formulated in terms of standards that are more explicit and precise than the desiderata we have considered. It also leaves room for the idea that the desiderata, which were taken here as reflecting the goal of scientific research or the idea of scientific progress, should be viewed, not as fixed once and for all, but as themselves subject to change in the course of the evolution of science.[49]

But I think it clear even at the present stage that scientific inquiry can be characterized by methodological principles which, while reflecting epistemological values, impose on scientific procedures and claims certain constraints of an objectively oriented, though partially vague, kind that preclude a view of science as an enterprise in which 'anything goes'.

NOTES

1. See, for example, Weber's essays, "The Meaning of 'Ethical Neutrality' in Sociology and Economics" and "'Objectivity' in Social Science and Social Policy" in Shils and Finch (1949). For a recent critical discussion of Weber's ideas, see Stegmüller (1979).
2. Grünbaum (1972), p. 61; italics cited.

3. Grünbaum (1980), p. 81; italics cited.
4. Kuhn (1970b), p. 264.
5. Dewey (1938), p. 9; italics cited.
6. See, for example, the articles Hempel (1943) and (1945), which offer a logical analysis of the qualitative concept "E confirms H" for formalized languages of a simple kind; and Carnap's writings (1950; 1952; 1971a), which develop a comprehensive logical theory of the quantitative concept of degree of confirmation, or inductive probability.
7. See, for example, Popper (1959), especially Appendix *ix; (1962), pp. 57–58; (1979), passim.
8. See, for example, Hempel (1965), pp. 257–258 and pp. 469–472.
9. Cf. Carnap (1956); Hempel (1973).
10. Popper (1959), p. 44; italics cited.
11. Ibid., p. 52.
12. Ibid., p. 53; italics cited.
13. Ibid., p. 55.
14. Ibid.
15. See, for example, Popper (1962), pp. 33–39.
16. Popper (1979), p. 356; italics cited.
17. I have here been concerned only with some general characteristics of Popper's methodology and have not considered his ideas of corroboration and of verisimilitude and other special issues. For a provocative discussion of the cognitive status of Popper's methodology, cf. Lakatos (1974) and Popper's reply (1974).
18. See Carnap (1950), chapter I.
19. Ibid. p. 7; italics cited.
20. Carnap (1928).
21. Carnap (1956), pp. 72–73.
22. See, for example, Carnap (1963a), p. 967.
23. Carnap (1963a); (1971a), pp. 13–16.
24. Carnap (1963a), pp. 978–979. The precise formulation of the relevant intuitive judgments and of their justificatory role is a rather subtle technical problem whose solution is based on results established by de Finetti, Kemeny, Shimony, and others. For details, see Kemeny (1963).
25. Goodman (1955), p. 67; italics cited.
26. Carnap (1963b), p. 990. See also Carnap's interesting remarks on p. 994 on differences and changes in intuitions concerning rational credibility.
27. In these respects, the explication of rational credibility by Carnap's precise theory of inductive probability is quite analogous to the explication of the concept of a correctly formed English sentence by a theory of English grammar. Such an explication must surely take account of the linguistic intuitions or dispositions of native speakers; but descriptive faithfulness has to be adjusted to the further objectives of constructing a grammar which is reasonably precise, simple, and general. For example, the idea that the conjunction of two well-formed sentences is again a well-formed sentence, and the idea that there must be an upper bound to the length of any well-formed sentence both have a certain intuitive appeal; but they are logically incompatible. Here, systematic-theoretical considerations will then decide which, if either, of the two is to be retained.
28. Carnap (1963a), p. 970; words in brackets supplied.

29. In this context, see Glymour's lucid and thought-provoking book (1980); it offers critical arguments against holism construed in the very comprehensive sense here adumbrated, and it propounds in detail a more restrictive account of the ways in which a hypothesis may be confirmed by empirical evidence.
30. Kuhn (1971), p. 144.
31. Kuhn (1970b), p. 237; italics cited.
32. Ibid.
33. Kuhn (1971), p. 144.
34. Kuhn (1977), pp. 321–322; also (1970a), pp. 205–206; (1970b), pp. 245–246. Van Fraassen (1980) discusses desirable characteristics of this kind under the heading of "virtues" of theories; see, for example, pp. 87–89.
35. Kuhn (1977), p. 324.
36. Ibid., p. 321. The entire essay presents the ideas here referred to very suggestively. For other passages concerning those issues, see Kuhn (1970b), pp. 241, 245–246, 261–262; (1970a), pp. 199–200.
37. See Holton (1978). That essay deals with the fascinating controversy between Millikan and the physicist Felix Ebrenhaft. The latter had, in similar experiments, found a large number of cases that did not agree with Millikan's hypothesis and therefore rejected the latter. Millikan himself discusses the issue in his book (1917), especially chapter VIII.
38. See Millikan (1917), pp. 165–172; also Holton (1978), p. 69.
39. Holton (1978), pp. 70–71.
40. Kuhn notes this point, for example, in (1977), p. 338.
41. Kries (1886), pp. 26, 291ff.; Nagel (1939), pp. 68–71; Carnap (1950), pp. 219–233.
42. Kuhn (1977); pp. 332–333 makes some remarks in a similar vein. It may be of interest to recall here that Reichenbach [cf. (1938), section 43] proposed a principle of simplicity not just as a criterion for the *appraisal* of a given hypothesis, but as a rule for the inductive *discovery* of laws. Briefly, he argued that in the search for a law connecting several quantitative variables (such as temperature, pressure, and volume of a gas), the method of always adopting the *simplest* hypothesis fitting the experimental data available at the time would, as the body of data grew, lead to a sequence of quantitative hypotheses which would mathematically converge on the law that actually connected the variables in question—provided there was such a law at all. In this justificatory argument for inductive reasoning to the simplest hypothesis, Reichenbach uses a rather precise characterization of the simplest curve, or surface, etc., through given data points. His method does not, however, take full account of the subtle and elusive questions concerning the simplicity of theories.
43. Some such factors are briefly discussed in Kuhn (1977), p. 325; Laudan (1977) devotes a great deal of attention to considerations of this kind.
44. It is broadly in this sense, I think, that Kuhn has recently characterized the application of the desiderata as a matter "of judgment, not taste" (1977, p. 337).
45. Chapin (1935), chapter XIX.
46. See, for example, Kuhn (1974).
47. Hempel (1979a; 1979b).
48. Hempel (1979b), section 8.
49. Detailed arguments in favor of this conception have been offered by Laudan (1977); Kuhn (1977), too, allows for some changes of this kind; my discussion in (1979a), section 6 deals briefly with this issue.

REFERENCES

Carnap, Rudolf. 1928. *Der Logische Aufbau der Welt*. Berlin-Schlachtensee: Weltkreis-Verlag. (English edition 1967, *The Logical Structure of the World and Pseudoproblems in Philosophy*. Berkeley: University of California Press.)

Carnap, Rudolf. 1950. *Logical Foundations of Probability*. Chicago: The University of Chicago Press.

Carnap, Rudolf. 1952. *The Continuum of Inductive Methods*. Chicago: The University of Chicago Press.

Carnap, Rudolf. 1956. "The Methodological Character of Theoretical Concepts." In Feigl, H. and M. Scriven (eds.), *Minnesota Studies in the Philosophy of Science*, vol. 1, pp. 38–76. Minneapolis: University of Minnesota Press.

Carnap, Rudolf. 1963a. "My Basic Conceptions of Probability and induction." In Schilpp (1963), pp. 966–979.

Carnap, Rudolf. 1963b. "Ernest Nagel on Induction." In Schilpp (1963), pp. 989–995.

Carnap, Rudolf. 1971a. "Inductive Logic and Rational Decisions." In Carnap and Jeffrey (1971), pp. 5–31.

Carnap, Rudolf. 1971b. "A Basic System of Inductive Logic, Part I." In Carnap and Jeffrey (1971), pp. 35–165.

Carnap, Rudolf and Richard C. Jeffrey (eds.). 1971. *Studies in Inductive Logic and Probability*. Berkeley: University of California Press.

Chapin, F. Stuart. 1935. *Contemporary American Institutions*. New York and London: Harper and Brothers.

Dewey, John. 1938. *Logic: The Theory of Inquiry*. New York: Holt.

Fraassen, Bas C. van. 1980. *The Scientific Image*. Oxford: Clarendon Press.

Glymour, Clark N. 1980. *Theory and Evidence*. Princeton, N.J.: Princeton University Press.

Goodman, N. 1955. *Fact, Fiction and Forecast*. Cambridge, Mass.: Harvard University Press.

Grünbaum, Adolf. 1972. "Free Will and Laws of Human Behavior." In Feigl, H., K. Lehrer, and W. Sellars (eds.), *New Readings in Philosophical Analysis*, pp. 605–627. New York: Appleton-Century-Crofts.

Grünbaum, Adolf. 1980. "The Role of Psychological Explanations of the Rejection or Acceptance of Scientific Theories," *Transactions of the New York Academy of Sciences*, Series II, 19, 75–90.

Hempel, Carl G. 1943. "A Purely Syntactical Definition of Confirmation," *Journal of Symbolic Logic* 8, 122–143.

Hempel, Carl G. 1945. "Studies in the Logic of Confirmation," *Mind* 54, 1–26, 97–121.

Hempel, Carl G. 1965. *Aspects of Scientific Explanation*. New York: The Free Press.

Hempel, Carl G. 1973. "The Meaning of Theoretical Terms: A Critique of the Standard Empiricist Construal." In Suppes, P. et al. (eds.), *Logic, Methodology and Philosophy of Science* IV, pp. 367–378. Amsterdam: North-Holland Publishing Company.

Hempel, Carl G. 1979a. "Scientific Rationality: Analytic vs. Pragmatic Perspectives." In Geraets, Th. F. (ed.), *Rationality Today/La rationalité aujourd'hui*, pp. 46–58. Ottawa: The University of Ottawa Press.

Hempel, Carl G. 1979b. "Scientific Rationality: Normative vs. Descriptive Construals." In Berghel, H., et al. (eds.), *Wittgenstein, the Vienna Circle, and Critical Rationalism*, pp. 291–301. Vienna: Hoelder-Pichler-Tempsky.

Holton, Gerald. 1978. "Subelectrons, Presuppositions, and the Millikan-Ehrenhaft Dispute." In Holton, G., *The Scientific Imagination: Case Studies*, pp. 25–83. Cambridge: Cambridge University Press.

Kemeny, John G. 1963. "Carnap's Theory of Probability and Induction." In Schilpp (1963), pp. 711–738.

Kries, Johannes von. 1886. *Die Principien der Wahrscheinlichkeitsrechnung*. Freiburg: Akademische Verlagsbuchhandlung; 2nd ed. Tübingen: Mohr, 1927.

Kuhn, Thomas S. 1970a. *The Structure of Scientific Revolutions*. 2nd ed. Chicago: The University of Chicago Press.

Kuhn, Thomas S. 1970b. "Reflections on My Critics." In Lakatos, I. and A. Musgrave (eds.), *Criticism and the Growth of Knowledge*, pp. 231–278. Cambridge: Cambridge University Press.

Kuhn, Thomas S. 1971. "Notes on Lakatos." In Roger C. Buck and Robert S. Cohen (eds.), *PSA 1970, Proceedings of the 1970 Biennial Meeting, Philosophy of Science Association*, pp. 137–146. *Boston Studies in the Philosophy of Science*, vol. 8. Dordrecht: D. Reidel Publishing Company.

Kuhn, Thomas S. 1974. "Second Thoughts on Paradigms." In Suppe, F. (ed.), *The Structure of Scientific Theories*, pp. 459–482. Urbana: University of Illinois Press.

Kuhn, Thomas S. 1977. "Objectivity, Value Judgment, and Theory Choice." In Kuhn, T. S., *The Essential Tension*, pp. 320–339. Chicago: The University of Chicago Press.

Lakatos, Imre. 1974. "Popper on Demarcation and Inductive." In Schilpp (1974), pp. 241–273.

Laudan, Larry. 1977. *Progress and Its Problems*. Berkeley: University of California Press.

Millikan, Robert Andrews. 1917. *The Electron*. Chicago: The University of Chicago Press. Facsimile edition: Chicago: The University of Chicago Press, 1963.

Nagel, Ernest. 1939. *Principles of the Theory of Probability*. Chicago: The University of Chicago Press.

Popper, Karl R. 1959. *The Logic of Scientific Discovery*. London: Hutchinson.

Popper, Karl R. 1962. *Conjectures and Refutations*. New York: Basic Books.

Popper, Karl R. 1974. "Lakatos on the Equal Status of Newton's and Freud's Theories." In Schilpp (1974), pp. 999–1013.

Popper, Karl R. 1979. *Objective Knowledge*. Rev. ed. Oxford: Clarendon Press.

Quine, W. V. 1969. *Ontological Relativity and Other Essays*. New York and London: Columbia University Press.

Reichenbach, Hans. 1938. *Experience and Prediction*. Chicago: The University of Chicago Press.

Schilpp, Paul A. (ed.), 1963. *The Philosophy of Rudolf Carnap*. La Salle, Illinois: Open Court.

Schilpp, Paul A. (ed.). 1974. *The Philosophy of Karl Popper*. 2 vols. La Salle, Illinois: Open Court.

Shils, Edward A. and Henry A. Finch, translators and editors. 1949. *Max Weber on the Methodology of the Social Sciences*. Glencoe, Illinois: The Free Press.

Stegmüller, Wolfgang. 1979. "Weltfreiheit, Interessen und Objectivität." In W. Stegmüller, *Rationale Rekonstruktion von Wissenschaft und ihrem Wandel*, pp. 175–203. Stuttgart: P. Reclam.

A Bibliography of Carl G. Hempel

[*Editor's note*: Special thanks to Richard Jeffrey for providing the editor with a bibliography of Hempel's work, which has been slightly revised for publication here. Those pieces reprinted in *Aspects*, the Jeffrey collection, and this volume have been identified below to facilitate cross-reference.]

1. *Beitraege zur logischen Analyse des Wahrscheinlichkeitsbegriffs*. Ph.D. thesis, 1934. 72 pp. University of Berlin, June 1934. [Subsequently translated into English by the author as "contributions to the logical analysis of the concept of probability."]
2. "On the Logical Positivists' Theory of Truth," *Analysis* 2 (1935), pp. 49–59. [Reprinted in Jeffrey.]
3. "Analyse logique de la psychologie," *Revue de Synthese* 10 (1935), pp. 27–42. (English translation, under the title "The Logical Analysis of Psychology," in H. Feigl and W. Sellars, eds., *Readings in Philosophical Analysis* (New York: Appleton-Century-Crofts, Inc., 1949), pp. 373–384). [Reprinted in translation in Jeffrey.]
4. "Zur Frage der wissenschaftlichen Weltperspektive," *Erkenntnis* 5 (1935/36), pp. 162–164.
5. "Ueber den Gehalt von Wahrscheinlichtkeitsaussagen", *Erkenntnis* 5 (1935/36), pp. 228–260. [Reprinted in translation under the title, "On the Content of Probability Statements," in Jeffrey.]
6. "Some Remarks on 'Facts' and Propositions," *Analysis* 2 (1935), pp. 93–96. [Reprinted in Jeffrey.]
7. "Some Remarks on Empiricism," *Analysis* 3 (1936), pp. 33–40. [Reprinted in Jeffrey.]
8. (with P. Oppenheim) *Der Typusbegriff im Lichte der Neuen Logik* (Leiden: A. W. Sijthoff, 1936). (Wissenschaftstheoretische Untersuchungen zur Konstitutionsforschung und Psychologie)
9. (with P. Oppenheim) "L'importance logique de la notion de type," *Actes du Congrés International de Philosophie Scientifique*, Paris, 1935, Vol. II (Paris: Hermann et Cie, 1936), pp. 41–49.
10. "Eine rein topologische Form nichtaristotelischer Logik," *Erkenntnis* 6 (1937), pp. 436–442.
11. "A Purely Topological Form of Non-Aristotelian Logic," *The Journal of Symbolic Logic* (1937), pp. 97–112.
12. "Le problème de la verité," *Theoria* (Goeteborg) 3 (1937), pp. 206–246. [Reprinted in translation under the title, "The Problem of Truth," in Jeffrey.]

13. "Ein System verallgemeinerter Negationen," *Travaux du 19e Congres International de Philosophie*, Paris, 1937, Vol. VI (Paris: Hermann et Cie, 1937), pp. 26–32.
14. "On the Logical Form of Probability-Statements," *Erkenntnis* 7 (1938), pp. 154–160. [Reprinted in Jeffrey.]
15. "Transfinite Concepts and Empiricism," Unity of Science Forum, *Synthese* 3 (1938), pp. 9–12.
16. "Supplementary Remarks on the Form of Probability Statements," *Erkenntnis* 7 (1939), pp. 360–363.
17. "Vagueness and Logic," *Philosophy of Science* 6 (1939), pp. 163–180.
18. Articles "Whole," "Carnap," "Reichenbach," in D. Runes (ed.), *Dictionary of Philosophy* (New York: Philosophical Library, 1942).
19. "The Function of General laws in History," *The Journal of Philosophy* 39 (1942) pp. 35–48. [Reprinted in *Aspects*.]
20. "A Purely Syntactical Definition of Confirmation," *The Journal of Symbolic Logic* 8 (1943), pp. 122–143.
21. "Studies in the Logic of Confirmation," *Mind* 54 (1945), pp. 1–26 and pp. 97–121. [Reprinted in *Aspects*.]
22. "Geometry and Empirical Science," *The American Mathematical Monthly* 52 (1945), pp. 7–17. [Reprinted in this volume.]
23. Discussion of G. Devereux, "The Logical Foundations of Culture and Personality Studies," *Transactions of the New York Academy of Sciences*, Ser. II, Vol. 7, No. 5 (1945), pp. 128–130.
24. (with P. Oppenheim) "A Definition of Degree of Confirmation," *Philosophy of Science* 12 (1945), pp. 98–115. [Reprinted in Jeffrey.]
25. "On the Nature of Mathematical Truth," *The American Mathematical Monthly* 52 (1945), pp. 543–556. [Reprinted in this volume.]
26. "A Note on the Paradoxes of Confirmation," *Mind* 55 (1946), pp. 79–82.
27. (with P. Oppenheim) "Studies in the Logic of Explanation," *Philosophy of Science* 15 (1948), pp. 135–175. [Reprinted in *Aspects*.]
28. (with P. Oppenheim) "Reply to David L. Miller's Comment," *Philosophy of Science* 15 (1948), pp. 350–352.
29. "Problems and Changes in the Empiricist Criterion of Meaning," *Revue Internationale de Philosophie*, No. 11 (1950), pp. 41–63.
30. "A Note on Semantic Realism," *Philosophy of Science* 17 (1950), pp. 169–173.
31. "The Concept of Cognitive Significance: A Reconsideration," *Proceedings of the American Academy of Arts and Sciences*, Vol. 80, No. I (1951), pp. 61–77.
32. "General System Theory and the Unity of Science," *Human Biology* 23 (1951), pp. 313–322.
33. *Fundamentals of Concept Formation in Empirical Science* (Chicago: University of Chicago Press, 1952), 93 pp. Volume II, No. 7, of *The International Encyclopedia of Unified Science*, Vol. II, No. 7. Spanish edition: *Fundamentos de la formación de conceptos en Ciencia empirica*. (Madrid: Alianza Editorial, S.A., 1988).
34. "Problems of Concept and Theory Formation in the Social Sciences," in *Science, Language, and Human Rights*. American Philosophical Association, Eastern Division, Vol. I (Philadelphia: University of Pennsylvania Press, 1952), pp. 65–86. German translation, "Typologische Methoden in den Sozialwissenschaften," in E. Topitsch (ed.) *Logik der Sozialwissenschaften* (Köln und Berlin: Kiepenheuer und Witsch, 4. Auflage 1967.) [Reprinted under the title, "Typological Methods in the Natural and the Social Sciences," in *Aspects*.]

35. "Reflections on Nelson Goodman's *The Structure of Appearance*," *Philosophical Review* 62 (1952), pp. 108–116.
36. "A Logical Appraisal of Operationism," *Scientific Monthly* 79 (1954), pp. 215–220. [Reprinted in *Aspects*.]
37. "Meaning," *Encyclopedia Britannica*, Vol. 15 (1956 ed.), p. 133.
38. "Some Reflections on 'The Case for Determinism," in S. Hook (ed.), *Determinism and Freedom in the Age of Modern Science* (New York: New York University Press, 1958), pp. 157–163.
39. "The Theoretician's Dilemma," in H. Feigl, M. Scriven, and G. Maxwell (eds.), *Minnesota Studies in the Philosophy of Science*, Vol. II (Minneapolis: University of Minnesota Press, 1958), pp. 37–98. [Reprinted in *Aspects*.]
40. "Empirical Statements and Falsifiability," *Philosophy* 33 (1958), pp. 342–348.
41. "The Logic of Functional Analysis," in L. Gross (ed.), *Symposium on Sociological Theory* (Evanston, Ill., and White Plains, N.Y.: Row Peterson & Co., 1959), pp. 271–307. Italian translation published as a monograph: *La logica dell'analisi funzionale* (Trento: Istituto Superiore di Scienze Sociali, 1967). [Reprinted in *Aspects*.]
42. "Science and Human Values," in R. E. Spiller (ed.), *Social Control in a Free Society* (Philadelphia: University of Pennsylvania Press, 1960), pp. 39–64. [Reprinted in *Aspects*.]
43. "Inductive Inconsistencies," *Synthese* 12 (1960), pp. 439–469. Also included in B. H. Kazemier and D. Vuysje (eds.), *Logic and Language: Studies Dedicated to Professor Rudolf Carnap on the Occasion of his Seventieth Birthday* (Dordrecht: D. Reidel, 1962). [Reprinted in *Aspects*.]
44. "Introduction to Problems of Taxonomy," in J. Zubin (ed.), *Field Studies in the Mental Disorders* (New York: Grune and Stratton, 1961), pp. 3–23. (Also contributions to the discussion on subsequent pages.)
45. *La formazione dei concetti e delle teorie nella scienza empirica* (Milano: Feltrinelli, 1961). (Contains items 33 and 39, translated and with an introduction by Alberto Pasquinelli.)
46. "Meaning," *Encyclopedia Americana*, Vol. 18 (1961 ed.), pp. 478–479.
47. "Deductive-Nomological vs. Statistical Explanation," in H. Feigl and G. Maxwell (eds.), *Minnesota Studies in the Philosophy of Science*, Vol. III (Minneapolis: University of Minnesota Press, 1962), pp. 98–169. Czech translation in K. Berka and L. Tondl (eds.), *Teorie modelu a modelování* (Prague: Nakladatelství Svoboda, 1967), pp. 95–172). [Reprinted in this volume.]
48. "Explanation in Science and in History," In R. G. Colodny (ed.), *Frontiers of Science and Philosophy* (Pittsburgh: University of Pittsburgh Press, 1962), pp. 9–33. [Reprinted in this volume.]
49. "Rational Action," *Proceedings and Addresses of the American Philosophical Association*, Vol. 35 (Yellow Springs, Ohio: The Antioch Press, 1962), pp. 5–23. [Reprinted in this volume.]
50. "Carnap, Rudolf," *Colliers Encyclopedia*, Vol. 5 (1962 copyright), pp. 457–458.
51. "Explanation and Prediction by Covering Laws," in B. Baumrin (ed.), *Philosophy of Science: The Delaware Seminar*, Vol. 1 (1961–62) (New York: Interscience Publishers, 1963), pp. 107–133. [Reprinted in this volume.]
52. "Reasons and Covering Laws in Historical Explanation," in S. Hook (ed.), *Philosophy and History* (New York: New York University Press, 1963), pp. 143–163. [Reprinted in this volume.]

53. "Implications of Carnap's Work for the Philosopy of Science," in P. A. Schilpp (ed.), *The Philosophy of Rudolf Carnap* (La Salle, Illinois: Open Court; and London: Cambridge University Press, 1963), pp. 685–709.
54. *Aspects of Scientific Explanation and Other Essays in the Philosophy of Science* (New York: The Free Press; London: Collier-MacMillan, Ltd., 1965). [*Aspects*.]
55. "Empiricist Criteria of Cognitive Significance: Problems and Changes," in C. G. Hempel, *Aspects of Scientific Explanation and Other Essays in the Philosophy of Science* (New York: The Free Press; London: Collier-MacMillan, Ltd., 1965), pp. 101–119. [This is a conflation, with certain omissions and other changes, of items 29 and 31. First published in *Aspects*.]
56. "Fundamentals of Taxonomy," in C. G. Hempel, *Aspects of Scientific Explanation and Other Essays in the Philosophy of Science* (New York: The Free Press; London: Collier-MacMillan, Ltd., 1965), pp. 137–154. [This is a revision of item 44. First published in *Aspects*.]
57. "Aspects of Scientific Explanation," in C. G. Hempel, *Aspects of Scientific Explanation and Other Essays in the Philosophy of Science* (New York: The Free Press; London: Collier-MacMillan, Ltd., 1965), pp. 331–496. Japanese translation of title essay published as a monograph by Bai Fu Kan, Tokyo, 1967. [First published in *Aspects*.]
58. "Coherence and Morality," *The Journal of Philosophy* 62 (1965), pp. 539–542.
59. "Comments" (on G. Schlesinger's "Instantiation and Confirmation"), in R. S. Cohen and M. W. Wartofsky (eds.), *Boston Studies in the Philosophy of Science*, Vol. 2 (New York: Humanities Press, 1965), pp. 19–24.
60. "Recent Problems of Induction," in R. G. Colodny (ed.), *Mind and Cosmos* (Pittsburgh: University of Pittsburgh Press, 1966), pp. 112–134. [Reprinted in this volume.]
61. *Philosophy of Natural Science* (Englewood Cliffs: Prentice-Hall, Inc., 1966). Translations: Japanese, 1967; Italian and Polish, 1968; Swedish, 1969; Portuguese and Dutch, 1970; French, 1972; Spanish, 1973; German, 1974; Chinese, 1986.
62. "On Russell's Phenomenological Constructionism," *The Journal of Philosophy* 63 (1966), pp. 668–670.
63. "Scientific Explanation," in S. Morgenbesser (ed.), *Philosophy of Science Today* (New York: Basic Books, 1967), pp. 79–88.
64. "Confirmation, Qualitative Aspects," in *The Encyclopedia of Philosophy*, Vol. II (New York: The MacMillan Company and The Free Press, 1967), pp. 185–187.
65. "The White Shoe: No Red Herring," *The British Journal for the Philosophy of Science* 18 (1967/68), pp. 239–240.
66. "Maximal Specificity and Lawlikeness in Probabilistic Explanation," *Philosophy of Science* 35 (1968), pp. 116–133. [Reprinted in this volume.]
67. "On a Claim by Skyrms Concerning Lawlikeness and Confirmation," *Philosophy of Science* 35 (1968), pp. 274–278.
68. "Logical Positivism and the Social Sciences," in P. Achinstein and S. F. Barker (eds.), *The Legacy of Logical Positivism* (Baltimore: The Johns Hopkins University Press, 1969), pp. 163–194. [Reprinted in this volume.]
69. "Reduction: Ontological and Linguistic Facets," in S. Morgenbesser, P. Suppes, M. White (eds.), *Philosophy. Science and Method. Essays in Honor of Ernest Nagel* (New York: St. Martin's Press, 1969), pp. 179–199. [Reprinted in this volume.]
70. "On the Structure of Scientific Theories," R. Suter (ed.), *The Isenberg Memorial Lecture Series 1965–1966* (East Lansing: Michigan State University Press, 1969), pp. 11–38. [Reprinted in this volume.]

71. "On the 'Standard Conception' of Scientific Theories," in M. Radner and S. Winokur, (eds.), *Minnesota Studies in the Philosophy of Science*, Vol. IV (Minneapolis: University of Minnesota Press, 1970), pp. 142–163. Also some contributions to "Discussion at the Conference on Correspondence Rules," pp. 220–259. [Reprinted without the additional comments in this volume.]
72. "Formen und Grenzen des wissenschaftlichen Verstehens," *Conceptus* VI, Nr. 1–3, (1972), pp. 5–18.
73. "Rudolf Carnap, Logical Empiricist," *Synthese* 25 (1973), pp. 256–268. [Reprinted in Jeffrey.]
74. "Science Unlimited?", *The Annals of the Japan Association for the Philosophy of Science* 4 (1973), pp. 187–202. [Reprinted in this volume.]
75. "The Meaning of Theoretical Terms: A Critique of the Standard Empiricist Construal," in P. Suppes et al. (eds.), *Logic, Methodology and Philosophy of Science IV* (Amsterdam: North Holland Publishing Company, 1973), pp. 367–378. [Reprinted in this volume.]
76. "A Problem in the Empiricist Construal of Theories" (in Hebrew, with English summary) *Iyyun, A Hebrew Philosophical Ouarterly*, 23 (1972), pp. 68–81 and 25 (1974) pp. 267–268.
77. "Formulation and Formalization of Scientific Theories: A Summary-Abstract," in F. Suppe (ed.), *Structure of Scientific Theories* (Urbana: University of Illinois Press, 1974), pp. 244–254.
78. "Carnap, Rudolf," *Encyclopedia Britannica*, 15th Edition (1974); Macropedia, Vol. 3, pp. 925–926.
79. *Grundzüge der Begriffsbildung in der empirischen Wissenschaft* (Dusseldorf: Bertelsmann Universitatsverlag, 1974), 104 pp. German translation of item 33, but enlarged by an additional chapter, not previously published, on theoretical concepts and theory change titled "Theoretische Begriffe und Theoriewandel: ein Nachwort (1974)", pp. 72–89 and pp. 97–98.
80. "The Old and the New 'Erkenntnis'," *Erkenntnis* 9 (1975), pp. 1–4.
81. "Dispositional Explanation and the Covering-Law Model: Response to Laird Addis," in R. S. Cohen, C. A. Hooker, A. C. Michalos, and J. van Evra (eds.), *PSA 1974: Proceedings of the 1974 Biennial Meeting of the Philosophy of Science Association* (Dordrecht: D. Reidel Publishing Co., 1976), pp. 369–376.
82. "Die Wissenschaftstheorie des analytischen Empirismus im Lichte zeitgenössischer Kritik," *Kongressberichte des XI. Deutschen Kongresses für Philosophie, 1975* (Hamburg: Felix Meiner Verlag, 1977), pp. 20–34.
83. *Aspekte wissenschaftlicher Erklärung* (Berlin and New York: Walter de Gruyter, 1977). (Revised translation of item 57 of this list, with a new section on statistical explanation). [Translation of this new section on statistical explanation (by Hazel Maxian), under the title, "Postscript 1976: More Recent Ideas on the Problem of Statistical Explanation." First published in this volume.]
84. "Dispositional Explanation," in R. Tuomela (ed.), *Dispositions* (Dordrecht and Boston: D. Reidel, 1978), pp. 137–146. (This is a revised version of section 9 of item 57.)
85. "Selección de una teoría en la ciencia: perspectivas analíticas vs. pragmáticas," in *La filosofía y las revoluciones científicas. Segundo Coloquio Nacional de Filosofía, Monterrey, Nuevo Leon. México* (Mexico City, D.F.: Editorial Grijalbo, S.A., 1979), pp. 115–135.

86. "Scientific Rationality: Analytic vs. Pragmatic Perspectives," in T. S. Geraets (ed.), *Rationality To-Day/La Rationalité Ajiourd'hui* (The University of Ottawa Press, 1979), pp. 46–58. Also remarks in the discussion, pp. 59–66, passim.
87. "Der Wiener Kreis-eine persoenliche Perspektive," in H. Berghel, A. Huebner, E. Koehler (eds.), *Wittgenstein, the Vienna Circle, and Critical Rationalism*. Proceedings of the Third International Wittgenstein Symposium, August 1978 (Vienna: Hoelder-Pichler-Tempsky, 1979), pp. 21–26.
88. "Scientific Rationality: Normative vs. Descriptive Construals," in H. Berghel, A. Huebner, E. Koehler (eds.), *Wittgenstein, the Vienna Circle, and Critical Rationalism*. Proceedings of the Third International Wittgenstein Symposium, August 1978 (Vienna: Hoelder-Pichler-Tempsky, 1979), pp. 291–301. [Reprinted in this volume.]
89. "Comments on Goodman's 'Ways of Worldmaking'," *Synthese* 45 (1980), pp. 193–199.
90. "Turns in the Evolution of the Problem of Induction," *Synthese* 46 (1981), pp. 389–404. [Reprinted in this volume.]
91. "Some Recent Controversies Concerning the Methodology of Science" (Chinese translation), *Journal of Dialectics of Nature*, Peking, Vol. 3, No. 5 (1981), pp. 11–20.
92. "Der Wiener Kreis und die Metamorphosen seines Empirismus," in Norbert Leser (ed.), *Das geistige Leben Wiens in der Zwischenkriegszeit* (Wien: Oesterreichischer Bundesverlag, 1981), pp. 205–215. [Reprinted in translation under the title, "The Vienna Circle and the Metamorphoses of Its Empiricism," in Jeffrey.]
93. "Analytic-Empiricist and Pragmatist Perspectives on Science" (Chinese translation of a lecture given in Peking), *Kexue Shi Yicong* [Collected Translations on History of Science], No. 1 (1982), pp. 56–63, concluded on p. 67.
94. "Logical Empiricism: Its Problems and Its Changes" (Chinese translation of a lecture given in Peking), *Xian Dai Wai Guo Zhe Xue Lun Ji* [Contemporary Foreign Philosophy; People's Publishing House], Vol. 2 (1982), pp. 69–88.
95. "Schlick und Neurath: Fundierung vs. Kohärenz in der wissenschaftlichen Erkenntnis," in *Grazer Philosophische Studien*, Band 16/17 (for 1982, but published in 1983), pp. 1–18. [Reprinted in translation under the title, "Schlick and Neurath: Foundation vs. Coherence in Scientific Knowledge," in Jeffrey.]
96. "Valuation and Objectivity in Science," in R. S. Cohen and L. Laudan (eds.), *Physics, Philosophy and Psychoanalysis: Essays in Honor of Adolf Grunbaum* (Dordrecht, Boston, Lancaster: D. Reidel, 1983), pp. 73–100. [Reprinted in this volume.]
97. "Kuhn and Salmon on Rationality and Theory Choice," *The Journal of Philosophy* 80 (1983), pp. 570–572.
98. *Methodology of Science: Descriptive and Prescriptive Facets*, Pamphlet Number IAS 814-84 (in series of lecture texts published by The Mortimer and Raymond Sackler Institute of Advanced Studies, Tel Aviv University, 1984), 30 pp.
99. "Der Januskopf der wissenschaftlichen Methodenlehre," in Peter Wapnewski (ed.), *Jahrbuch 1983/84, Wissenschaftskolleg—Institute for Advanced Study—zu Berlin* (Siedler Verlag, 1985), pp. 145–157.
100. "Wissenschaft, Induktion und Wahrheit," in brochure published by Fachbereich Wirtschaftswissenschaft der Freien Universität Berlin: "Verleihung der Würde eines Ehrendoktors der Wirtschaftswissenschaft an Prof. Dr. Phil. Carl G. Hempel (University of Pittsburgh) am 10. Dezember 1984." Published 1985.

101. "Thoughts on the Limitations of Discovery by Computer," in K. F. Schaffner (ed.), *Logic of Discovery and Diagnosis in Medicine* (Berkeley: University of California Press, 1985), pp. 115–122.
102. "Prova e verità nella ricerca scientifica," *Nuova Civiltà Delle Macchine*, Anno IV, nn. 3/4 (15/16) 1986, pp. 65–71 (Roma). (Translation by Patricia Pincini of lecture given in June 1986 at Locarno Conference under the title "Evidence and Truth in Scientific Inquiry.") English summary, p. 149.
103. "Provisoes: A Problem Concerning the Inferential Function of Scientific Theoties," *Erkenntnis* 28 (1988), pp. 147–164. Reprinted in A. Grunbaum and W. C. Salmon (eds.), *The Limitations of Deductivism* (Berkeley: University of California Press, 1988), pp. 19–36. [Reprinted in Jeffrey.]
104. "Limits of a Deductive Construal of the Function of Scientific Theories," in E. Ullmann-Margalit (ed.), *Science in Reflection. The Israel Colloquium*, Vol. 3 (Dordrecht: Kluwer Academic Publishers, 1988), pp. 1–15. [Reprinted in this volume.]
105. "On the Cognitive Status and the Rationale of Scientific Methodology," in *Poetics Today*, Vol. 9, Nr. 1 (1988), pp. 5–27. [Reprinted in Jeffrey.]
106. "Las facetas descriptiva y valorativa de la ciencia y la epistemología," in E. Villanueva, compilador, *Segundo Simposio Internacional De Filosofía* (1981), Vol. I (Mexico City, 1988), pp. 25–52.
107. *Oltre il Positivismo Logico: Saggi e Ricordi. A cura di Gianni Rigamonti* (Roma: Armando), 1989. Italian translations of 12 selected essays and of a recorded interview with Richard Nollan, 1982. [The interview is published in the companion to this volume under the title, "An Intellectual Autobiography."]
108. "Ernest Nagel" (Memorial Note, *1989 Year Book of the American Philosophical Society* (Philadelphia: The American Philosophical Society, 1990), pp. 265–270.
109. "Il significato del concetto di verità per la valutazione critica delle teorie scientifiche," *Nuova Civiltà delle Macchine*, Anno VIII, N. 4 (32) (1990), pp. 7–12. (English text, "The signification of the concept of truth for the critical appraisal of scientific theories," pp. 109–113. (The second word in the submitted typescript, however, was "significance.") [Reprinted under the title, "The Irrelevance of the Concept of Truth for the Critical Appraisal of Scientific Claims," in Jeffrey.]
110. "Hans Reichenbach Remembered," *Erkenntnis* 35 (1991), pp. 5–10. [Reprinted in Jeffrey.]
111. "Eino Kaila and Logical Empiricism," in I. Niiniluoto, M. Sintonen, and G. H. von Wright (eds.), *Eino Kaila and Logical Empiricism* (*Helsinki*: Philosophical Society of Finland, 1992), pp. 43–51. (*Acta Philosophica Fennica* 52.)
112. "Empiricism in the Vienna Circle and in the Berlin Society for Scientific Philosophy. Recollections and Reflections," in Friedrich Stadler (ed.), *Scientific Philosophy: Origins and Developments* (Dordrecht/Boston/London: Kluwer Academic Publishers, 1993), pp. 1–9. [Reprinted in Jeffrey.]

Index of Names

Achinstein, Peter 65, 207, 233, 235,
Adler, Alfred 377
Alexander, H. G. 136, 143
Avogadro 58, 63, 227, 331
Ayer, A. J. xvii, xxxiii

Bar-Hillel, Y. 141, 144
Barker, Stephen 115, 136, 139, 142–143
Bartley, W. W. 84
Bergmann, Gustav 253, 271
Bernoulli, Daniel 139
Birkhoff, George 16
Boehmer, H. 287, 296
Bohnert, Herbert 234
Bohr, Nils 53–54, 61–62, 205, 212, 220, 228
Bolyai, Janos 22
Boyle, Robert 14, 279, 322
Braithwaite, R. B. 45, 135, 142–143, 208, 213, 224, 234
Brandt, Richard 310
Bridgman, Percy 63–65
Brodbeck, May 84–85, 135, 144
Bromberger, Sylvain 84–85, 94, 136
Brutus 302

Caesar, Julius 301–302
Callebaut, W. xxxii–xxxiii
Calusius, Rudolf 322
Campbell, N. R. 65, 134, 143, 208, 213, 221, 233–234
Cantor, George 370
Carnap, Rudolf xiii, xvi, xx, xxiii, xxix, xxv, xxvii, xxxi, xxxiii, 17, 29, 34, 38, 42, 47–48, 63, 66, 78–79, 85, 113–114, 118, 129, 136, 139–141, 143–144, 150, 163–164, 189, 206–208, 213, 216–217, 230–231, 234–235, 241, 249,
253–254, 259–270, 272–275, 280, 292, 296, 298, 319, 347, 355, 375, 377, 379, 380–382, 384, 387, 390, 394
Casius 302
Chapin, F. Stuart 390, 393–394
Charles, Jacques 322
Churchman, C. West 45, 48
Coffa, Alberto 175, 177–178, 184–185
Cohen, Morris R. 144
Copernicus 385
Craig, William 242, 300
Cramer, Harald 108–109, 138, 144

Davidson, Donald 163, 324, 326
De Finetti, Bruno 142, 144, 392
Dewey, John 276, 295, 366–367, 370–371, 374, 392, 394
Dray, William 82–83, 85, 134, 144, 290–293, 295–297, 300–310, 316–317, 321, 326
Du Bois-Reymond, Emil 336, 338–340, 342
Duhem, Pierre 130, 134–135, 144, 365–366, 368, 370, 382, 385

Eberle, Rolf 84–85
Ebrenhaft 393
Ehrenhaft, Felix 246
Einstein, Albert xxvi, 28, 32, 221, 259, 377
Euclid 18–20

Feather, N. 235
Feigl, Herbert xxix, 85, 135, 144, 208, 216–217, 253, 264–269, 272, 274–275
Fermat, Pierre de 34
Fetzer, James H. xxxii
Feyerabend, Paul 60, 65–66, 84–85, 199, 206–207, 211–212, 217, 235, 357, 363–364, 370–371

Index of Names

Finch, Henry 395
Fraassen, Bas C. van 393–394
Frank, Philipp 253, 271
Frankel, Charles 300
Frege, Gottlob 10
Freud, Sigmund 282–283, 295, 377

Galileo 60, 81, 93, 97, 135, 144, 199–200, 205, 278–279, 330
Gallie, W. B. 300
Gardiner, Patrick 144, 294, 296
Gauss, Karl Friedrich 26
Gerschenkron, A. 289, 296
Gibson, Quentin 309, 312, 315, 326
Glass, Bentley 162, 164
Glymour, Clark 393–394
Godel, Kurt 16 n. 7
Goldbach, Christian 34
Good, I. J. 47
Goodman, Nelson xxiv, xxvi, xxxiii, 38–41, 47, 135, 137, 142, 144, 158, 163, 176, 184–185, 273, 355, 380, 392, 394
Gottlob, Adolf 287, 296
Grandy, Richard 158–161, 163, 234
Grunbaum, Adolf xiii, 85, 234, 373, 386, 391–392, 394

Hahn, Hans 253
Hanson, N. R. 65, 85, 139–140, 144, 235
Hayek, F. 302
Hempel, Carl G. ii, vii, xiii–xiv, xviii–xxii, xxiv–xxv, xxix, xxvi–xxviii, xxxii–xxxiii, 48, 77, 66, 80, 83–85, 134–135, 144, 167, 185, 216–217, 295, 309–310, 342–343, 392–394, 397
Hepburn, Ronald W. 342
Hesse, Mary 66, 216–217
Hilbert, David 19, 23, 25, 209–210, 216
Hinshelwood, Cyril 339, 343
Hitler, Adolf 284
Holton, Gerald 246, 249, 393, 395
Hosiasson-Lindenbaum, Janina 47
Hume, David xv–xvi, xxvi, 29, 344
Humphreys, W. C. 152–154, 159, 164

Jarrett, J. L. 144
Jeffrey, Richard C. vii, xxvii, 128, 142, 144, 165–167, 175, 178, 180, 183, 185, 233, 348–350, 397–398, 401–403

Kalish, D. 264
Kant, Immanuel 55
Kaplan, David 84–85
Kekule, F. A. 32

Kemeny, John 139, 145, 200, 207, 216, 235, 392, 395
Kepler, Johannes 93, 134, 199–200, 205, 278–279, 330, 351
Keynes, John Maynard 113, 298
Kim, Jaegwon xiii, 84–85, 185
King George III 330
Kitcher, Philip xiii
Koch, Sigmund 264
Koertge, Noretta 173, 184–185
Kolmogoroff, A. 138, 145, 224, 235
Korner, Stephan 137, 145
Kries, Johannes von 387, 394–395
Kruger, L. 175–177, 184–185
Kuhn, Thomas S. vii, xiii, xxxi–xxxiii, 48, 207, 217, 235, 352–355, 357–358, 375–359, 361, 364–371, 374, 382–385, 391–393, 395
Kyburg, Henry 216–217, 223, 234–235

Lakatos, Imre 361, 371, 392, 395
Laplace, Pierre-Simon 336
Laudan, Laurence xiii, 353, 393
Lehman, H. 184–185
Lewis, C. I. 138–139, 145
Lobatschefskij, N. I. 22
Louis XIV 307
Luce, R. D. 48, 141–142, 145, 310, 326
Luther, Martin 287, 291

MacCorquodale, K. 264
MacLane, S. 16
Mandelbaum, Maurice 302, 309
Massey, Gerald 154–155, 163–164
Maxian, Hazel 183
Maxwell, Grover 85, 144, 233
McConnell, D. W. 295
McMurrin, S. M. 144
Meehl, Paul 264
Mill, John Stuart 4
Millikan, Robert A. 246, 386, 393, 395
Mises, Richard von 133, 143, 145, 171, 183–186, 224, 253
Montague, Richard 85
Morgenbesser, Sidney 367, 371
Morgenstern, Oskar 141, 145
Morris, Charles 254

Nagel, Ernest 50, 61–62, 65–66, 136, 144–145, 189, 199, 206–208, 216–217, 222, 228–229, 233–235, 296, 364, 370–371, 387, 395
Neumann, John von 141, 145, 234
Neurath, Otto xxix, 189, 253–261, 272

Newton, Isacc 83, 134, 199, 205, 245, 278, 352–353
Neyman, Jerzy 142, 145
Nicod, Jean 35, 40–41, 47–48
Niiniluoto, Ilkka 183–185
Nozick, Robert xiii
Nute, Donald E. xxxii

O'Brien, J. F. 342
Oppenheim, Paul 77–78, 80, 83, 87, 134, 139, 145, 200, 207, 295, 309

Pap, Arthur 249
Passmore, John 306, 310
Pavlov, Ivan 255
Peano, Giuseppe xxvi, 6, 8–10, 12–13, 15, 17
Pears, David 47
Peirce, Charles S. xxxii
Peters, R. S. 316, 326
Pitt, Joseph 85
Planck, Max 53, 220
Poincare, Henri 26–27, 221
Pope Boniface VIII 289–290
Pope Innocent III 287
Pope John VII 287
Pope Sixtus IV 288
Popper, Karl R. vii, xiii, xviii, xxiii, xxxiii, 32, 47, 84, 134, 137, 145, 224, 245, 253, 358, 375, 377–378, 384, 392, 395
Ptolemy 385
Putnam, Hilary 60, 65–66, 207, 212, 232–233, 236

Quine, W. V. O. xix, xxviii, xxxiii, 17, 47, 185, 194, 200–201, 205, 207, 210–211, 215, 217, 230–231, 235–236, 260, 273, 323, 326, 382, 395

Raffia, H. 48, 141–142, 145, 310, 326
Ramsey, F. P. 208, 213, 221, 233–234, 242, 300
Rankine, William 50
Reichenbach, Hans xiii, xx, xxxiii, 55, 65, 136, 140, 145, 151–152, 162, 208, 210, 221, 223–224, 234, 253, 393, 395
Reimann, Bernhard 22, 28
Rescher, Nicholas xiii, 98, 136, 145, 185
Ritchie, B. F. 264
Roentgen, Wilhelm 247
Rudner, Richard 347–350
Russell, Bertrand xiv, 8, 10, 12, 16–17,
Ryle, Gilbert 135, 145, 256–257, 265, 270, 275, 305, 310, 318–319, 326

Salmon, Wesley C. xxv, xxvii–xxviii, xxxiii, 163–165, 168–175, 178–179, 181, 183–185

Savage, Leonard 141–142, 145, 224
Scheffler, Israel 47, 76, 84, 86, 100–101, 136, 145, 234–235, 310
Schilpp, P. A. 395
Schlesinger, C. 184
Schlick, Moritz 135–136, 145, 210, 216–217, 253–254, 256
Schwiebert, E. G. 288, 296
Scriven, Michael xxvii, 75–81, 84, 86, 90, 99, 101, 137–138, 140, 144–145, 301, 309
Sellars, Wilfred 136, 145, 235
Shaffer, Jerome 193, 206
Shapere, Dudley 65, 207, 235
Shils, Edward A. 395
Shimony, Abner 392
Siegel, Sidney 324, 326
Sklar, Lawrence xiii
Smart, J. J. C. 207
Snell, Willebrord 279
Sommerfeld, Arnold 205, 212
Spector, M. 66
Spence, K. W. 271
Stegmuller, Wolfgang 165, 171, 176, 177–181, 183–185, 217, 395
Strawson, Peter 307
Suppes, Patrick 223–224, 234–235, 324, 326

Tarski, Alfred xv, xxxiii, 16
Tolman, E. C. 264
Toulmin, Stephen 136, 145
Tuomela, Raimo 183
Turner, F. J. 285, 286, 289
Tyndall, John 338–339, 342

van der Waal, J. D. 322
Vlastos, Gregory xiii

Waismann, Friedrich 253–254
Wald, A. 142, 145
Watson, G. 293–294, 296
Watson, J. B. 255–256, 272
Weber, Max 258, 336, 342, 372, 391
Weyl, Hermann 234
Whewell, William 32, 47
White, M. G. 236
Whitehead, Alfred 10, 17
Williams, D. C. 139, 145
Wittgenstein, Ludwig xiv–xv, xvii, xxxii–xxxiii, 342–343
Wojcicki, R. 155–156, 163–164
Wolfe, A. B. 46

Young, J. W. 16

Index of Subjects

"a causative factor in" 93
"a cause of" 93
a posteriori knowledge xvi–xvii
a priori knowledge xvi–xvii
"about" 101–102, 137 n. 25
absolute vs. relativized characterizations 175
absolute vs. relativized homogeneity 169–170
abstract calculi xxvi, xxviii
abstract calculus C 49, 51, 55–61, 221, 226
abstractive methods 387
abstractive theories 218
acceptance does not entail truth 215
acceptance in pure science 350–352
acceptance of hypotheses 32–33
acceptance, direct 122
acceptance, inferential 122
acceptance, problem of 346
accepted-information model of scientific knowledge 122, 129
accepting a statement as true 123
accepting an hypothesis 128
accepting an inductive conclusion 345
accepting hypotheses vs. acting on hypotheses 348–350
accepting hypotheses, decision-theoretical approach to 350–352
accidental statistical generalizations 137 n. 28
account of science, descriptive-explanatory 359
account of science, normative-prescriptive 359
actions 45
adaptive vs. rational 369–370
addition defined 7
adjusting rules to inferences and inferences to rules 355, 380

adopting a course of action 128
algorithmic solutions 389
"all that can be said about an individual case" 341
ambiguity of induction xxvi, 41–43
ambiguity of I-S explanations 147–148
ambiguity of statistical explanation xxvii, 107, 111, 167–168
ambiguity of statistical prediction 107, 111
ambiguity of statistical systematization 105–107, 111, 115–116
ambiguity, epistemic 148
ambiguity, ontic 148
analogies and models 221–222, 229
analytic *a posteriori* knowledge xvi–xvii
analytic *a priori* knowledge 5–6
analytic empiricism xxx, 357–370, 375–377, 379–382
analytic empiricism, historic-pragmatist construal as corrective to 354–356
analytic sentences xxvi
analytic/synthetic distinction xvi–xvii, xviii–xix, xxi–xxii, xxxi, 204–205, 231–232
analyticity xvii, 205
antecedent conditions xxv
"antecedent conditions" 136 n. 19
antecedent vocabulary 50–52, 55–61
antecedently available language xxviii
antecedently available vocabulary 212–213, 232–233
anthropomorphic metaphors 335
applied mathematics xxvi, 14
applied science xxx, 130
arationality 370, 391
argument form admissible but not necessary 167

409

arguments, probabilistic 71, 77–78
arguments, statistical explanations as 165–167, 181–182
"as if" clauses 61, 228–229
"as if" the conclusion were true 345
ascertaining vs. understanding 322
aspects of concrete events 302
Aspects of Scientific Explanation, Hempel's vii, passim
assertion that an event did occur 150–151
assigning probabilities to hypotheses 348–350
"astronomical knowledge" 336
atomic facts xiv
atomic sentences xiv
Avogadro's number 58, 63, 227, 231
axiom of choice 9
axiom of infinity 12–13
axiomatization xxvi, 12–13, 54–55
axiomatization of geometry 209–210
axiomatization, critique of 222–225
axiomatized calculus 49, 51, 54–61
axiomatized deductive system, mathematics as an 6
axiomatized deductive systems xxvi, passim
axiomatized theories 23
axioms 18
axioms of a theory 6

Balmer's formula 53
basic principles C 237–239
"beautiful" cases 166–167
"because" 133–134
behaviorism 255–256
"behavioristic" outlook on decision problems 142–143 n. 58
behavioristics 255, 257
belief and goal attributions 326 n. 13
beliefs 45, 320–321
beliefs not deductively closed 323
blind guessing 332
Bohr's theory of the atom 53–54, 61–62, 219–220, 228
Bohr's theory, bridge principles of 220
Bohr's theory, internal principles of 220
boundary conditions 93, 99
brain-states and pain-states 194
bridge principles B 50, 52–54, 63–65
bridge principles xxvi, xxviii, 50, 52–54, 63–65, 209, 218, 233
broadly dispositional analysis 305–306, 308–309, 310 n. 15
broadly dispositional traits 318–320
"by means of chance" 166

$c(h, e) = r$ 113, 123, 127, 129
C, abstract calculus 221, 226
C, basic principles 237–239
C, theoretical calculus 209
C, uninterpreted axiomatized formal system 209
Carnap's elaboration of physicalism 259–266
categorical probability statements 138–139 n. 37
category mistake xxxii n. 1
causal conditionals xxxii n. 2
causal connections 91–94
causal explanation 70, 141 n. 50, 76, 90–94, 278–279
causal generalizations 90–94
causal laws 185 n. 30
causality 133–134
"cause and effect" 90–91
causes and reasons xxix
ceteris paribus clauses 90–91, 244
change in beliefs 204
change in meaning 204
"changed in meaning" 207 n. 19
classes of objects 10
classes of things 191, 193
classical mechanics 59
coextensive by law 267
coextensive, theoretically 267
cognitive significance xvii–xvix, xxvi, xxix, 332–333
cognitive status of methodology of epistemology 372
cognitive status of methodology of science 372
combined motive and belief causation 270–271
complete explanations 282–284, 301–303, 309 n. 6
complete understanding 337–338
completeness assumptions 244–245
concealing results 383
conclusions of statistical arguments 111–112
concrete events 284, 302
conditional truths 20
conditionals, counterfactual xxiii–xxiv
conditionals, material xx
conditionals, subjunctive xx, xxiii–xxiv
conditions of adequacy 299–300
  for explananations 89–90
  for explications 379
  for scientific explanations 74–77
  formal definition of 84 n. 13
  truth 100
confirmation xviii

Index of Subjects 411

confirmation c(h, e) 42–43
confirmation, measures of 48 n. 17
confirmed explanations 90
confirmed systematizations 99
confirming evidence 31
conflicting explanations 161–162
conflicting I-S arguments 147–149
conjectures 32
connecting principles 197–199
conscious choice as nonconsciously rational 324–325
consciously rational agents 323–325
consciousness 338–339
content measure function 125–126
context of discovery 30–33, 269, 375, 393 n. 42
context of justification 32–33, 269, 375, 393 n. 42
conventionalism concerning geometry 26–27
conventionalist stratagem 245, 377
conventionality 230
conventions of science 377
coordinating definitions xxvi
"coordinating definitions" 49, 63, 229
corpus of knowledge K 351–352
correspondence rules xxviii, 228, 265
correspondence rules R 221
counteracting (disturbing) forces 240–241
counterfactual conditionals xxiii–xxiv, 104, 137 n. 28
counterfactual predictions 100
covering laws 69
  as premises 95–97
  as rules of inference 95–97
  indispensability of 75
covering-law models xxv, 69–73, 83 n. 2, 87, 134 n. 3, 297–299
  as ideal types 281–282
  as idealized standards 281–282
(CR1) deductive closure condition 122
(CR2) logical consistency condition 122–123, 141–142 n. 54
(CR3) total evidence requirement 123
Craig's theorem 242–243
creative imagination 32
criteria of application 57–58
criteria of rationality 304
criteria of validation 34–35
criterion sentence 214
curve fitting 31, 393 n. 42
curve-fitting problems 345–346

decision-making 44–46, 324–325
decision-making experiments 324–325

decision-making under risk 314, 324
decision-making under uncertainty 314, 350
declarative sentences xvii
deduction xviii–xix, 20
deductive closure 323, 389
deductive inference 33–34
deductive nomological explanation 291
deductive predictions 98–99
deductive retrodictions 98–99
deductive rules of inference 39
deductive systematization 113
deductive validity 33–34, 47 n. 9
deductively valid arguments 111–112, 114–115
deductive-nomological (D-N) explanation xxvii, 87–102, 146, 276–278, 297–298
deductive-nomological model xxv, 69–70
deepanalysis, Stegmuller's conception of 165, 171–179
definability xxix
definability or derivability as reduction 191–192, 196–198
defined terms 6
definiendum xix
definiens xix
definition by postulates 209–211
definitions xix, xxi
definitions of dispositions xix–xxiv, 213, 262–264
"degree of factual support" 139 n. 39
"dependent variable" 94–95
Der logische Aufbau der Welt, Carnap's 260
derivability condition of adequacy 89–90
descriptions 74
descriptive predicates 51–52
descriptive principles 305
descriptive rationality, Kuhn's 382–383
descriptive-explanatory acccount of science 359
desiderata for theory choice 384–388
desiderata of scientific inquiry 352–356
desiderata of theory choice 359, 363–365
desiderata of theory choice in science 352–356
desideratum of accuracy 385–386
desideratum of simplicity 385
detachability of conclusions 167–168
deterministic explanations 170–171
deterministic laws 99
deterministic systems 93–94
deterministic theories 93–94
"deterministic with respect to the relevant notion of state" 195
dictionary definitions xix

dictionary, N. R. Campbell's notion of 213
dilemma of universal determinism 195–196
direct acceptance 122
disconfirmation xviii
disconfirming evidence 31
disposition to act rationally 307
dispositional analysis, broadly 305–306, 308–309, 310 n. 15
dispositional definitions 262–264
dispositional explanations 291–293, 305, 318
dispositional predicates xix–xxiv
dispositional properties xvii, xxiii
dispositional terms 96
dispositional traits, broadly 318–320
dispositions xxiii, 257, 262, 318
dispositions, definitions of xix–xxiv, 213, 262–264
dispositions, higher-order 318
disturbing (counteracting) forces 240–241
D-N explanation (*see also* deductive-nomological explanation) 276–278
D-N explanations 173
D-N vs. I-S explanations 175, passim
Duhem thesis 365–366
Duhem-Quine argument 382

elementary possibilities 107–108
elimination programs 242–244
ellipical explanations 282–283, 300
elliptical geometry 22, 28
emotional security as desideratum 363–364
empirical assertions 332–333
"empirical basis" 16 n. 2
empirical generalizations, mathematical theorems as 4–5
empirical hypotheses (relating laws to the world) 201
empirical interpretations xxvi, xxviii
empirical laws xxi–xxii
empirical meaning xvii
empirical science, fallibility of 332
empirical testability 332–333
empirical uncertainty vs. logical uncertainty 137 n. 26
empirical vocabulary presupposition 209
empiricial significance 265
empiricist criterion of meaning 261–262
  liberalized 265
  narrower 261–262
  wider 261–262
Empirische Soziologie, Neurath's 259
entelechies 73–74
entrenched predicates 176

entrenchment xxiv, 39–41, 163 n. 14, 184 n. 22
epistemic interdependence xxx
epistemic interdependence of goal and belief attributions 320–321
epistemic interdependence of motive and belief attributions 270–271
epistemic relativity of I-S explanation 167
epistemic relativization 169–170
epistemic utilities 44–46, 124–126, 350
epistemic values 372–373
epistemic values vs. moral values 351
epistemic vs. ontic completeness 244–248
epistemological valuations 378
equivalence condition 36
essential incompleteness 335–338
essential reference to particular individuals 106
esthetic satisfaction as desideratum 363–364
estimate of relative frequency 118
ethical values 45
Euclidean geometry xxvi, 21–24, 331
events of kinds 79, 91–92, 163 n. 9, 302
events under a description 302
"everything that is the case" 362
evidence, confirming 31, passim
evidence, disconfirming 31, passim
evidential relevance 31
exclamatory sentences xvii
exercise in epistemology xxii
exercise in ontology xxii
existential generalizations xviii
existentialism xiv
"explain", meanings of 79–83
explaining how vs. explaining why 335
explananda of statistical explanations 173–174
explanandum xxiv–xxv, 70, 78, 88, 277, 298, 334
"explanandum" 277
explanandum-event 277, 301–302
explanandum-event already known to have occurred 75
explanandum-phenomenon 70, 77, 88, 301–302
explanandum-sentence 70, 77, 88, 277
explanandum-statement 301–302
explanans xxiv, 70, 88, 277, 297, 334
"explanans" 277
explanation (*see also* explanations, D-N explanation, and I-S explanation) 237
  and prediction xxvii–xxviii
  and understanding 276
  as subsumption 134 n. 2, 295 n. 2

causal 70, 278–279
deductive mode of 87
explications of xxiv–xxvi
inductive mode of 87
meanings of 79–83
nonpragmatic conceptions of 82–83
of general uniformities 88–89
of scientific practices xxxi
of singular events 88 (*see also* covering-law models)
pragmatic conceptions of 80–83
pragmatics of 100–101
rational 316–317
sketch 282–284
teleological 73–74
theoretical 278
vs. justification 307
vs. justification of scientific procedure 360
why an event occurred 150–151
explanation-demanding questions 173–174, 180, 183 n. 7
explanations (*see also* explanation)
actual xxvii
by disposition 305
by motives 270–271
causal 90–94
confirmed 90
dispositional 291–293, 318
historical 300–301
of regularities 93–94
potential xxvii, 89
probabilistic 279–281
rational 289–294
true 90
with low probability 341–342
explanatory incompleteness 335–338
explicandum 379–380, 389
explications (*see also* rational reconstructions) xx, 365, 367–368, 379–382, 389, 390, 392 n. 27
conditions of adequacy for 379
of explanation xxiv–xxvi
explicatum 379–382, 390
explicit definitions xxviii
explicit linguistic specification, requirement of 209, 215–216
extensional connectives 263–264
extensional formulations xxi–xxii
extensional logic xxi, xxiii–xxiv

"fact" 185 n. 23
facts 256
"facts" xv, 272 n. 9
fair betting quotient 118

faking findings 383
fallibility of empiricial science 332
falsifiability xviii, 375
falsifiability criterion 367–368
falsifiability-in-principle 358
finite samples and single cases 110–111
fittest way to practice science 369
flagpole counterexample 94–95
"formal mode of speech" 190
frequency interpretation of statistical probabililty 108
fudging evidence 383
functional analysis 73–74
"functionally equivalent" theories 243
fundamental statistical laws 185 n. 30
Fundamentals of Concept Formation in Empirical Science, Hempel's xx

Geisteswissenschaften (or *Kulturwissenschaften*) 254–255
general laws 277–279, passim
general laws vs. principles of action 306
general laws, statements of 88
general theory of relativity 27–28
generalizations 96–97
generalizations, statistical 103
genetic explanations 287–289
geometrical figures 19
geometrical intuition 19
geometrical proofs 331
geometrical terms, implicitly defined 209–210
geometry and experience 18
ghost in the machine 257, 319–320
goal and belief attributions 320–321
goals 320–321
Goodman's linguistic reformulation xxvi
Goodman's new riddle of induction 38–40, 346

"had to occur" 70
half-life of radon 116–117, 132, 183 n. 9
Hempel's importance to philosophy of science vii, xiii
Hempel's methodology xiii
Hempel's students xiii
high probability 334
high probability requirement xxvii–xxviii, 179, 181–182
high probability requirement abandoned 166
higher-order dispositions 318
historical explanations 284–295, 300–301
historic-pragmatist conception 357–370
historic-pragmatist construal as corrective to analytic empiricism 354–356

historic-pragmatist construal of scientific inquiry 352–356
homogeneity condition 183–184 n. 10
　absolute vs. relativized 169–170
　relative to K 174–175
　relative to the class K 178
　relative to the class T 178
homogeneous reference class 171
homunculus 256
how an event took place 166
how did the event occur 180
how vs. why events occur xxx
hyperbolic geometry 22
hypotheses, acceptance of 32–33
hypotheses, invention of 32–33
hypothesis of universal psychophysical association 193
hypothetical probability statements 138–139 n. 37
hypothetical theories 218
hypothetico-deductive method 70
hypothetico-deductive model 237–239

I interpretative sentences 237–239
"ideal law" 322
ideal types xxix
ideal types, covering-law models as 281–282
idealized explanatory models 322–325
idealized standards xxix
idealized standards, covering-law models as 281–282
identity theory 193–194
identity theory, mind/brain 340
identity theory, physicalistic 267
idiosyncratic factors 388–391
"Ignorabimus" 338
"Ignoramus" 338
imagination and invention 34
imperative sentences xvii
implicit definition xxviii, 9–10, 61–62, 200–201, 216, 224, 235 n. 16, 381
"implicit definitions" 209–210
"incommensurability" 201, 203
incomplete explanations 282–283
incompleteness 16–17 n. 7
inconsistent theories 200
independent variable 94–95
indeterministic causal laws, L. Kruger on 175–176
index verborum prohibitorum, Neurath's 255–258
indirect procedures of testing 25–26
individual events 301–302

"individual events" 163 n. 9
induction xxvi, xviii–xix, xxx
　and valuation 44–46
　ambiguity of 41–43
　Goodman's new riddle of 38–40, 346
　justification of xxvi, 29–30
inductive (or logical) irrelevance 114, 119–121, 143 n. 61
inductive (or logical) probability 113, 280, 298–299
inductive (or probabilistic) explanations 298–299
inductive (or theoretical) ascent 239–240, 248
inductive acceptance rule for pure science 351
inductive ascent xxix
inductive degrees of support 34
inductive inference 29–30
inductive inference, central concept of 113
inductive logic xix, 123, 129, 387
inductive probability 71–72, 77–78, 112–113, 151, 392 n. 27
inductive projection, theory of 39
inductive rules of inference 39–40
inductive support 34, 112–113, 330–331
inductive systematization 113
inductive uncertainty xxx
inductive uncertainty of scientific knowledge 330–331
inductive validity 47 n. 9
"inductive" in a narrow sense 32–33
"inductive" in a wider sense 32–33
inductive-probabilistic explanation xxvii
inductive-probabilistic model xxv, 70–73, 78–79
inductive-statistical explanation 146–147
inference tickets 135 n. 11
inference, deductive 33–34
inference, nondemonstrative 29–30
inferential acceptance 122
"influence" 175–177
initial conditions xxv
instrumental rationality 366–367
instrumentalism 242–244
instrumentalist (or pragmatist) model 126–130
intensional logic xxiii–xxiv
interchangeability xix
internal principles xxvi, xxviii, 209, 218, 225–226, 233
internal principles I 50, 52–54, 55–63
internal principles, models as parts of 229
International Encyclopedia of Unified Science xx, xxxi, 254, 258

interpretation of an axiomatic system 6
"interpretation" 211
interpretations of Peano's primitives 9–10
interpretative sentences I 237–239
interpretative sentences R 209, 211–214
interpretative sentences, form of 213–214
interrogatory sentences xvii
intersubjective agreement 239
intersubjective reliability xxii
intersubjectivity 212, 260–261, 375–376
introspection 264, 267, 269–270
invention of hypotheses 32–33
i-predicate in K 160–161
I-S explanation (see also inductive-statistical explanation)
  conditions for 162
  epistemic relativity of 148
  of basic form 147
  vs. D-N explanations 175
isolated systems 93, 99

"Janus headed" conception of scientific methodology 360–361, 368, 376, 383
justication for scientific practices xxxi
justification of induction xxvi, xxx, 355–356
justification vs. explanation 307
justification vs. explanation of scientific procedure 360

K, a knowledge situation 148–153
K, an epistemic context xxvii
K, corpus of knowledge 351–352
K, deductive closure of 363
K, logical consistency of 363
K, set of accepted statements 362–363
kinds of events 79, 91–92, 163 n. 9, 191, 193, 302
kinds of knowing 81
kinds of phenomena 352–353
kinetic theory of gases 219, 226–227
knowing, kinds of 81
Kulturwissneshaft 266

language of science 266–267
Laplace's Spirit 336, 338
latent functionality 369–370
latent functionality 391
"lawlike propositions", Ryle's sense of 270
"lawlike sentence" 135 n. 7
lawlike sentences xxiii–xxiv, xxxii n. 3, 89, 137 n. 28, 153–154, 156–158
  probabilistic 71–72
  statistical 71–72, 103, 176–177
  universal 70–71

lawlikeness 163 n. 14
laws as analytic (definitions of kinds of systems) 201
laws (see also lawlike sentences)
  fundamental statistical 169
  nature of 70–73
  of coexistence 94, 135 n. 10
  of probabilistic-statistical form 279–280, 334
  of strictly universal form 279
  of succession 94, 135 n. 10
  of universal form 334
  strictly universal 169
leaving a statement in suspense 123
"legislative postulation" 215
Leibniz condition 179, 181–182, 185 n. 31
limits of scientific knowledge 330–341
limits of scientific understanding 341–342
linguistic aspect of physicalism 268–269
linguistic reformulation, Goodman's xxvi
linguistic turn 266
logical (or inductive) irrelevance 114, 119–121, 143 n. 61
logical (or inductive) probabilities 113, 280, 298–299
logical analysis 357, 379–381
logical atomism xiv–xv, xxii, xxxi
logical behaviorism 319–320
logical consistency 389
logical empiricism xviii–xvix, xxviii, xxxi, 208, 216, 232, 239, 253, 266, 269, 272 n. 2
Logical Foundations of Probability, Carnap's 347
logical function of explanations 80–81
logical meaning xvii
logical positivism xv, xix, xxii, xxxi, 253, 271, 357
logical positivist position xvi
logical probabilities (see also c(h, e) =r) 112–113, 151
logical probability related to statistical probability 118
logical probability, [r] as a xxv, xxvii, 146–162, passim
logical reconstructions 17 n. 9
logical structure of explanations 81–83
logical truths xvii, xix
logicism 13
lottery paradox 131–133, 347, 351
low-probability explanations 341–342

magnetism, theory of 237–242
material conditionals xx

material mode of speech 266
"material mode of speech" 190
material rules of inference 135–136 n. 11
mathematical certainty 18, 20–21
mathematical demonstrations 18
mathematical induction 7
mathematical theorems as analytic *a priori* truths 5–6
mathematical theorems as empirical generalizations 4–5
mathematical theorems as self-evident 3–4
mathematical truth 3
mathematics as a branch of logic 13
mathematics as an axiomatized deductive system 6
maximal specificity vs. total evidence 150–152
maximal specificity, epistemic relativity of xxviii
maximal specificity, requirement of xxvii, 149
"maximally specific i-predicate" 161
maximizing expected utility 314, 324
maximizing maximum utility 314
maximizing minimum utility 314, 350
maximum homogeneous reference classes 171–173
maximum reference classes as investigative maxim 184 n. 16
meaning analyses xix
meaning of theoretical terms 208
meaning postulates xix, xxviii
meaningful xviii
"meaningful" behavior 262–263
meaningless xviii
measure of epistemic utility, tentative 125
mechanical routines 34
mechanism in biology 190–192
mechanistic determinism 336
membership criterion 222–223
Mendelian genetics 116
metamathematics 375
method of inductive acceptance (MIA) 344
methodological anarchy 363
methodological conventions 377
methodological holism 382, 385–385, 393 n. 29
methodological individualism, Neurath on 259
methodological naturalism xxxi, 374
methodological pragmatism 374
methodological rationalism xxxi, 374
methodological unity of empirical science 269–271, 295
(MIA) method of inductive acceptance 344

Millikan oil-drop experiments 386
mind/body reduction 193–195
minimax principle 127–128, 142 n. 56, 142 n. 57
"model", senses of 228–229
"model of a theory" 61–63, 227–229
models and analogies 221–222, 229
models as components of theories 61–63
models as parts of internal principles 229
modified thesis of physicalism 264
molecular facts xiv
molecular sentences xiv
motivating reasons 74
motive and belief attributions, epistemic interdependence of 270–271
motive and belief causation, combined 270–271
motive and belief explanations 289–294
motives, explanations by 270–271
multiplication defined 8
mutual irreducibility 200

(N1) non-finite extension of nomic predicates 153, 163 n. 8
(N2) non-logical truth of nomic predicates 156
narrower reference class 149
narrowest reference class 120–121, 140–141 n. 47, 141 n. 50, 151–152, 162–163 n. 6
natural affinities 335
"natural kinds" 177–178
natural numbers 10–11
naturalistic account of theory choice 359
naturalistic facet of scientific methodology 378
Naturwissenschaft 266
Naturwissenschaften 254–255
necessary conditions alone non-explanatory 75
neovitalism 74
Neurath on prediction as objective of science 258–259
Neurath on theoretical understanding 258–259
Neurath's *index verborum prohibitorum* 255–258
new riddle of induction, Goodman's 346
Newtonian mechanics 59
Newtonian mechanics, application of 241–245
Nicod's criterion 35–36, 40–41, 47 n. 10, 48 n. 15
nomic expectability xxvii, 148–149, 173, 182, 299–300, 334

nomic expectability, [r] as a degree of xxvii, 146–162
nomic predicates 156
   non-finite extension of (N1) 153, 163 n. 8
   non-logical truth of (N2) 156
nomic relevance 175–178, 244, 247–248
nomic relevance, Coffa's conception of 175
nomically coextensive 267
nominal definitions xix
"nomological insight" 182
nomological predicates, Stegmuller on 176
nomological statements 89
nonalgorithmic desiderata 388
nonambiguity of D-N explanations 148
nonconjunctiveness of statistical systematization 131–133
nonconscious rationality of conscious choice 324–325
nondeliberate actions 308–309
nondemonstrative inference 29–30
non-Euclidean geometry xxvi, 21–24
nonextensional conditionals xxxi
nonextensional logic xxiii–xxiv
nonisolated systems 93, 99
nonobservable language xvii (see observable/theoretical distinction)
nonpragmatic conceptions of explanation 82–83
nonprimitives 9
nonpurposive teleological behavior 310 n. 21
normal science 246–247
normative or prescriptive approaches 358
normative principles 303–304
normative principles of action 290–292
normative rationality, Kuhn's 383
normative-prescriptive account of science 359
novel concepts 32
numbers 10–11

objectivity as intersubjectivity xxii, 376, 378
objectivity of science 389–391
objectivity of scientific inquiries xxii
objectivity of scientific inquiry 374, 376
object-language/metalanguage distinction xiv–xv
observability 51, 239
   of attributes of things 52
   of thing-like entities 52
observable predicates xxii
observable properties xxiii
observation sentences xxii, 257, 375
"observation sentences" 211–212

observation term 375
"observational predicate" 212
observational vocabulary 239
observational/nonobservational terms 51–52, 55–61, 219–220
observational/theoretical distinction xviii, xxii–xxiii, xxxi, 46 n. 2, 51–52, 55–61
observations as theory-laden xxiii, xxxi
ontic vs. epistemic completeness 244–248
ontological aspect of physicalism 268–269
"open ended" theoretical concepts 233
operational definitions 49, 63, 229, 230, 238
operational criteria of application 264
operationalism 56–58, 203–205, 226–227
ordinary language philosophy xiv

"p because p" nonexplanatory 337
P provisos 243–246
"$p(G, F) = r$" 104, 108, 146
$p(h, e)$ 380
$p(h, e) = r$ 346–347, 350
pain-states and brain-states 194
paradigms xxxii, 358, 365
paradox of analysis xix–xx
paradox of the explanation of the improbable 179–180
paradox, grue-bleen 346
paradox, lottery 347, 351
paradoxes of confirmation xxvi, 35–38
"paradoxical" in a logical sense 36
"paradoxical" in a psychological sense 36
paresis counterexample 75, 300–301
partial explanations 282–284, 303
particular conditions, statements of 88
particular event 79
Peano arithmetic 6–13
"P-equipollent" 262
phenomenalism vs. physicalism 260–261
phenomenalistic language xxii
phenomenology xiv
Philosophical Investigations, Wittgenstein's xv
physical geometry 24–25, 55, 202–204, 210
physical interpretation of a mathematical theory 25
physical interpretations of geometry 27
physical meanings 24–25
physical properties xxxi
physicalism 273–274 n. 33
   Carnap's elaboration of 259–266
   Feigl's version of 267–268
   linguistic aspect of 268–269
   linguistic reformulation of 266
   modified thesis of 264

Index of Subjects

physicalism (*continued*)
  ontological aspect of 268–269
  thesis of 261
  vs. phenomenalism 260–261
physicalistic identity theory 267
physicalistic language xxii
"physicalistic" vocabulary 255–257
picture theory of language xv
"place selection" 183 n. 6
Planck's constant 53
positive instances of type I 35, 37–38
positive instances of type II 35, 37–38
positive instances of type III 36–37
"postdictability" 136 n. 19
postulates 18, 23
postulates of a theory 6
potential explanations 89
potential explanations and predictions 115
potential systematizations 99
pragmatic conceptions of explanation 80–83
pragmatic phenomena of language xvi
pragmatic utilities 127
pragmatics of explanation 100–101
pragmatics of prediction 100–101
pragmatics of retrodiction 100–101
pragmatism xiv
pragmatist (or instrumentalist) model 126–130
preanalytically nonscientific theories 378
preanalytically scientific theories 378
"predictability" 136 n. 19
"predicting" 84 n. 4
prediction 237
prediction and control 276
prediction and explanation xxvii
prediction as objective of science for Neurath 258–259
prediction, pragmatics of 100–101
predictions 97–102, 331–332
predictions as arguments 76–77
predictions as statements 76
predictions from theories 3, 25
predictive arguments 76, 97, 100–102
predictive arguments not always explanatory 76–77
prescriptive or normative approaches 358
presuppositions of theoretical inference 240–241
pretheoretical vocabulary 52, 56–61, 219
primitive terms 6
primitives 9
principle of practical certainty 118
principle of testability xix

principle of verifiability xix
principles of action vs. descriptive generalizations 317
principles of action vs. general laws 306
"principles of action" 303–304
"principles of action", Dray's sense of 290–294
probabilistic (or inductive) explanations 71–72, 77–79, 162, 279–281, 298–299
probabilistic arguments 71, 77–78
probabilistic interpretative sentences 241–242
probabilistic lawlike sentences 153–154, 156–157
probabilistic nomological explanation 291
probabilistic-statistical laws 279–280, 298
probabilistic-statistical laws, form of 334
"probability", meaning of xx, 224
  as abstract property of sets 137–138 n. 29
  in empirical hypotheses xx
    as dispositions 137 n. 27, dispositions 177
    as frequencies xxv, 104, 137–138 n. 29, 176–177, 298
    as propensities (*see also* as dispositions) xxv, 137 n. 27
  in inductive reasoning xx
  inductive (*see also* probability, logical) 71–72, 78–79, 280
  logical (or inductive) 71–72, 78–79, 280, 298–299
  subjective 324–325
probability, statistical 71, 78–79, 151–152, 280
problem of induction 32, 46, 344, 349, 388
problem of induction, Hume's 344
problem of provisos xxix, 240–241, 248
Progress and Its Problems, Laudan's 353
projectibility 39–41, 184 n. 22
projections xxiv
proof sketches 282
proof theory 299–300
proof theory in metamathematics 282
proofs in geometry 331
propensity interpretation of probability 137 n. 27
protocol sentences 257
provisos 91, 93, 99
provisos P 243–246
provisos, problem of xxix, 240–241, 248
provisos, violations of 246–247
psychoanalysis 325
psychoanalytic theory 373
psychological novelty 14, 21

Index of Subjects   419

psychological traits as dispositional 262–265, 270–271
psychological traits as theoretical 265
Psychopathology of Everyday Life, Freud's 282–283
psychophysical connections 338–341
psychophysical reduction 193–195
pure geometry 24–25, 210
pure mathematics xxvi
pure science xxx, 130, 142–143 n. 58
pure science vs. applied science 349
pure scientific research 45–46, 329
purely deductive systems 15
purely scientific utility 46
purely theoretical utility 46
purposive action with conscious deliberation 293
purposive action without conscious deliberation 293
puzzle of "about" 102

quasi-theoretical connections 319–321

[r] as a degree of nomic expectability xxvii, 146–162
[r] as a logical probability xxv, xxvii, 146–162
[r] as an inductive probability xxvii, 146–162
R, correspondence rules 221
R, interpretative sentences 209, 211–214
R, rules of correspondence 209, 211–214, 229–231
radioactive half-life 341
Ramsey sentence 242–243
random experiments 108–111, 143 n. 61, 324–325
rational acceptability 121–131
rational action as a normative notion 311
rational action as explanatory concept 322–325
rational action, constraining norms of 313
rational action, information basis of 312, 323
rational action, total objectives of 312–313, 323
rational agents xxix, 305–306, 317–319, 321, 323–325
rational credibility 114, 121–131, 358, 381, 392 n. 27
"rational credibility" 346
rational decision-making 128–129, 311–315
rational expectability 173, 180–181
rational expectations for singular events 118
rational explanation 289–294, 303–305, 316–317, 325–326

rational reconstructions (*see also* explications) xxix, 357, 365, 367–368, 379–381
rationality as an appraisal 311, 316
rationality as an explanation 311, 316
rationality as descriptive-psychological concept 317–318
rationality as normative-evaluative concept 317
rationality of action vs. belief xxx
rationalization vs. explanation 293–294
rationally acceptable explanations and predictions 115
"real reasons", Gardner's sense of 294
realism, theoretical 244
reasons and causes xxix
recursive definition 7
reducibility to observables xxix
reduction xxviii
  as approximation, Smart and Putnam on 202–203
  as definability or derivability 191–192, 196–198
  Kemeny and Oppenheim account of 201–202
  linguistic conception of 189–190
  ontological conception of 189
  the linguistic turn in the study of 190
reduction sentences xxi, 63, 213, 216, 230–231, 238, 263–264, 292
reference class homogeneity 169–171
referential opacity 323
rejecting a statement as false 123
relative proofs 21
relevance 31
relevance, nomic 175–178
relevance, statistical 171–172
relevant antecedent conditions 92
reliability of scientific knowledge 332
reliable knowledge 357, 363
repeatable events 153–154, 157–158, 163 n. 9
repeatable experiments 166
representative sentences 63, 292
requirement of explicit linguistic specification 209, 215–216
requirement of high probability xxvii–xxviii, 179, 181–182
requirement of maximal specificity (RMS) xxvii, 149, 150–152
requirement of maximal specificity vs. requirement of total evidence 150–152
requirement of maximal specificity* (RMS*) 159–162
requirement of maximal specificity* (RMS*) revised 174–175

requirement of total evidence xxvii, 42–43, 72, 114–115, 117, 119–121, 133, 139 n. 40, 150–152, 154, 174–175, 244
retrodiction 237, 331–332
retrodiction, pragmatics of 100–101
reversible processes 99
revolutionary science 247
"riddles of the material world" 338
(RMS) requirement of maximal specificity 150–152
   G. Massey's objection to 154–155
   R. Grandy's objection to 158–161
   R. Wojcicki's objection to 155–156
   vs. (RMS*) 160–161
   W. Humphreys' objection to 152–154
(RMS*) requirement of maximal specificity* 159–162
(RMS*) requirement of maximal specificity* revised 174–175
rough criterion of evidential adequacy 119–121, 150–152
rule for inductive acceptance, tentative 126
rule of detachment 346
rule of inductive acceptance for pure science 351
rule of maximizing estimated utility 124
rules of application 43–44, 114
rules of correspondence 62–65, 241–242
rules of correspondence R 49, 209, 211–214
"rules of correspondence" 49, 229–231
rules of inductive acceptance 347–348
rules of inductive evaluation 347–348
rules of inductive inference 43
rules of inductive support 43
rules of inference 39–40, 43–44, 135–136 n. 11
rules or standards of scientific inquiry 357–370

Salmon and Stegmuller, comparison of 180–181
"same cause, same effect" 278–279
sameness of concepts 60
Schema I (for theories) 49–50, 54–55, 65 n. 4
Schema II (for theories) 49–50, 54–55, 64–65
Schema R (for the thing to do) 290
Schema R (for what was done) 317, 321
Schema R' (for what was done) 291
science xxxii
science of ethics 45
science of science xxxii
Science, Explanation, and Rationality, Fetzer's vii

scientific (theory) change, deductive aspects of 199–206
scientific community xxxii, 354, 365
scientific community, values of 359
scientific expertise 248
scientific inquiry, narrow inductivist view of 30–33
scientific inquiry, rules or standards of 357–370
scientific knowledge xxx, 329–333
   inductive uncertainty of 330–331
   reliability of 332
   scope of 332, 338–341
scientific materialism 338–339
scientific methodology, naturalistic facet of 378
scientific methodology, valuational facet of 378
scientific objectivity 389–391
scientific rationality xxx–xxxi, 357–370
scientific revolutions 384
scientific standards xxxii
scientific theories xxvi, xxviii–xxix, passim
   Hempel's conception of xxvi, xxviii
   semantic conception of, implicit 201
   standard conception of xxvi, xxviii, 208, 216
scientific theory T 49–50
scientific training 390
scientific understanding 278, 295, 329–330, 333–335, passim
scientific worldviews 388
scientists xxxii
scientists qua scientists make value judgments 347–351
scope of scientific knowledge 332, 338–341
Selected Philosophical Essays, Jeffreys's vii
"self-evidence" 3–4, 24
self-evident, mathematical theorems as 3–4
semantic conception of theories, implicit account of 201
semantical interpretation of a mathematical theory 24–25
semantical models of language xvi
sentences xvii
simple model world 98–99
simplicity 393 n. 42
single cases 166
single cases and finite samples 110–111
single cases with high probabilities 109
single cases with low probabilities 109
"single cases" 151–152
social sciences, Neurath on 254–259
socioeconomic status indicators 390

sociological-historical approach xxx
soundness, deductive 42
soundness, inductive 42
standard conception of scientific theories 208, 216
standards of evaluation 388
state of a system at a given time 93
states of affairs xiv, xxii
"states-under-a-theoretical characterization" 194
statistical covering laws xxv
"statistical deepanalysis" 179–182
statistical explanation, ambiguity of 107, 111
statistical explanation, conception of 165–166
   R. Jeffrey's objection to 165–168
   W. Salmon's objection to 168–175
   W. Stegmuller's objection to 171–179
statistical explanations as arguments 165–167, 181–182
statistical generalizations 103
statistical irrelevance 171–172
statistical lawlike sentences 153–154
statistical laws xxv, 103, 298–299
statistical prediction, ambiguity of 107, 111
statistical probability 151–152, 224, 280, 298
statistical probability hypotheses 71, 103–104, 107–111, 153–154, 156–157
statistical probability, frequency interpretation of 108
"statistical proof" 179–180
statistical relevance model xxv
statistical relevance, Hempel's conception of 160
"statistical syllogisms" 115–116
statistical systematization admits of degrees 131
statistical systematization, ambiguity of 105–107, 111
Stegmuller and Salmon, comparison of 180–181
stipulations 64
stochastic processes 166–167
subjective confidence or conviction 332
subjective factors 388–391
subjective probabilities 324–325
subjectivity 4
subjunctive conditionals xxi, xxiii–xxiv, xxxii n. 2, 104, 137 n. 28, 156–157, 176
subsumption, explanation as 134 n. 2
"sufficiently large" 130–131
"sufficiently small" 130–131
symmetry thesis 74–77, 97–102, 139–140 n. 44

synonymy xix
synonymy of terms 60
syntactical models of language xvi
synthetic *a priori* knowledge xvi–xvii
systematization, deductive 113
systematization, inductive 113
systematization, statistical 102–121
systematizations, confirmed 99
systematizations, potential 99
systematizations, true 99
systems, isolated 93, 99
systems, nonisolated 93, 99

$T = (C, I)$ 237–239
$T = (C, R)$ 221
$T = (I, B)$ 218–219
$T = c(I \cup B)$ 219
tautologies xvii
technical language xxii–xxiii
teleological explanation 73–74, 268
temporal intervals 64, 230
tendencies 285
tentative measure of epistemic utility 125
tentative rule for inductive acceptance 126
terminological conventions 214
testability 377–378
testability-in-principle 358, 360
testing theories of physical space 27
that an event took place 166
"the cause of X" 76
The Concept of Mind, Ryle's 256, 318
The Logical Structure of the World, Carnap's 379–380
"the only cause" 76
The Structure of Science, Nagle's 189
The Structure of Scientific Revolutions, Kuhn's xxxi, 48 n. 23
the thing to do vs. the thing that is done 303–304
"the thing to do", Dray's sense of 290–294, 316–317
theorems 23
theorems of geometry 18–20
theorems of mathematical probability as nonexplanatory 141 n. 49
theoretical (or inductive) ascent 239–240, 248
theoretical calculus C 209
"theoretical change" 227
theoretical explanation 134–135 n. 4, 278, 309 n. 3, 334
theoretical juice extractor 14
theoretical language xvii, passim
theoretical novelty 14, 21

theoretical predicates xxii, passim
theoretical principles 265, 319
theoretical properties xvii, xxiii, passim
theoretical realism 244
theoretical reduction 189–206
theoretical scenario 50, 52–61, 209, 218, 223
theoretical terms, meaning of 208
theoretical understanding, Neurath on 258–259
theoretical vocabulary 50–51, 55–61, 219, 237–239
theoretically coextensive 267
theories xxii–xxiii, passim
    as calculi plus interpretations 49–50
    as internal plus bridge principles 50
    as internal principles plus bridge principles 218–219
    embedded in models 228
    Hempel's conception of 49–50, 218–219
    semantic conception of (implicit account of) 201
    standard conception of 49–50, 54–61, 220–222, 237–239
theory change in science, deductive aspects of 199–206
theory choice in science 358
theory choice in scientific inquiry 352–356
theory of magnetism 237–242
theory-laden observations xxiii, xxxi
thesis of logicism 13
thesis of physicalism 261
"thing-like" entities 51–52
tool-for-optimal-action model of scientific knowledge 129
total body of scientific knowledge K 121–122
total evidence vs. maximal specificity 150–152
total evidence, requirement of xxvii, 42–43, 72, 114–115, 117, 119–121, 133, 139 n. 40, 150–152, 154, 244
total state of the universe 196
Tractatus Logico-Philosophicus, Wittgenstein's xiv
"transcendent riddles" 338–341
transitivity of identity 5, 7–8
translatability 268
"true by convention" 231
"true by definition" 6
true explanations 90
true sentences xv
true systematizations 99
truth as an epistemic utility 44–46
truth as an objective of science 44–46
truth by convention 215

truth condition of adequacy 89–90, 100
truth in virtue of facts 205
truth in virtue of meanings 205
truth of a sentence in a language xv–xvi
truth vs. confirmation 89
truth-functional logic xxi
truths of mathematics xxvi
"Two Dogmas of Empiricism", Quine's xix, passim

"unchanged in meaning" 207 n. 19
"under a description" 157
"under a specific description" 191
understanding 97–98, 180, 299–300, 329, 357
understanding via explanation 333–335
understanding vs. ascertaining 322
"understanding" 233
undesirably broad reference classes 171–173
undesirably narrow reference classes 171–173
uniformity of usage 216
uninterpreted axiomatized formal system C 209
unitary physicalist language 258, 260
unity of empirical science, methodological 269–271
universal covering laws xxv, passim
universal determinism, dilemma of 195–196
universal generalizations xviii, passim
universal laws 279, 297–298
universal laws, form of 334
universal psychophysical association hypothesis 193
universals xxiii
utilities, epistemic 125–126
utilities, pragmatic 127–128
utility measure 124

validity, deductive 33–34, 47 n. 9, passim
validity, inductive 47 n. 9, passim
valuational facet of scientific methodology 378
valuational presuppositions 45
valuations as dispositions 373
valuations, epistemological 378
values in science 372–373
values of scientists as determining theory choice 353–356
"verbalistic" outlook on decision problems 142–143 n. 58
verifiability xviii, 275, passim
verifiability criterion xvii, 367

verifiability criterion of meaningfulness xviii–xix
verification 330–331, passim
Vienna Circle 254, 256, 260
violations of provisos 246–247
"virtues" of theories 393 n. 34
vitalistic explanations 268
vocabulary, antecedent 50–52, 55–61
vocabulary, antecedent (or pretheoretical) 219
vocabulary, observational 239
vocabulary, pretheoretical (or antecedent) 219
vocabulary, theoretical 50–51, 55–61, 219, 237–239

"was to be expected" 299, 334
weight 162–163 n. 6
"weight" 151–152
what goes on 329
what the world is like 248
what there is 248
why an event took place 166, 276, 329
why in fact the agent acted as he did 306–307
why is there something rather than nothing? 340–342
why vs. how events occur xxx, 79–80
"working hypotheses" about causal relations 92–93
worldviews, scientific 388